Forest Fires

BEHAVIOR AND
ECOLOGICAL EFFECTS

Forest Fires

BEHAVIOR AND ECOLOGICAL EFFECTS

Edited by

EDWARD A. JOHNSON

*Department of Biological Sciences
and Kananaskis Field Stations
University of Calgary
Calgary, Alberta
Canada*

KIYOKO MIYANISHI

*Department of Geography
University of Guelph
Guelph, Ontario
Canada*

ACADEMIC PRESS

A Harcourt Science and Technology Company

San Diego   San Francisco   New York   Boston   London   Sydney   Tokyo

Front cover picture: Boreal forest crown fire estimated to have a frontal intensity of about 30,000 kW m^{-1}. The fire is about 20 km away, burning away from the camera. At least five convective updrafts can be identified by their darker smoke indicative of incomplete combustion. The wind was about 45 km h^{-1}. Notice the suggestion of shearing of the convective updrafts. Tornadic fire whirls were observed immediately in front of this advancing fire line. The fire eventually burned 48,000 ha.

This book is printed on acid-free paper. ∞

Copyright © 2001 by ACADEMIC PRESS

All Rights Reserved.
No part of this publication may be reproduced or transmitted in any form or by any means, electronic or mechanical, including photocopy, recording, or any information storage and retrieval system, without permission in writing from the publisher.

Requests for permission to make copies of any part of the work should be mailed to:
Permissions Department, Harcourt Inc., 6277 Sea Harbor Drive,
Orlando, Florida 32887-6777

Academic Press
A Harcourt Science and Technology Company
525 B Street, Suite 1900, San Diego, California 92101-4495, USA
http://www.academicpress.com

Academic Press
Harcourt Place, 32 Jamestown Road, London NW1 7BY, UK
http://www.academicpress.com

Library of Congress Catalog Card Number: 00-110672

International Standard Book Number: 0-12-386660-x

Transferred to Digital Printing 2007

00 01 02 03 04 05 SB 9 8 7 6 5 4 3 2 1

CONTENTS

7 Fire Plumes

G. N. Mercer and R. O. Weber

8 Coupling Atmospheric and Fire Models

Mary Ann Jenkins, Terry Clark, and Janice Coen

9 Surface Energy Budget and Fuel Moisture

Kenneth E. Kunkel

10 Climate, Weather, and Area Burned

M. D. Flannigan and B. M. Wotton

11 Lightning and Forest Fires

Don Latham and Earle Williams

12 Statistical Inference for Historical Fire Frequency Using the Spatial Mosaic

W. J. Reed

CONTRIBUTORS

Numbers in parentheses indicate the pages on which the authors' contributions begin.

TERRY L. CLARK (257), National Center for Atmospheric Research, Mesoscale and Microscale Meteorology, Boulder, Colorado 80307 (clark@ncar.ucar.edu)

JANICE COEN (257), National Center for Atmospheric Research, Mesoscale and Microscale Meteorology; Boulder, Colorado 80307 (janicec@ncar.ucar.edu)

MATTHEW B. DICKINSON (477), Kananaskis Field Stations, University of Calgary, Calgary, Alberta T2N 1N4, Canada (mdickins@ucalgary.ca)

FRANCIS E. FENDELL (171), Space and Technology Division, TRW Space and Electronics Group, Redondo Beach, California 90278 (frank.fendell@trw.com)

MICHAEL D. FLANNIGAN (351), Canadian Forest Service, Edmonton, Alberta T6H 3S5, Canada (mflannig@nrcan.gc.ca)

MARY ANN JENKINS (257), Department of Earth and Atmospheric Science, York University, Toronto, Ontario M3J 1P3, Canada (maj@science.yorku.ca)

EDWARD A. JOHNSON (1,477), Department of Biological Sciences and Kananaskis Field Stations, University of Calgary, Calgary, Alberta T2N 1N4, Canada (johnsone@ucalgary.ca)

KENNETH E. KUNKEL (303), Midwestern Climate Center, Illinois State Water Survey, Champaign, Illinois 61820 (k-kunkel@uiuc.edu)

DON LATHAM (375), Intermountain Fire Sciences Laboratory, USDA Forest Service, Rocky Mountain Research Station, Missoula, Montana 59802 (dlatham@fs.fed.us)

DAVID L. MARTELL (527), Faculty of Forestry, University of Toronto, Toronto, Ontario M5S 3B3, Canada (martell@smokey.forestry.utoronto.ca)

GEOFFREY N. MERCER (225), School of Mathematics and Statistics, University College, University of New South Wales, Australian Defence Force Academy, Canberra 2600, ACT, Australia (gnm@adfa.edu.au)

KIYOKO MIYANISHI (1,437), Department of Geography, University of Guelph, Guelph, Ontario N1G 2W1, Canada (kmiyanis@uoguelph.ca)

RALPH M. NELSON, JR. (79), USDA Forest Service, NFSNC Supervisor's Office, Asheville, North Carolina 28801 (rnelson01@fs.fed.us)

WILLIAM J. REED (419), Department of Mathematics and Statistics, University of Victoria, Victoria, British Columbia V8W 3P4, Canada (reed@math.uvic.ca)

KOZO SAITO (11), Department of Mechanical Engineering, University of Kentucky, Lexington, Kentucky 40506 (saito@engr.uky.edu)

DAROLD E. WARD (55), Fire Chemistry Research Work Unit, USDA Forest Service, Intermountain Research Station, Missoula, Montana 59807 (pyroward@aol.com)

RODNEY O. WEBER (151, 225), Department of Mathematics, University College, University of New South Wales, Australian Defence Force Academy, Canberra 2600, ACT, Australia (r.weber@adfa.edu.au)

EARLE R. WILLIAMS (375), Department of Earth, Atmospheric and Planetary Sciences, Massachusetts Institute of Technology, Cambridge, Massachusetts 02139 (earlew@ll.mit.edu)

MICHAEL F. WOLFF (171), Space and Technology Division, TRW Space and Electronics Group, Redondo Beach, California 90278 (michael.wolff@trw.com)

B. MICHAEL WOTTON (351), Canadian Forest Service, Great Lakes Forestry Centre, Sault Ste Marie, Ontario P6A 5M7, Canada (mwotton@nrcan.gc.ca)

PREFACE

Traditional approaches to the study of fire behavior and fire ecology (by foresters and ecologists) have been strongly rooted in a descriptive tradition in which fire behavior has been described in a very informal way, using loosely defined terminology such as "hot/cool fires" or "flammability" with no dimensions or units of measurement given. With this approach, the coupling of fire behavior and fire effects on individuals, populations, or communities could be, at best, qualitative or categorical. Furthermore, even when more quantitative measures of fire behavior have been attempted, many studies have used less than effective measures for attempting to understand the relationship between fire behavior and fire effects (e.g., the use of color-changing temperature-recording paints to correlate temperature with tree mortality). To effectively relate fire behavior to ecological effects, it is necessary to show that such instruments or measures are meaningfully related to the way that organisms are affected by fire. In other words, such studies should incorporate some model of heat transfer to the organisms or particular living tissues (e.g., to serotinous cones or to the tree cambium through bark).

In wildfires, variation in the response of the organism can result from variation in the appropriate measurements of fire behavior variables and variation in the organism's physical characteristics which affect heat transfer processes from the fire to living tissues. Many ecological studies cannot separate these components of variation and thus cannot make useful statements about the causes. On a different scale, studies of the spatial pattern of tree mortality within large burns (i.e., unburned "islands") have often looked for explanations of these patterns in differences in usual environmental factors such as topography, hydrology, and vegetation composition. While such differences may account for some unburned patches, in many cases these unburned islands do not appear to be related to any of these factors. Again, an understanding of the physical behavior of fire (e.g., coupling of fire and three-dimensional convection above the fire) might provide a useful approach to understanding some of these unburned islands.

Since the 1950s, the body of literature on fire behavior has been growing in journals of engineering, geophysics, meteorology, etc. Foresters and ecologists have not used much of this literature on the physical aspects of fire behavior to understand fire effects on individual organisms, populations, or communities. Part of the reason may be that their interests lie in the ecological responses to fire rather than in the physical processes of fire itself. Nevertheless, as indicated above, an understanding of the ecological effects of fire will necessarily involve some understanding of the processes involving the fire in order to ask appropriate research questions and to measure meaningful fire behavior variables. The *purpose* of this book is to provide an introduction to this literature and to convince those working in fire ecology of the importance of understanding these physical fire processes. The book is not meant to be a definitive coverage of the topic of fire behavior and ecological effects; some important topics are not covered because of space limitations. Possible future editions of the book may include other aspects of fire behavior and ecological effects.

This book attempts to bridge the gap between elementary texts on fire for foresters, ecologists, and other environmental specialists (e.g., Brown and Davis, 1973; Luke and McArthur, 1978; Pyne, 1984; Trabaud, 1989) and the technical combustion and heat transfer literature (e.g., *Journal of Fluid Mechanics, Symposium on Combustion, Combustion and Flame, Combustion Science and Technology*). We assume that the reader has an elementary knowledge of physics, chemistry, and biology and an introduction to calculus. Derivations of equations are shown when necessary for an understanding of the processes and relationships between variables. All fields have conventions (i.e., terminology, symbols, ways of presenting relationships) that are often difficult for outsiders to penetrate. Also all fields tend to take for granted that the reader knows why certain things are being done. For individuals outside the field, this is rarely the case. The contributors come from a wide range of different fields and they have tried to explain the conventions and approaches used as they arise in each chapter.

Although the contributors were instructed in writing their chapters to keep in mind the primarily non-physical-science background of the intended audience, we also wanted to provide enough detail and depth to allow the reader to grow into the material. Thus, the book may be read on different levels. A first reading of a chapter may provide some readers with a general intuitive grasp of the topic and its relevance without necessarily a complete understanding of the details and equations. We hope this first reading will entice the reader to delve further into the topic and to use some of the suggested readings to develop an understanding at a deeper level.

Development of this book was aided immeasurably by a workshop attended by all of the contributors in which first drafts of each chapter were discussed in terms of the general objective of the book, the intended audience for the book, and overall coordination of the connections between the chapters. The workshop was held at and funded by the National Center for Ecological Analysis and Synthesis in Santa Barbara. We also gratefully acknowledge Louis Trabaud and Paul Zedler, who, as representatives of the intended audience, provided invaluable feedback to all of the contributing authors.

REFERENCES

Brown, A. A., and Davis, K. P. (1973). "Forest Fire: Control and Use," 2nd ed. McGraw–Hill, New York.

Luke, R. H., and McArthur, A. G. (1978). "Bushfires in Australia." Australian Government Publishing Service, Canberra.

Pyne, S. J. (1984). "Introduction to Wildland Fire: Fire Management in the United States." Wiley, New York.

Trabaud, L. (1989). "Les Feux de Forêts: Mécanismes, Comportement et Environment." France-Selection, Aubervilliers Cedex, France.

E. A. Johnson
K. Miyanishi

ACKNOWLEDGMENTS

We thank the following people for reading and providing helpful comments on individual chapters: Frank Albini, Kerry Anderson, George F. Carrier, Terry L. Clark, Matthew B. Dickinson, Francis Fujioka, Robert Haight, Donald A. Haines, Stan Heckman, Rodman R. Linn, Tilden Meyers, Ralph M. Nelson, Tom Ohlemiller, Luis A. Oliveira, James Quintiere, Jon Regelbrugge, Jeffery S. Reid, Walter Skinner, Louis Trabaud, Charles E. Van Wagner, Neil R. Viney, Richard H. Waring, David R. Weise, Forman A. Williams, Zong-Liang Yang, and Paul H. Zedler. We gratefully acknowledge the National Center for Ecological Analysis and Synthesis in Santa Barbara, California, for providing funding and support for a workshop that greatly facilitated collaboration and interaction among the contributing authors. The Center is funded by NSF (Grant no. DEB-94-21535), the University of California-Santa Barbara, the California Resources Agency, and the California Environmental Protection Agency. We also acknowledge the contributions of Louis Trabaud and Paul H. Zedler in reading and commenting on all of the chapters as well as participating in the NCEAS workshop.

Marie Puddister in the Department of Geography, University of Guelph, did an admirable job of converting all of our graphics (rough hand drawings, photocopied diagrams, spreadsheet charts, and graphics files in numerous formats) into uniformly formatted figures. We greatly appreciate her skill and patience in working with authors scattered across North America and Australia.

Finally, we thank our editors, Charles Crumly, Donna James, Danielle Cummins, and Joanna Dinsmore, for their help in producing this book.

CHAPTER 1

Strengthening Fire Ecology's Roots

EDWARD A. JOHNSON
Department of Biological Sciences, University of Calgary, Calgary, Alberta, Canada

KIYOKO MIYANISHI
Department of Geography, University of Guelph, Guelph, Ontario, Canada

I. INTRODUCTION

Research on the connection between wildfires and ecological systems goes back to the early discovery that natural disturbances were a recurrent phenomenon in ecosystems and, as such, required an understanding of their effects on ecosystem structure and function. However, connecting wildfires to ecological systems has proceeded slowly. This is probably because forestry and ecology, the two fields primarily interested in wildfire effects on ecosystems, have been sidetracked by their traditional approach to studying ecological systems. Foresters are mostly interested in extinguishing or eliminating wildfires or in managing burns to produce certain effects in the forest (e.g., reduced competition between certain trees or creation of wildlife habitat). Ecologists have been interested in how fires change the composition and structure of ecological systems. The approach that has been taken to investigate these issues has, in general, involved

describing patterns of fire effects and correlating these to environmental factors.
This approach does not immediately lead to a concern with the mechanism of
interaction between fire processes and ecosystem processes. One could read the
ecological literature for a long time before finding any discussions that indicate
an appreciation that *the processes of combustion and heat transfer lie at the heart
of fire ecology.*

Fire ecology is a part of both environmental biophysics and ecology. Consequently, it is concerned with combustion, transfer of heat, mass and momentum
in wildfires, heat transfer between the fire and the organism, and finally how
these physical processes affect ecological processes. In the following chapters,
we will not be able to deal with these concerns in any complete manner. Consequently, we have chosen to accent certain topics which we feel ecologists and
foresters have not traditionally addressed and about which we believe they
should have more information. This means that most of the book is devoted to
presenting information on fire processes. The information on ecological effects
of fire is more limited than what most ecologists might hope to see. However,
this reflects the limitation of our understanding of combustion and heat, mass,
and momentum transfer. Particularly lacking are well accepted process-based
models for the connection of fire to ecological processes. This has been an on-
going problem noted by several authors over the last decades (Trabaud, 1989;
Johnson, 1992; Whelan, 1995; Bond and van Wilgen, 1996).

Readers may be asking at this point: As an ecologist/forester, do I really have
to understand all this about combustion and heat transfer? In fact, the purpose of this book is to convince readers of the necessity for understanding the
relevant processes involved in the ecological effects of fire. In the absence of a
process-based approach, the discipline of fire ecology is left with a collection
of case studies, that is, site-specific, species-specific, and also often method-
specific correlations between some ecological response and traditionally used
fire variables. Thus, there is little justification for the selection of variables
and relationships and no way to distinguish between relationships that are
physically and ecologically based and those that are statistically significant but
spurious.

Central to the approach advocated in this book is that we must invest in a
basic understanding of wildfire *materials* and *processes* and their effects on ecological (biological) materials and processes. Materials have properties and are
assembled into structures. Thus, in terms of heat transfer from a fire to the living tissues of a tree and its effect on tree mortality, an important material property of bark is its thermal diffusivity. The structure could be considered the
changing thickness of bark on different parts of the tree stem. Processes are
(natural) phenomena marked by a series of operations, actions, or mechanisms
which explain a particular result—in other words, how something works. Some
of the wildfire processes are ignition, extinction, flaming and glowing combus-

tion, rate of spread, and fuel drying. Some of the ecological processes affected by fire processes are population dynamics (i.e., birth, death, immigration, and emigration processes), nutrient cycling, and productivity.

II. PROCESSES

Briefly, this book is concerned with the following topics, most of which involve processes. Combustion is considered primarily as the means of generating heat and combustion products such as *smoke* (Chapter 3). *Flames* (Chapter 2) are one of the principal forms of combustion, smoldering being the other. Flames are also the main source of heat by which most wildfires spread. *Fire spread* (Chapters 5 and 6) brings together the different processes in the fire and its environment to explain why it moves. The *plume* (Chapter 7) above the flame is, as often as the flame, the cause of death of organisms. Since *fuel moisture* (Chapters 4) plays such a central role in coupling fire behavior and other environmental variables, it is given special treatment. Although the role of weather and climate has always been recognized as important in wildfires, recent decades have seen some significant advances in our understanding of all scales of interaction of wildfires and the weather. The *coupling of a wildfire and the atmospheric convection* (Chapter 8) above it has only begun but may help us understand better not only fire spread but also some phenomena such as unburned remnants left by fires. *Mesoscale meteorology* (Chapter 9) has shown how fuel moisture is affected by the energy exchange with the regional environment above the ecosystems. *Synoptic scale meteorology and climatology* (Chapter 10) have the longest history in fire ecology, but understanding in recent decades has helped us understand why wildfires occur only under limited kinds of synoptic conditions and how these conditions persist because of midtropospheric circulation patterns in the atmosphere. *Lightning* (Chapter 11) is the major natural cause of wildfires; however, until the advent of lightning locating systems, there was little understanding of the relationship between lightning and wildfire ignition. *Smoldering combustion* (Chapter 13) is a form of combustion that occurs in materials with particular properties (e.g., the partially to well decomposed organic matter (duff) that accumulates on top of the soil in some ecosystems). The process of smoldering is particularly important in ecosystems in which duff removal is necessary for plant regeneration. The *heat transfer to plants* (Chapter 14) uses many of the processes described earlier as input for the transfer into the plant to cause death of the whole plant or necrosis of parts. Many forest landscapes are mosaics of different ages resulting from the overburning of past fires. Determining the *frequency of fires* from this kind of spatial data has proved to be a sophisticated statistical problem (Chapter 12). *Fire management* has undergone changes in past decades from being primarily prevention and suppression to more ecological and economic concerns of understanding and mimicking the fire regime (Chapter 15).

III. TRANSFER RATES AND BUDGETS

The method for studying the preceding processes is generally by use of transfer rates and budgets of heat, mass, and momentum. Heat transfer results from the difference between objects in their average kinetic energies. The mechanisms of heat transfer are radiation, conduction, convection, and latent heat. Radiation is dependent on the temperature of the object. An object's temperature is considered to be proportional to its average kinetic energy. All objects above absolute zero radiate to their surroundings. When fireplaces were an important source of heat in homes, chairs had screens or wings to block the radiation from the fireplace. Thus, radiative transfer depends on the view of the radiating object. Conduction is heat transferred by the motion of adjacent molecules. Solid objects feel warm or cold to us because of conductive transfer between them and our skin. Convection is the mass transfer of heat in a fluid such as water or air as illustrated by hot-air heating systems in homes. Latent heat is the heat absorbed or released when water changes state; for example, evaporation of water absorbs 2450 J g^{-1}. In their simplest form, these mechanisms consist of the difference in temperature between objects times some coefficient which measures the resistance to transfer.

Mass transfer is the movement of materials and involves diffusion in which different classes of particles have different velocities. Diffusion has two modes: molecular and turbulent. Molecular diffusion is caused by the random motion of molecules, while turbulent diffusion is disorderly movement, often in eddies. Like heat, mass transfer is along a gradient but in this case a gradient of concentration or density rather than temperature. As with heat transfer, diffusion is modified by a resistance term.

Momentum transfer occurs because of resistance to flow in gases and liquids generated by collisions among rapidly moving molecules. Rates of convective heat transfer and of mass transfer are proportional to the viscosity or resistance of the fluid in which they occur. Perhaps the most important issue in momentum transfer discussed in the coming chapters will be boundary layers—the relatively still layer of fluid near solid surfaces in which heat and mass transfer are carried out by molecular diffusion or by boundary layer streaks.

The transfer rates are assembled into budgets which determine the input-output or storage of heat, mass, or momentum based on the law of conservation of energy, mass, and momentum. These budget equations form simple ways of understanding the dynamics of the processes of interest. Ecologists should be familiar with this approach since population dynamics (birth, death, and migration) are budgets which tell us how the numbers in populations come about.

We believe that the reason fire ecologists have not been recruited to the approaches given in the following chapters is the lack of tools to address process questions. This explains why seemingly obvious (to physical scientists) approaches are not taken. Consequently, the approaches used in the following chapters should be of as much interest as the contents.

Some tools are familiar to fire ecologists (e.g., careful measurements, lab and field experiments), but others, such as dimensional analysis (but see Gurney and Nesbit, 1998) and reasoning that seeks to interpret phenomena in light of physical principles, are not. Dimensional analysis helps define what variables should be incorporated and which ones are unnecessary. It also might provide a method for assembling these variables into functional relationships. Experiments (in the widest sense) supply the numerical constants and check the correctness against independent data. Biologists and foresters develop algebraic equations from empirical data by statistical curve fitting, usually with little physiological or biological justification. These functional relationships are simply mathematical descriptions of the data. They often make no dimensional sense. For example, what are the units of fire severity? Experiments are thus used to test classifications of a phenomenon, not the processes involved.

IV. EXAMPLES OF TRADITIONAL VS. PROPOSED APPROACH

Finally, perhaps a comparison between the traditional approach to a problem in fire ecology and the approach advocated in this book would be useful. A tree can be killed by a surface wildfire that has heated its base long enough to kill the cambium beneath the bark. Clearly this process of killing the cambium is affected by the temperature of the surface of the tree, how long this temperature is maintained, and how well the heat is transferred through the bark to the cambium. The question usually asked by ecologists in this situation is: What is the tolerance of different species to surface fires?

The traditional approach by ecologists in studying this fire tolerance has been to heat the base of the tree and record differences in temperature at the cambium (Uhl and Kauffman, 1990; Hengst and Dawson, 1994; Pinard and Huffman, 1997). The tree is heated with a torch, radiant heater, or a burning fuel-soaked rope which is tied around the tree. An unshielded thermocouple is exposed at the surface of the bark, and another is placed in the cambial layer beneath the bark. The temperature course of both thermocouples is recorded during heating.

The time-temperature curve of the outer bark surface (Figure 1) is compared between species, with differences in maximum temperature used to "demonstrate that external bark characteristics can affect heat absorption" (Uhl and Kauffman, 1990). It is not clear how heat absorption could be measured by a

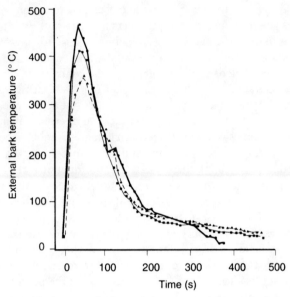

FIGURE 1   Temperature of bark surface during wick (experimental, tree surface) fires of *Tetragastris altissima, Jacaranda copaia,* and *Manilkara huberi* at Vitoria Ranch near Paragominas, Para, Brazil. From Uhl and Kauffman (1990), with permission.

thermocouple at the surface of the bark. Any differences in temperature at the bark surface are probably a result of the experimental setup related to variation in the wick fire and perhaps to flames from the burning bark if this is occurring.

The time-temperature curve of the thermocouple at the cambium surface (inside the bark) is usually related to bark thickness (Figure 2). Time-temperature curves increase more slowly in general with thicker bark species. Notice in Figure 2 that Uhl and Kauffman (1990) use the term temperature flux when the graph gives only temperature. Finally a relationship (Figure 3) is usually given between the maximum cambium temperature and bark thickness (for all species). The equation is then used to calculate the bark thickness at which a tree's cambium will exceed 60°C and cambium death is presumed to happen. Thus, bark thickness is used to infer fire tolerance.

A number of important issues are not considered in such studies. For example, heat transfer is never considered; only temperature is considered. There is no consideration of which heat transfer processes are operating. The transient nature of the heating is never incorporated into the results. Finally, the effect of the material properties of the bark on heat transfer is not considered.

This approach is indeed unfortunate since a more coherent and rigorous approach is almost always given in the references cited by these studies. However,

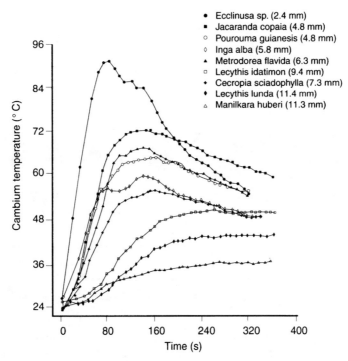

- Ecclinusa sp. (2.4 mm)
- Jacaranda copaia (4.8 mm)
- Pourouma guianesis (4.8 mm)
- Inga alba (5.8 mm)
- Metrodorea flavida (6.3 mm)
- Lecythis idatimon (9.4 mm)
- Cecropia sciadophylla (7.3 mm)
- Lecythis lunda (11.4 mm)
- Manilkara huberi (11.3 mm)

FIGURE 2   Mean temperature flux [sic] at the cambium surface during wick (experimental, tree surface) fires for two individuals (between 20 and 30 cm dbh) in each of nine taxa: *Eclinusa sp.*, *Jacaranda copaia, Metrodorea flavida, Pourouma guianensis, Inga alba, Cecropia sciadophylla, Lecythis idatimon, Lecythis lurida,* and *Manilkara huberi* at Vitoria Ranch near Paragominas, Para, Brazil. From Uhl and Kauffman (1990), with permission. Each species is labeled with the mean bark thickness (mm).

the significance of the heat transfer model (transient heat flow in a semi-infinite solid) given in Spalt and Reifsnyder (1962) does not seem to be recognized. The model (described in detail in Chapter 14) is as follows. Assume that the flame is heating primarily by conduction and that the surface (boundary layer) resistance is minimal, considering the proximity of the flame. The fire suddenly increases the surface temperature. The tree is further assumed to be large enough in diameter so that heating from the opposite side does not affect the cambium (i.e., the main conductive heat transfer is occurring perpendicular to the surface of the bark). The transient heat flow can then be described by

$$\frac{T - T_f}{T_i - T_f} = \text{erf}\left(\frac{x}{2\sqrt{\alpha\tau}}\right) \tag{1}$$

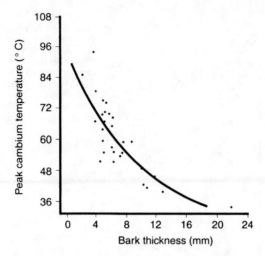

FIGURE 3 The relationship between bark thickness and peak temperature of the cambium during wick fires of 30 individuals distributed among 15 species at Vitoria Ranch near Paragominas, Para, Brazil. From Uhl and Kauffman (1990), with permission.

The left side of Eq. (1) gives the rise in temperature; the numerator gives the difference between the temperature at which the cambium is killed (T) and the flame temperature (T_f), and the denominator gives the difference between the initial (before heating) ambient temperature (T_i) and the flame temperature (T_f). The temperature rise is thus given relative to the initial difference between the flame and tree temperatures. Notice that the gradient of heat is defined by this term. On the right side of the equation, erf is the error function, a value which can be easily looked up in tables for the values in parentheses. The heat transfer is directly proportional to the bark thickness (x) (i.e., the depth for which the temperature gradient is being determined) and inversely proportional to the square root of the thermal diffusivity (α) (i.e., how the bark material affects the flow of heat), and the time it takes for the lethal temperature (T) to be reached (τ). Thus, if we solve for τ, we should be able to see how either trees of different α but same x or different x but same α have different fire tolerances. Thermal diffusivity (α) contains the relevant bark characteristics such as bark density, moisture, and conductivity that influence fire tolerance.

    The model we have given is the simplest, and more complicated ones can be formulated (e.g., Costa *et al.*, 1991). However, even this simple model illustrates an approach to the study of fire tolerance that is based on the process (heat conduction) by which the cambium is heated to lethal temperature. This model also provides a rationale for the choice of variables and shows how the variables interact.

REFERENCES

Bond, W. J., and van Wilgen, B. W. (1996). "Fire and Plants." Chapman and Hall, London.

Costa, J. J., Oliveira, L. A., Viegas, D. X., and Neto, L. P. (1991). On the temperature distribution inside a tree under fire conditions. *Int. J. Wildl. Fire* 1, 87–96.

Gurney, W. S. C., and Nisbet, R. M. (1998). "Ecological Dynamics." Oxford University Press, New York.

Hengst, G. E., and Dawson, J. O. (1994). Bark properties and fire resistance of selected tree species from the central hardwood region of North America. *Can. J. For. Res.* 24, 688–696.

Johnson, E. A. (1992). "Fire and Vegetation Dynamics: Studies from the North American Boreal Forest." Cambridge University Press, Cambridge.

Pinard, M. A., and Huffman, J. (1997). Fire resistance and bark properties of trees in a seasonally dry forest in eastern Bolivia. *J. Trop. Ecol.* 13, 727–740.

Spalt, K. W., and Reifsnyder, W. E. (1962). "Bark Characteristics and Fire Resistance: A Literature Survey." Occasional Paper No. 103. USDA Forest Service, Southern Forest Experiment Station.

Trabaud, L. (1989). "Les Feux de Forêts: Mécanismes, Comportement et Environment." France-Selection, Aubervilliers Cedex, France.

Uhl, C., and Kauffman, J. B. (1990). Deforestation, fire susceptibility, and potential tree responses to fire in the eastern Amazon. *Ecology* 71, 437–449.

Whelan, R. J. (1995). "The Ecology of Fire." Cambridge University Press, Cambridge.

CHAPTER 2

Flames

K. Saito

Department of Mechanical Engineering, University of Kentucky, Lexington, Kentucky

I. INTRODUCTION

This chapter intends to help ecologists, forest fire researchers, and fire-fighting strategists understand some of the fundamental aspects of combustion and think scientifically about their forest fire problems. As with any discipline, the best approach to understanding forest fires is to grasp the fundamentals—in this case, the structure and behavior of flames. Explanation and discussion are based on the physical and chemical aspects of combustion, with little emphasis on strict mathematical treatment of equations. The chapter addresses basic knowledge on the structure of diffusion flames and scaling laws, premixed flames, ignition, diffusion flame extinguishment, spreading flames, and the mechanism of diffusion flame anchoring. Some general background in combustion research is introduced at the beginning.

Fire is the common name we give to high-temperature gaseous combustion. This combustion can occur in open land or an enclosed environment. In either case, regardless of the type of fire, two things are certain: heat is released, and the fire may spread. Flame is the fundamental element of fire that produces the heat and the combustion by-products (such as CO_2, H_2O, CO, and smoke) by means of chemical reactions that occur between fuel and oxygen—called combustion.

Modern combustion research is composed of four theoretical supporting structures: thermodynamics, chemical kinetics, fluid mechanics, and transport processes (Williams, 1992). Additionally, three basic tools (experiment, theory, and computation) are available to help researchers approach combustion problems. Each tool has strengths and weaknesses.

Experiments have two roles: "insightful observation" and verification of assumptions that theory and computational methods employ (Hirano and Saito, 1994). The term "insightful observation" means to see things with unbiased minds using imagination and intuition. The second role of an experiment is well explained elsewhere in the scientific literature, while the first role is rarely emphasized, although it is important. In scientific research, the first and second roles often play together. For example, a simple experiment designed to verify assumptions also offers additional information to researchers. When a researcher conducts an experiment, two different points of view need to be kept in her or his mind: confirmation of assumptions and insightful observation. When the researcher's mind is entirely occupied by the confirmation role only, she or he may miss the opportunity for discoveries and inventions through insightful observation. Our history proves that insightful observation is the source of discoveries and inventions (Ferguson, 1993).

Computation, also referred to as computer experiments, can provide detailed results and even a virtual reality by simulation. The combination of two or more of these tools will increase not only the accuracy of the results but also the chance of getting correct answers, especially during the early stages of a research and development project (Wilson, 1954). Computational methods not only save time and energy of human calculation but also provide details of virtual reality under ideal initial and boundary conditions, which may be difficult to achieve by experiments.

II. BASIC ASPECTS OF COMBUSTION IN FOREST FIRES

"Combustion is exothermic (heat releasing) chemical reactions in flow with heat and mass transfer" (Williams, 1985; Linan and Williams, 1993). To be more specific, we can divide combustion into two types, based on the kind of chemical reaction and the flow regime (laminar and turbulent): high speed and low speed. Good examples of high-speed combustion are gas turbines, rocket motors, and internal combustion engines. Examples of low-speed combustion include forest fires and candle flames.

We can be more specific still by considering the type of flame in low-speed combustion. Forest fires, for example, burn with what are called diffusion flames. Diffusion flames can be maintained by a feedback loop; in the gas phase, heat is released by exothermic combustion of the secondary gaseous fuels, while in a condensed phase some of this heat is transferred back causing gasification of the primary fuel. Gasification is an endothermic (heat-absorbing) process that requires gaseous combustion in order to occur. This feedback of heat from the gas phase flames to the condensed phase fuels is an essential mechanism for sustaining the combustion process.

A. GOVERNING EQUATIONS

Governing equations or conservation equations can provide mathematically strict relationships among various quantities. In combustion science, conservation equations are partial differential equations expressing conservation of mass, momentum, energy, and chemical species. Derivation of conservation equations are detailed by Williams (1985) and Linan and Williams (1993) and are beyond the scope of this chapter; only the concept of these equations together with mathematical formulae for only mass and momentum conservation (because of their simple formulae) are explained.

By taking a volume element of flame fluid and assuming fluid velocity to be much smaller than the speed of light, we can conceptualize the conservation equations for mass, momentum, energy, and chemical species as follows.

Mass Conservation:

$$\left[\begin{array}{c} \text{Rate of accumulation} \\ \text{of mass} \end{array}\right] + \left[\begin{array}{c} \text{Rate at which mass flows} \\ \text{into the volume element} \end{array}\right] = 0.$$

The mathematical expression is $\varpi \partial/\varrho \partial t + \nabla(\varrho\bar{u}) = 0$, where ϖ = ratio of flow time to evolution time, ϱ = density over characteristic density, and $\hat{u}$ = velocity vector over characteristic velocity.

Momentum Conservation:

$$\left[\begin{array}{c} \text{Rate of increase} \\ \text{of momentum} \end{array}\right] = \left[\begin{array}{c} \text{Inertia} \\ \text{force} \end{array}\right] + \left[\begin{array}{c} \text{Pressure} \\ \text{force} \end{array}\right] + \left[\begin{array}{c} \text{Body} \\ \text{force} \end{array}\right] + \left[\begin{array}{c} \text{Viscous} \\ \text{force} \end{array}\right].$$

The mathematical expressions is $\varpi \partial(\varrho\bar{u})/\partial t = -\nabla(\varrho\overline{uu}) - \nabla p/M^2 + F/Fr + \nabla Ts/Re$, where p = pressure over characteristic pressure, M = Mach number, F = body force, Fr = Froude number, Re = Reynolds number, and Ts = shear stress tensor.

Energy Conservation:

$$\left[\begin{array}{c} \text{Thermal} \\ \text{energy} \end{array}\right] + \left[\begin{array}{c} \text{Chemical} \\ \text{energy} \end{array}\right] + \left[\begin{array}{c} \text{Kinetic} \\ \text{energy} \end{array}\right]$$

$$+ \left[\begin{array}{c} \text{Energy lost by conduction} \\ \text{and radiation} \end{array}\right] + \left[\begin{array}{c} \text{Work done on} \\ \text{surroundings} \end{array}\right] = \text{Constant}$$

Chemical Species Conservation:

$$\left[\begin{array}{c} \text{Accumulation} \\ \text{rate of given} \\ \text{chemical species} \end{array}\right] = - \left[\begin{array}{c} \text{Convection of the} \\ \text{chemical species out} \\ \text{of the volume element} \\ \text{by fluid motion} \end{array}\right] + \left[\begin{array}{c} \text{Diffusion of} \\ \text{the species into} \\ \text{the volume} \\ \text{element} \end{array}\right]$$

$$+ \left[\begin{array}{c} \text{Production of the} \\ \text{chemical species} \\ \text{by chemical} \\ \text{reactions} \end{array}\right]$$

The conservation (mass and energy) equations are often applied to the condensed phase.

When these conservation equations are solved under specified boundary and initial conditions, they can provide predictions on changes of pressure, velocity, temperature, and chemical species as a function of time and space. Linan and Williams (1993) point out that no combustion problems require the full description of all the terms in the conservation equations and recommend the simplification of combustion phenomena. For combustion of solids, which includes most forest fires, for example, the phenomenological derivation is more satisfying than the first principle approach of using these complete conservation equations. Here, phenomenological derivation means to come up with empirical equations using experimental data and physical principles, but not starting from the basic governing equations.

B. Adiabatic Flame Temperature and Soot Formation

The most desirable condition for combustion would be to reach the adiabatic flame temperature, the ideal maximum temperature that a combustion system can attain. In such a case, we would have no heat loss from the system. However, real systems don't reach this point because the actual combustion process consists of finite-rate chemical reactions with heat loss that occurs in the forms of convection, radiation, and conduction. All these losses contribute to leading the combustion system to incomplete combustion. Incomplete combustion produces products of incomplete combustion (PICCs). Common examples of PICCs include CO, NO_x, SO_x, dioxin, all kinds of intermediate hydrocarbons (C_nH_m, where n and $m = 1, 2, 3, \ldots$) and soot (solid particulates). The following diagram describes the generic process of combustion:

$$[Fuel] + [Oxidizer] \rightarrow [PCC] + [PICC] + Q \qquad (1)$$

where PCC are the products of complete combustion. When complete combustion occurs in an adiabatic system, no PICCs are formed, the heat of combustion Q reaches a maximum, and the system achieves adiabatic flame temperature. For example, adiabatic flame temperature for a methane + air reaction is 1875°C and that for most hydrocarbon + air systems falls between 1800°C and 2200°C. The detailed procedure to calculate adiabatic flame temperature is straightforward (e.g., Glassman, 1996) using the JANAF Thermochemical Tables (1985).

Flame temperature for an incomplete combustion system can also be calculated by assuming incomplete combustion products. Computer programs (e.g., NASA's Gordon and McBride and STANJAN, see Glassman, 1996) are available to calculate the flame temperature and equilibrium compositions. Equilibrium compositions, defined as species concentrations at a specified temperature, can be calculated using the equilibrium constant that is available in the JANAF Table. For example, an arbitrary second-order reaction can be written as

$$A_R + B_R \rightarrow C_P + D_P \qquad (2)$$

where A_R and B_R are reactants and C_P and D_P are products. A simple example is

$$CH_4 + 2O_2 \rightarrow CO_2 + 2H_2O \tag{3}$$

The reaction rate describing how quickly a given reactant, A_R, for example, disappears and converts to the product can be written using the Arrhenius approximation:

$$-d[A_R]/dt = -k[A_R][B_R] = d[C_P]/dt = -A \exp(-E/R_0 T)[A_R][B_R] \tag{4}$$

where [ ] means concentration, t is time, k is specific reaction rate constant, A is preexponential factor, E is activation energy, R_0 is universal gas constant, and T is temperature. The values of k, E, and A have been studied for many different types of hydrocarbon-air combustion systems and are available in the literature (e.g., Glassman, 1996). For further information regarding chemical kinetics, see Benson (1960) and Glassman (1996).

For combustion efficiency, the ratio of the actual heat release to the maximum ideal heat of combustion ranges from 50 to 95% (Pyne *et al.*, 1996). The combustion efficiency can be defined more specifically as the ratio of the actual carbon contained in the emissions of carbon dioxide compared to that theoretically possible, assuming that all the carbon was released as carbon dioxide. Under such combustion efficiency, many PICCs will be formed. As a result, for such a system, the temperature drops a few hundred degrees below the adiabatic flame temperature. However, there is a critical extinction temperature beyond which flame won't exist. A universal value of extinction, approximately 1300°C (Rashbash, 1962), was found for cellulosic fuels.

Soot is a typical PICC and has a large impact on the environment when it is formed in the case of forest fires by degrading ambient air quality, impairing visibility, worsening regional haze, and causing significant health problems (Pyne *et al.*, 1996, and Chapter 3 in this book). Therefore, a brief explanation of soot formation and properties will be given here. When soot is formed in a high-temperature flame zone, it emits a luminous yellow (black body) radiation; when it escapes from the flame, it can be seen as a cluster of small black particles. Smoke is a rather large cluster of soot particles (100 μm or larger) and is usually large enough for human eyes to recognize, while no clear-cut distinction is made between smoke and soot.

Scanning electron microscopy (SEM), transmission electron microscopy (TEM), electron spectroscopy for chemical analysis (ESCA), and laser desorption mass spectrometry (LDMS) have been used to investigate the physical and chemical structures of soot.

Soot can be collected directly from the hot flame zone by different direct sampling techniques (Fristrom and Westenberg, 1985): (1) a filter paper to collect soot through a vaccum line, (2) a small quartz needle and a stainless-steel mesh-screen to be inserted into a flame to deposit soot on them (Saito *et al.*, 1991), and (3) a thermophoretic sampling probe to collect soot (Dobbins and Megaridis, 1987). Technique (1) is suitable for measuring an average amount of soot produced in a relatively long time period (roughly an order of 10s or greater). The two other techniques are suitable for detailed chemical analysis. The residence time of technique (2) is roughly an order of a few seconds or greater, and that of technique (3) is roughly 50 milliseconds. Technique (2) provides a layer of soot deposit to be analyzed with little effect of heterogeneous catalysis on the soot and probe material. Technique (3) can provide individual soot particles that are collected by inserting a thermophoretic probe directly into a flame. Soot in the flame is attracted by copper and other metal grids by thermophoresis.

Most of the soot studies were conducted for laboratory scale candlelike laminar diffusion flames because they allow researchers easy access to laser optics and other measurement systems that require well-controlled laboratory conditions. These studies found that soot is actually a cluster consisting of small primary soot particles. The average diameter of the mature primary soot particle produced by small laminar diffusion flames was identified by a scanning electron microscope and a transmission electron microscope to be approximately 20 to 50 nm (1 nm = 10^{-9} m). The average molecular weight of soot is 10^6 g/mole, and the C/H ratio is 8 to 10 (Glassman, 1996). The mechanism of primary soot-particle formation, however, is not well understood. Two competing mechanisms that have been proposed recently include: (1) a gradual buildup of a carbonaceous product via formation and interaction of polycyclic aromatic hydrocarbons (PAHs) or acetylene (C_2H_2) molecules through dehydrogenation and polymerization processes, and (2) a direct condensation of liquidlike carbonaceous precursor materials to form a solidlike precursor soot particle as a result of fuel pyrolysis reaction, polymerization, and dehydrogenation.

Regarding the soot and smoke formation mechanisms, there is a belief that, because soot is formed in the gas phase during the process of fuel pyrolysis, difference in fuel type (either gas or liquid or solid) has little influence on soot formation processes. This may be true for the physical shape of primary soot particles which are spherical. However, a recent laser-desorption mass-spectroscopy study (Majidi *et al.*, 1999) showed that mature soot collected from several different laminar hydrocarbon-air diffusion flames have different in-depth chemical structures. This is interesting because, if it is proved true for a wide range of fuels, soot could be used to identify its parent fuel, helping in fire investigations. No LDMS data are available for soot collected from forest fires.

A few studies focused on smoke generated from liquid pool fires. Fallen smoke particles were collected on the ground (Koseki *et al.*, 1999; Williams and Gritzo, 1998) or sampled at a post flame region by an airborne-smoke-sampling package (Mulholland *et al.*, 1996). They used crude oil fires whose diameters ranged from 0.1 to 20 m and sampled smoke particles generated by these fires; primary smoke particles from these crude oil fires looked spherical, and the primary particle diameters varied between 20 and 200 nm depending on the sampling location and diameter of the pool. These studies found a general trend: the larger the fire, the larger the primary smoke particle, suggesting the observed result may be attributable to the longer resident time.

The smoke generation rate from many different types of solid fuels was tested using a laboratory scale apparatus. When small solid samples were exposed under constant heat flux, they increased their temperature, released pyrolysis products and eventually emitted smoke. The amount of smoke can be measured by a light-scattering device or can be collected on a filter paper for weight measurement. When a pilot ignitor is placed over the sample, the sample can be ignited to study material flammability and flame spread characteristics (Fernandez-Pello, 1995; Kashiwagi, 1994; Tewarson and Ogden, 1992; Babrauskas, 1988). The flame spread occurs in horizontal, upward, and downward directions with and without an external (radiation) heat source. These kinds of experimental apparatus require relatively small samples (e.g., 10 cm wide × 30 cm long for flame spread tests, Saito et al., 1989), but some of them are designed for large scale upward spread tests (Orloff et al., 1975; Delichatsios et al., 1995). Effects of sample size and geometrical shape on ignition, flammability and flame spread rate are not well understood (Long et al., 1999). Caution should be taken when the laboratory test results are applied to evaluate behavior of forest fires that involves many different types, kinds and sizes of trees, and bushes and duff in the forest.

III. TEMPERATURE, VELOCITY, SPECIES CONCENTRATION, AND FLAME HEIGHT

Measurements of flame temperature, velocity, height, and chemical species concentrations are important for understanding the structure and characteristics of forest fires. The first three quantities can be measured at a point and thus allows determination of profiles. Height can be measured as the time-averaged maximum height of the yellow flames since it is most relevant to radiant heat released from the flame.

For the first three quantities, profile data can provide more information than point data, but any (one-, two-, and three-dimensional) profile data require more human effort and better (often more sophisticated) equipment than point measurement. The profile data provide us with the relative change of the measured quantity (such as temperature), thus helping us better understand fire phenomena.

A. Temperature Measurement

Temperature measurement with thermcouples is the most commonly used flame temperature measurement technique. It is an intrusive point-by-point measurement and is affected by radiation (q_{rad}), convection (q_{cov}), and conduction (q_{cod}) heat losses. These heat losses can be described in terms of heat flux: $q_{rad} \sim \varepsilon_b \sigma T_b^4 (1 - a)$; $q_{cov} \sim \dot{\eta}(T_b - T_a)$; $q_{cod} \sim \lambda(T_b - T_a)/l_T$. Here, ε_b = emissivity of the thermocouple bead, a = absorption coefficient of the gas surrounding the thermocouple, σ = Stefan-Boltzmann constant, T_b = thermocouple bead temperature, T_a = gas temperature surrounding the thermocouple bead, l_T = a characteristic length related to the high-temperature gradient in the gas, $\dot{\eta}$ = heat transfer coefficient, and λ = thermal conductivity of gas. To minimize these heat losses, a thermocouple with a small bead and a small wire diameter is advantageous. To minimize the conduction heat loss, the wire can be placed along an isotherm in the flame. Radiation heat loss proportional to the fourth power of temperature may become dominant above 1000°C. Soot deposit on the thermocouple junction changes the thermocouple output reading due to the change in emissivity of the junction. A sheathed thermocouple is available to minimize the radiation effect. When a bare thermocouple is used, caution should be taken for the measurement on sooty flames because the fire-generated smoke and soot can coat the thermocouple bead quickly and change its emissivity, increasing the radiation heat loss. An extrapolation to zero insertion time can be used to eliminate the soot-coating effect on the thermocouple (Saito et al., 1986). When thermocouples are used for field tests, a lead wire that connects the thermocouple and its detector should be as short as possible and lifted above ground to avoid a short circuit due to an accidental step or contact with water. The lead wire should be covered by a heat-insulating material to prevent convective heat loss when significant convective heat loss is expected.

The infrared thermograph technique involves use of an infrared camera and an image recording and analysis system capable of obtaining a two-dimensional thermal image from a remote location. This technique was applied to measure upward flame spread rate on wood slabs (Arakawa et al., 1993) and temperature profiles in forest fires (Clark et al., 1998; 1999). Radke et al. (1999) applied an airborne imaging microwave radiometer to obtain a real-time airborne remote sensing. These techniques can be very useful in identifying the hot spot or pyrolyzing condensed-phase front which can be related to the active flame spread front. Note that the infrared technique may not represent exactly the actual temperature of the flame or condensed phase because the phase emissivities vary from location to location and because absorption by smoke, H_2O, and CO_2 will influence the infrared signal. Arakawa et al. (1993) suggest the use of a band-pass filter (10.6 ± 0.5 μm) to eliminate the absorption effects of these gases.

A color video camera may be used to obtain relative flame temperature change in forest fires because the flame color can be roughly related to the flame temperature. The color video camera technique is simple and useful for seeing relative change throughout the flame, but it is more qualitative than the infrared thermograph. Combination of the thermocouple, infrared thermograph, and color video camera techniques may be most ideal for reseachers to assess flame temperature correctly.

Other temperature measurement procedures that are mainly suitable for laboratory-scale flames include the holographic interferometer (Ito and Kashiwagi, 1988), two-color pyrometer (Gaydon and Wolfhard, 1979), spectrum-line reversal method (Gaydon and Wolfhard, 1979), and coherent antistoke Raman spectroscopy (Demtroder, 1982). The two-color pyrometer employs two different wavelength band-pass filters to eliminate emissivity of the object. It is suitable to measure from a remote location a local temperature of sooty-yellow flames whose profiles are fairly uniform along the direction of measurement. Typical examples are combustion spaces of internal combustion engines and boilers. All other techniques are based on a laser optic system which requires a dark room and accurate alignment. These systems are sensitive to environmental change and not suited to field measurements.

B. Velocity Measurement

A pitot tube velocity measurement is a commonly used velocity measurement technique. It is an intrusive and point-by-point measurement (Sabersky *et al.*, 1989). Pitot tubes can accurately measure velocity in the range of a few meters per second to above 100 m/s, covering most of the wind velocity range in forest fires. The pitot tube is relatively simple and can measure a one-dimensional velocity component, but it also can be applied to a three-dimensional measurement by changing the direction of the pitot tube head when the flow is at steady state.

A simple velocity measurement technique with practical use is a video camera. Recording the fire behavior in a video camera and reviewing it on a monitor screen, researchers may be able to trace the trajectory of smoke and fire brand particles moving in and over the fire area and calculate approximately the local velocity as well as two dimensional velocity profiles. This technique is qualititive at best because the smoke and particles may not exactly follow the flow stream and also the flow may be complex and three-dimensional. Wisely using the video image, however, researchers can gain valuable information on the approximate flow velocity in relation to the overall fire behavior.

Other available velocity measurement techniques, all of which are laser-based methods and only suitable to laboratory experiments, include laser doppler velocimetry (LDV) and particle image velocimetry (PIV) or laser sheet particle tracking (LSPT). LDV is a point-by-point measurement, and PIV and LSPT are capable of measuring two-dimensional, and possibly three-dimensional velocity profiles. All three of these techniques require seeding of small trace particles, laser optics, and a data acquisition system. An example of these techniques applied to flame base structure is shown later.

C. SPECIES CONCENTRATION MEASUREMENT

Gas chromatography is the most commonly used method of analyzing the concentration of species, such as CO_2, CO, H_2O, N_2, O_2, and many different hydrocarbons. A sampling probe collects sample gases by means of either a batch or continuous flow system. The batch sampling requires transport of the sample to a laboratory where the samples will be analyzed. The continuous flow system offers an on-line analysis at the site. The continuous flow system is better than the batch sampling system because the on-line method eliminates the possibility of sample contamination during transportation and storage. For field experiments, portable gas analyzers can be used for the on-line measurement of concentrations of some or all the following species: O_2, N_2, CO_2, H_2O, CO, NO_x, and other hydrocarbons. These instruments require frequent calibration. Their sensitivity and response time should satisfy the required accuracy of each experiment.

A quartz microsampling probe is commonly used for sampling of laboratory-scale flames. It consists of a tapered quartz microprobe with a small sonic orifice inlet which accomplishes rapid cooling and withdrawal of the sampling gas by adiabatic expansion (Fristrom and Westenberg, 1985). During the measurement of water concentration, the sampling line should be heated above the dew point to prevent water condensation; this is called wet-base sampling. Another method is dry-base sampling in which all water is removed before analysis. The dry-base technique requires a water condensation unit in the sampling line. A water-cooled stainless-steel sampling probe is often used to provide dry-base sampling of large scale fires. When an uncooled probe is used for dry-base measurement, a water condensation unit is needed to be sure all water is removed before analysis. The sampling line that connects the probe and the batch (or the probe and the analytical instrument) should be free from contamination by leftover gas or air leaks. The commonly used sampling line materials are Teflon, quartz, and stainless steel.

The previously mentioned sampling techniques are intrusive and provide time-average species concentrations within a specific sampling volume. Physical disturbances due to the probe may be negligible for forest fires, but they may become important for small-scale laboratory flames. For futher information, see Fristrom and Westenberg (1985).

D. FLAME HEIGHT MEASUREMENT

The purpose of flame height measurement is to estimate radiation intensities from the flame, an important parameter for assessing hazards to personnel and the rate of flame spread. The most widely accepted interpretation of flame height is the height where the flame achieves the maximum temperature. Another definition that has more practical application to forest fires is the vertical distance from the flame base to the yellow visible flame tip. These two definitions may not always give the same flame height. For a small candle flame the maximum temperature is near the visible yellow flame tip, while for a 1-m diameter crude oil fire the maximum temperature is achieved at about two thirds of the time-averaged visible flame height, probably because a fire-induced strong turbulent air flow cools the upper portion of the 1-m diameter fire.

Here, two different methods of determining flame height will be introduced. The first is to measure the visible flame tip, and the second is to identify the location of the maximum flame temperature. The values obtained by these techniques are valid to within ±10% uncertainty.

To measure the time-average visible yellow flame height, a motor-driven 35mm still camera or a motion color video camera can be used. Both are capable of measuring temporal flame behavior from which time-averaged flame height can be measured. The latter equipment has an advantage in accuracy because it offers the measurement of flame height over the slow motion video screen. For a smoke-covered flame, however, neither technique is applicable because the flame tip is invisible. For these types of fires, an infrared camera can be used to penetrate the smoke layer and measure the maximum flame temperature location. Caution is required when the infrared camera is used to identify the maximum flame temperature and/or visible flame tip because emissivity of the flame is unknown, and it varies with fuel and location in the flame.

Figure 1 is a simplified plot of dimensionless flame height (visible flame height divided by a horizontal scale of fire) as a function of dimensional parameter measuring Froude number (the ratio of inertia force to buoyancy force) (Williams, 1982). The error bars in the correlation are due to different measurement techniques, different fuels, and different flame height definitions. However, the correlation itself is reasonably high over the 15 orders of magnitude in Froude number and the 6 orders of magnitude in the ratio of flame height to the fuel-bed dimension.

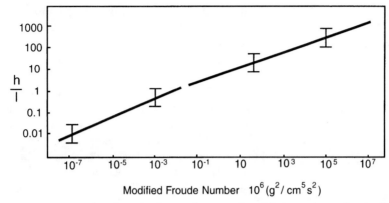

FIGURE 1   Dimensionless flame height as a function of modified Froude number (Williams, 1982). Quantity l is a horizontal scale of the fire.

IV. PREMIXED AND DIFFUSION FLAMES

The idea of categorizing flames into two different types, premixed and diffusion flames, and then studying each type thoroughly has proven to be successful. Many studies have contributed to a thorough understanding of these two types of flames. Based on these studies, much progress has been made in the field of combustion (Williams, 1985; Linan and Williams, 1993; Glassman, 1996; Hirano, 1986). Although diffusion flames can represent forest fires, the premixed flame allows us to better understand the diffusion flame by comparing its structure with that of premixed flames. Thus, in this chapter, characteristics of both diffusion and premixed flames are explained. However, note that these concepts are by no means perfect and may not be applicable to all forest fire phenomena. For example, the microstructure of a flame's leading edge can be interpreted by using a different concept (Venkatesh et al., 1996).

A diffusion flame is defined as any flame in which the fuel and the oxidizer are initially separated. Diffusion flames are represented by candle flames, match flames, wood fires, and forest fires. A premixed flame is defined as any flame in which the fuel and the oxidizer are initially mixed. A good example of a premixed flame is a uniform combustible mixture contained inside a tube that is ignited by a spark creating a flame spreading along the tube (Williams, 1985). The most significant difference between diffusion and premixed flames is that in the former the fuel-oxidizer ratio varies throughout the flame, whereas in the latter it is constant everywhere in the flame. Some flames possess both characteristics, and they will be called partially premixed flames.

According to the flow regime, both diffusion and premixed flames can be classified as laminar, turbulent, or transitional. Laminar flames have smooth, steady flow characteristics in and around the flame, whereas turbulent flames have irregular, disorganized flow characteristics in and around the flame. The use of Reynolds number (the ratio of viscous force to the inertia force) for judging flames to be either turbulent or laminar is not always straightforward because of uncertainty involved in determining characteristic parameters; a visual observation is often good enough to make this judgment. Some flames may be laminar in one section and turbulent in the other section (e.g., the flame shape in a 1-m diameter crude oil fire is turbulent except at the flame base where the flame anchors and flame shape is laminar). In the following discussion, premixed flames and four different types of diffusion flames will be explained.

A. The Premixed Flame and Its Burning Velocity

One important characteristic of the premixed flame is the burning velocity. Assuming the pressure p constant, and with the specific heat c_p of the mixture taken as constant, the overall energy conservation shows that the energy per unit mass added to the mixture by the combustion, H_0, is

$$H_0 = c_p(T_\infty - T_0) \tag{5}$$

where T_0 and T_∞ are the initial and final temperature, respectively. This energy can be determined by the rate of conversion of reactant mass to product mass. Using Eq. (4), this conversion rate (mass per unit volume per unit time) can be written as

$$w = A Y_F^m Y_O^n \exp(-E/R_0 T) \tag{6}$$

where the subscripts F and O identify fuel and oxidizer, the mass fractions Y_F and Y_O are proportional to the concentrations (the number of molecules per unit volume) of the reactants, and m and n are the overall reaction orders with respect to fuel and oxidizer, respectively. In terms of the thickness δ of the flame, the chemical energy released per unit area per unit time may be approximated as

$$H_0 w \delta = \lambda(T_\infty - T_0)/\delta \tag{7}$$

where λ is the thermal conductivity.

Using Eqs. (5) and (7) and solving for δ, we obtain

$$\delta = [\lambda/(c_p w)]^{1/2} \tag{8}$$

which shows that the flame thickness varies inversely with the square root of the reaction rate. The mass of reactant converted per unit area per unit time is $\rho_0 v_0$, where ρ_0 is the initial density of the mixture and v_0 is the laminar burning velocity. By conservation of fuel mass in the steady flow, we have

$$\rho_0 v_0 = w\delta \tag{9}$$

Using Eqs. (8) and (9), the laminar burning velocity takes the form

$$v_0 = (1/\rho_0)[(w\lambda)/c_p]^{1/2} \tag{10}$$

Equation (10) shows that the laminar burning velocity is proportional to the square root of the ratio between the diffusivity, $\lambda/(c_p\rho_0)$, and the reaction time, ρ_0/w. Equation (10) agrees with laminar flame spread experiments using a vertical tube filled with a fuel and oxidizer mixture. When the mixture is ignited, a steady laminar flame spread occurs from the top to the bottom. The flame spread rate can be measured fairly accurately using a motor-driven camera or a high-speed video camera or an array of thermocouples placed along the inner surface of the tube.

B. IGNITION

Ignition can be achieved either by heating the system containing a combustible mixture with thermal energy or by creating chain reactions with autocatalytic reactions. Chemical reactions whose reaction rate is controlled by the concentration of initially present reactants are called thermal, whereas reactions whose rate is affected by the concentrations of the intermediate and final products are called autocatalytic.

Thermal ignition can be achieved by supplying an external source of heat to the system. An electric arc, a spark plug, or a pilot flame can be the external heat source. The external heat source provides excess heat to the system and raises its temperature continuously if the heat input exceeds the heat loss. At some point, the temperature rise will become highly accelerative and a high heat evolution will occur. This is called ignition.

Autocatalytic ignition can be achieved by increasing the rate of chain carrier generation over the termination reaction of chain carriers. The initiation of chain carrier generation may be achieved by thermal energy. After the initiation reaction has started, the reaction may become self-accelerative even after the external heat is removed. Then the system will ignite by satisfying the self-sustaining condition. Lightning may be a good example of autocatalytic ignition.

An important topic in forest fires is determining the condition under which given combustible forest materials can ignite. There are two types of ignition—

spontaneous (unpiloted) ignition and piloted ignition. Unpiloted ignition can be achieved by raising the temperature of the system with hot boundaries and adiabatic compression. Piloted ignition can be achieved by adding heat to the system with an external heat source. Ignition temperature (explained later) for unpiloted ignition is higher than for piloted ignition. For example, ignition temperature for piloted ignition of cellulose is approximately 350°C, whereas that for unpiloted ignition is approximately 500°C (Williams, 1982).

Other important aspects of ignition related to forest fires include ignition temperature, flammability limits, and minimum ignition energy.

C. Ignition Temperature

Ignition temperature can be defined as the critical temperature that the condensed phase needs to achieve for burning to begin (Williams, 1985). The ignition temperature is a useful criterion to assess the flammability of materials and estimate flame spread rate. However, there are some variations among different ignition conditions. For example, some researchers use a black body as the heating source, and others use a halogen or infrared lamp (Babrauskas, 1988).

Ignition tests conducted in an inert atmosphere under various external heat flux conditions showed that the ignition temperature indeed varies as a function of external radiant heat flux. The higher the external heat flux, the higher is the ignition temperature (Delichatsios and Saito, 1991). Thus, some caution should be taken when laboratory ignition temperature data are applied to forest fire problems.

D. Flammability Limits

If there is too little or too much fuel in a fuel-air mixture, ignition may not occur even if a pilot flame provides sufficient heat to the mixture. For example, a propane + air mixture with fewer than 2.2 mole of propane or more than 9.5% of propane can't ignite. The maximum and the minimum concentrations of fuel within which flame can spread are called, respectively, rich and lean limits of flammability. Flammability limits depend on pressure and temperature; such data for gases are well established (Zabetakis, 1965). Flammability of solid fuels can be linked to ignition of solids that release pyrolysis products under external heating. These pyrolysis products will mix with air forming a flammable mixture whose flammability can be tested. Thus, knowledge gained through studies on flammability limits for gases can be applied to understand the flammability of solids.

E. MINIMUM IGNITION ENERGY

Williams (1985) offers a simple and practically useful definition of minimum ignition energy. "Ignition will occur only if enough energy is added to the gas to heat a slab about as thick as a steadily propagating adiabatic laminar flame to the adiabatic flame temperature." Thus, the minimum ignition energy, H, can be described by

$$H = (\tilde{A}\delta)\rho_0[c_p(T_\infty - T_0)] \qquad (11)$$

where $\tilde{A}$ is the cross-sectional area of the slab, δ is the adiabatic laminar flame thickness, ρ_0 is the initial density of the mixture, c_p is the average specific heat at constant pressure, T_∞ is the adiabatic flame temperature, and T_0 is the initial temperature of the mixture. Using $\delta \approx \lambda/c_p\rho_0 v_0$ (Williams, 1985), Eq. (11) can be written as

$$H/\tilde{A} = \lambda(T_\infty - T_0)/v_0 \qquad (12)$$

where λ is the average thermal conductivity of the gas and v_0 is the laminar burning velocity discussed earlier.

V. EXTINCTION OF DIFFUSION FLAMES

There are two main flame extinction criteria: the Damköhler extinction criterion and the Zel'dovich extinction criterion, both of which were detailed by Williams (1985) and Linan and Williams (1993). Here a summary of the Damköhler extinction criterion is presented.

Extinction can be defined as a balance between a chemical time, t_c, and a residence time, t_r, in the gas-phase flames. A nondimensional parameter known as Damköhler number, Da, is defined as the ratio of a flow (or diffusion) to chemical reaction times in the gas:

$$\text{Da number} = (\text{A flow or air diffusion time})/(\text{Chemical reaction time}) = t_r/t_c \qquad (13)$$

Using the chemical time, $t_c = AY_FY_O \exp(E/R_0T)$, and the residence time, $t_r = l/u$ or l^2/α, Eq. (13) can be written as

$$\text{Da number} = (l/u)(AY_FY_O)^{-1} \exp(-E/R_0T)$$
$$\text{or} \quad (l^2/\alpha)(AY_FT_O)^{-1} \exp(-E/R_0T) \qquad (14)$$

where u is the characteristic flame velocity, l is the characteristic length, α is the thermal diffusivity of the gas, and $m = n = 1$ denoting the first-order chemical reaction.

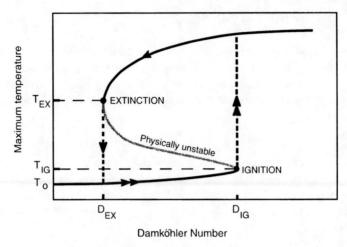

FIGURE 2   Schematic illustration of the dependency of the maximum temperature on the reduced Damköhler number for diffusion flames. Here, for cellulose-based fuels, T_{EX} (extinction temperature) = 1300 ± 100°C, T_{IG} (ignition temperature) = 350 ± 50°C, and T_0 (initial fuel temperature) = 20°C.

The maximum temperature can be taken as an important parameter to control extinction. Figure 2 shows temperature as a function of the Da number, a well-known S-shaped curve. Variation of parameters that control fires causes the Da number to change continuously. For example, if the Da number based on T_0 is increased beyond the critical Da number marked as ignition (D_{IG}), then a discontinuous change to a much higher temperature will occur (this passage is shown in a double arrow). If the upper branch of the Da number is decreased continuously to the critical Da number marked as extinction (D_{EX}), then a discontinuous change to a lower temperature will occur (this passage is shown in a single arrow). At the lower branch temperature, there is no combustion, whereas at the upper branch temperature, active combustion is taking place. The middle branch is unstable and not physically relevant. The Damköhler extinction criterion is useful in obtaining overall chemical kinetic parameters for gas-phase combustion from experimental data.

There are eight different strategies to achieve extinction: isolate the fuel, isolate the oxidizer, cool condensed phase, cool the fuel, cool the gas, inhibit the chemical reaction homogeneously, inhibit chemical reaction heterogeneously, and blow the flame away (Williams, 1982).

The Zel'dovich extinction criterion, derived by assuming a large activation energy asymptote (Linan and Williams, 1993), is beyond the scope of this book. Readers who are interested in the Zel'dovich extinction criterion may consult Linan and Williams (1993).

VI. DIFFUSION FLAMES
AND SCALING ANALYSIS

Diffusion flames can be classified based on the forces that control the flow and mixing processes. The major forces acting on the diffusion flames may be assumed to include momentum, viscous, and buoyancy forces. Molecular diffusion driven by concentration differences is also assumed to play an important role in the mixing process. Based on these assumptions, scaling analysis (Williams, 1969; Emori and Schuring, 1977) can be conducted to categorize the diffusion flames into four different types. Scaling laws are important in fire research as described in the scaling law section later in this chapter. By defining u_0 = the fuel exit velocity, d = the burner port diameter, D = the binary diffusion coefficient of a fuel into air, g = the acceleration due to gravity, and ν = the kinematic viscosity, the following three pi-numbers (dimensionless numbers) can be formed when a fire burns in calm air:

- Peclet (Pe) number = [Fuel mass transferred by momentum of incoming fuel flow]/[Fuel mass transferred by concentration difference] = $u_0 d/D$
- Reynolds (Re) number = [Momentum force of incoming fuel flow]/[Viscous force of air and fuel] = $u_0 d/\nu$
- Froude (Fr) number = [Momentum force of air induced by the flame]/[Buoyancy force in the flame] = u_0^2/dg

Here four different types of flames, each identified by different combinations of the preceding three pi-numbers, will be discussed.

A. Re NUMBER-CONTROLLED LAMINAR
DIFFUSION FLAMES WITH LOW Pe NUMBER

Consider the case of a hypodermic needle held vertically and in which a hydrocarbon fuel constantly flows through the bottom end to the top. When the fuel is ignited, a small laminar diffusion flame with flame height of a few millimeters is established on the port of the hypodermic needle. This flame is called a microdiffusion flame. The existence of this type of flame was predicted by Williams (1985) and proven experimentally by Ban et al. (1994). It is governed by the balance between the momentum of the incoming fuel flow and the molecular diffusion (both in axial and radial directions) of the gaseous fuel and air. The shape of a microdiffusion flame is spherical and different from the more commonly observed candlelike diffusion flame.

The momentum and species equations are coupled, and a similarity solution has been obtained for the coupled system. Microdiffusion flame experiments on CH_4, C_2H_2, and C_2H_4 were conducted under ambient atmospheric pressure conditions. Flame shape was photographically recorded and compared well with the theoretical predictions. Figure 3 (see color insert) shows a photograph and a schematic of an ethylene-air microdiffusion flame established on a 0.25-mm-diameter hypodermic needle (Ban *et al.*, 1993).

B. Re Number-Controlled Turbulent Jet Diffusion Flames with High Pe Number

By increasing the incoming fuel velocity in the previously mentioned hypodermic needle, the flame will become unstable, and at some point the flame can no longer be sustained at the burner port. This is because the characteristic time of the incoming fuel velocity is smaller than the time required for the flame to complete the chemical reaction. By increasing the diameter of the burner, a stable flame with a higher Re number can be achieved. For example, a hydrogen jet diffusion flame can be stabilized on a 1.43-mm-diameter nozzle burner up to Re = 4800, and a methane jet diffusion flame can be stabilized on a 9.45-mm-diameter nozzle burner up to Re = 8440 (Takahashi *et al.*, 1996). Figure 4 (see color insert) shows a hydrogen-air turbulent jet diffusion flame stabilized on the 1.43-mm-diameter burner (Takahashi *et al.*, 1996). For turbulent jet diffusion flames, the flame will be established as the result of the mixing of the incoming fuel jet with quiescent air. The momentum force controls the fuel jet, and the inertia and viscous forces of both the air and the fuel control the mixing process. This type of flame (high Pe and Re numbers) is stabilized by a turbulent shear mixing, but not by molecular diffusion (Takahashi *et al.*, 1996; Venkatesh *et al.*, 1996).

C. Re and Fr Number-Controlled Laminar Diffusion Flames with Low Pe Number

Applying a relatively slow fuel flow rate (approximately a few centimeters per second) to the burner with a diameter of a few millimeters to a few centimeters, a candlelike flame can be established over the burner port. This is the regime of buoyancy-controlled laminar diffusion flames. Buoyancy forces acting on the hotter, less-dense gas in the region of the flame often distort the shape of the flame sheet. Figure 5 shows a candlelike laminar methane-air diffusion flame established on a 1.6-cm-diameter Pyrex burner. Faraday (1993, originally published in 1861) was the pioneer to identify this type of diffusion flame as the most basic type and conducted a series of noble experiments. After Faraday,

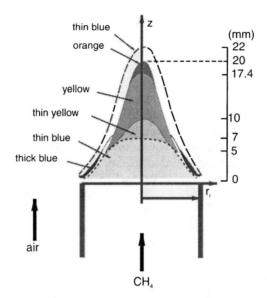

A small laminar methane - air diffusion flame

FIGURE 5   A schematic of a small methane-air diffusion flame established on a 1.6-cm diameter Pyrex tube (the average fuel flow rate at the burner port is 2.2 cm/s and the air flow rate is 9.3 cm/s).

many researchers have used this type of diffusion flame to further understand their chemical, thermal, and fluid mechanical structures. These flames are also ideal for optical and chemical species sampling measurements because they are stable, steady, and large enough to obtain accurate measurements.

The structure of laminar diffusion flames in this category is also of current interest for a number of reasons. There have been recent demands for understanding the mechanisms of turbulent flames. A promising approach to the analysis of turbulent flames is to introduce conserved scalars (Linan and Williams, 1993) so that data on laminar flames can be used. Correlations of laminar flame structures on the basis of mixture fractions allow turbulent flame properties of interest to be calculated with relative simplicity. The mixture fraction, Z, for the two-feed system (oxidizer and fuel) can be defined as

$$Z = (Y_i - Y_{i2})/(Y_{i1} - Y_{i2}) \tag{15}$$

Here Y_i is the mass fraction of the i-chemical element; the subscripts 1 and 2, respectively, refer to the composition in the fuel and the oxidizer flows. There are indications that the correlations can be approximately applied even for soot and chemical species that are known to be out of the chemical equilibrium region (Koylu and Faeth, 1991). If the binary diffusion coefficients of all pairs of species are approximately equal, then the concentrations of all species can be uniquely related to Z through algebraic equations. If this condition is not satisfied, then different diffusion rates can cause Z for different elements (N, O, C, H) to take different values (Linan and Williams, 1993).

Here, structure of a laminar coflow methane-air diffusion flame will be introduced because methane is the simplest saturated hydrocarbon, yet it can represent the general nature of hydrocarbon-air combustion. A 20-mm-high methane-air diffusion flame, established on a 16-mm-diameter Pyrex burner, was probed by a microsampling probe, and 15 different steady chemical species were separated; concentration of each of these 15 species was measured with a gas chromatograph. Two-dimensional profiles of steady flame temperature were measured by traversing a fine thermocouple. Figure 5 shows a schematic of the laminar methane-air diffusion flame that was probed using the experimental apparatus shown in Figure 6 (Saito et al., 1986). The main elements of the apparatus include a quartz coflow burner with Pyrex chimney that can produce a stable and steady laminar diffusion flame, a quartz microsampling probe that can probe chemical species from the flame, fuel and air supply and control sys-

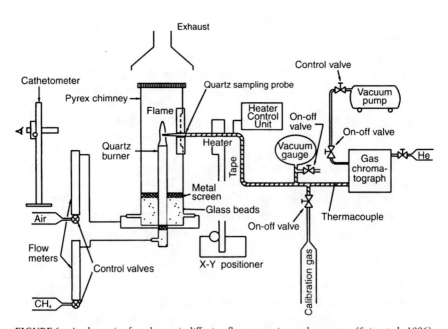

FIGURE 6   A schematic of methane-air diffusion flame experimental apparatus (Saito et al., 1986).

tem, a gas chromatograph, and a cathetometer for flame height measurement. The sampling line that connects the quartz probe and gas chromatograph is wrapped with a heating tape to prevent condensation of water and heavy molecular species.

The distinctive color differences shown in the Figure 5 flame can be explained as follows. A thick blue near the burner port is associated with chemiluminescence from exited CH and CO molecules. Above that thick blue, there is a thin blue (chemiluminescence from CO) covering the entire flame. Inside the thin blue, there is yellow-soot region and on the top of that yellow-soot region, a narrow orange-soot region is formed. The yellow- and orange-soot regions are clearly separated. The two soot-color zones were found by Saito et al. (1987). At the yellow region, young soot particles are formed, and the soot particle temperature is higher than the gas temperature because the young soot particles can actively react with the gas molecules, causing a strong exothermic reaction on the soot particle surface. However, at the orange-soot region, young soot becomes mature and has lost its surface reactivity, bringing the soot particle and gas temperatures into equilibrium.

The centerline profiles of chemical species of the flame is shown in Figure 7. The structure of the methane-air diffusion flame, based on results from careful measurements such as Figure 7, is illustrated in Figure 8 (Linan and Williams, 1993). In Figure 8, Z_c is the stoichiometric mixture fraction, and the one-step rate function, Eq. (6), is used. At the point, $Z = Z_c$, the fuel and oxidizer are consumed at an infinite rate, releasing the maximum heat (and achieving the maximum temperature) due to the complete combustion on the stoichiometric fuel-oxidizer mixture. The mixture is fuel-lean for $Z < Z_c$ and fuel-rich for $Z > Z_c$; for both cases, temperature drops from the maximum. At or around $Z = Z_c$, an active chemical reaction is taking place by balancing with diffusion of chemical species. This region is called the reactive-diffusive zone. There are two separate zones in each side of the reactive-diffusive zone. One is the $Z < Z_c$ side where oxygen diffuses into the flame and the other is the $Z > Z_c$ side where fuel diffuses into the flame. In these (convective-diffusive) zones, no chemical reaction takes place and convection and diffusion terms control the structure.

Kinetic models have been proposed to calculate concentrations of species for methane-air flames. These models often consist of more than 200 chemical reaction steps, but simplifications are needed to understand the structure. A reduced kinetic mechanism has been proposed by carefully selecting only essential reactions. Williams proposed to use the minimum sets that can describe the main feature of the flame and apply steady state approximations in this set to obtain a reduced mechanism. Table 1 shows the proposed four-step mechanism (Linan and Williams, 1993) consisting of fuel consumption, water-gas shift, recombination, and oxygen consumption and radical production. The Table 1 results were obtained from guesswork on 14 fundamental reaction steps offered

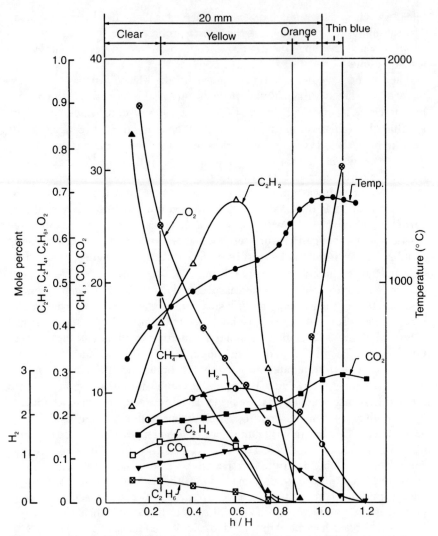

FIGURE 7 The centerline profiles of temperature and chemical species for a laminar methane-air diffusion flame established over an overventilated 1.6-cm-diameter coflow burner (Saito *et al.,* 1986).

by Linan and Williams (1993) and applying steady state approximation for the intermediaries, O, OH, HO_2, CH_3, CH_2O, and CHO. The usefulness of this four-step mechanism has been demonstrated. Even simpler three- or two-step mechanisms were also proven to be sufficient to explain many aspects of flame structure.

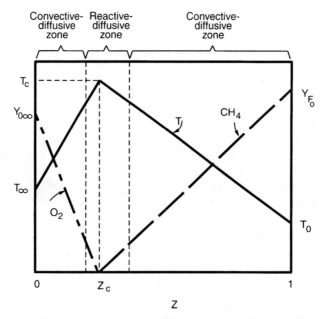

FIGURE 8 Thermal and chemical structures of the methane-air diffusion flame plotted against mixture fraction Z.

TABLE 1 Four-Step Mechanism for the Methane-Air Flame

Fuel consumption	$CH_4 + H \rightarrow CH_3 + H_2$
	$CH_3 + O \rightarrow CH_2O + H$
	$CH_2O + H \rightarrow CHO + H_2$
	$CHO + M \rightarrow CO + H + M$
	$H + OH \rightarrow O + H_2$
	$H + H_2O \rightarrow OH + H_2$
	$CH_4 + 2H + H_2O \rightarrow CO + 4H_2$
Water-gas shift	$CO + OH \Leftrightarrow CO_2 + H$
	$H + H_2O \Leftrightarrow OH + H_2$
	$CO + H_2O \Leftrightarrow CO_2 + H_2$
Recombination	$O_2 + H + M \rightarrow HO_2 + M$
	$OH + HO_2 \rightarrow H_2O + O_2$
	$H + H_2O \rightarrow OH + H_2$
	$2H + M \rightarrow H_2 + M$
Oxygen consumption and radical production	$O_2 + H \Leftrightarrow OH + O$
	$O + H_2 \Leftrightarrow OH + H$
	$OH + H_2 \Leftrightarrow H_2O + H$
	$OH + H_2 \Leftrightarrow H_2O + H$
	$O_2 + 3H_2 \Leftrightarrow 2H_2O + 2H$

From Linan and Williams, 1993.

D. Fr NUMBER-CONTROLLED TURBULENT
DIFFUSION FLAMES WITH HIGH Pe NUMBER

When the velocity of incoming fuel jet is relatively low compared to the velocity of hot gas accelerated in the turbulent flame zone by buoyancy, a buoyancy-controlled turbulent flame will be obtained. Good examples of this type of flame are turbulent liquid pool fires, which can be created by igniting a liquid fuel placed in a relatively large diameter open top pan (say larger than 1 m for most hydrocarbon fuels and a lesser diameter for alcohol fuels). When the diameter of the pan is less than a few centimeters, the flame shape will be laminar. When the diameter is approximately between a few centimeters and 1 m, the flame will be in a transient regime between laminar and turbulent. When the diameter is larger than 1 m, the flame exhibits fully developed turbulence and often emits black smoke; sometimes the entire flame will be covered by a thick dense smoke.

Figure 9a (see color insert) shows a diesel oil fire of a 15-m^2 open-top container, conducted in an open field with a wind of approximately 9–12 m/s; a thick dense smoke covers almost the entire flame making the flame height measurement difficult. An infrared thermograph technique, described in the temperature measurement section, was applied to obtain a two-dimensional temporal map of flame temperatures from a remote location (Figure 9b; see color insert). Interestingly, the infrared image in Figure 9b shows the highest temperature zone for such a large-scale pool fire to be near the flame base. A similar result was found in a recent large-scale crude oil fire where an infrared (IR) thermograph was used to measure temperature distributions of a smoke-covered flame (Koseki et al., 1999). It is not clear whether the result is false due to the change of emissivity of the flame or real. If it is real, it may be attributed to the enhanced cooling effect by buoyancy-induced convection of outside air in the upper portion of the flame where a large-scale puffing motion takes place.

E. SCALING LAWS

Forest fires spread nonuniformly and nonsteadily over nonhomogeneous materials. To understand complicated mechanisms of forest fires, it may be necessary to design a laboratory-scale test without changing the essential chemistry and physics governing the forest fires and measure specific aspects of forest fire behavior under well-controlled laboratory conditions. Prediction models then can be developed against these experimental data. Scale modeling is an experimental technique to realize these needs. The purpose of scale modeling is well described by Johnstone and Thring (1957): "commit your blunders on a small scale and make your profits on a large scale." The role of scale modeling in the study of forest fires is certainly important because many forest fire phenomena are so large that full-scale experiments are difficult to conduct due to safety conditions, economical and time constraints, as well as technical difficulty in controlling the parameters. In scaling combustion and fire phenomena, the same temperature for both the model and the full-scale may be required to make the same chemical reactions for both.

The rules that researchers must follow in setting up and interpreting the results of scale model experiments are called scaling laws or model rules (Emori and Schuring, 1977). To give a reliable forecast of full-scale phenomena, the development of scaling laws is very important, and indeed the use of scale models can be justified only if they are able to provide information relevant to the full-scale phenomena.

Governing equations can be applied to obtain strict scaling laws shown below.

$$\pi_i = \pi_i' \tag{16}$$

where π_i (pi-numbers) represent dimensionless numbers involved in the full-scale (or prototype) phenomena ($i = 1, 2, 3, \ldots$), and the prime represents the scale model. Williams (1969) showed us that the strict governing equations in combustion systems yield 29 pi-numbers. Thus, it is difficult to satisfy the Eq. (16) requirement. For example, take π_1 as Reynolds number and π_2 as Froude number. Then π_1 requires the characteristic velocity to be proportional to (Length scale)$^{-1}$, whereas π_2 requires the characteristic velocity to be proportional to (Length scale)$^{1/2}$, resulting in a scaling conflict. However, if we can find reasons to neglect one of them, then the scaling problem can be resolved. Thus, the essential problem of scale modeling is to select only a few groups of pi-numbers and discard others to realize the scaling laws.

Scaling laws were derived for the last type of flames, referred to as pool fires, by employing the following assumptions (Emori and Saito, 1983): (1) the pressure is atmospheric for both model and the full-scale fire, (2) the dominant heat input to the fuel surface is given by radiation from the optically thick flame, (3) conducted heat through the rim of the pan is negligibly small, (4) inertia and buoyancy terms in the motion of gas and air are dominant compared with viscous term, and (5) the flame pulsation phenomenon is governed by the inertial and buoyant effect of combustion gas and air.

With the same ambient conditions for both the scale model and the full-scale model and with the use of the same fuel under the same ambient conditions, scaling laws were reduced to a functional relationship of the form

$$\Phi\{u_1/u_2,\ h/d,\ df/u_1,\ dg/u_2^2,\ E_r/(\rho_f q_f V_f)\} = 0 \tag{17}$$

where u_1 is the lateral wind velocity, u_2 is the buoyancy-induced velocity, h is the visible flame height, d is the diameter of the liquid pool, f is the frequency of the flame pulsation, g is acceleration due to gravity, E_r is the radiant heat flux received by a radiometer at geometrically similar points, scaled according to the size of the fuel bed, V_f is the rate of decrease of height of fuel surface by combustion, Φ is an arbitrary function, ρ_f is the density of the fuel, and q_f is the heat value per unit mass of liquid fuel.

It was found experimentally that the regression rate V_f does not depend on the diameter of the fuel container. The burning time $t_b = h_f/V_f$, where h_f is the depth of liquid fuel in the container, does not depend on the size of container if the initial fuel thickness is the same. As a result, the condensed phase is controlled by $E_r/(\rho_f q_f V_f)$ in Eq. (17), whereas the flame behavior is governed by four other pi-numbers, u_1/u_2, h/d, df/u_1, dg/u_2^2. This is the major difference between pool fires and crib fires (to be explained in the following section), where the common group of pi-numbers governs both condensed and gas phases.

Equation (17) can be used to design scale models and determine specific experimental conditions for scale model experiments. Using h/d and dg/u_2^2, for example, the model experiments are to be performed with wind velocities proportional to the square root of the fuel bed diameter. The scaling then predicts variations of flame height and vertical velocity in the flame. Wind tunnel fire experiments can be designed to verify these predictions (Emori and Saito, 1983).

F. WOOD CRIB FIRES

In contrast to the pool fires where the regression rate of the fuel is the same in winds of different velocities, the burning rate of wood crib fires can easily be increased by flowing air on it. Thus, scaling laws on crib fires employ the same assumptions—(1), (3), (4), and (5)—as employed for the pool fires. Assumption (2), however, is modified to account for energy transfer by buoyancy. Thus, assumption (2) for crib fires includes both radiation and convection as the dominant heat input. Figure 10 shows the configuration of a crib fire bed and a Japanese cedar crib fire. The configuration of the crib promotes three-dimensional turbulence so that the flow remains turbulent in cribs at smaller diameters than in pool fires, suggesting that the geometrical configurations play an important role in combustion.

With the use of the same crib materials and under the same ambient conditions, the pi-numbers obey a functional relationship of the form

$$\Psi\{u_1/u_2, h/l_c, l_c/u^2 t, l_c g/u_2^2, E_r t/(\rho_c l_c q_c)\} = 0 \qquad (18)$$

In addition to the symbols defined for pool fires, here l_c is the length of the crib stick, t is the characteristic time, Ψ is an arbitrary function, ρ_c is the density of the crib stick, and q_c is the heat value per unit mass of the crib material.

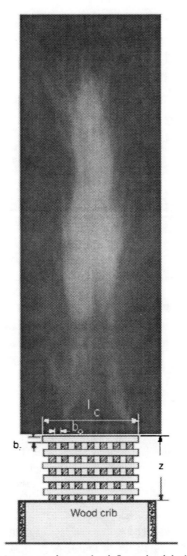

FIGURE 10   A Japanese cedar wood crib fire and crib bed configuration.

Experimental data are available to support the proposed scaling laws (Emori and Saito, 1983). Figure 11 shows that the temporal changes of irradiance from pool fires of 10 m (prototype) and 1.2 m (the scale model) in diameter become the same at corresponding times after the burning. This is consistent with the scaling law prediction, $E_r/(\rho_f q_f V_f)$, which yields $E_r/V_f = E_r'/V_f'$ and $V_f = V_f'$ for

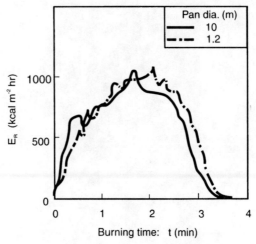

FIGURE 11 Irradiance measured at geometrically similar points as a function of time after ignition for hexane pool fires (Emori and Saito, 1983).

pool fires using the same fuel. For crib fires, the temporal change of total irradiance, both irradiance and time, are scaled to the size of the crib with $l_c = 90$ cm $(= l_3)$ according to the scaling laws l_c/u^2t, l_cg/u_2^2, and $E_rt/(\rho_cl_cq_c)$. Figure 12, temporal change of total irradiance from crib fires of three different sizes, proves the scaling law predictions [Eq. (18)].

Byram and Nelson (1970) and Porscht (1975) confirmed the prediction from df/u_1 and dg/u_2^2 that the pulsating frequency becomes inversely proportional to the square root of characteristic length (e.g., diameter) of the fuel bed. Blinov and Khudyakov (1957) and Emori and Saito (1983) showed that irradiance E_r and burning time t are independent of pool diameter. Emori and Saito verified that the burning rate of a 1.2-m-diameter hexane pool fire is little influenced by lateral wind of up to 10 m/s and that the flame inclination is the same if dg/u_2^2 is kept constant.

The concept of pool and crib fires can be used to categorize the currently existing various types of fires. In forest and urban fires (Emori and Saito, 1982; Soma and Saito, 1991), for example, the heat release rate and the velocity of air is known to be respectively scaled by (Characteristic length)$^{5/2}$ and (Characteristic length)$^{1/2}$. In enclosure fires (Quintiere, 1988), such as a fire in the room of a house, the burning rate is known to depend on ventilation from openings, and the Fr number can scale the velocity of air. These fires, therefore, can be categorized as crib fires. On the other hand, a fire on a railroad passenger car was best characterized as a combination of pool and wood crib fires (Emori and Saito, 1983).

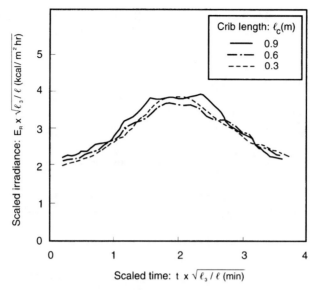

FIGURE 12   Scaled irradiance measured at a geometrically similar location as a function of scaled time for crib fires (Emori and Saito, 1983). The l is characteristic length of crib bed, and l_3 is the crib length of 90 cm. The square root relationships are based on Fr number scaling described in Eq. (18).

VII. SPREADING FLAMES

In forest and urban fires, both laminar and turbulent flame could spread either in one-dimensional, two-dimensional, or three-dimensional fuel configurations with steady or unsteady modes of spread. However, most of the fire spread modes will be turbulent with strong or minor interactions with wind. Flame spread direction can be either concurrent or opposed to the wind direction. Because of the difficulty associated with the unsteady nature of turbulence, most of the detailed measurements on spreading fires are done for a two-dimensional laminar flame with a steady state spreading rate over a one-dimensional or two-dimensional fuel surface. Theoretical models developed range from simple one-dimensional laminar steady state models to complex three-dimensional turbulent spread models (Williams, 1976; Wichman, 1992). These models, derived for and tested in well-controlled laboratory conditions, address the fundamental nature of flame spread.

The structure of spreading flames over condensed fuels, either liquid or solid, has been studied both theoretically and experimentally. A difficulty, however, lies in the transient measurement technique, which is far more difficult than the steady-state measurements because of the highly transient nature of flame spread behavior. It has been a challenge to obtain the two-dimensional and three-dimensional profiles of chemical species concentration, velocity, and temperature for the spreading flames. Progress has been made recently by an international collaborative effort (Ito *et al.*, 1999) to use simultaneously laser sheet particle tracking (LSPT), single-wavelength (HI), or dual-wavelength (DWHI) holographic interferometer and infrared thermograph techniques. This simultaneous measurement technique made transient two-dimensional measurements of velocity, fuel concentration, and temperature for spreading flames possible. A dual-wavelength hologram was obtained for a spreading flame over propanol that was placed in a Pyrex tray 30 cm long and 1 cm wide (Ito *et al.*, 1999). Two different lasers—argon-ion and He-Ne—were used as the light source. The dual-wavelength hologram contains two different data—temperature and fuel concentrations. For the temperature holograms, each colored line represents an isothermal contour, whereas for the concentration holograms it represents equal fuel density contours. From the second hologram, the fuel vapor concentration profiles in the upstream from the spreading flame-front was obtained.

Fire spread can occur if there is communication between the burning region and the virgin fuel. With the definitions of v = the flame spread rate, q = the net energy per unit area per second transported across the surface of fire inception, and Δh = the difference in thermal enthalpy per unit mass between the fuel at its ignition and the virgin fuel, the energy balance can be written as

$$\rho_f v \Delta h = q \tag{19}$$

Flame spread can be classified based on its spreading direction (i.e., downward, horizontal and upward). Using this classification, for example, flame spread over a 15-degree angle up-slope hill can be treated as the combination of upward and horizontal spread.

Many other parameters can influence the spread rate in forest fires. Those parameters include the type of fuel (thermally thick or thermally thin solid fuel, char-forming or non-char-forming, porous, liquid, etc.), the fuel conditions (moisture contents, temperature, and aging), the size and geometry (discrete or continuous) of the fuel bed, and the ambient (wind, temperature, and humidity) conditions. Rothermel (1972) extended the flame spread equation (19) to forest fires by adding two important components, wind and fuel bed slope effects:

$$v = q/\rho_f \Delta h = I_R \xi (1 + \Phi_w + \Phi_s)/\rho_f \omega Q_{ig} \tag{20}$$

where I_R = reaction intensity, ξ = propagating flux ratio (the proportion of the reaction intensity that heats adjacent fuel particles to ignition), Φ_w = a dimensionless multiplier that accounts for the effect of wind in increasing the propagating flux ratio, Φ_s = a dimensionless multiplier that accounts for the effect of slope in increasing the propagating flux ratio, ω = the effective heating number (the proportion of a fuel particle that is heated to ignition temperature at the time flaming combustion starts), and Q_{ig} = ignition energy. Here $\Delta h = \omega Q_{ig}$. Further explanation of Eq. (20) is not offered here because it is available in Rothermel (1972) and Pyne et al. (1996).

Williams (1982) stresses that developing a general formula to predict the spread rate including all the preceding parameters is not practical nor the best avenue. He rather suggests developing the spread rate formula for special cases and studying each case thoroughly. Taking his advice as a guide, four types of fire spread that have application to forest fires are summarized here (Williams, 1976, 1982). Note that for downward and horizontal spread, radiation is believed to become important, whereas for upward spread both radiation and convection become important.

A. Downward and Horizontal Flame Spread

Types of forest fuel may be categorized as thermally thin or thermally thick. For thermally thin fuels, temperature rises across the entire cross section of the virgin fuel prior to arrival of the flame. For thermally thick fuels, the fuel interior may not be heated appreciably prior to arrival of the flame, and the temperature won't rise throughout the entire cross section of the virgin fuel prior to arrival of the flame. For downward and horizontal spread, the spread direction is opposed to the fire-induced flow direction, resulting in a slow and steady spread rate (≈ 1 cm/s or less).

Downward and horizontal spread may be associated with an early stage of flame spread in forest fires. A small circulation that was found in the gas phase just ahead of the spreading flame's leading edge may play an important role in providing heat from the flame to a virgin fuel surface (Ito et al., 1999). Gas-phase heat conduction and radiation may also be important.

For forest fires, radiation from the flame may be a dominant heat transfer mode of flame spread (Albini, 1986). Applying $q = (\varepsilon \sigma T^4 h) \sin \theta / \ell$ to Eq. (19), Eq. (21) will be obtained for thermally thin fuels:

$$v = (\varepsilon \sigma T^4 h) \sin \theta / (\ell \rho_f \Delta h) \tag{21}$$

where h is the visible flame height, θ is the angle between the flame (usually the vertical dimension) and the exposed virgin fuel surface ($\theta > \pi/2$), and ℓ is fuel bed thickness. When the fuel is thermally thick, it won't be heated throughout its depth prior to ignition. Then, using thermal diffusivity α ($= \lambda / c_p \rho_f$), the fuel thickness that is heated by the flame, $(\alpha h/v)^{1/2}$, can be used instead of the fuel thickness ℓ in Eq. (21). Thus, for thermally thick fuels,

$$v = (\varepsilon\sigma T^4)^2 h \sin^2 \theta / [\alpha\rho_f^2(\Delta h)]^2 \tag{22}$$

B. TURBULENT UPWARD FLAME SPREAD

Upward flame spread occurs with a large heat release rate created by rapid ($\approx$ 1 m/s or more) and acceleratory spread modes because the flame comes in contact with the fuel in the downstream direction effectively preheating the fuel surface by convection and radiation. Upward spread rate is much faster than downward and horizontal spread rates where buoyancy-induced flow has either negative or no effect. Saito *et al.* (1985) conducted a study of upward flame spread over vertically suspended (PMMA, wood and particle board) wall samples (0.3 m wide × 1.2 m high × 0.015 m thick) and formulated the spread rate. In the forest, upward flame spread will occur when a tree trunk ignites near the root and flame spreads along the trunk. Figure 13 shows a schematic of upward flame spread over a vertical wall. The spread rate is given by

$$v = 4q^2(x_f - x_p)/[\pi\lambda\rho_f c_p(T_p - T_0)^2] \tag{23}$$

where x_f is the visible flame tip height, x_p is the pyrolysis flame tip height, T_p is the ignition temperature, and T_0 is the initial fuel temperature. For a wall burning in building fires, a typical value of $q = 2$–2.5 W/cm². The value q may be considerably larger for forest fires where intense radiant heat is generated from the surrounding burning trees.

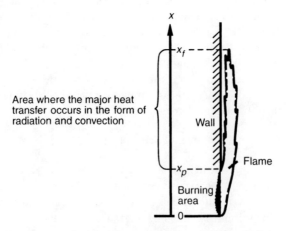

FIGURE 13 A schematic of upward flame spread over a vertical wall (Saito *et al.*, 1985).

C. Flame Spread through Porous Fuel Beds

Forests can be treated as porous fuel beds consisting of layers of dead leaves and pine needles and thickets of trees. When these fuels are burned, flame spread occurs through the porous media fuel beds. When radiation from the burning fuel element dominates the heat transfer, Eq. (19) becomes

$$v = \varepsilon\sigma(T_f^4 - T_0^4)/f_p\rho_f(\Delta h) \tag{24}$$

where T_f is the burning fuel surface temperature, T_0 is the initial fuel temperature, and f_p is the porosity of the fuel beds. Progress has been made recently using a two-dimensional numerical calculation model to predict the maximum spotting distance in fires over porous media (McDonough et al., 1998). This numerical model can accommodate variable geometrical shape of terrain and wind speed. The model prediction was roughly verified against qualitative observations made during forest fires.

Wind-aided flame spread is important in forest fires because of the enhancing effect of winds on spread rate in running crown fires (Pyne et al., 1996), firebrands and spotting fires (Albini, 1979), and fire whirls and fire storms (Emori and Saito, 1982; Soma and Saito, 1991). Detailed explanations and discussions appear in Chapters 5 and 6 in this book and are also available from Albini (1981), Carrier et al. (1991), Wolff et al. (1991), and Pyne et al. (1996), so no further discussion is provided here.

VIII. STRUCTURE OF FLAME BASE

One of the important aspects of diffusion flames is the mechanism of flame anchoring. A commonly raised question is why and how diffusion flames are anchored. Study of the anchoring mechanism of diffusion flames is important because of its connection to fire extinguishment, control of fire spread, and flame instability. Here, two different mechanisms of flame anchoring will be explained: one sustains Fr number controlled pool fires and the other sustains Re number-controlled turbulent jet diffusion flames. Figure 14 shows an LSPT experimental apparatus applied to a small-scale pool fire. Flame-induced flow near and around the flame leading edge where the pool fire is anchored was visualized by the LSPT technique. Figure 15 shows a two-dimensional flow structure visualized by the LSPT technique for a 45-mm-diameter propanol pool fire (Venkatesh et al., 1996).

For the Re number-controlled turbulent jet diffusion flames, the mechanism of anchoring depends on the formation of shear-stress-related circulation zones. Because of the shearing between the fuel jet, whose flow speed may be much higher than the flame propagation speed, and the oxidizer, a stagnant circulation

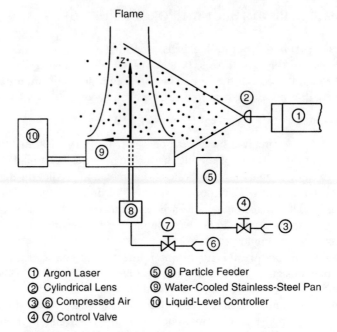

FIGURE 14 Schematic of a laser sheet particle tracking system applied to a small-scale propanol pool fire.

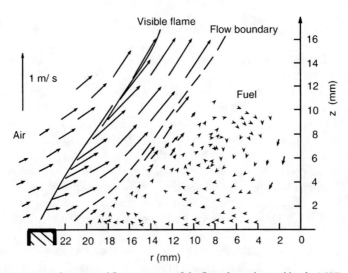

FIGURE 15 A two-dimensional flow structure of the flame base obtained by the LSPT technique shown in Figure 14. Center of the flame base is at $r = 0$.

zone is developed where mixing occurs and the flame is allowed to anchor. For pool fires, the shear stresses at the rim are lower by two orders of magnitude compared to the jet diffusion flame, demonstrating the insignificance of shear-stress-induced mixing and suggesting the presence of a molecular-diffusion-mixing zone due to finite-rate chemistry. Nevertheless, both anchoring mechanisms require the generation and sustained presence of a mixing zone, where fuel and oxidizer can intermix. The mechanisms by which these two flames produce these zones are quite distinct.

The structure of the flame leading edge where the pool fire is anchored is similar to that of a spreading flame. Because of finite-rate chemistry influences, the flame sheet has a finite thickness. Because the diffusion flame is bounded on either side by two premixed flames (Figure 8), the concept of the premixed flame propagation speed (described earlier) can be applied. The direction of the flame propagation is opposite to the tangential component of the air entrainment vector, V_t. If V_t is greater than the flame propagation speed near the base, the flame will not be able to anchor or stabilize. The LSPT measurements, shown in Figure 15, confirmed that V_t in the convective-air entrainment zone was lower than the stoichiometric flame speeds for the anchored pool fire. For the flame with V_t higher than the stoichiometric flame speed, however, the flame was no longer able to anchor the burner port resulting in lift-off or blow-off, which led to extinction. A schematic of structure of diffusion flames shown in Figure 8 also provides additional explanation. The flame is able to spread when the oxidizer and fuel mixture is in the reactive-diffusive zone; the spread rate is maximum at $Z = Z_c$ and becomes slower both at $Z < Z_c$ and at $Z > Z_c$.

IX. CONCLUSIONS

Topics introduced in this chapter cover a small portion of flame studies that have been accomplished in the past. Some interpretations given here are the author's view and subject to further discussion. The intention of the author is to discuss fundamentals of combustion and to stimulate readers to think scientifically when they deal with forest fire problems. References are provided for readers who are interested in advancing their understanding in flames and combustion.

ACKNOWLEDGMENTS

Ralph Nelson read the manuscript and offered many valuable comments. This author is largely indebted to Forman Williams and Alvin Gordon who taught principles of combustion and have been the constant source of education. This work is sponsored both by Kentucky-NASA EPSCoR program and Trinity Industrial Corporation through a general education gift.

NOTATION

ROMAN LETTERS

A	pre-exponential factor	s^{-1}
$\tilde{A}$	cross-sectional area of the slab	m^2
D	binary diffusion coefficient of a fuel into air	$m^2\,s^{-1}$
E	activation energy	$J\,mol^{-2}$
E_r	radiant heat flux received by a radiometer of geometrically similar points	$W\,m^{-2}$
F	body force	N
H	minimum ignition energy	J
H_o	energy per unit mass added to the mixture by the combustion	$J\,g^{-1}$
I_R	reaction intensity	$W\,m^{-2}$
PCC	product of complete combustion	
PICC	product of incomplete combustion	
Q	heat of combustion	J
Q_{ig}	ignition energy	J
T	temperature	K
T_a	gas temperature surrounding the thermocouple	K
T_b	thermocouple bead temperature	K
T_o	the initial temperature	K
T_f	burning fuel surface temperature	K
T_p	ignition temperature	K
T_∞	the final temperature	K
T_S	shear stress tensor	P_a
V_f	the rate of decrease of height of fuel surface by combustion	$m\,s^{-1}$
V_t	the tangential component of the air entrainment vector	$m\,s^{-1}$
Y_F	mass fraction of fuel	

Y_i	mass fraction of the i-chemical element	
Y_{i1}	mass fraction of the i-chemical species in the fuel	
Y_{i2}	mass fraction of the i-chemical species in the oxidizer	
Y_O	mass fraction of oxidizer	
Z	mixture fraction	
Z_c	stoichiometric mixture fraction	
a	absorption coefficient of gas surrounding the thermocouple	
c_p	specific heat at constant pressure	$J\ K^{-1}\ mol^{-1}$
d	burner port diameter or diameter of liquid pool	m
f	frequency of the flame pulsation	s^{-1}
f_p	porosity of the fuel beds	
g	acceleration due to gravity	$m\ s^{-2}$
h	visible flame height	m
h_f	depth of liquid fuel	m
Δh	the difference in thermal enthalpy per unit mass between the fuel at its ignition and the virgin fuel	$J\ g^{-1}$
k	specific reaction rate constant	s^{-1}
l	characteristic length	m
ℓ	fuel bed thickness	m
l_3	90 cm	
l_c	crib stick length	m
l_T	a characteristic length related to the high temperature gradient in the gas	m
m, n	reaction order	
p	pressure over characteristic pressure	
q	the net energy per unit area per second transported across the surface of fire inception	$W\ cm^{-2}$
q_c	heat value per unit mass of the crib material	$J\ g^{-1}$
q_{cod}	conduction heat loss	$W\ m^{-2}$
q_{cov}	convection heat loss	$W\ m^{-2}$
q_f	heat value per unit mass of liquid fuel	$J\ g^{-1}$
q_{rad}	radiation heat loss	$W\ m^{-2}$
t	characteristic time	s

t_b	h_f/V_f	s
t_c	chemical time	s
t_r	residence time	s
u_0	fuel exit velocity	m s^{-1}
v	flame spread rate	m s^{-1}
v_0	laminar burning velocity	m s^{-1}
u_1	lateral wind velocity	m s^{-1}
u_2	buoyancy-induced velocity	m s^{-1}
$\hat{u}$	velocity vector over characteristic velocity	
w	production rate (mass per unit volume per unit time) of chemical species by chemical reaction	$\text{g m}^{-3}\text{s}^{-1}$
x_f	the visible flame tip height	m
x_p	the pyrolysis height	m

GREEK LETTERS

Φ	an arbitrary function	
Φ_W	a dimensionless multiplier that accounts for the effect of wind in increasing the propagating flux ratio	
Φ_S	a dimensionless multiplier that accounts for the effect of slope in increasing the propagating flux ratio	
Ψ	an arbitrary function	
α	thermal diffusivity	$\text{m}^2\,\text{s}^{-1}$
δ	adiabatic laminar flame thickness	m
ε_b	emissivity of the thermocouple bead	
λ	thermal conductivity of gas	$\text{W m}^{-1}\,\text{K}^{-1}$
ν	kinematic viscosity	$\text{m}^2\,\text{s}^{-1}$
π	pi-number	
ρ_c	density of the crib stick	g m^{-3}
ρ_f	density of fuel	g m^{-3}
ρ_0	the initial density of the mixture	g m^{-3}
ϱ	density over characteristic density	
ϖ	ratio of low time to evolution time	
ω	the effect of heating number	

ξ propagating flux ratio, the portion of the reaction
         intensity that heats adjacent fuel particles to ignition

$\overset{\backslash}{\eta}$ heat transfer coefficient $W\ m^{-2}\ K^{-1}$

DIMENSIONLESS GROUPS

D_{IG} critical Da number for ignition

D_{EX} critical Da number for extinction

Fr Froude number

M Mach number

Pe Peclet number

Re Reynolds number

CONSTANTS

σ Stefan–Boltzmann constant

R_0 universal gas constant

REFERENCES

Albini, F. A. (1979). "Spot Fire Distance from Burning Trees—A Predictive Model." General Technical Report INT 56, Intermountain Forest and Range Experiment Station, Forest Service, USDA, Ogden.

Albini, F. A. (1981). A model for the wind-blown flame from a line fire. *Combustion and Flame* **43**, 155–174.

Albini, F. A. (1986). Wildland fire spread by radiation—A model including fuel cooling by natural convection. *Combustion Science and Technology* **45**, 101–113.

Arakawa, A., Saito, K., and Gruver, W. A. (1993). Automated infrared imaging temperature measurement: With application to upward flame spread studies. *Combustion and Flame* **92**, 222–230.

Babrauskas, V. (1988). "Handbook of Fire Protection Engineers." SFPE, Boston.

Ban, H., Venkatesh, S., and Saito, K. (1994). Convection-diffusion controlled micro-diffusion flames. *J. Heat Transfer* **116**, 954–959.

Benson, S. W. (1960). "The Foundations of Chemical Kinetics." McGraw-Hill, New York.

Blinov, V. I., and Khudyakov, G. N. (1957). Certain laws governing diffusion flames in liquids. *Academiia, Nauk, USSR, Daklady* **113**, 1094–1098.

Byram, G. M., and Nelson, R. M., Jr. (1970). The modeling of pulsating fires. *Fire Tech.* **6**, 102–110.

Carrier, G. F., Fendell, F. E., and Wolff, M. F. (1991). Wind-aided fire spread across arrays of discrete fuel elements. I. Theory. *Combust. Sci. Tech.* **75**, 31–51.

Clark, T. L., Coen, J. L., Radke, C. L., Reeder, M., and Packham, D. (1998). Coupled atmosphere-fire dynamics. *In* "Third International Conference of Forest Fire Research," pp. 67–82. Coimbra, Portugal.

Clark, T. L., Radke, L., Coen, J., and Middleton, D. (1999). Analysis of small-scale convective dynamics in a crown fire using infrared video camera imagery. *J. Appl. Meteorol.* 38, 1401–1420.

Delichatsios, M. A., and Saito, K. (1991). Upward fire spread: Key flammability properties, similarity solutions and flammability indices. *In* "Fire Safety Science—Proceedings of Third International Symposium" (G. Cox and B. Langford, Eds.), pp. 217–226. Elsevier Applied Science, London.

Delichatsios, M. A., Delichatsios, M. M., Lougheed, G. D., Crampton, G. P., Qian, C., Ishida, H., and Saito, K. (1995). Effect of external radiant heat flux on upward fire spread: Measurements on plywood and numerical predictions. *In* "Fire Safety Science—Proceedings of the Fourth International Symposium" (T. Kashiwagi, Ed.), pp. 421–432. International Association for Fire Safety Science, Washington.

Demtroder, W. (1982). "Laser Spectroscopy." Springer-Verlag, New York.

Dobbins, R. A., and Megaridis, C. M. (1987). Morphology of flame-generated soot as determined by thermophoretic sampling. *Langmuir* 3, 254–259.

Emori, R. I., and Schuring, D. J. (1977). "Scale Models in Engineering." Pergamon Press, London.

Emori, R. I., and Saito, K. (1982). Model experiments of hazardous forest fire whirl. *Fire Tech.* 18, 319–327.

Emori, R. I., and Saito, K. (1983). A study of scaling laws in pool and crib fires. *Combust. Sci. Tech.* 31, 217–230.

Faraday, M. (1993). "The Chemical History of a Candle." Cherokee Pub., Atlanta.

Ferguson, E. S. (1993). "Engineering and the Mind's Eye." The MIT Press, Cambridge.

Fernandez-Pello, A. C. (1995). The solid phase. *In* "Combustion Fundamentals of Fire" (G. Cox, Ed.). Chapter 2, pp. 31–100. Academic Press, San Diego.

Fristrom, R. M., and Westenberg, A. A. (1985). "Flame Structure," 2nd ed. McGraw-Hill, New York.

Gaydon, A. G., and Wolfhard, H. G. (1979). "Flames: Their Structure, Radiation and Temperature," 4th ed. Chapman Hall, London.

Glassman, I. (1996). "Combustion," 3rd ed. Academic Press, San Diego.

Hirano, T. (1986). "Combustion Physics." Kaibundo, Tokyo. (In Japanese.)

Hirano, T., and Saito, K. (1994). Flame spread phenomena: The role of observation in experiments. *Progr. Energy Combust. Sci.* 20, 461–485.

Ito, A., and Kashiwagi, T. (1988). Characterization of flame spread over PMMA using holographic interferometry: Sample orientation effects. *Combustion and Flame* 71, 189–204.

Ito, A., Konishi, T., Narumi, A., Tashtoush, G., Saito, K., and Cremers, C. J. (1999). The measurement of transient two-dimensional profiles of velocity and fuel concentration over liquids. *J. Heat Transfer* 121, 413–419.   •

JANAF. (1985). "Thermochemical Tables," 3rd ed. (M. W. Chase, Jr., *et al.*, Eds.). J. Physical and Chemical Reference Data.

Johnstone, R. E., and Thring, M. W. (1957). "Pilot Plants, Models, and Scale-up Methods in Chemical Engineering." McGraw-Hill, New York.

Kashiwagi, T. (1994). Polymer combustion and flammability—Role of condensed phase. *In* "Twenty-fifth International Symposium on Combustion," pp. 1423–1437. The Combustion Institute, Pittsburgh.

Koseki, H., Iwata, Y., Natsume, Y., Takahashi, T., and Hirano, T. (1999). Tomakomai large scale crude oil fire experiments. *Fire Tech.* 36, 24–38.

Koylu, U. O., and Faeth, G. M. (1991). Carbon monoxide and soot emission from liquid-fueled buoyant turbulent diffusion flames. *Combust. Flame* 87, 61–76.

Linan, A., and Williams, F. A. (1993). "Fundamental Aspects of Combustion." Oxford University Press, New York.

Long, R. T., Jr., Torero, J. L., Quintiere, J. G., and Fernandez-Pello, A. C. (2000). Scale and transport considerations on piloted ignition of PMMA. *In* "Fire Safety Science—Proceedings of Sixth International Symposium." The International Association for Fire Safety Science, Washington.

Majidi, V., Saito, K. Gordon, A. S., and Williams, F. A. (1999). Laser-desorption time-of-flight mass-spectrometer analysis of soot from various hydrocarbon fuels. *Combust. Sci. Tech.* **145**, 37–56.

McDonough, J. M., Garzon, V. E., and Saito, K. (1998). "A Porous Medium Model of Turbulent Wind-driven Wildland Fires." Department of Mechanical Engineering Internal Report, University of Kentucky, Lexington, KY.

Mulholland, G., Liggett, W., and Koseki, H. (1996). The effect of pool diameter on the properties of smoke produced by crude oil fires. *In* "Twenty-sixth International Symposium on Combustion," pp. 1445–1452. The Combustion Institute, Pittsburgh.

Orloff, L., de Ris, J. N., and Markstein, G. H. (1975). Upward turbulent flame spread and burning of fuel surface. *In* "Fifteenth International Symposium on Combustion," pp. 183–192. The Combustion Institute, Pittsburgh.

Porscht, R. (1975). Studies on characteristic fluctuations of the flame radiation. *Combust. Sci. Tech.* **10**, 73–76.

Pyne, S. J., Andrews, P. L., and Laven, R. D. (1996). "Introduction to Wildland Fire," 2nd ed. John Wiley & Sons, New York.

Quintiere, J. G. (1988). "Handbook of Fire Protection Engineers." SFPE, Boston.

Quintiere, J. G. (1998). "Principles of Fire Behavior." Delmar Pub., Albany, NY.

Radke, L. F., Clark, T. L., Coen, J. L., Walter, C., Riggan, P. J., Brass, J., and Higgans, R. (1999). Airborne remote sensors look at wildland fires. *In* "Fourth International Airborne Remote Sensing Conference and Exhibition/ 21st Canadian Symposium on Remote Sensing." Ottawa.

Rashbash, D. J. (1962). The extinction of fires by water sprays. *Fire Res. Abstr. Rev.* **4**, 28–53.

Rothermel, R. C. (1972). "Mathematical Model for Predicting Fire Spread in Wildland Fuels." Res. Pap. INT-115, Intermountain Forest and Range Experiment Station, Forest Service, USDA, Ogden.

Sabersky, R. H., Acosta, A. J., and Hauptman, E. G. (1989). "Fluid Flow," 3rd ed. Macmillan, New York.

Saito, K., Quintiere, J. G., and Williams, F. A. (1985). Upward turbulent flame spread. *In* "Fire Safety Science—Proceedings of the First International Symposium," pp. 75–86. Hemisphere Pub., Washington.

Saito, K., Williams, F. A., and Gordon, A. S. (1986). Structure of laminar co-flow methane-air diffusion flames. *J. Heat Transfer* **108**, 640–648.

Saito, K., Gordon, A. S., and Williams, F. A. (1987). A study of two-color soot zone for small hydrocarbon diffusion flames. *Combust. Sci. Tech.* **51**, 291–312.

Saito, K., Wichman, I. S., Quintiere, J. G., and Williams, F. A. (1989). Upward turbulent flame spread on wood under external radiation. *J. Heat Transfer* **111**, 438–445.

Saito, K., Gordon, A. S., Williams, F. A., and Stickle, F. A. (1991). A study of the early history of soot formation in various hydrocarbon diffusion flames. *Combust. Sci. Tech.* **80**, 103–119.

Soma, K., and Saito, K. (1991). Reconstruction of fire whirls using scale models. *Combust. Flame* **86**, 269–284.

Takahashi, F., Vangsness, M. D., Durbin, M. D., and Schmoll, W. J. (1996). Structure of turbulent hydrogen jet diffusion flames with or without swirl. *J. Heat Transfer* **118**, 877–884.

Tewarson, A., and Ogden, S. D. (1992). Fire behavior of polymethylmethacrylate. *Combustion and Flame* **89**, 237–259.

Venkatesh, S., Ito, A., Saito, K., and Wichman, I. S. (1996). Flame-base structure of small scale pool fires. *In* "Twenty-Sixth International Symposium on Combustion," pp. 1437–1443. The Combustion Institute, Pittsburgh.

54   K. Saito

Wichman, I. S. (1992). Theory of opposed-flow flame spread. *Progr. Energy Combust. Sci.* **18**, 553–593.

Williams, F. A. (1969). Scaling mass fires. *Fire Res. Abstr. Rev.* **11**, 1–23.

Williams, F. A. (1976). Mechanisms of flame spread. *In* "Sixteenth International Symposium on Combustion," pp. 1281–1294. The Combustion Institute, Pittsburgh.

Williams, F. A. (1982). Urban and wildland fire phenomenology. *Progr. Energy Combust. Sci.* **8**, 317–354.

Williams, F. A. (1985). "Combustion Theory." Benjamin/Cummings, San Francisco.

Williams, F. A. (1992). The role of theory in combustion science. *In* "Twenty-Fourth International Symposium on Combustion," pp. 1–17. The Combustion Institute, Pittsburgh.

Williams, J. M., and Gritzo, L. A. (1998). In situ sampling and transmission electron microscope analysis of soot in the flame zone of large pool fires. *In* "Twenty-Seventh International Symposium on Combustion," pp. 2707–2714. The Combustion Institute, Pittsburgh.

Wilson, E. B., Jr. (1954). "An Introduction to Scientific Research." McGraw-Hill, New York.

Wolff, M. F., Carrier, G. F., and Fendell, F. E. (1991). Wind-aided fire spread across arrays of discrete fuel elements. II. Experiments. *Combust. Sci. Tech.* **77**, 261–289.

Zabetakis, M. G. (1965). "Flammability Characteristics of Combustible Gases and Vapors." U.S. Bureau of Mines, Bulletin #627, Washington.

CHAPTER 3

Combustion Chemistry and Smoke

D. WARD

*Fire Sciences Laboratory, Rocky Mountain Research Station, USDA Forest Service,
Missoula, Montana*

I. INTRODUCTION

Fires burning in vegetative fuels, whether prescribed or wildfires, emit a complex mixture of particles and gases into the atmosphere. The diversity in composition of combustion products results from wide ranges in fuel types, fuel chemistry, and fire behavior. This chapter discusses the state of knowledge concerning the effect of these variables on the characteristics of smoke. In addition, the chemical and physical characteristics of the vegetation affect the rate of combustion and influence the overall fire behavior.

The need to bring fire back into ecosystems that, historically, have been dependent on fire has created a great deal of concern over the smoke released and the effect of the smoke on air quality values in both urban and nonurban settings. This has resulted in a need to develop methods for estimating the rate of release of smoke and for methods of burning to minimize smoke production consistent with achieving goals of the managers. Air Quality Regulations have been the main drivers for improving guidelines needed for reducing smoke emissions and managing the dispersion of smoke in the atmosphere. Many of the principles discussed in detail in this chapter have direct application in reducing smoke from wildland fires or understanding the toxicity of the smoke to humans.

Relatively little biomass is burned in the United States in comparison to the rest of the world. The total biomass consumed globally is currently estimated to be on the order of 5–8 Pg (1 Pg equals 10^{15} g) per year (Crutzen and Andreae, 1990). In comparison, 1988 was one of the most extreme "fire years" in recent history in the United States with almost 2 million ha of land burned by wildfires, including the large fires of Yellowstone National Park. If we consider the average fuel consumption per ha to be 45 t (1 metric tonne is 10^6 g), the total fuel consumed by wildfires in the United States in 1988 would have been ~90 Tg (1 Tg is 10^{12} g). In addition, Ward *et al.* (1993) estimated that prescribed fires for 1989 consumed about 40 Tg of fuel in the United States. The total of ~130 Tg of fuel consumed per year is ~2% of the total burned globally. It is interesting that more than 80% of all biomass burning takes place in tropical countries (Hao and Liu, 1994).

The discussion in this chapter will focus on the chemistry of natural ecosystem fuels important from an emissions production standpoint; then combustion processes will be discussed; and finally a general discussion of the chemistry of smoke from fires in natural vegetation fuels will be presented. Process level mechanistic approaches have not been as useful for explaining the range and mix of emissions resulting from the burning of biomass in the open environment as has proven beneficial for studying the mix of emissions from the combustion of hydrocarbon and coal (Glassman, 1977). For wildland fires, the chemistry of fuels is very complex, the distribution of fuel elements is random but highly variable, and weather influences are nonlinear and exceptionally difficult to quantify with moisture gradients occurring throughout the fuel complex. Outstanding use has been made of controlled environment combustion laboratory experiments to simulate field conditions (Yokelson *et al.*, 1996). Correlation and linear regression modeling techniques have been used almost exclusively in establishing the influence of independent variables on smoke production. Generally, approaches used in studying smoke emissions can be broken into four levels of investigation based on scale and control of variables. These are:

1. Highly controlled microcombustion or pyrolysis devices (Ward, 1979; Clements and McMahon, 1980; Susott, 1980),
2. Laboratory experiments (Lobert *et al.*, 1991; Reinhardt and Ward, 1995; Yokelson *et al.*, 1997),

3. Field-based tower experiments, generally used for studying smoke emissions from small plots (Ward and Hardy, 1991; Ward et al., 1992; Hoffa et al., 1999), and

4. Airborne studies of full-scale fires burning under ambient atmospheric conditions (Radke et al., 1991; Nance et al., 1993; Reid et al., 1998; Yokelson et al., 1999).

The characterizations of fuels and resulting emissions discussed in the following sections draws heavily from all four scales of research.

II. FUEL CHEMISTRY AND COMBUSTION

A. FUEL

Plant material consists of polymeric organic compounds, generally described by the chemical formula $C_6H_9O_4$ (Byram, 1959), although we know that this generalized chemical formula varies depending on whether we are interested in the carbon content of grass or wood or material that is undergoing decay. Regardless, plant tissue is approximately 50% carbon, 44% oxygen, and 5% hydrogen by weight. The content of most wood varies between 41% and 53% cellulose, between 15% and 25% hemicellulose, and between 16% and 33% lignin (Browning, 1963). The lignin content of wood is much higher (up to 65%) in decaying (punky) wood, in which the cell wall polysaccharides are partially removed by biological degradation.

Dead plant material accumulates on the forest floor or in lakes and bogs. This material is rich in nutrients and is transformed by microorganisms into humus in which individual plant parts are no longer identifiable. In the process, sulfur and nitrogen accumulate in the organic matter because many of the microorganisms require these elements for growth and reproduction. In addition, tree needles and leaves generally have a higher composition of nitrogen and sulfur than woody stems and limbs. Allen (1974) reported sulfur content values between 0.08% and 0.5% for plant material and 0.03% and 0.4% for organic soils. Nitrogen is one of the most dominant of the macronutrients and is of primary concern because of the large number of nitrogen-based compounds produced when biomass is burned. Nitrogen makes up as much as 0.2% of the older wood of some species, 1% of needles of some pines, and up to 2.7% of fallen hardwood leaves (Clements and McMahon, 1980). We will discuss the nitrogenous products of combustion in Section III.C under the topic of emissions of trace gases (Ward et al., 1996).

B. Combustion Chemistry

The generalized representation of biomass (or fuel) does not explain the diversity of compounds produced in smoke as a result of burning the material. Combustion efficiency is a term used to describe the overall conversion of carbon to carbon dioxide (Ward and Hardy, 1991). Byram (1959) illustrated the case for complete combustion of plant fuel with the following chemically balanced oxidation reaction:

$$4C_6H_9O_4 + 25O_2 + [0.322MH_2O + 94.0N_2]$$
$$\rightarrow 18H_2O + 24CO_2 + [0.322MH_2O + 94.0N_2] + 11.6 \times 10^9 J \quad (1)$$

Moisture in the fuel and water and nitrogen in the air are shown as bracketed quantities because they do not take part in the combustion reaction. It should be noted that the combustion of plant material releases the moisture contained with the fuel shown in brackets for this case and the water of combustion which is 18 kg moles of water for each 4 kg moles of plant material [(18*18 g)/(4 * 145) = 0.559]. For each kg of plant material (dry weight basis), 0.559 kg of water is released. Therefore, if one considers the moisture content of the fuel, it is readily seen that the rate of release of water is equal to the water of combustion plus the contained moisture (0.559 + 0.01 M). It should further be recognized from Eq. (1) that 4 kg moles of plant material release 11.6×10^9 J of heat energy. If we divide this heat release by the molecular weight and number of kg-mol of fuel, the high heat of combustion can be derived as 20,000 kJ/kg. The high heat of combustion is defined as the heat released from a unit mass of fuel without accounting for the heat required to vaporize the moisture of combustion or moisture content of the fuel (see Byram (1959) for a more complete treatment).

The combustion of plant material from open fires is seldom if ever 100% efficient; hence, products of incomplete combustion are produced that are of a major concern from an air pollution standpoint. Carbon monoxide is the major carbonaceous product of incomplete combustion, whereas carbon dioxide and water are the major products of complete combustion. Carbon dioxide contains as much as 95% of the carbon released in smoke from biomass fires where the combustion efficiency is high (e.g., grassland fires). On the other hand, the sum of carbon in other compounds of incomplete combustion makes up less than 5% (Figure 1). For fires of lower combustion efficiency, the percentage of carbon released as carbon dioxide would be lower and the products of incomplete combustion higher than shown in Figure 1. Minor constituents such as nitrogen, phosphorus, and sulfur affect the mix of pollutants generated by burning plant material. Other factors are important as well and contribute to the diversity of combustion products. For example, most plant material contains classes of compounds known as extractables, consisting of aliphatic and aromatic hydro-

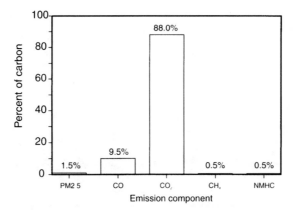

FIGURE 1   The average distribution of carbon between the primary products of combustion for broadcast burns of logging slash in the Pacific Northwest in the United States.

carbons, alcohols, aldehydes, gums, and sugars. These extractables, as a group, have a higher heat value than cellulose, lignin, and hemicellulosic substances. Other extractables from forest fuels contain a complex mixture of terpenes, fats, waxes, and oils. Not all of the extractable substances have low boiling points. For example, some of the extractives were shown by Shafizadeh *et al.* (1977) to evaporate or pyrolyze only at temperatures above 300°C.

C. PYROLYSIS

Pyrolysis is defined as the chemical breakdown of solid fuel under the influence of heat and usually in an oxygen-deficient environment. The pyrolytic decomposition of cellulose is generally believed to follow one of two paths dependent on whether the pyrolysis is occurring under high-temperature or low-temperature conditions. Usually under low-temperature conditions (200–280°C), the cellulose undergoes dehydration with the evolution of char, H_2O, CO_2, CO, and other compounds. Under higher temperature conditions (280–340°C), the pyrolysis proceeds in the production of levoglucosan, a volatile fuel that supports a gas-phase flame (Kilzer and Broido, 1965).

Thermal gravimetric analysis methods have been used by Susott (1980) and Susott *et al.* (1979) to evaluate the evolution of pyrolysis gases from solid fuels as a function of temperature during heating. The evolution of pyrolysis gas using this technique exhibits a spectrum reflecting the thermal stability of the fuel components as shown in Figures 2 and 3. Each component released can have a

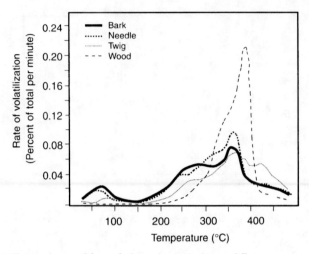

FIGURE 2 Different classes of forest fuels tend to volatilize at different rates as a function of temperature. The rates of volatilization were determined using thermal gravimetric measurement techniques.

different molecular weight and chemical form which can have significant implications regarding the formation of emissions. These materials pass through the flame structure or are released directly into the atmosphere. Oxidation may or may not occur at the solid fuel interface. For woody biomass, most researchers have studied the decomposition of cellulose and hemicellulose. Little is known about the thermal decomposition of bark, of which suberin—a polymer of long-chain hydroxy fatty acids esterified to phenolic acids—is one substance. The degradation products in the 425°C peak form a sizable fraction, but the chemical content may be different than for the woody carbohydrate products (Figure 2).

D. PHASES OF COMBUSTION

Preheating, flaming, smoldering, and glowing combustion compete for available fuel (Figure 3) and are markedly different phenomena that contribute to the diversity of combustion products. The fuel characteristics (including arrangement, distribution by size class, moisture, and chemistry) dominate in affecting the duration of flaming and smoldering combustion phases and combustion efficiency. Open combustion occurs through a diffusion flame process in which the fuel from the interior of the flame (oxygen-deficient area) diffuses outward, and the oxygen from the free-air diffuses inward. This results in a narrow envelope

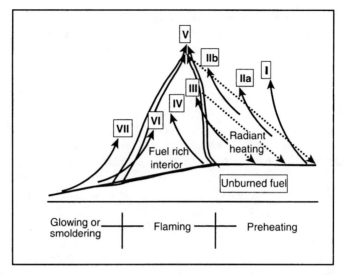

FIGURE 3   Diffusion flame model of pyrolysis and vaporization of volatiles during the preheating, flaming, glowing, and smoldering phases of combustion. The more highly volatile oleoresins are evaporated from the surface fuels in advance of the flame front for an advancing fire moving from left to right. The diagram above is a cross-section through a line of fire with the line of fire extending through the paper. There are several features describing the thermal environment and the effect of changing heat load on the early evaporation and later pyrolysis of fuel:

I.   Evaporation of the highly volatile compounds. No condensation or polymerization is expected for the compounds released that are not in close proximity to the flame envelope.

IIa. Evaporation of the highly volatile compounds increases in this zone and the potential for polymers to form increases as bond rupturing takes place for the terpenes.

IIb. Abundant evaporation of oleoresins occurs with partial oxidation due to the intense thermal environment and extreme bond rupturing resulting from the high-intensity radiant energy source.

III. Introduction of evaporated oleoresins into a high-temperature oxygen-deficient environment with considerable oxidation occurring as molecules diffuse through the flame envelope. Those not undergoing complete oxidation may be fragmented into ethylene units and/or free radicals.

IV. Very high-temperature environment with oxygen depletion occurring as the carbon fragments reach the tip of the flame envelope. The amount of compound oxidation is dependent on the depth of the flame envelope and the amount of ventilation (turbulence). As the flames become taller, the heat feedback to the solid fuels becomes less, and the radiant energy loss becomes greater. The result is that particle formation/polymerization reactions may be increased because of the loss of heat within the fuel-rich zone.

V.   Recombination takes place with the formation of compounds not found during the evaporation phase or inside the flame envelope due to pyrolysis. Aromatic hydrocarbon molecules are synthesized during this phase of transport.

VI. Products of pyrolysis and glowing combustion are transported across the flame surface.

VII. Transport of products of pyrolysis and glowing combustion completely miss the flame envelope and enter the atmosphere with no additional oxidation.

around the fuel-rich zone where the oxygen and fuel are mixed in stoichiometric proportions to produce rapid oxidation chemical reactions which result
in the visible emission of light called a flame. The chemical reactions also cause
the rapid liberation of heat which feeds back to the interior of the flame envelope and ahead of the flame envelope (Figure 3). This causes further vaporization of fuels with low vapor pressures and pyrolysis of solid fuels.

Initial heating of unburned fuel releases the more volatile components by
distillation which then leads to pyrolysis and fragmentation of polymers and
the release of oxygenated organic compounds (Yokelson *et al.*, 1996). Flaming
is initiated when the fuel-to-oxygen mixture reaches flammable proportions and
there is a source of ignition. Flaming and smoldering combustion are reasonably
distinct combustion processes that involve different chemical reactions and are
quite different in appearance (see also Chapters 2 and 13 in this book). Flaming
combustion dominates during the startup phase, with the fine fuels and surface
materials supplying the volatile fuel required for the rapid oxidation reactions
to be sustained in a flaming environment. The heat from the flame structure
and the diffusion and turbulent mixing of oxygen at the surface of the solid
fuel promote generation of the heat required to sustain the pyrolysis processes.
Early in the flaming phase, the more volatile hydrocarbons are vaporized from
the fuels. Later the cellulosic and lignin-containing cellular materials decompose through pyrolysis. These processes produce the fuel gases that sustain the
visible flame structure (Figure 4).

Once carbon and ash begin to build up on the solid fuel surfaces, the pyrolytic reactions no longer produce sufficient fuel gases to maintain the flame
envelope. For combustion to continue, oxygen must diffuse to the surface of the
fuel. Diffusion of oxygen and the availability of oxygen at the fuel surface are
enhanced through turbulence in the combustion zone and through premixing
by oxygen transport at ground level. This allows oxidation to take place at the
solid fuel surface and provides for heat evolution and heat feedback to accelerate the pyrolytic reactions and volatilization of the fuel gases from the solid
fuel. The process ultimately leads to the production of charcoal for which the
only combustion occurring is of the glowing type—a surface reaction of oxygen with carbon.

III. SMOKE PRODUCTION

The smoldering combustion phase produces large amounts of particulate matter
and carbon monoxide. Fires of very low intensity (those in which the flaming
combustion phase is barely sustained) produce proportionately higher amounts
of particulate matter. Heading fires generally are associated with conditions that
produce two to three times as many particles (by weight) as backing fires. The

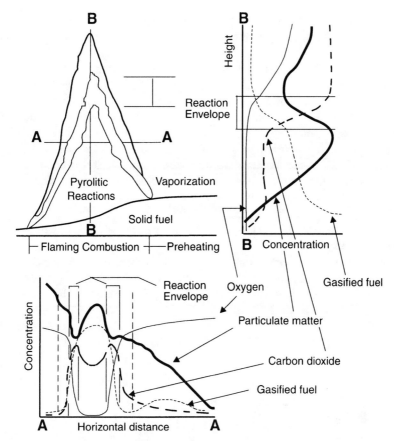

FIGURE 4   Relative concentrations of basic combustion gases with horizontal distance, A–A, and height, B–B, cross-sectioning a typical diffusion flame resulting from the oxidation of pyrolyzed forest fuels.

formation of particles results primarily from two processes: (1) the agglomeration of condensed hydrocarbon and tar materials and (2) mechanical processes which entrain fragments of vegetation and ash.

A. RELEASE OF CARBON

When vegetation is burned, carbon is released in the form of carbon dioxide, carbon monoxide, hydrocarbons, particulate matter, and other substances in decreasing abundance (Figure 1). A carbon mass-balance procedure is often employed by researchers in developing emission factors for different fuel conditions (Ward and Hardy, 1984). An emission factor is defined as the mass of a specific combustion product released per unit dry mass of fuel consumed and usually is expressed in pounds per ton or grams per kilogram. Emission factors are highly variable, but generally a large part of the variance can be explained through using a measurement of combustion efficiency for the independent variable. The combustion efficiency is a measure of the overall oxidation capacity for the combustion of fuel under a given set of weather and fuel conditions. It is a ratio of the mass of carbon released in the form of carbon dioxide to the mass of carbon in the original fuel and ranges from 0.98 for flaming combustion of fully cured grass vegetation with virtually no smoldering combustion to 0.75 for 100% smoldering combustion of deep duff. To simplify the calculation of combustion efficiency, Ward *et al.* (1996) adopted the use of modified combustion efficiency (MCE), or the ratio of the carbon released as carbon dioxide to the sum of the carbon released in the form of carbon dioxide and carbon monoxide. MCE is illustrated in Figure 5 for a number of different fuel types and is often used to predict emission factors for different compounds and particulate matter (Ward and Hardy, 1991; Ward *et al.*, 1996). For example, Ward *et al.* (1996) showed that MCE can be predicted for savanna fuels by using the

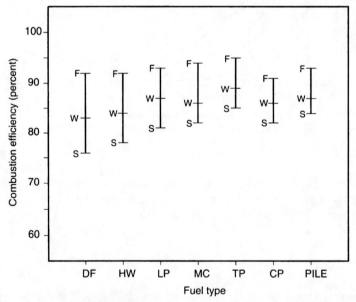

FIGURE 5   Range in combustion efficiency (the ratio of carbon released as CO_2 to the sum of carbon released as CO_2 and CO) for fires burning in seven different fuel types as follows: DF, Douglas Fir; HW, Hardwood; LP, ponderosa pine and lodgepole pine (long-needled conifers); MC, Mixed Conifer; TP, Tractor Piles; CP, Crane Piles; PILE, Combined TP and CP.

ratio of the mass of standing grass to the total mass of standing grass and litter using:

$$MCE = 0.85 + 0.111 \text{ Grass}/(\text{Litter} + \text{Grass})^{0.34} \qquad R^2 = 0.68 \qquad (2)$$

The model was developed for African savanna conditions of completely cured grass (no green grass). When the grass is lying flat (i.e., grass = 0) the predicted MCE is 0.85. When the grass is standing and there is no litter, the MCE is predicted to be 0.96. A summary of algorithms for predicting emission factors for several vegetation types, including the African savanna ecosystem, will be provided.

B. FORMATION OF PARTICLES

Wildland fires are a complex form of the diffusion flame process where pyrolysis of solid fuels produce vapors that interdiffuse with oxygen from the atmosphere. Turbulent mixing of fuel and oxygen is important, and as the turbulence increases, the flame characteristics and the chemical processes occurring in the flame zone change. Although this has never been completely substantiated, it is believed that there is a level of fire intensity where MCE reaches its highest level (perhaps 0.96 to 0.98 for many fuel types) (Ward, 1983) and particulate matter production is lowest (Nelson and Ward, 1980). For very high-intensity fires, some of the pyrolyzed fuel may no longer pass through an active oxidation zone. At times, even in lower intensity fires, pockets of unburned, partially oxidized gaseous fuels escape the combustion zone or undergo delayed ignition. The influence of flame turbulence on combustion efficiency is not fully understood; however, as the intensity of the fire increases and the zone of complete mixing of gaseous fuel and oxygen moves farther from the solid fuel, combustion efficiency is believed to decrease and the production of pollutants to increase. Because of the increased depth and height of the flame zone, heading fires and area fires create an extended reducing environment in which continued pyrolysis and synthesis of hydrocarbon gases and fragmented particles can occur under conditions of reduced oxygen content. In addition, heat reradiated from the particles to the atmosphere can slow down the reactions as the unburned gases and particles are convected away from the active combustion zone. Combustion of the particles requires prolonged exposure at high temperatures ($>800°C$) in a zone with near ambient concentrations of oxygen (Glassman, 1977). Greater premixing seems to reduce the production of fine particles ($< 2.5 \mu m$ diameter).

Fuel chemistry plays an important role in the formation and release of particles to the atmosphere. For example, it is well known that fuels high in oleoresins smoke profusely when burned. The pitch has a high terpene content; the terpenes are similar to aromatic compounds in structure (five carbon atom building blocks instead of six carbon atoms such as occurs in benzene and some other highly aromatic compounds). These highly volatile compounds have a low oxygen content and during pyrolysis produce ethylene units which tend to polymerize and form long-chain compounds which, in turn, condense into tar-like substances that coat most particles from fires. Ward (1979) found that emission factors for particulate matter are inversely related to the oxygen content of the fuel molecules and suggested that cellulose and hemicellulose may produce less particulate matter when burned (due to their high oxygen content) than the very low oxygen content oleoresins.

Smoke particles have been measured using sophisticated instruments ranging from Differential Mobility Particle Sizer (DMPS) to methods for sizing particles based on their aerodynamic properties. These instruments have been used onboard aircraft to cover the broad range of particle sizes from 0.01 to over 43 μm (Radke et al., 1991; Reid and Hobbs, 1998). The results suggest a very pronounced number concentration peak at a diameter of 0.15 μm. The volume distribution, when one assumes a unit mass density for smoke particles, represents the mass distribution and exhibits a bimodal distribution with peaks at about 0.3 μm and greater than 43 μm (Figure 6).

Emission factors have been measured for a number of different combinations of forest fuel and weather influences (Lobert et al., 1991; Ward and Hardy, 1991; Nance et al., 1993; Andreae et al., 1996; Ward et al., 1996; Ferek et al., 1998;

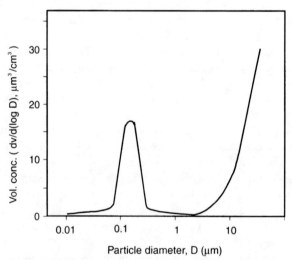

FIGURE 6   Bimodal distribution of particles on a volume basis. The percentage of particles larger than 10 μm typically increases for the flaming phase of combustion due to the turbulence generated in the combustion zone and the lofting of ash and partially burned biomass fragments.

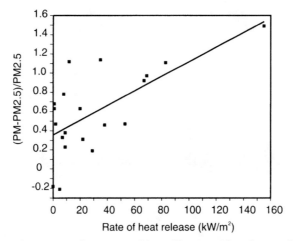

FIGURE 7   Higher intensity fires entrain additional large particles relative to fine particles.

Hoffa *et al.*, 1999; Yokelson *et al.*, 1999). Empirical data suggest that emission factors for particulate matter range from 4 to more than 40 g/kg for particles less than 10 μm in diameter (EFPM10); for particles without regard to size (EFPM), the range may be 20% larger, and emission factors for particles less than 2.5 μm diameter (EFPM2.5) are typically 10% smaller than EFPM10. However, the difference between EFPM and EFPM2.5 is highly dependent on fire intensity as illustrated by Figure 7. For savanna grassland ecosystems, Ward *et al.* (1996) found that EFPM2.5 can be predicted from the following equation:

$$EFPM2.5 = 87.65 - 88.51\ MCE \qquad R^2 = 0.56 \qquad (3)$$

The content of particulate matter varies between flaming and smoldering combustion. Between 1 and 10% of a particle's mass may consist of trace elements of potassium, chlorine, sulfur, phosphorus, and sodium. On a percentage basis, the mass of trace elements contained in particles from smoldering combustion is 10–20% of that from the flaming phase. Figure 8 for logging slash fires in the Pacific Northwest illustrates this dramatic difference between flaming and smoldering combustion (Ward and Hardy, 1989).

Emissions of graphitic and organic carbon are especially important because of the increased absorption of light by smoke particles that are high in graphitic carbon content. Absorption of light is due primarily to the black carbon content of the particles (Patterson *et al.*, 1986). For logging slash fires of the Pacific Northwest, emission factors for graphitic carbon ranged from 0.46 to 1.18 g/kg of fuel consumed. For fuel beds burned under laboratory conditions, the emission factors ranged to 5.4 g/kg of fuel consumed. Though graphitic or black carbon is produced proportional to the intensity of the fire, it is generally true

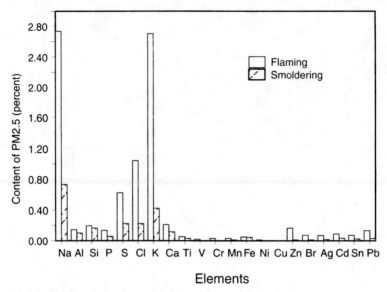

FIGURE 8 Difference in percent composition of particles for flaming and smoldering combustion reflecting the difference in ash content of the fine particles.

that emission factors for PM2.5 are inversely proportional to higher intensity burns. The organic fraction of the particles smaller than 2.5 μm is as much as 50–70% of the mass of the particulate matter for the smoldering phase, but can be lower for the flaming phase emissions.

Reid *et al.* (1998) conducted a very intensive study of "young" smoke (less than 2 hours old) in comparison to "aged" smoke (from 2 to 4 days old) from biomass fires in the very humid Brazilian Amazon region. Among other findings, they hypothesized that hydrocarbons of fewer than 11 carbon atoms were depleted over time and converted to CO_2, CO, and reactive molecular species and likely removed through dry deposition and/or by conversion to particulate matter. Although somewhat contradictory, the argument concludes that individual particle mass increased over time, and it was estimated that 20–45% of the mass concentration of the particles was due to the condensation of organics. It is interesting to note that most of the active "new" fires are lit between 1:00 and 3:00 P.M. local time and active flaming is nearly complete in 3 to 4 hours with smoldering combustion then dominating. Smoldering combustion continues for several hours to several days (Ward *et al.*, 1992). The organic content of particles produced during smoldering combustion of biomass is approximately 20% higher than the organic content of particles produced during the flaming combustion phase (Ward and Hardy, 1988) (likely dominating the emissions for the "young" fire plumes of Reid *et al.*, 1998). There are still many issues to be resolved before it can be conclusively demonstrated that condensation of hydrocarbons occurs at a higher rate than volatilization in contributing to the mass of particulate matter once the smoke is more than a few minutes old.

C. EMISSION OF TRACE GASES

In this section, fuel chemistry and combustion efficiency effects are discussed as they affect the production of trace gases from combustion sources. In particular, studies are reviewed that have examined the production of gases containing nitrogen, sulfur, and chlorine. Other gases produced, such as methane, nonmethane hydrocarbons (NMHC) and oxygenated hydrocarbons, also are discussed in this section.

1. Nitrogen Gases

Generally, temperatures within flame structures of vegetation fires do not exceed 1000°C, which suggests that molecular nitrogen gas (N_2) from the atmosphere is not dissociated to combine with free radicals within the combustion zone to form oxides of nitrogen (NO_x). Several studies have reported the production of NO_x from the burning of biomass.

Evans et al. (1977) measured NO_x in forest fire plumes in Australia. Their findings showed NO_x concentrations as high as 0.024 ppm. The conclusion was that NO_x is produced proportional to fire intensity. The same relationship has not been found for fires in the United States.

Results of Clements and McMahon (1980) suggest that the production of NO_x strongly increases with the nitrogen content of the fuel burned. Conversion of fuel-bound nitrogen to NO_x can occur readily in oxygen-depleted air. Simple linear regression equations were developed with coefficients of determination greater than 0.70 for predicting the production of NO, NO_x, NO_2, as functions of the nitrogen content of the fuel. The quantity of fuel nitrogen converted to NO_x was found to range from 6.1 to 41.7% for wood and organic soil, respectively.

Proximate analysis for the nitrogen content is an important measurement to make in interpreting differences between NO_x emissions for different ecosystems. The emissions of N_2O from combustion of forest fuels showed a small variance for two extreme fuel types—the boreal forest of Canada and the chaparral fuels of California. However, for some measurements of NO_x the difference between boreal and chaparral NO_x emissions was quite large (Hegg et al., 1989).

Hegg et al. (1989) measured emission factors for NH_3 of 0.1 to 2.0, for NO_x of 0.81 to 8.9, and for N_2O of 0.16 to 0.41 g/kg of fuel consumed. Nitrogen loss from the biological site was significant. In addition, a fraction of the nitrogen is contained within the particulate matter. NH_4 is contained within the particles and may contribute as much as 0.05 g/kg of fuel consumed.

Volatilization and release of nitrogen to the atmosphere can be significant on a total mass basis per unit area burned. Yokelson et al. (1996) demonstrated a relationship between the NH_3/NO_x ratio and MCE where NO_x dominates the emissions during flaming combustion and NH_3 dominates during smoldering combustion as follows:

$$NH_3/NO_x = 14(1 - MCE) \qquad R^2 = 0.96 \qquad (4)$$

The total of NH_3 and NO_x accounts for approximately 50% of the fuel nitrogen. Therefore, if the nitrogen content of the fuel is known and the MCE for the combustion is known, then the emissions of NH_3 and NO_x can be estimated.

Crutzen et al. (1985) found an average ratio of N_2O to CO_2 of 1.5×10^{-4} for tropical biomass burning. If one sums the carbon released as CO, CO_2, CH_4, and NMHC and assumes an emission rate of 1% for particulate matter (Ward and Hardy, 1991), the conversion of biomass to CO_2 is 77.8% efficient. This compares favorably with average rates of combustion efficiency for logging slash fires of 82.8% in the western United States (Ward and Hardy, 1984). The emission factor for N_2O was computed from Crutzen et al. (1985), and the result is 0.22 to 0.44 g N_2O per kilogram of fuel consumed. The emission factor for burning of tropical biomass is quite similar to that for boreal, chaparral, and coniferous biomass in North America as measured by Hegg et al. (1989). Ozone is not a byproduct of biomass combustion but forms as a product of secondary chemical reactions once the combustion products enter the atmosphere.

2. Sulfur Emissions (Carbonyl Sulfide)

Along with nitrogen, sulfur is one of the essential nutrients required in the synthesis of plant amino acids and other physiologically important substances (Grier, 1975; Tiedemann, 1987). Hence, the volatilization and loss of these important nutrients is of extreme interest in sustaining the productivity of ecosystems. Nitrogen can be replaced through symbiotic N-fixation, whereas sulfur is replenished mainly through atmospheric deposition. Very little work has been done in identifying the form of the sulfur- or nitrogen-containing emissions released during the combustion of biomass fuels.

Crutzen et al. (1985) measured carbonyl sulfide (COS) emission ratios to CO_2 of approximately 4 to 8×10^{-6} (5.71 to 11.42 mg/kg of fuel consumed). Ward et al. (1982) measured emission factors for COS of 0.18 to 2.36 mg/kg of fuel consumed from controlled experiments in a combustion laboratory-burning hood facility. The experiments were for fuels of varying sulfur content ranging from 0.55% (by weight) for organic soil to a low of 0.065% for pine needles. Ward et al. (1982) calculated emission factors from Crutzen et al. (1979) of 16.2 and 35.3 mg/kg of fuel consumed for fires in the Rocky Mountain area of the United States. The results of Ward et al. (1982) can be used to predict emission factors for COS as a function of the rate of heat release per unit area (I_R, kW/m^2) as follows: $EF_{COS} = 0.732 - 0.0065 I_R$ where the coefficient of determination is 0.99. This would suggest that the production of COS is independent of the sulfur content of the fuel—a result that probably is not valid. However, it may be concluded that COS production is extremely sensitive to the thermal environment reflected through the apparent dependence on I_R. Other sulfur-containing compounds were measured as well including H_2S, $(CH_3)_2S$, CS_2, CH_3SSCH_3, and other unknown mercaptan compounds. The sulfur quantified made up less than 0.25% of the total sulfur released during the combustion experiments. The balance of the sulfur may be partially explained by measurements of Yokelson et al. (1996) who estimate the molar ratio of SO_2 emitted to that of CO_2 as $(7.3 \, ! \, 3.0) \times 10^{-4}$. As with nitrogen compounds, the sulfur content of the fuel is highly variable and becomes the dominant characteristic in estimating the release of sulfur compounds.

3. Methyl Chloride

Methyl chloride has been suggested as a natural tracer unique to the combustion of biomass fuels (Khalil et al., 1985) and seems to work well for apportioning the impact of residential wood combustion. For open burning, Ward (1986) found that methyl chloride is produced in much greater quantities in the smoldering combustion phase than in the flaming phase. Thus, unless the actual ratio of flaming to smoldering is known for smoke impacting a site, the apportionment may be in error.

Ward (1986) found an inverse relation between chlorine content of fine particles and the amount of methyl chloride released and that emission factors for methyl chloride are inversely proportional to the rate of heat release. For chaparral fuels, methyl chloride emission factors ranged from 16 to 47 mg/kg of fuel consumed (Ward and Hardy, 1989). Reinhardt and Ward (1995) found that most, if not all, of the methyl chloride is produced from the smoldering combustion process.

4. Carbon Monoxide

Carbon monoxide is the second most abundant carbon-containing gas released during the combustion of biomass (Figure 1). Combustion efficiency and MCE are nearly perfectly correlated with the ratio of the production of carbon monoxide relative to carbon dioxide. Ward and Hardy (1991) found particulate matter concentration to be strongly correlated with carbon monoxide concentration ($r = 0.89$). Reinhardt and Ward (1995) found the concentration of formaldehyde to be correlated with the concentration of carbon monoxide ($r = 0.93$). Generally, emission factors for carbon monoxide on a mass basis are ten times greater than for the fine particle fraction. Emission factors for carbon monoxide range from 40 to over 300 g/kg of fuel consumed. Ward *et al.* (1996) found for fires in savanna grassland ecosystems that the following relationship can be used for computing emission factors for carbon monoxide (EFCO):

$$EFCO = 1834(1 - MCE) * 0.64 \tag{5}$$

5. Methane and Nonmethane Hydrocarbons

Methane is produced in much larger quantities during the smoldering combustion phase than in the flaming phase (Ward and Hardy, 1991). Emission factors are about two to three times greater for the smoldering phase than for the flaming phase. For one study, however, they ranged from 5.7 to 19.4 g/kg of fuel consumed for smoldering combustion and the flaming phase emission factors ranged from 1 to 4.2 g/kg. The volatile hydrocarbons are well correlated with methane, and emission factors for the hydrocarbons are well correlated with MCE. Ward *et al.* (1992) and Hao and Ward (1993) demonstrated that linear relationships between CH_4 emission factors and MCE are quite different for forest and savanna fires. In addition to the different slopes of the linear regression models, typical MCE values were found to be 0.84 for slash fires in primary forests of Brazil, 0.90 for second-growth deforestation burns in the same area, and 0.94 for cerrado savanna ecosystems of Brazil. Ward *et al.* (1996) found for fires in savanna ecosystems that the following equations can be used for calculating emission factors for CH_4 and NMHC:

$$EFCH_4 = 60.76 - 62.41 \, MCE \qquad R^2 = 0.84 \tag{6}$$

$$EFNMHC = 45.5 - 45.8 \, MCE \qquad R^2 = 0.96 \tag{7}$$

IV. MINIMIZING SMOKE PRODUCTION

One way to reduce the amount of air pollutants released to the atmosphere is to adjust the fire return interval to minimize the total smoke production. Here, an example is presented using information for a pure grassland ecosystem in Africa (Ward *et al.*, 1996). The decision to be made is whether it should be burned annually or biennially or a fire return interval of 3 or more years should be used to minimize the release of pollutants to the atmosphere. To assist in making this decision, we can apply the equations presented in the previous sections to first of all estimate the MCE using Eq. (2) which in turn can be used to estimate the emission factors needed for evaluating the emissions of PM2.5, CO, CH_4, and NMHC. We first need to know something about the rate of litter accumulation. For our Africa example, Ward *et al.* (1996) assumed that the area would be burned biennially and that half of the first year's standing grass decomposes naturally and the other half becomes incorporated into the litter layer during the 2-year period (i.e., $L = 0.5G$ in the second year). The ratio under this set of circumstances for the standing grass to the sum of standing grass plus the litter would be 0.67. Even though the total amount of fuel consumed is less, the MCE declines as a result of the increase in litter so that the emission factors are larger. The total products of incomplete combustion are shown in Table 1 to be more than the amount that would be produced were the area burned annually. The amount of nonmethane hydrocarbons and PM2.5 is 8 and 12% greater and the CH_4 released is 60% greater than would occur with annual burning. Other examples could be developed based on higher or lower decomposition rates for the litter and longer periods between burns.

TABLE 1 Example of Application of Algorithms for Assessing the Optimal Burning Interval for Minimizing the Release of Pollutants to the Atmosphere from the Use of Prescribed Fire in a Pure Grassland Ecosystem

	1 year	2 years
Available fuel	1.000	1.500
$G/(G + L)$	1.000	0.667
MCE	0.961	0.947
$EFCO_2$	1762	1736
EFCO	45.8	62.6
$EFCH_4$	0.8	1.7
EFNMHC	1.5	2.1
EFPM2.5	2.6	3.9
Emissions		
CO_2	1762	2604
CO	45.8	93.8
CH_4	0.78	2.51
NMHC	1.49	3.21
PM2.5	2.59	5.79
Ratios to one year		
CO_2	1.00	1.48
CO	1.00	2.05
CH_4	1.00	3.21
NMHC	1.00	2.16
PM2.5	1.00	2.23
The total emissions relative to burning every year		
CO_2		0.74
CO		1.02
CH_4		1.60
NMHC		1.08
PM2.5		1.12

After Ward et al., 1996.

TABLE 2   Algorithms and Ratios Developed for Logging Slash of the Pacific Northwest in the United States and Used for Computing Emission Factors (g of emission per kg of dry fuel consumed)

	Algorithm	Ratio model	Flaming	Smoldering	Units
MCE			0.92	0.75	
Carbon dioxide		MCE * 1834	1687	1376	g/kg
Carbon monoxide		(1 − MCE) * 1834 * 0.64	94	293	g/kg
Methane	42.7 − MCE * 43.2		3.0	10.3	g/kg
Nonmethane hydrocarbons	0.760 + EFCH$_4$ * 0.616		2.6	7.1	g/kg
PM2.5	67.4 − MCE * 66.8		5.9	17.3	g/kg
PM10		1.18 * EFPM2.5	7.0	20.4	g/kg
PM	93.3 − MCE * 90.5		10	25	g/kg
Benzene		0.00592 * EFCO	0.56	1.74	g/kg
Benzo[a]pyrene		13 * EFPM	131	331	µg/kg

After Ward *et al.*, 1993.

Ward *et al.* (1993) used similar algorithms developed for vegetation types in the United States to estimate the total criteria pollutant and air toxic emissions released from the prescribed use of fire nationally for the year 1989. The system they employed identified the fuel consumed by National Fire Danger Rating Fuel Model types and assigned a ratio of flaming to smoldering for each type. Then emission factors were computed for flaming and smoldering based on using an MCE of 0.92 for flaming and 0.75 for smoldering. Table 2 lists the algorithms and the emission factors for flaming and smoldering combustion used in this study. It immediately follows that a weighted emission factor can be computed by knowing the amount of fuel consumed during flaming versus smoldering combustion. For example, for 100% flaming a valid emission factor for particulate matter for particles less than 2.5 μm diameter would be 5.9 g of PM2.5/kg of fuel consumed on an oven dry weight basis. On the other hand, for a fuel complex where we expect about 30% of the fuel to be consumed during flaming and 70% during smoldering, we would weight the emission factor by multiplying the flaming emission factor by 30% and the smoldering emission factor by 70%, add the two together, and derive a weighted emission factor of 13.9 g of PM2.5/kg of fuel consumed.

V. CONCLUSIONS

The combustion of biomass in the open environment is a source of pollution and greenhouse gases to the atmosphere. Incomplete combustion, primarily from the smoldering phase, results in a very complex mixture of gaseous and particulate matter called smoke. The mixture is somewhat dependent on the elemental composition, size distribution, and moisture content of the fuel particles at the time of burning. The environmental conditions prevailing at the time of the fire and the type of fire (whether heading, backing, flanking, or an area fire) are important as well. Combustion efficiency can be calculated and/or estimated for determining the magnitude of emission factors for different compounds and particulate matter for different combinations of fuel and fire. The outcome of these calculations can be used for assessing management strategies in developing environmental impact assessments. By knowing fuel loading, available fuel, and ratio of fuel consumed by phase of combustion, the amount and rate of release of smoke can be estimated.

REFERENCES

Allen, S. E. (1974). "Chemical Analysis of Ecological Materials." Wiley, New York.

Andreae, M. O., Andreae, T. W., Annegarn, H., Beer, J., Cachier, H., Canut, P., Elbert, W., Maenhaut, W., Salma, I., Wienhold, F. G., and Zenker, T. (1996). Airborne studies of aerosol emissions from savanna fires in southern Africa. 2. Aerosol chemical composition, *J. Geophys. Res.* 103, 32,119–32,128.

Browning, B. L. (1963). "Methods of Wood Chemistry. Vol I." Wiley, New York.

Byram, G. M. (1959). Combustion of forest fuels. *In* "Forest Fire Control and Use" (K. P. Davis, Ed.), pp. 155–182. McGraw Hill, New York.

Clements, H. B., and McMahon, C. K. (1980). Nitrogen oxides from burning forest fuels examined by thermogravimetry and evolved gas analysis. *Thermochim. Acta* 35, 133–139.

Crutzen, P. J., and Andreae, M. O. (1990). Biomass burning in the tropics: Impact on atmospheric chemistry and biogeochemical cycles. *Science* 250, 1669–1678.

Crutzen, P. J., Delaney, A. C., Greenberg, J., *et al.* (1985). Tropospheric chemical composition measurements in Brazil during the dry season. *J. Atmos. Chem.* 2, 233–256.

Crutzen, P. J., Heidt, L. E., Krasnec, J. P., Pollock, W. H., and Seiler, W. (1979). Biomass burning as a source of the atmospheric gases CO, H_2, N_2O, NO, CH_3Cl, and COS. *Nature (Lond)* 282, 253–256.

Evans, L. F., Weeks, I. A., Eccleston, A. J., and Packham, D. R. (1977). Photochemical ozone in smoke from prescribed burning of forests. *Environ. Sci. Tech.* 11, 896–900.

Ferek, R. J., Reid, J. S., Hobbs, P. V., Blake, D. R., and Liousse, C. (1998). Emission factors of hydrocarbons, halocarbons, trace gases and particles from biomass burning in Brazil. *J. Geophys. Res.* 103, 32,107–32,118.

Glassman, I. (1977). "Combustion." Academic Press, New York.

Grier, C. (1975). Wildfire effects on nutrient distribution and leaching in a coniferous ecosystem. *Can. J. Soil Sci.* 5, 599–607.

Hao, W. M., and Liu, M. H. (1994). Spatial and temporal distribution of tropical biomass burning. *Global Biogeochem. Cycles* 8, 495–503.

Hao, W. M., and Ward, D. E. (1993). Methane production from global biomass burning, *J. Geophys. Res.* 98, 20,657–20,661.

Hegg, D. A., Radke, L. F., Hobbs, P. V., *et al.* (1989). Emissions of some biomass fires. *In* "Proceedings of the 1989 National Air and Waste Management Association Meeting," Paper No. 089-025-003. Air and Waste Management Association, Pittsburgh.

Hoffa, E. A., Ward, D. E., Hao, W. M., Susott, R. A., and Wakimoto, R. H. (1999). Seasonality of carbon emissions from biomass burning in a Zambian savanna. *J. Geophys Res.* 104, 13,841–13,853.

Khalil, M. A. K., Edgerton, S. A., and Rasmussen, R. A. (1985). Gaseous tracers for sources of regional scale pollution. *J. Air Pollut. Control Assoc.* 35, 838–840.

Kilzer, F. J., and Broido, A. (1965). Speculations on the nature of cellulose pyrolysis, *Pyrodynamics* 2, 151–163.

Lobert, J. M., Scharffe, D. H., Hao, W. M., Kuhlbusch, T. A., Seuwen, R., Warneck, P., and Crutzen, P. J. (1991). Experimental evaluation of biomass burning emissions: Nitrogen and carbon containing compounds. *In* "Global Biomass Burning: Atmospheric, Climatic, and Biospheric Implications" (J. S. Levine, Ed.), pp. 289–304. MIT Press, Cambridge.

Nance, J. D., Hobbs, P. V., Radke, L. F., and Ward, D. E. (1993). Airborne measurements of gases and particles from an Alaskan wildfire. *J. Geophys Res.* 98, 14,873–14,882.

Nelson, R. M., Jr., and Ward, D. E. (1980). Backfire particulate emissions and Byram's fire intensity. Research Note SE-290. USDA Forest Service, Southeastern Forest Experiment Station, Asheville.

Patterson, E. M., McMahon, C. K., and Ward, D. E. (1986). Absorption properties and graphitic carbon emission factors of forest fire aerosols. *Geophys. Res. Lett.* 13, 129–132.

Radke, L. F., Hegg, D. A., Hobbs, P. V., Nance, J. H., Lyons, J. H., Laursen, K. K., Weiss, R. E., Riggan, P. J., and Ward, D. E. (1991). Particulate and trace gas emissions from large biomass fires in North America. *In* "Global Biomass Burning: Atmospheric, Climatic, and Biospheric Implications" (J. S. Levine, Ed.), pp. 209–224. MIT Press, Cambridge.

Reid, J. S., and Hobbs, P. V. (1998). Physical and optical properties of young smoke from individual biomass fires in Brazil. *J. Geophys. Res.* 103, 32,013–32,030.

Reid, J. S., Hobbs, P. V., Ferek, R. J., Blake, D. R., Martins, J. V., Dunlap, M. R., and Liousse, C. (1998). Physical, chemical, and optical properties of regional hazes dominated by smoke in Brazil. *J. Geophys. Res.* 103, 32,059–32,080.

Reinhardt, T. E., and Ward, D. E. (1995). Factors affecting methyl chloride emissions from burning forest biomass. *Environ. Sci. Tech.* 29, 825–832.

Shafizadeh, F., Chin, P. S., and DeGroot, W. F. (1977). Effective heat content of green forest fuels. *For. Sci.* 23, 81–89.

Susott, R. A. (1980). Thermal behavior of conifer needle extractives. *For. Sci.* 26, 347–360.

Susott, R. A., Shafizahdeh, F., and Aanerud, T. W. (1979). A quantitative thermal analysis technique for combustible gas detection. *J. Fire Flamm.* 10, 94–104.

Tiedemann, A. R. (1987). Combustion losses of sulfur from forest foliage and litter. *For. Sci.* 33, 216–223.

Ward, D. E. (1979). "Particulate Matter and Aromatic Hydrocarbon Emissions from the Controlled Combustion of Alpha Pinene," PhD dissertation. University of Washington, Seattle.

Ward, D. E. (1983). Particulate matter emissions for fires in the palmetto-gallberry fuel type *For. Sci.* 29, 761–770.

Ward, D. E. (1986). Characteristic emissions of smoke from prescribed fires for source apportionment. *In* "Proceedings of the Annual Meeting of the Pacific Northwest International Section Air Pollution Control Association."

Ward, D. E., and Hardy, C. C. (1984). Advances in the characterization and control of emissions from prescribed fires. *In* "Proceedings of the 78th Annual Meeting of the Air Pollution Control Association," Paper No 84-363. Air Pollution Control Association, Pittsburgh.

Ward, D. E., and Hardy, C. C. (1988). Organic and elemental profiles for smoke from prescribed fires. *In* "Receptor Models in Air Resources Management" (J. G. Watson, Ed.), pp. 299–321. Air and Waste Management Association, Pittsburgh.

Ward, D. E., and Hardy, C. C. (1991). Smoke emissions from wildland fires. *Environ. Int.* 17, 117–134.

Ward, D. E., Hao, W. H., Susott, R. A., Babbitt, R. E., Shea, R. W., Kauffman, J. B., and Justice, C. O. (1996). Effect of fuel composition on combustion efficiency and emission factors for African savanna ecosystems. *J. Geophys. Res.* 101, 23,569–23,576.

Ward, D. E., Hao, W. H., Susott, R. A., Kauffman, J. B., Babbitt, R. E., Cummings, D. L., Dias, B., Holben, B. N., Kaufman, Y. J., Rasmussen, R. A., and Setzer, A. W. (1992). Smoke and fire characteristics for cerrado and deforestation burns in Brazil: BASE-B experiment. *J. Geophys. Res.* 97, 14,601–14,619.

Ward, D. E., McMahon, C. K., and Adams, D. F. (1982). Laboratory measurements of carbonyl sulfide and total sulfur emissions from open burning of forest biomass. *In* "Proceedings of the 75th Annual Meeting of the Air Pollution Control Association." Air Pollution Control Association, Pittsburgh.

Ward, D. E., Peterson, J., and Hao, W. H. (1993). An inventory of particulate matter and air toxic emissions from prescribed fires in the USA for 1989. *In* "Proceedings of the 1993 Air and Waste Management Association Meeting." Air and Waste Management Association, Pittsburgh.

Yokelson, R. J., Goode, J. G., Ward, D. E., Susott, R. A., Babbitt, R. E., Wade, D. D., Bertschi, I., Griffith, D. W. T., and Hao, W. M. (1999). Emissions of formaldehyde, acetic acid, methanol, and other trace gases from biomass fires in North Carolina measured by airborne Fourier transform infrared spectroscopy. *J. Geophys. Res.* 104, 30,109–30,125.

Yokelson, R. J., Griffith, D. W. T., and Ward, D. E. (1996). Open-path Fourier transform infrared studies of large-scale laboratory biomass fires. *J. Geophys. Res.* 101, 21,067–21,080.

Yokelson, R. J., Susott, R., Ward, D. E., Reardon, J., and Griffith, D. W. T. (1997). Emissions from smoldering combustion of biomass measured by open-path Fourier transform infrared spectroscopy. *J. Geophys. Res.* 102, 18,865–18,877.

CHAPTER 4

Water Relations of Forest Fuels

RALPH M. NELSON, JR.

USDA Forest Service, Rocky Mountain Research Station, Fire Sciences Laboratory, Missoula, Montana

I. INTRODUCTION

Earlier chapters have described forest fire behavior and effects, the general nature of forest fire flames, and the processes of combustion and smoke production in forest fuels. It was pointed out that moisture in these fuels acts to retard the rate of combustion. Among the fire behavior factors affected are the preheating and ignition of unburned fuels, rate of fire spread (or fire growth), rate of energy release, and production of smoke by burning and smoldering fuel. If we are to improve our understanding of these and other aspects of fire behavior, we must be able to quantify fuel moisture content within reasonable bounds. Moisture content, expressed as a fraction, is the mass of water held by unit mass of ovendry fuel and is determined primarily by fuel type and weather. It also may be expressed as a percentage of the fuel ovendry weight by multiplying by 100%.

For our purposes, fuels will be considered as living or dead and, within these broad categories, as individual particles or a collection of particles making up a fuel complex, or stratum. In many cases, of course, a fuel complex is composed of a mixture of live and dead particles from various fuel types whose moisture contents may vary over a wide range. The open grasslands of the southwestern United States exemplify a single fuel stratum, whereas the palmetto-gallberry-southern pine fuels of northern Florida exhibit at least three strata—a mat of pine needles in varying stages of decay intermixed with various living grasses and forbs, an intermediate layer of shrubs, and a pine overstory. The trunks of standing individual trees greater than 5–10 cm in diameter usually are scorched or charred by fire but do not burn and are not considered part of the fuel complex. In the conifer forests of the western United States and Canada, needle ladders (sometimes referred to as needle drape) often extend from the canopy to the ground so that fire can climb into the tree crowns. Similar behavior in some other conifer species such as spruce is caused by branches extending to the ground. When the crown fuels are drier than normal and winds are strong, these transitions from surface fire to crown fire occur with a large increase in energy release rate. The fire appears to acquire a violent character, and its increased energy often renders useless any attempt to control it.

The present chapter begins with a review of the influence of moisture on the combustion of forest fuels and how fuel characteristics determine the moisture content level of these fuels. It summarizes current understanding of the amount of water these fuels can hold, gain and loss of this water, and how the governing processes have been described mathematically. Only fuels associated with or originating from vascular plants (grasses, shrubs, and trees as opposed to mosses and worts that lack internal structure for transporting water) are considered. The discussion of live fuels emphasizes the physiological aspects of water transfer, but only a few studies related to mechanisms of water transport in these fuels are discussed. For example, the effects of photosynthesis, respiration, and growth on water potential and water movement are not described. Brief mention is made of soil water transport which, itself, has been a prominent topic in many soil physics and hydrology texts. In the case of dead fuels, a description of what the author believes are the more relevant studies is given. Only a few of the published studies of moisture content change in wood and forest fuels have dealt with diurnal change; a glimpse at prediction models for diurnal moisture variation is provided. Near the end of the chapter, several methods of measuring the moisture content of live and dead forest fuels are briefly discussed, and the reader is referred to works related to the fuel moisture aspects of current fire behavior and fire danger rating systems. Information on the general topic of forest fuel moisture relationships may be found in Luke and McArthur (1978), Chandler et al. (1983), Pyne (1984), and Pyne et al. (1996).

II. FOREST FUELS

The behavior of a spreading fire is determined by factors such as weather, topography, fuel quantity, and fuel moisture content. In practice, the burning characteristics of individual forest and wildland fuel particles are difficult to describe, but the influence of moisture content, particle physical properties, and particle arrangement on the burning characteristics of vegetation layers or litter and duff is even more complex and not well understood. This section discusses the effects of fuel moisture content on fuel burning rates and the fuel characteristics related to moisture content change.

A. FUEL CLASSIFICATION

Because forest fires usually are categorized according to the location of the uppermost fuel stratum through which they burn (ground, surface, or crown fires), it makes sense to think of fuel classification in the same way. Thus fuels may be classified as ground fuels, surface fuels, or crown fuels. Ground fuels consist of the highly decomposed organic material in contact with the inorganic layer and include duff, roots, peat, and rotten wood or bark coming from downed twigs and branches. Next to the ground fuel is a layer of surface fuels consisting of recently fallen and partially decomposed tree leaves (and/or conifer needles), fallen twigs, bark and branches, live or dead grasses, forbs, and shrubs less than 1.8 m tall (Davis, 1959). In moist climates, the fuels above mineral soil have three components: the recently cast *litter layer,* the partially decomposed *fermentation layer,* and the well-decomposed *humus layer.* Stocks (1970) refers to duff as all material above the upper surface of the mineral soil. A distinction between litter and duff is made in this chapter, however, with "litter" referring to the litter and fermentation layers and "duff" referring to the humus layer. The characteristics of duff layers are discussed further in Chapter 13 in this book. Dead ground and surface fuels contain both free and bound liquid water and are sensitive to precipitation and to changes in atmospheric relative humidity and temperature. Crown fuels include the canopies of most conifers and the chaparral and pocosin shrub types typical of the southwestern and southeastern United States. The latter two fuel types, which some investigators would categorize as surface fuels, grow to a height of 5 m or more. The moisture content of these live fuels is determined primarily by environmental and physiological factors. The canopies of deciduous species also are crown fuels, but fires rarely spread through these strata because the crowns are relatively sparse and have higher moisture contents than conifers. Exceptions occur under dry, windy conditions and when the fuels contain highly volatile and flammable substances, as in various species of *Eucalyptus.*

B. Fuel Burning Rates
and Moisture Content

The effect of fuel moisture on fire behavior is to slow the rate of burning, or rate of fuel consumption. For a spreading line of fire, the average burning rate can be computed from the mass of fuel consumed per unit area of ground divided by the time required to consume fuel on the area. Thus high fuel moisture content retards the rate of fuel consumption per unit of burning area (kg m^{-2} s^{-1}) by decreasing the mass of fuel consumed and increasing the particle burning time, or *particle residence time* (the time during which flame resides on individual particles in the fuel layer combustion zone). In addition, it increases the fuel preheating time. With respect to the thermal history of unit mass of fuel, the various phases of combustion include the processes of preheating, volatilization, charring, smoldering, and glowing; these are described in Chapters 3 and 13 in this book.

Consider first the fuel preheating time (or ignition time). The heat required for the onset and completion of volatilization of the fuel (volatilization begins at about 200°C and is assumed complete when the fuel temperature reaches 400°C) is called the heat of ignition Q_T (kJ kg^{-1}) by Wilson (1990) and is computed from the defining equation

$$Q_T = Q_f + MQ_M \tag{1}$$

where Q_f (kJ kg^{-1}) is the heat required to raise unit mass of dry fuel from ambient temperature to 400°C, Q_M (kJ kg^{-1}) is the energy to heat unit mass of water to 100°C and then vaporize it, and M is the fractional moisture content (a fuel particle or layer average expressed on an ovendry weight basis). Thus, an increase in M increases the amount of heat required to raise the temperature of unit mass of fuel from ambient temperature to 200°C and increases the preheating time. Various investigators have observed increases in ignition time with increasing M in dead and live fuels (Fons, 1950; Xanthopoulos and Wakimoto, 1993). The analytical studies of Albini and Reinhardt (1995) suggest that the increase in ignition time is due to increases in particle thermal conductivity and volumetric heat capacity with increasing M. A different aspect of ignition deals with what happens when firebrands, either short or long range, are blown or dropped into unburned fuels. Again, preheating and ignition are slowed by fuel moisture content, and the heat demand may be estimated by Eq. (1). Blackmarr (1972) has shown that the probability of ignition in slash pine litter is a strong function of moisture content and firebrand characteristics.

A second effect of fuel moisture on the burning rate involves a decrease in fuel consumption owing to the interaction among various fire behavior characteristics. The available fuel loading, defined here as the mass of fuel consumed per unit area of ground during flaming combustion, depends in a complicated way on moisture content, flame temperature, and the mass fractions of volatiles and char produced during combustion of unit mass of original fuel. The fuel chemical composition and rate of thermal decomposition are two factors affecting the relative amounts of volatiles and char produced which, in turn, determine the heat of combustion and the fraction of fuel available for flaming combustion. These relationships are discussed in detail by Albini (1980) and Susott (1982). The presence of moisture also causes a reduction in flame temperature because some of the heat generated in combustion is used to heat the inert water vapor in the products of combustion and because oxygen in the air is diluted by water vapor leaving the heated fuel (Byram, 1959). The reduced temperature retards the rate of decomposition and tends to drive the combustion process toward production of char rather than toward production of high-temperature volatiles. Results from burns of *Eucalyptus* leaves in a flow calorimeter (Pompe and Vines, 1966) suggest that water in fuel promotes smoke formation, reduces the heat of combustion, lowers the rate of temperature rise of air flowing in the calorimeter, and leads to much less intense burns. The effect of moisture on fuel consumption was estimated by Van Wagner (1972a) who used data from experimental fires spreading in jack pine, red pine, and white pine stands to develop a semiempirical relationship showing that duff consumption decreases as M increases. His equation for the weight of duff consumed, W (kg m^{-2}), is given by

$$W = 0.941(1.418 - M)/(0.1774 + M) \qquad (2)$$

suggesting that consumption in these fuels approaches zero when M approaches 1.418. The origin of Eq. (2) is discussed in Chapter 13 in this book. In this model, downward transfer of heat by radiation in the fuel layer combustion zone drives off moisture and then raises the temperature of the dry duff to the ignition point. Closure of the model requires a decrease in the emissivity of the flaming front with increasing M, suggesting that reduced radiative transfer to unburned fuel constitutes yet another means of slowing the rate of fuel consumption. King (1973) discusses the reduction of flame emissivity due to presence of moisture in terms of a reduction in soot concentration and an increase in carbon monoxide concentration according to the water gas reaction in which water vapor and carbon (soot) combine to form carbon monoxide and hydrogen.

The third effect of fuel moisture on the rate of burning is an increase in the fuel particle residence time. Albini and Reinhardt (1995) have shown with experimental and theoretical studies that, to the extent that flaming combustion is fueled by a simple sublimation process, the characteristic particle burning time is proportional to Q_T from Eq. (1). Because this demand for heat is satisfied by radiative and convective heat transfer within the combustion zone, moisture also will increase the burning time by reducing radiation to the particles (King, 1973). It is noted that the effects of M on particle ignition time and particle burning time are sometimes difficult to separate because a fraction of M is lost in preheating and the remainder is lost during particle consumption by flaming combustion. Research has not yet clarified how moisture change takes place while a fuel layer burns, but the relative amounts of water lost during preheating and flaming combustion seemingly would depend on particle size, shape, and arrangement and on the initial value of M (i.e., on the amounts of liquid and/or adsorbed water present).

C. Fuel Characteristics Affecting Moisture Content

If the effects of weather are disregarded, the most significant factors affecting the amount of water held and transported in woody and vegetative particles are chemical composition, internal structure, and physical properties. In the case of a fuel bed, several layer characteristics may be significant also. The amount of moisture held in the cell walls of fuel particles is related to composition and crystalline structure of the walls, whereas the liquid water held in the cell cavities is determined by the larger scale capillary structure.

1. Chemical Composition

The chemical constitution of wood is similar to that of foliage. Stamm and Harris (1953) give the cell wall composition of wood on a percentage of dry weight basis as cellulose (40–55%), hemicellulose (15–25%), lignin (15–30%), and extraneous and extractive matter (2–15%). Extractives include various organic compounds such as resins, sugars, and fatty acids that may be soluble in either alcohol, water, or organic solvents (e.g., xylene or ether). The only mineral constituent they cite is ash with a content between 0.1 and 4%. Thus, the holocellulose content (cellulose plus hemicellulose) ranges from about 55 to 80%. When ranked according to their hygroscopicity (or affinity for water), these components are ordered as hemicellulose, cellulose, lignin, and extractives. On the other hand, foliage contains the same components but in different amounts. According to Susott (1980), the range in composition of old and new needles of three western conifers is holocellulose (35–44%), lignin (18–19%), and extractives (37–47%). The higher percentage of extractives in needles than in wood tends to reduce water takeup and rates of moisture exchange (Anderson,

1990a, 1990b). Yet, the experimental data of Anderson (1990b) tend to confirm the results of earlier investigators (Dunlap, 1932; Blackmarr, 1971), showing that forest fuels are slightly more hygroscopic than wood under identical conditions of air temperature and relative humidity. This could be the case if the tendency toward smaller moisture contents due to smaller hemicellulose content of the fuels is more than offset by the fact that the hygroscopicity of hemicellulose exceeds that of wood cellulose by 50% (Browning, 1963); alternatively, the extractives in some conifer needles and deciduous leaves may be more hygroscopic than wood holocellulose. As litter fuel weathers over time, water-soluble extractives are leached from the surface and interior of the particles, and the holocellulose content is reduced due to consumption by microorganisms. This loss of dry mass can be approximated as an exponential decrease with time (Olson, 1963).

2. Internal Structure

The instantaneous level of moisture content in forest fuels is strongly influenced by the internal structure of the material and by whether the fuel is alive or dead. The factors controlling water movement in living plants are osmotic forces due to intercellular differences in plant water concentration and capillary tension forces created by transpirational demand at the external surfaces of the leaves (the term "leaves" includes conifer needles also). This demand for water must be satisfied primarily through absorption by the roots. In dead fuel particles, the most significant factors are two transport properties: (1) the permeability of the fuel to liquid water (bulk flow in cell cavities by capillarity), and (2) the moisture diffusivity (molecular flow in cell walls by bound water diffusion or in air spaces by diffusion of water vapor). An additional factor in litter and duff layers is liquid water drainage due to gravitational forces.

a. Live Fuels

The cells in living fuels contain protoplasm, cell wall material, and one or more vacuoles. The living protoplasm is largely made up of water, protein, various dissolved organic compounds and salts (cytoplasm), and a nucleus that controls inheritance and the metabolic activity of the cell through the genes it contains. The vacuole is located within the cytoplasm and is filled with cell sap (about 98% water plus various dissolved compounds). The watery solution surrounded by the nonliving primary cell wall is called the *protoplast*. The wall itself is mainly composed of cellulose, hemicellulose, and pectin; the water it contains is referred to as the *apoplast*. As the cell matures, the vacuole enlarges and eventually occupies most of the cell (Figure 1). Water movement in the cell is accomplished primarily by osmosis, for which the thin outer layer of cyto-

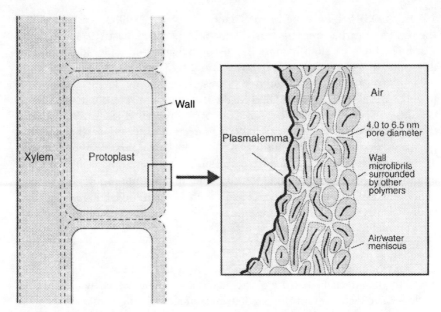

FIGURE 1 Mature cell wall showing the xylem through which water flows and the protoplast in the vacuole. The enlarged inset shows the apoplast, including air-water interfaces between the polymer microfibrils. The plasmalemma is pressed against the wall from the inside and pulled in the same direction by tension from the outside. This is the source of plant turgor. Adapted from Kramer and Boyer (1995).

plasm in contact with the cell wall interior, the *plasmalemma,* functions as the semipermeable membrane. In general, any given cell is surrounded by other cells or by a solution containing various solutes. Hence cell A with higher solute concentration than cell B will have a lower water concentration than cell B so that water will move from B to A by osmosis. Cell A will become turgid and cause the cytoplasm to press against the cell wall. This is the origin of "turgor pressure." In the opposite sense, when water leaves cell B, the cytoplasm shrinks from the cell wall, and if carried to the extreme, cell B dies. In addition to osmosis, liquid imbibition (or absorption) within the cell wall occurs in live or dead cells. The influence of the fine structure of the wall on this process is discussed in the next section.

In living fuels, liquid soil water absorbed by the roots moves by capillary flow through the water-conducting xylem of the stem, into the leaves, and then into the atmosphere through the leaves as water vapor. The processes of liquid absorption and vertical ascent are highly significant because they can limit the rate of water supply to the leaves by reducing plant turgor and causing stomatal closure. Figure 2A shows the structure of a deciduous leaf and pathways for

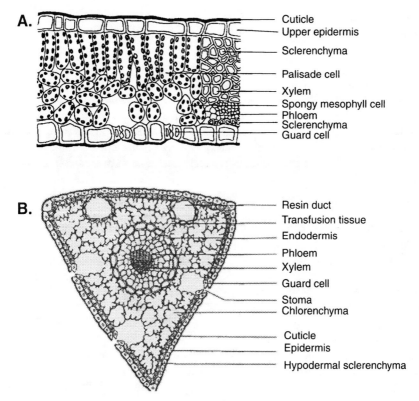

A.
— Cuticle
— Upper epidermis
— Sclerenchyma
— Palisade cell
— Xylem
— Spongy mesophyll cell
— Phloem
— Sclerenchyma
— Guard cell

B.
— Resin duct
— Transfusion tissue
— Endodermis
— Phloem
— Xylem
— Guard cell
— Stoma
— Chlorenchyma
— Cuticle
— Epidermis
— Hypodermal sclerenchyma

FIGURE 2 Structure of leaves from a deciduous tree and a conifer. Water vapor loss is primarily through the internal air spaces and guard cells, with a small amount of vapor diffusing directly through the epidermis and cuticle: (A) species not identified, (B) eastern white pine. From Kramer and Kozlowski (1979) and Kozlowski and Pallardy (1997), respectively.

liquid transfer. Food, water, and nutrients are supplied to and removed from the leaf through veins (not shown) that penetrate the mesophyll (palisade and spongy parenchyma) cells. These cells, in which photosynthesis occurs, are surrounded by a network of small intercellular spaces; other spaces just behind the lower epidermis are relatively large. These air spaces create a large amount of internal surface from which evaporation can occur. The vapor then diffuses toward and through the guard cells surrounding the stoma and into the atmosphere. This process is referred to as *stomatal transpiration*. A small amount of vapor also exits the leaf through the epidermal layers but is reduced by surface cuticle—*cuticular transpiration*. The stomates are found on both sides of the leaves but generally are more numerous on the underside. Conifers, with the exception of pines, also produce needles in which the mesophyll consists of palisade and spongy parenchyma cells. The pine needle cross section in Figure 2B shows sclerenchyma cells which provide mechanical support and chlorenchyma cells that contain chloroplasts. The transfusion tissue is involved in solute transport.

If the supply of water is not a limiting factor, the two major sources of resistance to evaporation are associated with the leaf and the atmosphere. The former resistance primarily is due to covering of the leaf surfaces by a relatively waterproof layer of cutin and wax (cuticle) and to stomatal resistance related to size of the aperture. In pine species, wax also is present in stomatal chambers. Aperture size of the stomates strongly depends on plant stresses and environmental factors such as sunlight, air temperature, and relative humidity. Atmospheric resistance is due to the leaf boundary layer, a thin region adjacent to the surfaces in which the factors affecting transport of vapor into the atmosphere are changing from leaf values to atmospheric values. This resistance, determined by particle size and shape and by wind speed, decreases as air flow over the particle increases. More information on water transport as related to the internal structure of living fuels may be found in texts such as Slatyer (1967), Kramer and Boyer (1995), and Kozlowski and Pallardy (1997).

b. Dead Fuels

The amount of moisture held in the cell walls of live or dead fuels (as discussed earlier in connection with cell wall imbibition) is related to the fine-scale structure of the walls. Approximately 50 individual cellulose chains are held together in parallel chance groupings by hydrogen bonds and various other types of secondary bonding to form a crystalline region, or crystallite. These elements are thought to be about 0.06 μm long—much shorter than a cellulose chain—due to the frequent occurrence of amorphous regions between the crystallites. The amorphous regions are regions in which the chains are highly disordered. Thus individual chains pass through several crystalline and amorphous regions but are linked at only a few points in the amorphous regions so that most of the active hydroxyl groups are available to take on water molecules. This water then forces the chains farther apart and swells the material. Because water is adsorbed throughout the amorphous regions but only on the surfaces of the crystallites, the water-holding ability of the cellulose varies directly with the proportion of the cellulose that is amorphous. The crystallites, with their associated amorphous regions, become further aggregated into fibrils that are embedded within the lignin-hemicellulose matrix materials to make up the various layers of the cell wall. In these fibrous materials, a secondary wall (the S_1, S_2, and S_3 layers) composed mainly of cellulose is found inside the primary wall and provides structural integrity (Figure 3A). In wood, the S_2 layer makes up 60–80% of the cell wall, thus controlling shrinkage, swelling, and other physi-

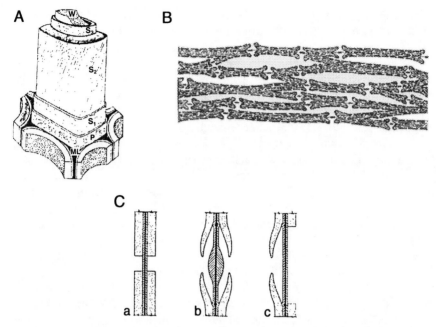

FIGURE 3 Structural features of a softwood. (A) View of a mature tracheid in which ML is the middle lamella that cements tracheids together, P is the primary wall, S_1, S_2, and S_3 are secondary walls, and W is the "warty" layer of unknown composition. From Coté (1967). (B) Pathways for continuous liquid or vapor flow through the large white cell cavities and bordered pits and for intermittent bound water/water vapor flow through cavities and the darker cell walls (the middle lamella between tracheids is not clear). From Stamm (1946). (C) Three kinds of pit pairs—a, simple pit with the dark middle lamella, cross-hatched primary wall, and dotted secondary wall; b, bordered pit with the torus midway in the pit membrane; c, half-bordered pit with no torus. Reproduced with permission of The McGraw-Hill Companies from Brown, H. P., Panshin, A. J., and Forsaith, C. C. (1949). "Textbook of Wood Technology, Vol. 1," 1st ed. McGraw-Hill, New York.

cal properties (Siau, 1995). References on details of the fine-scale structure of wood and cellulose are Browning (1963) and Panshin and de Zeeuw (1980).

The fractional moisture content M of dead fuel particles can reach maximum values between 2.5 and 3, depending on the wood or foliage particle specific gravity (Stamm, 1964). Both types of fuel exhibit sudden changes in moisture transport mechanisms and in electrical, mechanical, and thermal properties when M is in the range 0.21 to 0.35. These changes are associated with a value of M called the *fiber saturation point* that may be estimated for each fuel; the variability in its value can be due to differences in fuel temperature, mechanical stress, specific gravity, or chemical composition. The fiber saturation point, M_{fsp}, has been defined in various ways. It is most often defined as that M value obtained when isothermal equilibrium data for M are plotted as a function of increasing fractional relative humidity, H, and then extrapolated from $H = 0.95$ to $H = 1$ (Stamm, 1964). This procedure estimates the M value at which the fuel cell walls are saturated but no condensed water exists in the capillary structure. The method is an idealization, however, because there is no abrupt transition from a "saturated cell wall" region to a "liquid water" region; instead, the transition occurs over a range in H from about 0.9 to 0.995. Above the latter value, the fuel becomes increasingly saturated. The figure generally used for M_{fsp} by fire researchers is 0.30, independent of fuel type or state.

Gain or loss of moisture by dead fuel particles above M_{fsp} occurs by movement of liquid water through the fuel capillary structure in response to surface tension forces. The specific permeability (m^2) of a porous material is a measure of the ease with which fluids can move in bulk through the material in response to a pressure gradient in the fluid. In woody and vegetative particles, permeability to water depends on the internal structure of the material and extent of its swelling. Because forest fuels usually exchange moisture with the atmosphere at temperatures below 60°C, vapor pressure differences are not extremely large, and moisture transfer by gas flow often is considered negligible in comparison with that due to liquid flow. In softwoods, passage of liquid water is through fiber cavities in series with tiny holes in the pit membrane called pit membrane pores; on the other hand, continuous liquid movement through the permanent cell wall capillaries contributes little to the overall flow (Stamm, 1964). Figure 3B is a sketch of the gross capillary structure of a softwood and paths for liquid transport through fiber cavities and pit membrane pores. Darkened areas represent a network of lignocelluosic strands (or fibrils) intertwined to form the cell wall, whereas the white areas represent fiber (or tracheid) cavities and pit chambers within the bordered pits. Figure 3C shows three types of pit pairs found in softwoods and hardwoods. Bordered and half-bordered pits are common in softwoods, whereas simple pits occur in hardwoods. The enlarged region in the center of the pit membrane is called the torus; the pit becomes aspirated when the torus is sealed against a border by surface tension forces created in drying (Hart and Thomas, 1967). The pits also can become impervious to water because of encrustation by extractives.

The internal structure and liquid permeability of hardwoods is more complex and more variable than that of softwoods. The primary conducting elements, or vessels, are interspersed among tracheids and other fibers and interconnected by simple pits. Resistance to liquid flow is offered by the pits and by internal growths, or tyloses, lodged in the vessels of most species. In the pits, resistance is due to encrustation rather than aspiration because pits in hardwoods have no torus in the pit membrane. Furthermore, the openings in the pit membrane are about 10 times smaller in hardwoods than in softwoods, but not so small as to prevent water flow. One of the most permeable woods is red oak (no tyloses) with a longitudinal permeability of about 2×10^{-10} m^2. The sapwood and heartwood of many softwoods and hardwoods exhibit longitudinal permeabilities ranging from 10^{-12} to 10^{-15} m². Comstock (1970) measured longitudinal/tangential permeability ratios in softwoods ranging from about 500 to 8×10^4. Siau (1995) refers to an unpublished study of eight hardwoods by Comstock reporting longitudinal/tangential permeability ratios between 3×10^4 and 4×10^8. Additional information concerning flow pathways and the relation of internal structure to fluid flow in wood may be found in Stamm (1964) and Siau (1995).

The flow of liquid water in layers of deciduous or coniferous litter and duff is determined by the amount of moisture retained on and within the particles and in the interstices between particles. In a series of laboratory experiments, Stocks (1970) applied simulated rainfall amounts ranging from 12.7 to 50.8 mm to air-dry ponderosa pine duff layers 7.6 cm thick at a rate of about 25.4 mm h^{-1}. The results showed that the percentage of the applied rainfall retained by the duff decreased from 20 to 11% as rainfall amount increased. When the laboratory experiment was repeated using moist duff, water retention was halved and the percentages decreased from 10 to 5% as rainfall increased. Stocks (1970) also reported field and laboratory experiments suggesting that the moisture content of the duff in a thin sublayer adjacent to underlying mineral soil is smaller than in the sublayers just above it (Figure 4). These results are consistent with

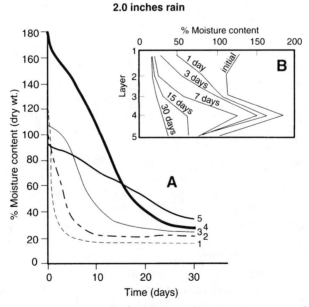

FIGURE 4   Laboratory drying rates for five separate sublayers within a 7.6-cm duff layer (primarily ponderosa pine) after 50.8 mm (2 inches) of simulated rain. "Layer 1" is at the top of the duff layer. (A) Moisture content versus time after rain. (B) The vertical distribution of moisture content for specific times after rain. Adapted from Stocks (1970).

the idea that the amount of rainfall retained by the layer depends on a balance among capillary forces across the duff-soil interface, capillary forces among the fuel particles, gravitational forces acting over the vertical depth of the layer, and attractive forces drawing water into the particles. Structural details of the layer affect this balance by determining the specific permeability of the duff. Fosberg (1977) measured the permeability of several western conifer duff layers to air and reported values ranging from about 10^{-8} to 5×10^{-11} m^2. These values are orders of magnitude larger than those for wood, reflecting the much larger sizes of the conducting pores.

Moisture diffusivity $(m^2\ s^{-1})$ is the property that governs exchange of moisture in dead fuel particles below M_{fsp}. Water movement is due to combined diffusion involving a gradient in bound water concentration across the cell walls and a gradient in partial vapor pressure through the fiber cavities and pit membrane pores. Stamm (1964) states that bound water diffusion differences between softwoods and hardwoods are small, so diffusion of vapor through the continuous void structure is the major source of variability among species. Stamm (1964) and Choong (1965) suggest, however, that in softwoods the percentage of the transverse moisture diffusion attributable to vapor diffusion is 22% or less at temperatures below 50°C. Thus structural differences between softwoods and hardwoods normally cause only small differences in diffusional flow. Though the circumstances under which diffusion operates in dead foliage particles are not well known, equilibrium and dynamic moisture relationships in these fuels are well represented in form, if not in magnitude, by research results for wood over a large range in M. This is verified to some extent in data reported by Van Wagner (1979) on drying of jack pine litter that are similar to those from studies on wood and paper (Nelson, 1969). Other data demonstrate, however, that diffusivities differ for particles of wood and foliage. Values of the integral moisture diffusivity (the average value for the particle or layer), D_{av}, observed by Linton (1962) for *Eucalyptus obliqua* wood ranged from 1.4 to 42 $\times$ 10^{-11} $m^2\ s^{-1}$ for twigs 6 mm in diameter and from 3.5 to 45 $\times 10^{-12}$ $m^2\ s^{-1}$ for twigs 3 mm in diameter. Linton also found that individual leaves from *Eucalyptus obliqua* and *Eucalyptus radiata* exhibited D_{av} values ranging from 0.24 to 11 $\times 10^{-12}$ $m^2\ s^{-1}$. Anderson (1990a) reported a comprehensive study in which grasses, hardwood leaves, conifer needles, and square softwood sticks initially in equilibrium at 26.7°C and H of 0.9 were subjected to a rapid drop in H to 0.2 at which they equilibrated and then were exposed to another step change in H back to 0.9. Weathered foliage fuels were studied also. Results showed that D_{av} of the wood ranged from about 0.1 to 2 $\times 10^{-10}$ $m^2\ s^{-1}$, in good agreement with measurements of transverse diffusion in softwoods and hardwoods for which D_{av} ranged from 0.07 to 1.7 $\times 10^{-10}$ $m^2\ s^{-1}$ (Stamm, 1960; Simpson, 1993a). On the other hand, Anderson's foliage diffusivities (for recently cast fuel) ranged from about 10^{-14} to 10^{-12} $m^2\ s^{-1}$. These values are in approximate agreement with the observations of Linton (1962), but even Anderson's values for weathered fuels (10^{-13} to 10^{-12} m^2 s^{-1}) are smaller than those for wood by two orders of magnitude.

3. Physical Properties

The differences in moisture diffusivities of dead wood and foliage just described are due, at least in part, to differences in mass density, particle size and shape, and extractive content. Moreover, the effects of weathering on moisture relationships operate primarily through these variables. The density of wood depends on its specific gravity and moisture content, ranging from about 300 to 800 kg m^{-3} (Simpson, 1993b); values for foliage fuels fall well within this range (Anderson, 1990a). The effect of density on diffusion has not been identified in experimental studies. Though he acknowledged the lack of conclusive data, Hart (1964) argued that two samples of wood drying in response to identical gradients in M will exhibit diffusivities inversely proportional to their densities. Particle density also is affected by water-soluble extractives that, because of their bulking action, leave the fiber cell wall in an expanded state when drying from the green condition takes place. For foliage, this effect may be overshadowed by the flow reduction due to wax on particle surfaces. Van Wagner (1969a) presents data showing that *response time* τ (s) (the time required for accomplishment of $1 - 1/e$, or 63.2%, of the total moisture content change due to a step change from a constant initial value to a constant final value) decreased by factors averaging about 5.5 and 2.5 for red pine needles and aspen leaves that had been pretreated with xylene to remove the waxes and resins. This result implies that fuel particles with surface wax removed by weathering will gain or lose moisture about three to six times faster than unweathered particles whose surface wax is intact.

The effects of particle size are best obtained from the theory of diffusion in solids (Crank, 1975). The diffusion theory, discussed briefly in Section III.D.2, indicates that the theoretical effect of minimum particle dimension d (m) on D_{av} is zero because the local diffusivity, D, is treated as a constant and therefore equals D_{av}. Dimension d usually is the radius of a cylinder (or sphere) or the half-thickness of a slab. It is well known, however, that D_{av} for wood is not constant, but a function of moisture fraction M and fuel temperature T_f.

The measurement of D_{av} is further confounded by the possible occurrence of two nondiffusional mechanisms that can be mistakenly interpreted as diffusion. The first of these involves the slowing of *sorption* [this term is used when reference is made to the combined processes of adsorption and desorption (see Section III.D.1) or when no distinction between the two processes is necessary] due to molecular rearrangements associated with relaxation of shrinkage or swelling stresses. These rearrangements are more pronounced during adsorption in thin particles ($<$ 3 mm thick) when fractional humidity H exceeds 0.5; they make new adsorption sites available at a slower rate than the initial diffusion rate (Christensen, 1965). The second mechanism is moisture transfer at the solid surface. The theoretical influence of this process is expressed in terms of the dimensionless mass transfer Biot number, Bi—the ratio of resistance to moisture transfer at the surface to that within the particle. This number is discussed further in Section III.D.2. Byram (1963, unpublished) expressed the results from classical theory in terms of response time and found τ proportional to d^2 when Bi approaches infinity. For smaller Bi, approaching zero, τ was predicted as proportional to d. Nelson (1969) tested Byram's predictions with laboratory drying experiments on cellulosic materials exposed to fixed step changes in H at constant temperature. Two values of τ were observed for most runs. The τ versus d^2 relationship was approximated for sawdust layers (τ depended on $d^{1.8}$ rather than d^2), square wooden sticks, and thin slabs of paper (during the later stage of drying) for initial values of M well above M_{fsp}. The approximate d^2 dependence observed in all fuels studied suggests that diffusivity D_{av} was nearly independent of d in these tests. Results for the paper slabs are presented in Figure 5. The data show that the predicted τ versus d result was observed in approximate form (τ depended on $d^{1.27}$ rather than d) during the early stage of drying when the resistance to transfer at the surface must have exceeded that within the slabs. For the two thinnest slabs, the tendency for τ to become independent of d indicates that the drying rate depended entirely on the external conditions. In other work, the effect of thickness on moisture change in *Eucalyptus regnans* sapwood specimens was studied by King and Linton (1963) who found that τ increased with increasing d, but not according to d^2 as would be expected from theory. Taken together, the data suggest that the predicted effects of d on τ will be approximately correct for forest fuels and that D_{av} is nearly independent of d in the diffusion-controlled range when external conditions are constant.

Particle shape theoretically affects moisture exchange by means of differences in constants derived in analytical solutions of the equations describing diffusion in the various particles. In practice, these effects due to shape operate partly through Biot number Bi and partly through the particle *surface-to-volume ratio* σ (m^{-1}) that influences the exchange of heat and mass between the particle and the surrounding fluid. Larger values of the ratio result in larger exchange rates because of increasing surface area per unit of particle volume. This ratio is one of the factors defining the openness of fuel layers which, in turn, may determine the rate-controlling dimension during moisture exchange. A second controlling factor, the dimensionless fuel bed *packing ratio,* β, represents the volume of fuel particles within unit volume of fuel bed. An interesting question not yet answered by research is the following: if a fuel layer of constant packing ratio β, vertical depth δ (m), and uniformly distributed cylindrical particles of radius d (or constant σ) is exposed to a step change in H at constant T,

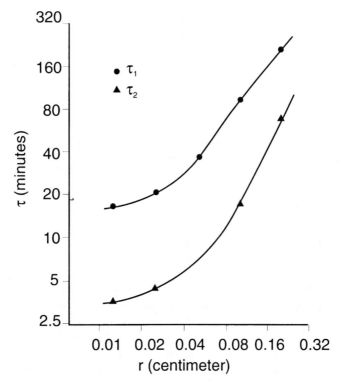

FIGURE 5   Logarithmic plot of response time versus total thickness of paper slabs drying at 27.2°C and 36% relative humidity, where τ_1 refers to early drying when surface effects were present but not dominant and τ_2 to the later stages when internal diffusion was controlling the process. For small values of slab thickness, response time tends to become independent of thickness due to surface control. From Nelson (1969).

is its rate of moisture exchange controlled by δ or d? Though the answer to this question partly depends on the imposed boundary conditions, in general one would expect that moisture gain or loss in tightly packed beds is controlled by δ, and that as β decreases and the beds become more open, the rates of moisture exchange would approach the rate for a single particle. The limited data of Anderson et al. (1978) for beds of ponderosa pine needles suggest this is so. In later work, Anderson (1990a) compared the diffusivities of layers with different degrees of packing with those of widely spaced needles drying in similar environmental conditions and found that D_{av} decreased as the fuel beds became deeper or more dense. Using the method of nonlinear least squares, Anderson also developed equations for predicting the effects of fuel particle and fuel bed variables. He produced four groups of equations—each group distinguished by the hardness of wax on the fuel surfaces or by the amount of surface cuticle. Within each group, equations were developed for recently cast and weathered fuels undergoing adsorption and desorption. All equations showed that response times were proportional to the quantity $\sigma^a \beta^b \delta^c$, where exponents a, b, and c differed among fuel species and sorption conditions. Anderson stated that increasing β and δ when σ is constant decreases D_{av} and increases τ, and that additional study is needed.

III. FUEL MOISTURE RELATIONSHIPS

The moisture content of living fuels involves free water and water vapor in plant void spaces and in the cell walls and surrounding tissues. For example, an equilibrium relationship between water content and water potential describes water retention in plant tissue, bulk capillary flow takes place in the stems of woody plants, and water exits the interior of leaves by vapor diffusion (Kramer and Boyer, 1995). Similar phenomena take place in dead fuels. Approaches for describing moisture held at equilibrium in hygroscopic porous solids range from empirical and semiempirical models to rigorous theories based on the physical chemistry of solutions or statistical thermodynamics. Water movement in porous bodies, including dead forest fuels, can be described mathematically in two ways. In the first, water transport is assumed to obey the conservation equations of engineering mechanics (Spolek and Plumb, 1980). The second approach involves the procedures of nonequilibrium thermodynamics (Katchalsky and Curran, 1965). The underlying foundation of this method rests in the conservation equations of mass and energy and in the thermodynamic laws but involves additional assumptions about the conservation of momentum and deviations from equilibrium. The primary driving force for mass flow is the spatial gradient of the partial molal Gibbs free energy; this idea has been accepted by soil and plant water researchers for many years and now is drawing the attention of wood-drying researchers (Cloutier and Fortin, 1991, 1993; Siau, 1995; Zhang and Peralta, 1999). The rapidly developing thermodynamic methods, which have been little used in forest fuel moisture research, are not discussed further because of space limitations.

A. WATER POTENTIAL

Researchers studying water relationships in soils and plants as separate thermodynamic systems have long recognized that water tends to move in these materials from regions of high Gibbs free energy to regions of lower free energy. This is consistent with the thermodynamic concept of equilibrium which tells us that two bodies (or two regions within a single body) are in equilibrium only when the Gibbs free energy difference between them is zero. Water moves across the soil–atmosphere, soil–plant, and plant–atmosphere interfaces in response to free energy gradients, so viewing the three separate domains as one unified thermodynamic system is a logical research approach. This single-system concept has been labeled SPAC to refer to the "soil–plant–atmosphere continuum" and has been accepted by many workers in the two subject matter areas.

The energy status of water in living or dead plants, soils, or the atmosphere measures the potential energy of the water in its current state when temperature, pressure, and chemical composition are the only variables. This energy is the partial molal Gibbs free energy, or chemical potential, μ (J mol^{-1}). The standard reference state in both soil and plant water research, denoted by μ_0, is taken as pure liquid water at atmospheric pressure and at the temperature of water in the material. The reference location is an arbitrarily chosen elevation. Because water is essentially incompressible (except at very small moisture fractions), its density (kg m^{-3}) and molecular weight (kg mol^{-1}) can be used to express μ in units of pressure, J m^{-3}. Thus water potential ψ is defined as

$$\psi = \rho_w(\mu - \mu_0)/M_w = (\mu - \mu_0)/V_w \qquad (3)$$

where ρ_w, M_w, and V_w are, respectively, the density, molecular weight, and molar volume of liquid water and μ and μ_0 are chemical potentials of water in the plant and in the reference state. For plant water in the reference state, $\psi = 0$. The water potential ψ (J m^{-3}), applicable in both soils and plants, is written as

$$\psi = \psi_p + \psi_m + \psi_s + \psi_g \qquad (4)$$

where ψ_p is the pressure potential owing to plant turgor (hydrostatic forces in excess of those due to atmospheric pressure acting on cell walls and internal membranes), ψ_m is the matric potential caused by capillary or binding forces at internal surfaces of cell walls, ψ_s is the osmotic potential due to a reduction in water vapor pressure by solutes and surface attraction, and ψ_g is the gravitational potential which originates from the difference in gravitational force acting on water at a given location and at the reference location. In general, ψ_g is taken to be negligible in the water of living plants, and ψ_p is considered to be zero in soil water. Cowan (1965) and Philip (1966) were among the early workers who attempted to describe water flow in living plants with physical coefficients like diffusivity and conductivity, but within the continuum framework. Lemon et al. (1971) and Riha and Campbell (1985) are other modelers who have used a similar approach. References summarizing the SPAC concept and research on water transport in plants are Kramer and Boyer (1995) and Kozlowski and Pallardy (1997).

B. CONSERVATION EQUATIONS

Though the conservation laws are equally applicable to live fuels, they are discussed here in the context of dead fuels because their application to water transport problems in these fuels is relatively simple. The solid fuel and its associated moisture comprise four separate phases: the dry solid, free liquid water, bound liquid water, and a gas phase consisting of a mixture of dry air and water vapor. Equations expressing the conservation of mass, momentum, and energy may be written for each phase. The intricate chemical and structural features of fuel particles or layers require the moisture exchange modeler to consider carefully the elementary volume selected as the basis for writing equations governing the process. Though a set of differential equations can be written to describe changes within each phase in the system, it is not possible to know which set to assign at each point in the medium. One approach to this problem is to define an elementary volume within the particle just large enough to contain all phases and satisfy the definitions of all system variables. Then a set of volume-averaged equations may be written that should be valid at all locations and the fuel-water system may be treated as a continuous medium (Spolek and Plumb, 1980). This procedure is known as *local volume averaging* and has been used by several researchers to describe heat and mass transfer in wood. Perre *et al.* (1990) pointed out that moisture transport in wood at temperatures between 60 and 120°C is best considered a low- to moderate-temperature process taking place with variable total pressure. On the other hand, an assumption of constant total pressure may be reasonable for open duff layers or for moderately thin fuel particles because within-particle temperatures, even in intense sunshine, seldom exceed 60°C, and the internal pressure does not differ greatly from atmospheric. Perre *et al.* also assume thermal equilibrium among all phases—an assumption that is commonly used because of the mathematical simplification it provides.

In the present discussion, we consider the unaveraged equations (equations for each phase) that would describe heat, mass, and momentum transfer in an elementary volume of fuel particle. Suppose that shrinking and swelling of the particle are ignored and gravitational body forces, kinetic energy changes, and the rate of doing work on the water are negligible. Then the equation that balances the transport of phase i in direction x (m) in a plane body (a slab, for example) may be written as

$$\partial(\rho_i\phi_i)/\partial t = -\partial(\rho_i\phi_i v_i)/\partial x - w_i = -\partial(J_i)/\partial x - w_i \qquad (5)$$

where ρ_i (kg m^{-3}), ϕ_i, v_i (m s^{-1}), and w_i (kg m^{-3} s^{-1}) are the density, volume fraction, velocity, and evaporation rate of phase i—all on a unit volume of particle basis—referred to a coordinate system on the stationary particle. The mass fluxes J_i (kg m^{-2} s^{-1}) take forms that depend on the transport mechanism for phase i. Subscript i ($i = s, w, b, g$) denotes the solid, free liquid, bound liquid, or gas phase, respectively. The dry air component of the gas phase, indicated by subscript a, can be described with its own mass conservation equation given by

$$\partial(\rho_a\phi_a)/\partial t = -\partial(\rho_a\phi_a v_a)/\partial x = -\partial(J_a)/\partial x \qquad (6)$$

where $w_a = 0$. In addition, the diffusion flux of air can be written as

$$J_a = -\rho_a\varepsilon D_v\partial[\ln(P_T - P_v)]/\partial x = -\rho_a\varepsilon D_v\partial[\ln(P_a)]/\partial x \qquad (7)$$

where εD_v $(m^2\ s^{-1})$ is the diffusivity of water vapor in air (or air in water vapor) corrected for presence of the solid (ε is the void fraction) and total gas pressure P_T depends on the partial pressures of air and vapor, P_a and P_v, according to

$$P_T = P_a + P_v \qquad (8)$$

Equality of temperatures T_a and T_v of the two gases and applicability of the ideal gas law are assumed.

For moisture change in forest fuels, it is usually not necessary to include equations for momentum conservation explicitly because flow rates are slow and fuel temperatures rarely exceed 60°C. Though the transport relationships describing J_i (Fick's law and Darcy's law) may be derived from approximate expressions of momentum conservation, their origin is empirical because of their discovery during 19th century experiments. The equations for J_i, where $i = w$, b, v (v denotes the water vapor component of the gas), may be written as

$$J_w = -(K_w/\eta_w)\partial(P_w)/\partial x = -(K_w/\eta_w)\partial(P_T - P_c)/\partial x \qquad (9)$$

$$J_b = -\rho_s m D_b\partial[\ln(P_b)]/\partial x \qquad (10)$$

$$J_v = -(K_g/\eta_g)\partial(P_T)/\partial x - \rho_v\varepsilon D_v\partial[\ln(P_v)]/\partial x \qquad (11)$$

where K_w (m^2), η_w $(m^2\ s^{-1})$, and P_w $(J\ m^{-3}$ or $kg\ m^{-1}\ s^{-2})$ in Eq. (9) are the specific permeability of the medium to liquid water and kinematic viscosity and pressure of the liquid phase. This equation is Darcy's law describing movement of liquid water in an unsaturated two-phase system by capillary forces. Capillary pressure P_c $(J\ m^{-3})$ is the pressure difference at a gas-liquid interface given by

$$P_c = (P_g - P_w) = (P_T - P_w) = 2\gamma/r \qquad (12)$$

where γ is the surface tension of the liquid and r is the mean radius of the interface (Siau, 1995). Equation (10) originates in the equation of motion for the adsorbed phase (Babbitt, 1950; Nelson, 1986b) and describes bound water diffusion in the cell walls. Quantities ρ_s $(kg\ m^{-3})$, D_b $(m^2\ s^{-1})$, m, and P_b $(J\ m^{-2})$ denote the density of the dry particle, bound water diffusivity, local moisture content fraction, and spreading pressure of water in the cell wall. The two-dimensional spreading pressure of a film adsorbed on the internal surface of a solid is analogous to the pressure of a three-dimensional gas in that adsorbed water molecules jump from one sorption site on the surface to another because of differences in film surface tension. Terms on the right side of Eq. (11) account for the capillary flow and diffusion of water vapor in the medium. Quantities K_g (m^2), η_g $(m^2 s^{-1})$, ρ_v $(kg\ m^{-3})$, and εD_v $(m^2 s^{-1})$ refer to the specific permeability of the medium to gas flow, kinematic viscosity of the gas, density of the vapor, and corrected vapor diffusivity. Equations (9)–(11) may be substituted into Eq. (5) to obtain the within-phase equations of moisture transport. When $i = w$ in Eq. (5), $w_i = w_w$; when $i = b$, $w_i = w_b$; when $i = g$, $w_i = w_g = w_v = -(w_w + w_b)$ and the interphase sink/source terms sum to zero.

The conservation of energy within phase i may be written as

$$\rho_i \phi_i c_i \partial T_i / \partial t + c_i J_i \partial T_i / \partial x + w_i \lambda_i = -\partial q_i / \partial x = \partial(k_i \partial T_i / \partial x)/\partial x \qquad (13)$$

where c_i $(J\ kg^{-1}\ K^{-1})$ and T_i (K) are the constant-pressure specific heat and temperature of phase i, λ_i $(J\ kg^{-1})$ is the heat of evaporation/condensation of phase i, k_i $(J\ m^{-1}\ s^{-1}\ K^{-1})$ is the thermal conductivity, and q_i $(J\ m^{-2}\ s^{-1})$ is the heat flux that accounts for all mechanisms of heat transfer to the phase. In Eq. (13), it is assumed that conduction is the only means of heat transfer, and use is made of another empirical relationship—the Fourier law of heat conduction. Equation (13) must be summed over all phases (assuming local thermal equilibrium) because all are capable of transferring heat. It is noted that $J_s = w_s = w_a = \lambda_s = \lambda_a = 0$; in addition, $\lambda_b = (\lambda_w + \lambda_d)$, where λ_d $(J\ kg^{-1})$ is the differential heat of sorption. A discussion of general wood–water thermodynamic relationships, including the differential heat of sorption, is given by Skaar (1988).

Consistent with their application and the desired level of complexity, Eqs. (5)–(13), or a subset of them, may be solved analytically or numerically in conjunction with fixed or variable boundary conditions involving one or more of the following variables: moisture content, temperature, relative humidity, wind speed, solar radiation, and net longwave radiation from various sources. Use of these equations requires still more information that includes expressions for void fraction, transport properties, thermal properties, and thermodynamic relationships. For multiple phases, relative amounts of space occupied by free water and water vapor are needed. Examples of the incorporation of this kind of information into models describing multiphase transport through volume averaging are the studies of Spolek and Plumb (1980), Perre *et al.* (1990), and Fernandez and Howell (1997).

C. LIVE FUEL MOISTURE

The understanding of water retention and movement in plants and in the soil supporting their growth is basic to development of models for predicting the moisture content of live surface and crown fuels. These models would help forest fire management and control personnel quantify the behavior of crown fires and anticipate the transition from surface fire to crown fire. Testing of a crown fire spread model that uses fuel moisture content as an externally supplied input parameter is underway (Alexander *et al.*, 1998).

1. Moisture Characteristic Curves

The equilibrium relationship between plant tissue water content and water potential at constant temperature is often referred to as the moisture characteristic curve by plant water researchers, and similar terminology is used in soil water studies. It is determined, in effect, by a balance between water potential of the solution in the vacuole of the cell (protoplast) and water in the cell walls (apoplast) and provides a mechanism by which liquid water moves into and through plants (see Figure 1). Though the cell wall is permeable to water and solutes, the membrane (or plasmalemma) surrounding the wall's inner edge is impermeable to solutes. Thus water moves into or out of the vacuole according to the difference in potential across the membrane. These differences, however, are usually so small that local equilibrium may be assumed (i.e., $\psi_{va} = \psi_{cw}$ where subscripts *va* and *cw* denote the vacuole and cell wall). Techniques are available to measure the components of ψ_{va} and ψ_{cw} separately—matric potential of the cell wall water $\psi_{m(a)}$, pressure (or turgor) potential of the solution in the vacuole, and the osmotic potential of both regions (Kramer and Boyer, 1995). The relationship between $\psi_{m(a)}$ and relative water content, RWC, is shown in Figure 6 for a *Taxus* branch suspended in a pressure chamber (McGilvary and Barnett, 1988; Kramer and Boyer, 1995). Plant water researchers express ψ in terms of bars or MPa rather than $J\ m^{-3}$, so their terminology is used here (1 bar = 0.1 MPa = $10^5\ J\ m^{-3}$). The osmotic component in the cell wall, $\psi_{s(a)}$ in Figure 6, contributes in a minor way to the total potential of the wall, which is the sum of the two components displayed. Quantity RWC is the water content of a material expressed as a percentage of the water content when the material is fully turgid (in equilibrium with pure water). This turgid state is a more convenient reference point than the dry state because it is experimentally reproducible and not as subject to error due to plant growth. Thus RWC is computed from

$$RWC = [(W_c - W_d)/(W_t - W_d)]100\% \tag{14}$$

where W_c, W_t, and W_d are the material weights in the current, turgid, and ovendry (dried at 100°C) states. Gardner and Ehlig (1965) are among the researchers who have studied the water potential-RWC relationship. They measured RWC, ψ, ψ_s, and ψ_p of cotton, pepper, sunflower, and birdsfoot trefoil leaves and found a linear relation between osmotic potential ψ_s and RWC. The total potential ψ was not linearly related to RWC because of a change in the modulus of elasticity of the cell wall when the turgor potential ψ_p approximated +2 bars. This change in elasticity was associated with the early stages of wilting.

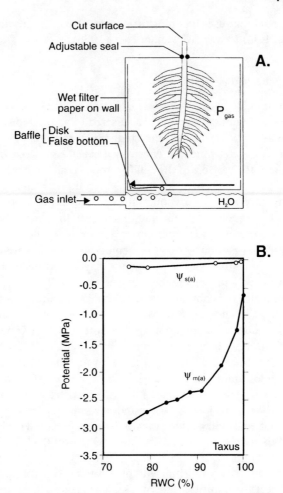

FIGURE 6 Determination of the moisture characteristic curve. (A) Pressure chamber in which incoming gas increases the pressure around a *Taxus* branch and forces xylem sap through the cut surface. The pressure is adjusted to maintain zero flow at the surface and provides an estimate of the internal tension on the apoplast. (B) Pressure chamber measurements show that water potential decreases as relative water content decreases and that a greater pull is being exerted by the leaves on water in the xylem as water content decreases. Osmotic water potential of the cell wall is given by $\psi_{s(a)}$. From Kramer and Boyer (1995).

2. Water Transport in the SPAC

The mechanisms governing transport of water through living fuels may be described by several methods. One is empirical and often used to fill knowledge gaps with information based on experimental data. In the second method, it is assumed that the water-conducting system is analogous to an electrical circuit and requires evaluation of flow resistances in series and/or parallel connections. A third method involves solution of the differential equations governing flow in the various soil–plant–atmosphere domains and uses conductivities and diffusivities rather than resistances. The domains most conducive to analysis with differential equations are those describing soil water flow and uptake by roots. Modelers often describe flow in the whole plant using some combination of these three methods (Federer, 1979; Riha and Campbell, 1985).

a. The Ohm's Law Analogy

The driving force for water movement in transpiring plants originates in the leaves and is transmitted to the roots through the cell sap. Transpiration reduces the leaf water potential, and this disturbance is propagated through the xylem of the leaves and stem to the roots where the reduced potential induces water flow toward the roots. This process occurs when water content of the soil is high or the transpirational demand is low. On the other hand, when the soil water content is low or the transpirational demand is great, leaf water potential falls to such a low value that the leaves lose their turgor and the stomates tend to close. Thus, the leaf resistance is greatly increased, the rate of transpiration is reduced, and further loss of water is controlled by the plant and soil rather than the atmospheric conditions (Cowan, 1965). In the SPAC method of analysis, the overall movement of water may be analyzed using an analog of Ohm's law which states that the flow of current in an electrical circuit is proportional to the potential difference across the circuit. The electrical analog shown in Figure 7 is a simplified representation of the paths taken by water from its entry via the water table to its exit through the plant leaves. The xylem portion of the circuit applies to the complete vascular system—xylem in the roots, stem, and leaves. The overall process may be crudely approximated with four subprocesses: water flow from soil to roots, from roots to stem, from stem to leaves, and from leaves to the atmosphere. Figure 8 illustrates the relative magnitudes of change in water potential for each component of the SPAC. Because the change in potential from root to leaf is only a few bars, the stem sometimes is omitted from considerations of overall flow.

The SPAC is nonstationary in character because of diurnal and seasonal variation and because of periods of hydration and desiccation. To eliminate some of this complexity, plant water researchers often assume a steady state rate of transport. Though this represents considerable simplification, investigators can work with a series of steady states instead. The steady state assumption implies that a disturbance of water potential in the roots (or leaves) is propagated instantaneously throughout the plant to the leaves (or roots) and that internal changes in water content during the process are small in comparison with the magnitude of the disturbance (Cowan, 1965). Jarvis (1975) and Waring and Running (1976) argue, however, that in plants with water sources of sufficient

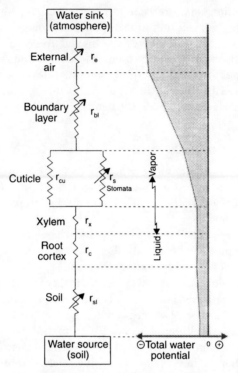

FIGURE 7 Electrical analog showing soil, root, vascular, leaf, and atmospheric flow resistances in the SPAC. Resistances r_{sl}, r_c, and r_e correspond to r_{so}, r_r, and r_a in Eq. (15); the overall resistance due to r_{cu}, r_s, and r_{bl} is represented by r_l in Eq. (15). The shaded area at the right indicates relative magnitudes of water potential for the various components. From Rose (1966).

size in the stem and branches and with radial (or horizontal) pathways to the stem xylem of low enough resistance, sufficient water may be withdrawn from these regions of storage to satisfy the transpirational demand for considerable periods. Jarvis points out that little is known about resistances to flow between these storage regions and the xylem and about the moisture characteristic of the cambium, phloem, or roots. Nevertheless, if steady state is assumed, stem resistance is omitted, and soil moisture is not limiting, the transpiration rate Q (m s^{-1}) equals the flow rate in all plant domains and

$$Q = (\psi_{so} - \psi_r)/(r_{so} + r_r) = (\psi_r - \psi_l)/(r_r + r_l) = (\psi_l - \psi_a)/(r_l + r_a)$$
(15)

where r (bar-s m^1) symbols denote resistances and subscripts *so*, *r*, *l*, and *a* refer to soil, root, leaf, and atmosphere. The last term in this equation is a rough approximation during the day because its value is influenced by temperature

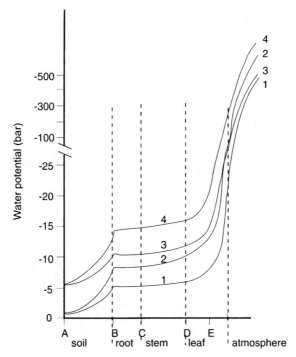

FIGURE 8   The distribution of water potential throughout a woody plant (not to scale). Curves 1 and 2 indicate extremely wet soil; curves 2 and 4 indicate a high atmospheric demand. The largest change in potential occurs in the leaf. From Hillel (1980b).

changes associated with solar radiation and periods of cloudiness. Moreover, the term often is written with differences in vapor pressure or vapor concentration replacing the water potential difference (Kramer and Boyer, 1995).

It is recognized by plant water researchers that the assumptions commonly associated with the SPAC approach, namely steady state flow and constant flow resistances, make this kind of analysis a crude approximation at best. For example, total resistances often must be evaluated by dividing the potential difference by transpiration rate Q. They are continuously changing, however, because transport can be controlled either by the soil, plant, or atmosphere. Numerous authors have studied the relative importance of r_{so} and r_r, and differences of opinion continue to exist. Results from these studies are briefly summarized by Kramer and Boyer (1995) who state "evidence seems to indicate that over a considerable range of soil water content and with average root density, root resistance exceeds soil resistance to water movement." Leaf resistance also is complicated and includes components due to liquid flow in the petiole, veins, and mesophyll cells as well as vapor diffusion through the cuticle, substomatal cavities, and stomates (Kreith and Sellers, 1975). Root and leaf resistances from several studies are reported by Slatyer (1967) and Cowan and Milthorpe (1968).

b. Soil–Root Models

Movement of soil water may be described by the Richards equation which is based on mass continuity and momentum conservation as expressed by Darcy's law. Derivations of the Richards equation for saturated and unsaturated flow may be found in soil physics texts such as Hillel (1980a). If the influence of osmotic potential on soil water flow is negligible and the system is isothermal, the Richards equation for vertical flow in unsaturated soils may be written as

$$\partial\theta/\partial t = C_h(\partial h/\partial t) = \partial[\kappa_h(\partial h/\partial z) + \kappa_h]/\partial z \tag{16}$$

where θ is the volumetric soil water content, t (s) is time, h (m) is the pressure head applicable to flow caused by capillary and attractive forces, C_h (m^{-1}) is the specific water capacity given by $(\partial\theta/\partial h)$, κ_h (m s^{-1}) is the hydraulic conductivity, and z (m) is the vertical distance from the reference location. Head h is given by $P_w/\rho_w g$, where P_w $(\text{J m}^{-3}$ or $\text{kg m}^{-1}\text{ s}^{-2})$ is the hydrostatic pressure, ρ_w (1000 kg m^{-3}) is the density of water, and g (9.8 m s^{-2}) is the gravitational acceleration. For horizontal flow, z must be interpreted as a horizontal distance, and the second term in the brackets is set to zero. In the case of saturated flow, conductivity κ_h may be regarded as the hydraulic conductivity of the saturated soil (a constant), and the derivative of the second term in brackets vanishes. Various empirical and theoretical relationships expressing conductivity κ_h in terms of θ or matric potential ψ_m and soil texture have been reported (Scholl, 1976; Jury et al., 1991; Alessi et al., 1992). Equation (16) may be expressed in terms of ψ_m (J kg^{-1}) by substituting the relationships

$$\psi_m = P_w/\rho_w = gh \qquad \text{and} \qquad \rho_w\kappa_h = g\kappa_m \tag{17}$$

into Eq. (16) to obtain

$$\rho_w(\partial\theta/\partial t) = \rho_w C_m(\partial\psi_m/\partial t) = \partial[\kappa_m(\partial\psi_m/\partial z) + g\kappa_m]/\partial z \tag{18}$$

where C_m $(\text{s}^2\text{ m}^{-2})$ is the slope of the soil-moisture characteristic curve, or specific water capacity $(\partial\theta/\partial\psi_m)$, and conductivity κ_m (kg-s m^{-3}) differs from κ_h. Equations (16) and (18) contain no source or sink terms to account for loss or gain of soil water by the roots. Riha and Campbell (1985) added a sink term to the right side of Eq. (18) to account for absorption of such water. Their term was of the form

$$U = \rho_t(\psi_{so} - \psi_r)/(r_{so} + r_r) \tag{19}$$

where U is the root sink strength (kilogram of water absorbed per second per cubic meter of soil) and ρ_r the root density (meter of root length per cubic meter of soil). Potentials are in units of J kg^{-1} and resistances in $\text{m}^3\text{ kg}^{-1}\text{ s}^{-1}$.

The uptake of soil water by roots may be modeled by two differing approaches in which the roots are treated as uniform cylinders of length L. The first method considers a single root of uniform radius surrounded by a cylindrical sheath of soil. When solute effects are neglected and equilibrium is achieved between liquid in the cell vacuole and that in the cell wall, radial transport of water into the root may be described by

$$\partial \psi_m / \partial t = D_h(\partial^2 \psi_m / \partial r^2) + (D_h/r)(\partial \psi_m / \partial r) \tag{20}$$

where D_h (m^2 s^{-1}) is the hydraulic diffusivity of the root tissue, r (m) is radial distance in the root, and potential ψ_m has the units J m^{-3} (Hillel, 1980b). The relationship between κ_h of Eq. (16) and D_h of Eq. (20) is

$$C_h D_h = \kappa_h \tag{21}$$

whereas Eqs. (16)–(18) may be used to write

$$C_h = g C_m \tag{22}$$

When the soil moisture contains a significant amount of solutes, differences in osmotic and matric potential between the sheath of soil surrounding the root and the root itself must be considered (Hillel, 1980a; Flowers and Yeo, 1992). The second approach is to consider the entire root system as a moisture sink which penetrates the soil to a known depth. This root zone depth may be subdivided into layers that contain different root masses per unit volume of soil. The major fault of this method is that it is based on gross averages of root density and water potential over the root zone.

c. Leaf–Atmosphere Models

The influence of the atmosphere on water movement in plants can be evaluated from the rate of evaporation from a free water film covering the leaf surface. This rate is referred to as the *potential rate of evaporation*, Q_0 (kg m^{-2} s^{-1}), and may be estimated from the simple "Ohm's law" equation

$$Q_0 = (\rho_{sat} - \rho_{va})/r_a \tag{23}$$

where ρ_{sat} (kg m^{-3}) is the concentration (or density) of water vapor at the leaf surface under saturated conditions and ρ_{va} the vapor concentration in the ambient air. Further discussion of Q_0 is given in Chapter 9 in this book. Resistance offered by the air to vapor diffusion from the leaf surface is r_a (s m^{-1}) and can be estimated by two methods. The first method, for continuous areas of more or less uniform vegetation, utilizes the logarithmic wind profile (see Chapter 9 in this book) and therefore considers such factors as vegetation height, zero plane displacement, and roughness length (Monteith, 1963a; Rutter, 1968). The second approach, for single leaves, makes use of boundary layer concepts and depends on leaf size and shape and on wind speed (Kreith and Sellers, 1975). When evaporation takes place from the interior leaf surfaces, the process is referred to as *transpiration*. In this case, the resistance to vapor diffusion occurring within the leaf, r_l, should be accounted for and r_a in Eq. (23) replaced by $(r_a + r_l)$. Resistance r_l includes cuticular resistance which is in parallel with the resistance to vapor diffusion. This diffusion, or stomatal, resistance is made up of resistances originating in the mesophyll cell walls, intercellular spaces, and stomatal pores linked in series. Schönherr (1976) studied the diffusion of water through cuticular membranes and found that the water permeability of the cuticle is completely determined by waxes rather than cutin in the cuticular layer. Cuticular resistance is 10 to 100 times greater than the stomatal resistance and generally can be ignored in calculations of r_l (Kozlowski and Pallardy, 1997).

The term *evapotranspiration* refers to loss of water from the soil due to the combined water loss from plants by transpiration and from the soil surface by evaporation (see Chapter 9). When all surfaces are moist, evapotranspiration proceeds at the potential rate and depends primarily on meteorological factors. In this case, the *potential evapotranspiration, Q_0*, may be calculated by at least two different methods. The first is outlined by Kreith and Sellers (1975) and is applicable to evaporation from large areas (an ecosystem or field). Their approach requires that temperatures of the evaporating surfaces be known. The second method utilizes the Penman equation—an equation which accounts for the effect of solar radiation on the evaporation from a water surface in such a way that knowledge of surface temperatures is not required. Various forms of the equation are widely used in agricultural applications to compute Q_0. Monteith *et al.* (1965) and Monteith (1980) modified the Penman equation to describe potential evaporation from whole plants or crops over an extended area. The modified model, referred to as the Penman-Monteith equation, is discussed in Chapter 9.

According to Monteith (1980), there are only two ways to compute the actual evapotranspiration rate, Q. One method is to utilize complex models within the SPAC framework (Lemon *et al.,* 1971; Waring and Running, 1976); the second is to apply models such as the Penman-Monteith equation to the entire plant canopy. This application, which assumes that stomatal resistance is uniform throughout the canopy and that temperature, humidity, and momentum profiles are similar within and above the canopy, was questioned by Philip (1966) and Tanner (1968). Subsequent research has shown that the Penman-Monteith equation is useful when applied to well-developed crop canopies but underpredicts transpiration from sparse canopies when soils are dry and overpredicts it when soils are wet. Refinements in the equation by agricultural researchers have led to compartment schemes in which evaporation from the soil surface and transpiration from the crop canopy are treated separately (Shuttleworth and Wallace, 1985; Choudhury and Monteith, 1988). This approach has produced better agreement with experimental measurements of transpiration because of improved modeling of soil heat and vapor fluxes (Wallace *et al.*, 1990; Tourula and Heikinheimo, 1998).

When the soil moisture or soil water potential falls below some critical value, the evapotranspiration rate falls below the potential value. Lopushinsky and Klock (1974) found that transpiration rates in several conifer seedlings began to decrease when the soil water potential fell below −2 bars (Figure 9). The transpiration rates of several pines were slower than those of fir and spruce by more than a factor of two, explaining the ability of pines to survive on fairly dry sites and the need for fir and spruce to occupy more moist sites. Lopushinsky and Klock attributed the decrease in transpiration rate to stomatal closure and noted that such closure takes place in all species when the leaf water potential ranges from about −14 to −25 bars. Brown (1977) studied the wilting process in two range grasses under controlled environmental conditions. He measured

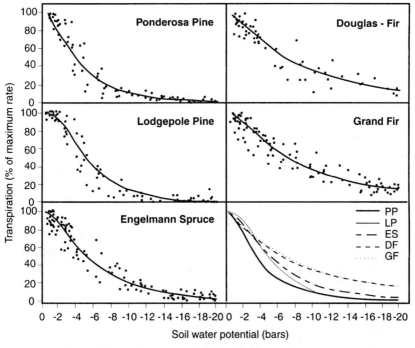

FIGURE 9 Transpiration rate as a function of soil water potential for seedlings of various conifers. Curves at the lower right represent regression equations through the experimental data. From Lopushinsky and Klock (1974). Reprinted from Forest Science (vol. 20, no. 2, p. 183) published by the Society of American Foresters, 5400 Grosvenor Lane, Bethesda, MD 20814-2198. Not for further reproduction.

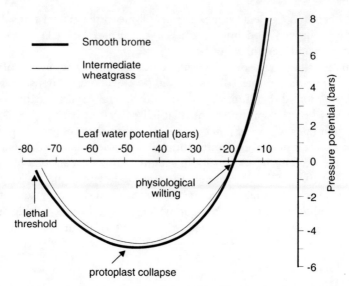

FIGURE 10   Relationship between pressure (or turgor) potential and leaf water potential for two range plants. From Brown (1977).

soil and leaf water potentials during a drying process in which the soil water potential decreased from 0 to about −65 bars. In the early stages of drying, the pressure potential (the difference between the leaf water and osmotic potentials) decreased to zero—a value labeled by Brown as the *physiological wilting point* (Figure 10). As drying continued, the pressure potential decreased to a minimum of about −5 bars (the point of *protoplast collapse*) and then increased toward zero where severe wilting occurred at a leaf water potential between −70 and −80 bars; Brown labeled this point (or range) the *lethal threshold* and associated it with the collapse of all plant cells. Brown (1977) stated that other unpublished research by him indicates the measurement techniques he used are equally applicable to trees, shrubs, forbs, and grasses growing in arid environments and that similar experiments can provide quantitative guidelines for evaluating the adaptability of range plants to dry conditions.

3. Canopy Moisture Content

The literature of forest physiology contains much information on evaporation and transpiration rates but is almost devoid of attempts to model the water content of understory and overstory canopies. It is likely that within a given fuel type there exist critical combinations of canopy moisture content, relative weights of live and dead fuel, and ambient wind speed which will permit the spread of fire in tree crowns, understory shrubs, or open grassland. The ability to predict canopy moisture content seasonally would be a help in understanding and anticipating the initiation and spread of fire in these fuel types.

Changes in the moisture content of live fuels are poorly understood—partly because plant physiologists have studied the response of plants to environmental stress in terms of water potential rather than moisture content (Brown *et al.*, 1989). These authors also state that seasonal variation in water content is more closely related to phenological changes such as leaf growth and maturation than to environmental factors. In earlier work, Kozlowski and Clausen (1965) pointed out that changes in plant tissue moisture content can result from changes in actual water content, tissue dry weight, or both. Their phenological study of leaves from several forest trees during the 1963 growing season showed that the moisture content of all deciduous leaves studied decreased until mid-June, after which time the rate of decrease slowed and became dependent on species. This result was attributed to increases in dry weight rather than changes in the mass of water present. On the other hand, moisture content of zero-age conifer needles decreased continuously, whereas in 1-year-old needles it increased until about mid-July and then became relatively constant. In this case, the dry weight changes were attributed to carbohydrate translocation into new needles and out of the older ones. Little (1970) conducted a year-long study of water, sugar, starch, and crude fat content in juvenile balsam fir needles. In contrast with the results of Kozlowski and Clausen (1965), he found that seasonal fluctuations in moisture content generally were caused by changes in actual water content rather than by dry weight changes. Little pointed out, however, that changes in moisture content due to dry weight changes increase with increasing age, and that increased carbohydrates may not have accounted for the total change in dry weight of his study—as they might have in the maturing balsam fir needles of the Kozlowski and Clausen study. Research on Engelmann spruce needles conducted by Gary (1971) showed that moisture contents were determined primarily by dry weight changes and that differences between needles from north and south slopes were small. Chrosciewicz (1986) sampled the foliar moisture content of several Canadian conifers in central Alberta. His sampling procedure was designed to minimize variation due to the diurnal cycle and sample location within the canopy. Figure 11 shows some of his results for new and old (needles 1, 2, and 3+ years old) foliage measured during 1974. The flushing of new foliage coincides with the moisture content minimum in the old foliage. Moisture content of the new foliage approaches that of the old in September.

In a study extending from January to October 1971, Hough (1973) measured extractive content, phosphorus content, and moisture content of old and new needles to determine their effects on the flammability of natural sand pine stands. A sharp rise in moisture content of new needles up to 10 months in age coincided with the initiation of new growth, whereas ether-soluble extractives

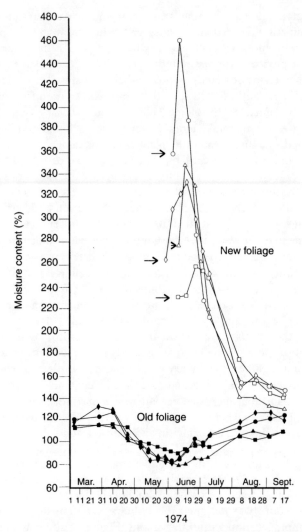

FIGURE 11 Seasonal moisture content variation in new and combined old foliage (1, 2, and 3+ years old) of four Canadian species. Right-pointing arrows indicate the start of flushing of new foliage. With the arrows as reference points, the species from top to bottom of the figure are white spruce, black spruce, balsam fir, and jack pine. From Chrosciewicz (1986).

of new and old (9–18 months in age) needles peaked in late February and then decreased until June. Phosphorus content of new needles decreased until March and then peaked in late May, while that of the old needles decreased until April and then slowly increased. Hough concluded that both weather and fuel variables affect fire occurrence in stands of sand pine in Florida and that crown fires would be most likely to occur in late February and early March. He stated that the severity, or intensity, of wildfires is not reduced until the new needles, with their high moisture content and low extractive content, make up most of the crown.

One naturally would expect understory vegetation to show patterns of variation in moisture, mineral, and extractive content similar to those in tree crowns (Hough, 1973), and the results of numerous experimental studies have shown this to be so. Fire researchers have documented the moisture content of chaparral and brush fuels in southern California for many years. Countryman (1974), for example, reported field data on the variation in moisture content of manzanita and snowbrush (Figure 12). New growth in both species typically begins in early June with moisture contents of about 200% of the fuel ovendry weight. The moisture content of old foliage increases more slowly, and both species reach a minimum level in September. Countryman stated that the percent of total fuel made up of living material usually ranges from 65 to 85% and that fire behavior in the chaparral types depends on both live fuel moisture content and relative amount of dead fuel present. According to Blackmarr and Flanner (1975), the understory shrubs growing on organic soil in the pocosin areas of eastern North Carolina cause a severe fire hazard during periods of drought. These authors sampled the moisture content of six shrub species native to the area for a 22-month period. Their 1964 and 1965 data for deciduous and evergreen species showed that all parts of the evergreen foliage were consistently drier (ranging from 250 to 100% during the summer) than those of the deciduous species (350 to 150%), tending to make them more flammable. In earlier work, Blackmarr and Flanner (1968) measured the diurnal moisture content variation in four North Carolina pocosin species over a 36-hour period during June 1965 and found differences ranging from about 25 to 40% moisture content.

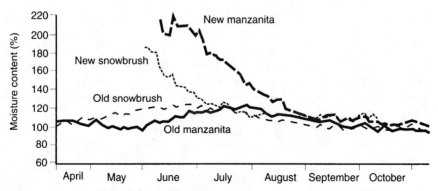

FIGURE 12   General trend of seasonal moisture content variation in two chaparral fuels in southern California. From Countryman (1974).

The moisture content of live fuels in open grasslands generally follows patterns similar to those of the understory shrub and crown fuels discussed earlier, but it also differs in some respects. These patterns are determined primarily by weather. The annual grasses usually have shallow root systems and therefore depend strongly on moisture availability near the soil surface for their growth. In the spring, these green grasses can attain fractional moisture content values larger than 3.5. As hotter weather approaches, the growth of annuals (roughly from May to July in the United States) is shortened by the absence of rainfall. The plants mature, produce seed, and then begin to cure (or dry). Their rate of curing depends on soil moisture and rainfall, resulting in curing times that may vary from three to eight weeks. As the plants flower and produce seed, the color of their various parts turns from green to yellow (or other colors such as purple, depending on species). According to Luke and McArthur (1978), this process begins in the annuals of Australia when the moisture content fraction has decreased to about 1.75. As the moisture fraction drops below 1.0, yellowing has affected about 30% of the plant surface, and the grass is said to be 30% cured; at a moisture fraction of about 0.5, the grass is 60% cured and has reached a state of dryness such that it cannot recover its moisture regardless of the amount of additional rainfall. At moisture fractions below 0.3, annuals are close to fully cured and become dead fuel now subject to changes in atmospheric relative humidity and temperature. The perennial grasses exhibit a similar, but longer, curing period (from May to September in the United States) because they have root systems that penetrate more deeply into the soil than those of annuals; this makes them less sensitive to short-term weather and surface soil moisture. In warm and humid climates, some parts of perennials cure and die whereas other parts remain alive but in a dormant state (Schroeder and Buck, 1970). In the United States, Brown (1977) reported water potential relationships he observed during the wilting of two range grasses (see Section III.C.2.c); Brown et al. (1989) measured the moisture content of perennial grasses under aspen stands in Wyoming. The latter authors noted a change in color of the green vegetation at a moisture fraction of about 1.0 (corresponding to 30% curing of Australian annuals) and found that these grasses green up and increase in moisture content in response to rainfalls greater than 6 mm. Similarly, Luke and McArthur (1978) reported a delay in the curing of Australian perennials due to 30 mm of rain. The moisture content of grassland fuels in Australia has been related to the Keetch-Byram Drought Index (see Section IV for several references on drought). Luke and McArthur (1978) state that Australian annuals are fully cured when the Keetch-Byram index exceeds 200 and that perennials are cured when the index is close to 600. In studies conducted in the United States (northeastern California), Olson (1980) observed that the foliar moisture content of bluebunch wheatgrass (a perennial) decreased exponentially as the Keetch-Byram index increased.

While the need for models that predict moisture content change in the canopies of grass or understory/overstory vegetation is generally recognized, the complexity of the task may have precluded all attempts to develop a reasonably complete physics- and physiology-based model (Tunstall, 1991). The required model should include compartments treating such topics as soil water transport to roots and directly to the atmosphere, water takeup by roots, water rise in the main stem, growth of roots, stem, and foliage as determined by photosynthesis and respiration, storage and translocation of water and food, water potential interactions, water vapor diffusion through leaves and into the atmosphere, and senescence (the process of wilting and dying). A practical solution to the problem may be achievable if we can eventually identify and model a few key controlling processes (Howard, 1978); alternatively, perhaps we can successfully modify existing models of growth and transpiration in conifers and understory vegetation to obtain reasonable predictions of canopy moisture content (Running, 1978, 1984a, 1984b).

D. Dead Fuel Moisture

The moisture content of dead forest fuels in the hygroscopic range is of considerable interest to fire managers. The equilibrium moisture content is useful in laboratory studies in which classical diffusion theory describes moisture change driven by step changes in the fuel's environment. More recently, equilibrium concepts have been combined with diffusion theory to develop new methods for estimating the moisture transport properties of fuels under field conditions.

1. Equilibrium Moisture Content

The fractional equilibrium moisture content M_c represents the constant value of M attained by dead forest fuels when they are exposed for an extended period in air of constant relative humidity and temperature and in which changes due to variability in wind, solar radiation, and barometric pressure are small. In laboratory experiments, equilibrium values of fuel moisture content depend on whether that state is attained following adsorption or desorption because of various mechanical and environmental factors that result in sorption hysteresis. The effects of hysteresis are not apparent in field studies because of a continuously changing environment. Models that describe the equilibrium state in fuels and soils over the entire range in relative humidity eventually will simplify the description of capillary and sorption processes in composite layers of soil, duff, and litter.

a. Adsorption and Desorption

When wood and foliage reach equilibrium at various values of H and constant T, the resulting M_c values, plotted versus H, form a sigmoid curve charac-

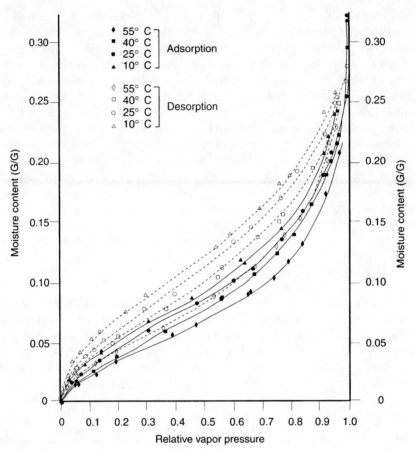

FIGURE 13 Sorption data for klinki pine wood at four temperatures. Note crossing of the 40°C adsorption isotherm over the 25°C curve at a relative vapor pressure of about 0.95. From Kelsey (1957). Reproduced by permission of CSIRO Australia.

teristic of most natural materials. If the successive H values increase, M_e increases, and the process is referred to as *adsorption;* when the H values decrease, *desorption* is occurring. The resulting curves are called adsorption and desorption *isotherms,* and the combined processes are often referred to as sorption. Figure 13 illustrates these isotherms for klinki pine wood at temperatures from 10 to 55°C. The region in H from 0 to about 0.2 is considered one in which most of the water in the wood is held by hydrogen bonding in a single layer. The slightly curvilinear region from 0.2 to about 0.9 involves water held in multiple layers within the cell walls but possibly also includes a small amount of capillary condensed water in the finest cell wall capillaries as H approaches 0.9.

Most of this water adds its volume to that of the cell walls and the composite is best thought of as a solid solution (water dissolved in cell wall substance). Because of swelling, the internal area available for adsorption due to forming cellulose-water bonds or water bridges between hydroxyl groups is about 1000 times greater than the area that would be available if water did not swell wood. True capillary condensation takes place for H between 0.9 and 0.95, but these values are rarely exceeded in commonly used experimental methods of measurement. Specialized measurements extending H to 0.999+ have been made on wood and wood pulp using the tension plate, pressure plate, pressure membrane, or nonsolvent water methods (Cloutier and Fortin, 1991; Zhang and Peralta, 1999).

The effect of previous history on sorption for a given temperature is evident in Figure 13 as the difference in M_e values for adsorption and desorption at constant H, referred to as *sorption hysteresis*. The two most generally accepted explanations of hysteresis are based on the availability of sorption sites in the material and on the rheological properties of wood. In the sorption site theory (Urqhart, 1960), it is supposed that during the initial desorption from the green condition, all available hydroxyl groups will have water molecules attached to them. As these molecules leave the sites and shrinkage occurs, many of the sites become mutually satisfied through formation of cellulose–cellulose bonds and a certain amount of rearrangement of the cellulose takes place; some unsatisfied hydroxyl groups remain, however, even when H is zero. Now as H is increased during adsorption, water molecules are adsorbed on these remaining sites, and nearly all sites will have one or more attached molecules, reducing the attraction for additional molecules. Thus a greater H is required during adsorption to attach the same number of water molecules (or achieve the same M_e value) as in desorption. This mechanistic theory of Urqhart (1960) remains, for many researchers, the most reasonable explanation of hysteresis.

Research has not yet shown whether Urqhart's theory conflicts with the thermodynamic arguments proposed by Barkas (1949) to explain hysteresis in wood based on rheological properties, but it seems likely that the two explanations will prove to be compatible. The rheological theory states that hysteresis appears because gels such as wood are not perfectly elastic and hydrostatic stresses caused by resistance to shrinking or swelling of the cell wall introduce M_e values differing from the stress-free (or perfectly elastic) value. On the one hand, fiber direction differences in the cell wall (Figure 3A) induce tension into wood undergoing desorption, causing a larger M_e than if the wood were stress free. On the other hand, interfibrillar bonding causes wood that is swelling during adsorption to be in compression, resulting in a smaller M_e than if the wood were stress free. Barkas (1949) shows that the area enclosed by a constant-temperature hysteresis loop (as in Figure 13) represents the net work loss during the entire sorption cycle. For spruce wood, Barkas calculates that this area equals 17% of the total work of desorption. According to Skaar (1988), this means that the ratio of M_c values for adsorption and desorption in spruce should come to 0.83 (1 − 0.17 = 0.83) when averaged over the entire range in H. This overall *hysteresis ratio* averages about 0.83 for both softwoods and hardwoods (Skaar, 1988). For woody fuels, the hysteresis ratio for specific values of H generally ranges from 0.75 to 0.85 over the entire range in H from 0 to 1 (Stamm, 1964). These discrete ratios for some foliage fuels are more sensitive to H, as indicated by values ranging from 0.61 to 0.86 (Nelson, 1984).

The immediate effect of temperature on water vapor sorption by forest fuels is a reduced value of M_e as T increases. The reduction is due to the strong dependence of saturation vapor pressure on T and the resulting tendency for more water molecules to leave their sorption sites as T increases but H remains constant. This reversible effect is illustrated in the desorption data of Figure 13. Van Wagner (1972b) reported a decrease in M_c averaging about 0.024 per 10°C increase for desorption in aspen and pine leaf litter and about 0.018 in aspen wood splints. On the other hand, King and Linton (1963) and Anderson *et al.* (1978) reported a decrease of about 0.01 per 10°C increase for sorption in assorted natural fuels and ponderosa pine needles, respectively. At high values of T and H, a second temperature effect, an irreversible reduction in hygroscopicity, occurs when exposure times are long. The wood becomes discolored, and the most hygroscopic component, the hemicellulose, decomposes first followed by the pure cellulose. A still different effect at the higher values of H is that the adsorption isotherms of wood and cellulosic materials at temperatures exceeding about 40°C may cross those for lower temperatures. The effect is present in Figure 13 in which the 40°C adsorption isotherm crosses the 25°C isotherm at about $H = 0.95$. Though the 55°C isotherm does not extend to sufficiently high H values to indicate crossing at that temperature, the data suggest that crossing can occur in once-dried wood at about 40°C. Unfortunately, there exist no extensive laboratory data illustrating crossover in wood. The phenomenon has been observed in American Uplands cotton (Urquhart, 1960) in which the 70 and 90°C isotherms crossed those for lower temperatures when H exceeded 0.85. Both Urquhart (1960) and Stamm (1964) attribute the crossover effect in cotton to plasticization of the cellulose fibers; the resulting molecular rearrangement relieves drying stresses introduced when the cotton was first dried, making new sites available for adsorption. Figure 13 also shows that for $H = 0.5$ the magnitude of hysteresis in klinki pine wood decreases by a factor of nearly 2 as T increases from 10 to 55°C; Skaar (1988) refers to German research indicating that sorption hysteresis in European spruce wood disappears between 75 and 100°C. In a study of water vapor sorption by conifer and hardwood litter, Van Wagner (1972b) noted that hysteresis decreased by a factor of about 2.5 at "medium" values of H when T increased from 16 to 49°C.

Sorption of water vapor by forest fuels in the field is strongly affected by environmental variables, and even the finest fuels rarely achieve equilibrium with the continuously changing H and T values of the ambient air. The effects of H and T on sorption are modified, however, by wind and by daytime solar radiation and nighttime longwave radiation to the sky. These factors cause H and T in the air immediately adjacent to the fuel surface to differ from their ambient values. Thus M_c for a given fuel is a continuously changing value that would be obtained if the fuel were exposed for a hypothetically infinite time to these different values of "fuel" temperature and fractional relative humidity, T_f and H_f. One way to estimate M_c is to assume that T_f and H_f are the same as T and H. In this case, approximating T_f with T might be acceptably accurate only if the fuel were beneath a dense forest canopy, and even then, perhaps only at night. The second, and more appropriate, way to obtain M_c under field conditions should utilize measurements or models to evaluate T_f and H_f. Byram and Jemison (1943) collected data on fuel temperature, air temperature and humidity, wind, and fuel moisture content in an artificial sun apparatus in which fuels (hardwood leaf litter or basswood slats) were exposed to simulated solar radiation and wind generated by light bulbs and fans. The data were used to develop constants in a mathematical expression describing the effects of radiation and convection on the temperature difference $(T_f - T)$. Byram and Jemison then used this difference to compute the ratio H_f/H by assuming that vapor pressures in the fuel and air were equal. Values of M_c were determined by applying T_f and H_f to tabulated equilibrium sorption data for wood. These figures were in good agreement with measured M_c, substantially verifying the substitution of T_f and H_f for T and H. Byram and Jemison also studied experimentally the effect of wind and solar radiation on M_c obtained following the drying of basswood slats containing free water. They found that M_c for irradiated slats was larger when ventilated than when wind was absent and attributed the difference to higher H_f and smaller T_f values caused by the cooling action of the wind. Van Wagner (1969b) repeated the field experiments of Byram and Jemison in a laboratory setting to test the form of their equation for T_f. He observed similar effects of wind and radiation on T_f for jack pine and aspen litter fuels, but the mathematical form of his equation was slightly different. Viney (1991) critiqued the Byram and Jemison (1943) approach and noted omission of the effects of longwave radiation and vertical gradients in vapor pressure on T_f and H_f.

b. Sorption Models

The study of sorption systems involving various solid–gas interactions has led to identification of five types of sorption (Stamm, 1964). Type I sorption occurs on all solids in a single molecular layer at temperatures above the critical temperature of the gas. Types II and III take place in swelling solids and exhibit sorption in multiple layers. In Type II, large quantities of heat are liberated due to strong interaction between the solid and the gas; in Type III, this interaction is weak. Types IV and V are special cases of Types II and III and describe sorption in solids with very fine pores that become filled before the saturation vapor pressure is reached. Researchers in the fields of wood products, textile manufacturing, leather processing, and food technology have found that water vapor sorption isotherms for their respective materials follow Type II sorption. One of the first isotherms derived for Type II adsorption was the Brunauer-Emmett-Teller (BET) isotherm—a model that did not represent water takeup in wood very well but spawned other theories that describe water takeup over the entire range in H. Prior to the BET theory, however, the Bradley isotherm had been developed on the basis of reduced dipole attraction between successive layers of adsorbed water vapor molecules. Anderson and McCarthy (1963) derived an equation similar in form to the Bradley isotherm. Their semiempirical theory was based on experimental data that suggested the natural logarithm of the partial molal enthalpy of adsorption is linearly related to M_e; it successfully described water vapor sorption data in silk, nylon, and cotton fibers over a range in H from about 0.1 to 0.85. A detailed discussion of sorption theories is beyond the scope of this chapter; the reader may consult the summary in Skaar (1988).

Despite its apparent success, the Anderson-McCarthy sorption equation must be regarded as approximate because it uses the highly improbable assumption that the entropy change of the adsorbed water always is zero. In an alternative approach, Nelson (1984) reported that for sorption in forest fuels the natural logarithm of the change in Gibbs free energy, $\ln \Delta G$, is linearly related to M_e over most of the hygroscopic range according to

$$\ln \Delta G = A + BM_e = \ln \Delta G_d[1 - (M_e/M_{est})] \tag{24}$$

where ΔG_d (J kg^{-1}) is the limiting value of ΔG as M_e and H approach zero, and A and B are constants. Quantity M_{est} is an estimate of the fiber saturation point M_{fsp} under desorption conditions; for adsorption, M_{est} approximates the αM_{fsp} product, where α is the hysteresis ratio. The decrease in Gibbs free energy when pure liquid water is adsorbed by forest fuels is defined as the difference between the Gibbs free energy of water vapor in equilibrium with pure liquid water (the reference state) and the energy of water in the adsorbed state. Thus at constant T,

$$\Delta G = G_0 - G = \int_p^{p_0} V_v dp = -\int_{p_0}^{p} (RT/M_w)dp/p = -(RT/M_w) \ln H \tag{25}$$

where G_0 refers to the reference state, V_v is the specific volume of water vapor at ambient temperature T, R is the universal gas constant ($8.32 \text{ J mol}^{-1} \text{ K}^{-1}$), and M_w is the molecular weight of water ($0.018 \text{ kg mol}^{-1}$). Vapor pressures in the adsorbed and reference states are p and p_0, respectively, and the vapor is assumed to be an ideal gas. The sorption isotherm may be written by combining Eqs. (24) and (25) to give

$$M_c = \{\ln[(RT/M_w)(-\ln H)] - A\}/B \qquad (26)$$

in which B is a negative number. Clearly, as H approaches 0, M_c approaches negative infinity, and as H approaches 1, M_c approaches infinity. Increasingly erroneous values of M_c are predicted as H decreases below about 0.1 or increases beyond 0.85. The model in Eq. (26) was used to describe water vapor sorption in forest fuels, including hardwood leaves, pine needles, and wiregrass (Nelson, 1984). The maximum difference between measured and modeled M_c values was about 0.02.

In a comprehensive study, Anderson (1990b) utilized Eq. (26) to describe sorption in recently cast and weathered (over one winter) litter collected from sites in northern Idaho and western Montana. The twofold purpose of the research was (1) to test applicability of the model for forest fuels and (2) to determine whether the model would aid in separation of the litter types studied into groups with similar M_c values. A total of 94 runs involving adsorption and desorption in fresh and weathered fuels and spanning a temperature range from 278 to 322 K was made. Anderson found that a minimum of 89% of the variation in $\ln \Delta G$ of Eq. (24) was explained by variation in M_c. He identified four distinct fuel groups for recently cast fuels on the basis of ± 0.02-wide bands within the M_c data; there were only a few species changes within the basic groups for the weathered litter. In order of increasing M_c, the four groups are referred to as grasses, spruce and fir needles, pine and cedar needles, and aspen and larch foliage. Encouraged by results of the study, Anderson stated that only a few fuel groups were needed to describe changes in M_c for a large assortment of fuel litters. The constants A and B of Eq. (26), documented in his Table 2, provide a wealth of information for researchers interested in modeling equilibrium relationships or water transport below M_{fsp} in foliage fuels.

c. Complete Sorption Isotherms for Wood and Soil

The isotherms of the previous section involve sorption of water below M_{fsp} and, therefore, are strictly applicable for $H < 0.9$. Soil and plant researchers have been interested in the region $H > 0.9$ for many years, but only a few wood and paper researchers have reported on sorption in this region since the mid-1960s. What is referred to here as a *complete isotherm* involves a range in M_c from the ovendry to water-saturated condition. Only when wood is totally saturated will $H = 1$ and water potential $\psi = 0$. For abiotic materials such as wood and soil, ψ must be interpreted as the sum of the osmotic and matric potentials because both of them cause a reduction of the vapor pressure of liquid water. Moreover, they are not easily separated experimentally (Cloutier and Fortin, 1991). Thus Eqs. (3) and (25) may be combined to give

$$\psi = \rho_w(\mu - \mu_0)/M_w = -\rho_w\Delta G = (\rho_w RT/M_w) \ln H \qquad (27)$$

so that water potential is negative, as it should be if the vapor pressure is reduced below that of the reference state. Capillary water within and between fuel or soil particles is assumed to form hemispherical menisci in which the liquid–gas interface separates liquid pressure P_w and gas pressure P_g. These pressures are related to the principal radii of curvature of the air–water interface, r_1 and r_2, and the surface tension of water, γ, by the equation

$$P_w - P_g = -\gamma[(1/r_1) + (1/r_2)] = -2\gamma/r = \psi \qquad (28)$$

when $r_1 = r_2 = r$ and the wetting angle of water on the solid material is assumed to equal zero. Figure 14 shows data relating M_c to ψ for western hemlock specimens 10 mm thick in the longitudinal direction at $T = 21°C$. The data at high M_c values were obtained in 1979 by Fortin (cited by Cloutier and Fortin, 1991) with the tension plate and pressure plate methods; the dashed lines represent data obtained with saturated salt solutions in the hygroscopic

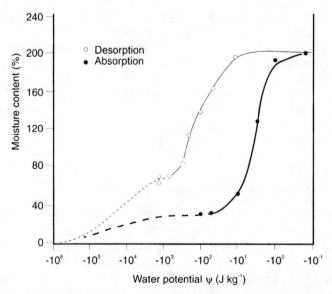

FIGURE 14   Moisture content of western hemlock sapwood as a function of water potential over the range from the nearly ovendry to saturated condition. Temperature is 21°C. From Cloutier and Fortin (1991).

region using Eq. (27). For this figure, ψ has units J kg^{-1}, but these are easily changed to J m^{-3} because the density of water is 1000 kg m^{-3}. The hysteresis prominent in the data of Figure 14 is attributed by Cloutier and Fortin (1991) to the "ink-bottle effect" in which water gain or loss is dependent on the distribution of capillary sizes and shapes. In wood, evaporation to a given water content is determined by radii of the smaller channels (pit membrane pores connecting cell cavities), whereas condensation approaching the same water content depends on the radii of the larger pores (the cell cavities). Thus Eqs. (27) and (28) indicate that both ψ and H are smaller for desorption than absorption at constant water content, in qualitative agreement with Figure 14.

In earlier years, soil water equilibrium relationships were expressed in terms of matric suction (a pressure difference across a differentially permeable membrane) and moisture content. Campbell et al. (1993) have simplified the presentation of such data by developing a complete isotherm similar in form to Eq. (26). The isotherm of Campbell et al., given by their Eq. (4), may be written (using symbols of this chapter for water activity and equilibrium water content) as

$$H = \exp\{-(M_w/RT)\exp[(1 - M_c/M_1)\ln(-\psi_d)]\} \qquad (29)$$

where M_1 is the extrapolated value of M_c when $\psi = -1$ J kg^{-1} and ψ_d is the water potential of ovendry soil, evaluated as the constant -10^6 J kg^{-1}. Because the quantities $-\psi_d$ and M_1 of Campbell et al. are equivalent to ΔG_0 and M_{cst} of Eq. (24), it is easy to show using Eqs. (27) that Eqs. (24) and (29) are identical in form. In addition to simplifying the mathematical modeling of soil drying, Eq. (29) may be used with Eq. (26) to characterize moisture equilibrium between mineral soil and the adjacent duff and litter layers. This possibility was pointed out by Siau (1995).

2. Laboratory Studies

The transport of moisture through a fuel particle is determined by the relative magnitudes of internal and external transfer of water as determined by particle properties and ambient conditions. However, in highly impermeable wood or in a composite fuel layer (litter, duff, and mineral soil, for example), internal evaporation and/or infiltration (gravitational flow of liquid water) also may take place. These processes initially were studied with diffusion theory, but advances in describing the flow of liquid water in wood and granular materials by capillarity now provide more realistic analyses of moisture transport.

a. Transport Mechanisms

The mechanisms by which water moves through the interior of dead forest fuel particles and layers are capillarity, diffusion, and infiltration. The latter process is described in soil physics texts such as Hillel (1980b). When gravitational forces have ceased to operate, liquid water is brought to the particle or layer surface where it evaporates into the atmosphere; water in vapor form either condenses on a cell wall or continues to diffuse until it passes into the environmental air. Spolek and Plumb (1980) stated that the permeability of wood to the gas phase can be neglected because many of the bordered pits become aspirated by surface tension forces associated with liquid flow. Thus for materials with fine capillaries, Darcy's law [Eq. (9)] is used to compute flows of liquid, and the flux of vapor [Eq. (11)] may be regarded as due to diffusion only. In analyzing liquid flow, Spolek and Plumb (1981) defined the local liquid saturation, S, as

$$S = (m - m_{fsp})/(m_{max} - m_{fsp}) \tag{30}$$

where m denotes local values of the moisture fraction and m_{max} is the maximum possible m calculated from the wood density (Stamm, 1964). Clearly, S can range from 0 to 1 when $m \geq m_{fsp}$. If the fuel temperature is below 60°C and the total pressure P_T in Eq. (9) is considered constant, the liquid water flux J_w may be expressed in terms of S or m by establishing a relationship between capillary pressure P_c [see Eq. (12)] and S. Spolek and Plumb (1980) reported a set of equations for this purpose based on the capillary flow model of Comstock (1970). In later work, Spolek and Plumb (1981) presented the empirical equation

$$P_c = 1240S^{-0.61} \tag{31}$$

where P_c has the units J m^{-3}. These authors implied that local values of S can replace the volume-averaged values they measured at several positions within each of their wood samples, so S is used in Eq. (31). When this equation is combined with Eq. (9) and P_T is assumed constant, the flux of liquid becomes

$$J_w = -756(K_w/\eta_w)S^{-1.61}(\partial S/\partial x) \tag{32}$$

where J_w is on a volume of whole wood basis. Combination of the relation $\rho_w \phi_w = \rho_f(m - m_{fsp}) = \rho_f(m_{max} - m_{fsp})S$ with Eqs. (5) and (32) leads to the liquid water transport equation

$$\rho_f(m_{max} - m_{fsp})\partial S/\partial t = 756(K_w/\eta_w)\partial[S^{-1.61}(\partial S/\partial x)]/\partial x - w_w \tag{33}$$

where ρ_f is density of the dry fuel.

The diffusion mechanism involves bound water and water vapor. Local diffusion of bound water when $m < m_{fsp}$ is due primarily to surface diffusion—a process in which water molecules hop from one sorption site to another in two dimensions. This diffusion is driven by a gradient in spreading pressure, P_b, and is described by Eq. (10). Practical use of Eq. (10), however, requires a known relationship between m and P_b (e.g., Nelson, 1986a). In addition, water vapor traverses the cell cavities by diffusion. The ideal gas law may be used to express Eq. (11) (with the diffusion term only) in terms of a gradient in vapor pressure.

Thus combined bound water and water vapor diffusion consists primarily of a series of steps involving evaporation from a cell wall, vapor diffusion across the adjacent cell cavity, condensation on the opposite wall, and bound water diffusion through the wall. Stamm and Nelson (1961) found from combined diffusion calculations based on an electrical conductance analogy utilizing premeasured wood structural features that 86% of the total flow in softwoods with a specific gravity of 0.4 drying at 50°C is due to this mechanism. Much of the remaining flow is continuous vapor diffusion through the cavities and pit membrane pores. Rather than solve separate mass conservation equations for bound water and water vapor, other modelers (Choong, 1965; Gong and Plumb, 1994) have used a combined diffusivity in place of D_b in Eq. (10). When this modified equation is substituted into Eq. (5), the diffusivity in the resulting diffusion equation becomes a complex function of fuel internal structure, air (or fuel) temperature, and local moisture fraction m. Numerically based prediction models combining capillary and diffusive mechanisms with energy conservation and volume averaging have been used to simulate the drying of wood (Spolek and Plumb, 1980; Perre et al., 1990; Gong and Plumb, 1994).

In some types of fuel, evaporation takes place within the material rather than on the surface. This retreating-surface mechanism occurs in thick layers of porous materials (Nissan and Kaye, 1957) and in wood in which the free water is relatively immobile due to little or no continuous void structure (Stamm, 1964). Nissan et al. (1959) analyzed this phenomenon in thick textiles wound on bobbins that were dried in a wind tunnel at about 50°C. The temperature-time data for wool in Figure 15 indicate a temperature plateau at 30°C, about 5°C greater than the wet-bulb temperature. This increase is due to a balance between heat arriving at the internal surface of evaporation and heat used to evaporate free water and produce outward vapor diffusion. Nissan et al. called this intermediate temperature the *pseudo-wet-bulb temperature* and described its calculation. The two major assumptions are that (1) temperature and vapor concentration gradients in the outer dry layer are similar and (2) all incoming heat is used to evaporate free water at the retreating free water surface. The heat balance equation derived by Nissan et al. is

$$k(\Theta - \Theta_{pwb}) = (\varepsilon D_v)_{pwb}\lambda_{pwb}(\rho_{pwb} - \rho)$$
$$= 0.00216(\varepsilon D_v)_{pwb}\lambda_{pwb}[(p_{pwb}/T_{pwb}) - (p/T)] \tag{34}$$

where subscript *pwb* denotes evaluation at the pseudo-wet-bulb temperature, absence of a subscript indicates ambient air values, k ($J\ m^{-1}s^{-1}K^{-1}$) is thermal conductivity of the material, Θ and T are Celsius and Kelvin temperatures, ρ ($kg\ m^{-3}$) is vapor concentration, λ ($J\ kg^{-1}$) is the latent heat of evaporation, and p ($J\ m^{-3}$) the vapor pressure. With this equation and a generalization of Eq. (43), it can be shown that response time τ should vary directly with d^2 and

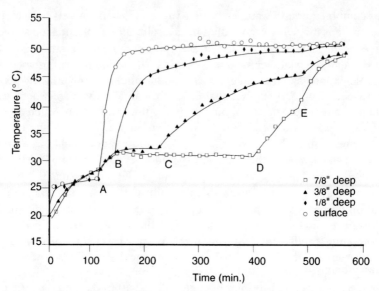

FIGURE 15 Temperature variation with time at various depths for a thick porous wool bobbin dried at 50.9°C and an airspeed of 4.35 m s^{-1}. Point A corresponds to the end of the constant-rate period; the pseudo-wet-bulb temperature, indicated by points B, C, and D, is about 30°C; point E represents the time at which the bobbin is essentially dry. Thermocouple depths in inches are indicated in the lower right corner. From Nissan and Kaye, reprinted by permission from *Nature* 180: 142–143, copyright 1957, Macmillan Magazines Ltd.

T/P_{sat}, but inversely with $(1 - H)$. Here, d is layer thickness, P_{sat} is the saturation vapor pressure at T, and H is the humidity fraction. Interestingly, approximations to these relationships were observed for layers of sawdust (Nelson, 1969) and jack pine litter (Van Wagner, 1979) that were losing water only through their upper surfaces. These results were further verified in unpublished studies by the writer with measured temperature distributions similar to those in Figure 15 and with predicted drying rates and plateau temperatures close to measured values for sawdust layers drying in air at temperatures from 14 to 32°C and humidity fractions from 0.25 to 0.79.

b. Data Analysis

The drying of solids under fixed external conditions (constant M_c) typically is divided into a constant-rate and two falling-rate periods (Van Wagner, 1979). In the constant-rate period, drying depends on external factors but is independent of the nature of the solid. This period is seldom seen in forest fuels except for short periods of time—when the exposed surface is covered with a water film. The decrease in M with time t is linear. When this free water has disappeared, or the rate of water supply from the fuel interior becomes smaller than the surface evaporation rate, the first falling-rate period is triggered as the surface begins to dry. Moisture loss depends on both internal and external factors; in most fuels, the decrease in $M - M_c$ with time approximates an exponential decay. A second falling-rate period often occurs when all free water has disappeared and the fuel temperature begins to rise. This second period, while differing in mechanism from the first, also may exhibit an exponential decay, but one that usually is preceded by a time of transition from the first exponential period to the second. This type of response was observed by Anderson (1990a) in experiments on drying and wetting of forest foliage below M_{fsp} and is shown in the semilogarithmic plot of Figure 16. Similar results were reported by Nelson (1969), Anderson et al. (1978), and Van Wagner (1979). The fraction of evaporable moisture remaining in the fuel at any time, denoted by E, is given by

$$E = (M - M_c)/(M_{in} - M_c) \tag{35}$$

where M is the fuel particle (or layer) average moisture fraction and M_{in} is the initial value. Note that E in Eq. (35) corresponds to $(1 - E)$ in the graph of Anderson (1990a) in Figure 16.

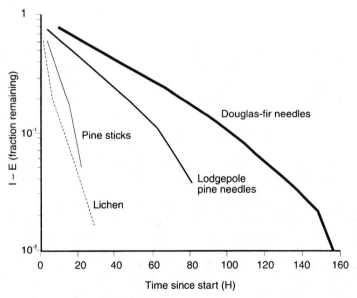

FIGURE 16 Variation in the fractional amount of evaporable moisture remaining as a function of drying time for several forest fuels. The quantity $(1 - E)$ in this figure corresponds to E in Eq. (35). The four fuels exhibit two or more drying stages. Pine sticks dry quickly when compared with pine and fir needle layers. From Anderson (1990a).

Laboratory drying (or wetting) results such as those in Figure 16 may be analyzed by means of classical diffusion theory (Crank, 1975). The theory assumes that the fuel is homogeneous and isotropic, drying is isothermal, and diffusivity D is constant. The equation for diffusion of moisture in a plane slab of thickness $2d$ (moisture exchange from both surfaces) is

$$\partial m/\partial t = D(\partial^2 m/\partial x^2) \tag{36}$$

and may be solved subject to the convective boundary condition

$$-D(\partial m/\partial x)_s = h_m(m_e - m_s) \tag{37}$$

where subscript s indicates a surface value, h_m (m s^{-1}) is the surface emission (or convective mass transfer) coefficient, and $m_e = M_e$, a constant. Though Eq. (36) strictly applies to movement below fiber saturation, the equation often is applied when free water is present for convenience. Such an analysis may fortuitously describe the moisture change in cases when the interior free water is highly mobile but drying is controlled by diffusion in the drier layers. Solving Eq. (36) subject to Eq. (37) leads to an infinite series solution that may be integrated from $x = -d$ to $x = d$ and divided by $2d$ to obtain E in the form

$$E = 2\mathrm{Bi}^2 \sum_{n=1}^{\infty} \{\exp(-\beta_n^2 \mathrm{Fo})/[\beta_n^2(\beta_n^2 + \mathrm{Bi}^2 + \mathrm{Bi})]\} \tag{38}$$

where $\mathrm{Bi} = h_m d/D$ is the mass transfer Biot number, $\mathrm{Fo} = Dt/d^2$ is the mass transfer Fourier number, and β_n is the nth positive root of the equation $\beta \, \mathrm{Tan} \, \beta = \mathrm{Bi}$. When Fo exceeds 0.1, the series converges rapidly and an approximate solution is found by retaining only the first term for $n = 1$. If the ratio of surface resistance to internal flow resistance is small, Bi effectively approaches infinity and β_1 approaches $\pi/2$. The moisture change is driven by a significant gradient in m. Thus E is written as

$$E = (8/\pi^2) \exp(-\pi^2 Dt/4d^2) = K \exp(-t/\tau) \tag{39}$$

where constant K is sometimes assumed equal to unity and the response time is

$$\tau = 4d^2/\pi^2 D \tag{40}$$

On the other hand, when the surface resistance is relatively large and the gradient in m is small, Bi approaches zero and β_n^2 approaches Bi. In this case, E corresponding to $n = 1$ becomes

$$E = \exp(-\mathrm{BiFo}) = \exp(h_m t/d) = \exp(-t/\tau) \tag{41}$$

and τ is given by

$$\tau = d/h_m \tag{42}$$

Equations (40) and (42) display the τ-versus-d dependence described in Section II.C.3. Similar analyses are possible for other fuel shapes—the cylinder, sphere, or square rod—and form the basis of the response time theory initially proposed by Byram (1963, unpublished). This analysis provides a convenient framework with which to interpret experimental data, but D is not a constant in fibrous materials and should be considered an integral diffusivity over the range of moisture content change. Procedures for calculating the bound water and water vapor diffusivities separately in terms of m and T (or T_f) and then combining them into an overall diffusivity are described by Nelson (2000).

The method used to evaluate integral diffusivity D_{av} from experimental data depends to some extent on the type and amount of data available. Crank (1975) shows that a semiinfinite solid dries or wets in proportion to the square root of time t and that plots of E versus $t^{1/2}$ according to Eq. (38) for constant Bi are effectively linear during roughly the first half of the exchange process. Thus when data are plentiful for short times, D_{av} for the slab may be found from a different solution of Eq. (36) written as

$$D_{av} = \pi d^2 E^2 / 4t = \pi d^2 s^2 / 4 \qquad (43)$$

where s is the slope of the linear portion of a plot of E versus $t^{1/2}$ (Crank, 1975). This equation is based on three assumptions: (1) the slab dries or wets as a semiinfinite solid early in the process of moisture change, (2) shrinkage/swelling is negligible, (3) the surfaces instantaneously attain the equilibrium value, M_c. McNamara and Hart (1971) suggest using Eq. (43) in the range $0.5 \leq E \leq 1$. When data are available for more widely spaced time intervals or when $E < 0.5$, Eq. (39) may be used to obtain an estimate of D_{av} by solving for D.

For a single experiment, the effect of surface transfer on the moisture exchange process cannot be separated from that due to internal diffusion because the mass transfer correlations applicable to saturated wood surfaces are not valid for unsaturated surfaces (Plumb *et al.*, 1985; Siau and Avramidis, 1996). The effects can be separated, however, with a procedure that requires two or more experimental runs with samples of differing thickness (Choong and Skaar, 1972). A summary of the results of various investigators utilizing this procedure reports a 100-fold variation in mass transfer coefficient h_m (Siau and Avramidis, 1996). A theory for predicting h_m in the hygroscopic range was reported by Morén *et al.* (1992), and an identical relation was developed by Siau and Avramidis (1996).

3. Field Studies

Linton (1962) used the theory of diffusion in solids to describe moisture change in fuel particles in terms of a constant diffusivity and evaluated this quantity with laboratory experiments. Byram (1963, unpublished) also applied theory to the problem, but his timelag concept is applicable only to step changes between fixed initial and final values of average moisture fraction M. The oscillating moisture content data obtained under field conditions differ from laboratory data which tend to decrease exponentially. Linton (1962) and Viney and Hatton (1989) discussed a different timelag that is useful in field studies. Their timelag is concerned with the lapse in time between the continuously varying equilibrium moisture content, M_c (defined as the M value determined by assuming equilibrium with the diurnally changing fuel temperature and relative humidity), and the actual M value for the fuel. Thus Viney and Hatton (1989) and Viney and Catchpole (1991) recommended that "timelag" be used only to denote the lag of M behind M_c and that "response time" be applied to the time needed for 63% of the total change between fixed moisture contents. Recognizing this distinction, Anderson (1990a) used "response time" to characterize the moisture exchange rates of fuels in his laboratory tests. The terminology of Viney and Catchpole is adopted in this chapter and is recommended for all future studies of fuel moisture relationships to minimize confusion.

a. Weather Factors

In addition to continuous variation in air temperature and relative humidity, other variables influencing moisture change in forest fuels include wind speed and direction, solar and longwave radiation, precipitation, canopy interception of rainfall and sunshine, filtration through fuel layers, deposition of dew, contact with fog, and source/sink effects of underlying soil. The models of Byram and Jemison (1943) describing fuel temperature and relative humidity fraction in terms of ambient values, solar radiation, and wind speed were briefly summarized in Section III.D.1. Using the iterative Newton method, Viney (1992a) presented coupled theoretical models for soil surface temperature and humidity fraction based on a surface energy balance and complex expressions for resistance to vapor transport. Viney compared predictions of T_f and H_f based on his surface balance models and on the Byram-Jemison models with values measured at the under surfaces of *Eucalyptus globulus* leaves and found that his models provided more accurate predictions of T_f and H_f than the simpler models. The mean absolute error in T_f predicted by Viney (1992a) was 0.52 K; the error for predicted H_f was 0.051.

Byram and Jemison (1943) pointed out the influence of wind on drying rate and final M_c values for forest fuels below M_{fsp}. Their data suggested that the cooling action of wind on the drying rate of irradiated fuels is greater than the tendency of wind to increase the evaporation rate. The data showed that desorption rates are slowed and M_c is increased, whereas both adsorption rates and M_c values are increased by the action of wind. Other data showed that the wind effect is reversed in the absence of radiation—i.e., drying rates for nonventilated fuels are slower than for ventilated fuels. Van Wagner (1979) conducted laboratory tests on the drying of jack pine litter at 27°C and 30% relative humidity

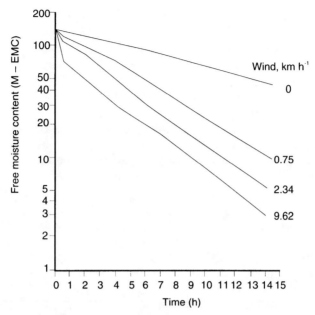

FIGURE 17 Change of free moisture content (that above the equilibrium moisture content, EMC) of jack pine litter with time at four wind speeds ranging from 0 to 9.62 km hr^{-1} (2.7 m s^{-1}). Even at the high moisture contents implied here, the effect of wind is evident only during the first hour of drying. From Van Wagner (1979).

in wind but without solar radiation; the results in Figure 17 show that the litter (which initially contained considerable free water) dried in two stages when the wind speed exceeded about 0.75 km hr^{-1} (0.2 m s^{-1}). These results indicate that differences in drying rate are established during the first hour of drying; though this first hour constitutes only a small fraction of the total drying time, a significant fraction of the moisture content change takes place during this period. For times exceeding one hour, the drying rate is roughly independent of wind speed.

The rise in average moisture content fraction, ΔM, due to rainfall has been studied by Stocks (1970) for duff (see Section II.C.2) and by Fosberg (1972) for wood cylinders. In his theoretical approach, Fosberg used numerical and analytical solutions of the diffusion equation to argue that water uptake by wood cylinders is limited by the rate at which the wood can adsorb water and that this limiting rate is of the order of 1 mm day^{-1}. He also stated that the theory predicts realistic rates of adsorption and desorption associated with precipitation and postprecipitation drying and that M increases exponentially with increasing duration of rainfall. In the empirical approach of Van Wagner (1987), ΔM

is expressed in terms of the initial value of M and the net rainfall (observed rainfall amount in the open less a correction of 0.5 mm for canopy interception). These and other treatments of ΔM due to rainfall have been critiqued by Viney (1991). Different aspects of rainfall are concerned with interception by the canopy (if present as understory or overstory) and infiltration through the litter and underlying soil but are not discussed here. These topics are treated by Helvey (1967) and Hillel (1980b).

Nocturnal condensation of water on plant surfaces can be caused by deposition of water from the atmosphere or by *distillation,* the upward transport of water vapor from the soil by turbulent diffusion (Monteith, 1963b). In a study of the effect of condensation on leaf litter moisture content, Viney and Hatton (1990) measured M for single *Eucalyptus globulus* leaves in contact with the ground at two sites in southeastern Australia. They developed a theory of condensation into which measured fluxes of net all-wave radiation, soil heat, and sensible and latent heat at the leaf surfaces were introduced to obtain a predicted ΔM (the change in average moisture fraction attributable to condensation). Good agreement between observed and predicted values is evident in Figure 18, but M is underpredicted by the model when fog is present. Viney and Hatton (1990) state that the model tends to predict a zero ΔM during fog episodes when, in fact, the leaves may have taken on moisture through direct con-

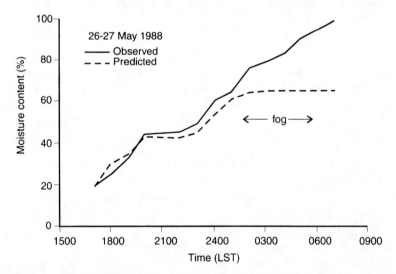

FIGURE 18   Comparison of predicted moisture contents of *Eucalyptus* leaves with those observed overnight on 26 May 1988 during experiments in eastern Australia. Moisture content change due to condensation of water vapor from the atmosphere and soil is predicted well, but the effects of fog apparently are not explained by the model. Reprinted from Agricultural and Forest Meteorology, Volume 51, Viney, N. R., and Hatton, T. J., Modelling the effect of condensation on the moisture content of forest litter, pages 51–62. Copyright 1990, with permission from Elsevier Science.

tact with fog droplets. The measured values of M during condensation also may be sensitive to fuel shape and orientation which could cause reductions in absorbed and surface-held water due to runoff (Viney, 1991).

Though the effects of soil moisture have been neglected in many studies of dead fuel moisture, at least two studies have presented evidence that the former affects the latter. Hatton *et al.* (1988) used a randomized block design to study the response of leaves of several species of *Eucalyptus* to wetted (20% dry soil weight or less) and untreated dry (8% dry weight or less) soil over a period of 48 hours. Within each cleared 8 × 8 m plot, leaves were arranged in a 2 × 1 × 0.16 m mound and also were uniformly scattered over the ground for full exposure to the weather. The leaves were held in place by nylon netting. Hatton *et al.* concluded that soil moisture influenced fuel moisture because the scattered leaves approached moisture contents of 100%, well beyond values attainable by vapor transport alone (dew was observed). In addition, leaves on top of the mound on wetted soil were about twice as moist as corresponding leaves on the dry soil. This difference was explained in terms of more water available for distillation from the wetter soil. In a different study conducted in aspen stands of central Alberta, Rothwell *et al.* (1991) measured the water potential and volumetric water content (in %) at locations 2–3 cm above and below the transition zone between the litter layer and mineral soil. The authors concluded, on the basis that water flows from higher to lower levels of water potential, that soil water affects fuel availability and fire behavior in stands of aspen because upward capillary flow from the soil kept the litter moist in their experiments.

b. Prediction Models

Viney and Hatton (1989) measured the moisture content of dead leaves, bark, and twigs of *Eucalyptus macrorhyncha* and *Eucalyptus rossii* in contact with the ground for the purpose of assessing models for predicting the moisture content of fine fuels. They reported significant differences in equilibrium moisture fraction M_e within the three fuel types of the study and stated that submodels for dew and distillation should be included in models to be used in open or lightly foliated areas. Viney and Catchpole (1991) and Viney (1992a, 1992b) developed a method for estimating fuel moisture response time and timelag (the lag of fuel moisture fraction M behind M_e) from field observations. The basis of their method is shown in Figure 19 in which it is assumed that both M and M_e follow a 24-hour sinusoidal cycle. In present notation, the equations describing such data are

$$M(t) = M_{av} + a \, \mathrm{Sin}[\omega(t - \varphi_m)] \tag{44}$$

$$M_e(t) = E_{av} + A \, \mathrm{Sin}[\omega(t - \varphi_e)] \tag{45}$$

where M_{av}, E_{av}, a, and A are constants obtained from least squares regression analysis, t is time (h), φ_m and φ_e (h) are phase lags relative to midnight, and ω

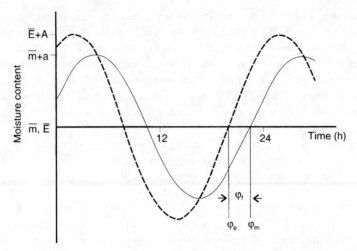

FIGURE 19   Idealized sinusoidal variation in equilibrium (dashed curve) and fuel (solid curve) moisture contents with time under field conditions. The average values of E and M shown in this figure are represented as E_{av} and M_{av}, respectively, in Eqs. (44) and (45). The timelag φ_f is the difference in phase for the two curves. From Viney (1992a, 1992b).

is the angular frequency given by $\omega = \pi/12 \text{ hr}^{-1}$. Hourly observations of M and M_e over a period of several days are filtered with a time series decomposition algorithm to eliminate long-term trends and random fluctuations and to obtain φ_m and φ_e from the two sets of data. As Figure 19 shows, the difference $\varphi_m - \varphi_e$ is the timelag (or phase lag) of the fuel, φ_f. The final step is to determine response time τ from

$$\tau = (1/\omega) \, \text{Tan}(\omega\varphi_f) \tag{46}$$

The derivation of Eq. (46) is based on three additional assumptions: (1) the resistance to internal moisture transport is small compared to that at the fuel surface, (2) τ is constant during the diurnal cycle, and (3) time since midnight of the day prior to the first day of the measurement period ($t = 0$) is large compared with the fuel's response time. These assumptions are reasonable for fine or poorly ventilated fuels in which moisture is exchanged with little or no gradient in moisture content. An alternative to evaluating φ_f with the sinusoidal assumptions is to represent the two data sets with cubic spline interpolations and then choose the value of φ_f that maximizes correlation between the two sets. In a further modification of these methods, Catchpole *et al.* (in press) combined spline interpolation with the sorption isotherm in Eq. (26) to find estimates of M_e and τ without assuming sinusoidal variation in M and M_e. According to Catchpole *et al.*, tests of the model with data collected in Tasmanian buttongrass moorland and Western Australian mallee shrubland have given them "some confidence in the robustness of the method and the appropriateness of the model." The methods described here should be used only for fuels with short (3 h or less) response times and would apply during periods of sorption-controlled moisture exchange rather than periods of rainfall or evaporation and condensation.

In larger fuels or fuels that dry in the presence of moisture content gradients, internal resistance exceeds that at the surface. Viney (1992a, 1992b) pointed out that such conditions require solution of the diffusion equation appropriate for the type of fuel particle under consideration. To evaluate the constant moisture diffusivity, D, for different particles, he solved the diffusion equations applicable to the semiinfinite solid (representing a layer of leaf or needle litter in contact with the ground), the slab (a flat leaf), the cylinder (a twig or needle), and the long square rod (a fuel moisture indicator stick)—all subject to sinusoidal surface boundary conditions. A parameter common to all four solutions is the dimensionless number, kd, where k is given by

$$k = (\omega/2D)^{1/2} \qquad (47)$$

and d is the depth of interest for the semiinfinite solid, thickness of the slab, diameter of the cylinder, or side length of the square rod. Though solutions for the average moisture fraction were derived analytically, their complexity precluded evaluation of kd, and hence D. Numerical representations of φ_f for the four fuel arrangements are shown in Figure 20. If the timelag is evaluated from

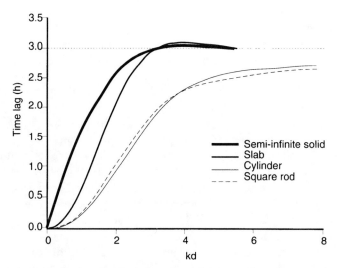

FIGURE 20   Relationship between the fuel timelag and dimensionless parameter kd for the four fuel shapes indicated. Quantity d represents the depth for which the timelag is desired in a semiinfinite solid, thickness for a slab, diameter for a cylinder, and side length for a square rod. From Viney (1992a, 1992b).

field data and d is known, then k may be determined from Figure 20 and used in Eq. (47) to compute D. Viney (1992a, 1992b) compared his calculated D values for leaf and twig data from the Viney and Hatton (1989) study with those reported by Linton (1962) and found good agreement for leaves and 6 mm twigs; his D for 3 mm twigs slightly exceeded Linton's values. In a different comparison, Viney's D for a lodgepole pine litter layer was within the range of corresponding values measured by Anderson (1990a).

Several investigators have developed diurnal prediction models from field data. Fosberg (1975) derived a model for conifer litter and duff from mass conservation equations for water vapor and the first law of thermodynamics. The physical properties of litter and duff horizons were taken from measurements on ponderosa pine and lodgepole pine layers or from the literature. Fosberg adapted his model to field conditions by introducing time-dependent boundary conditions consisting of step changes of arbitrary duration and magnitude. Though he did not compare his computed M-versus-t curves with data from laboratory or field experiments, Fosberg's predicted response times based on layer properties were close to several values in the literature. A predictive model of moisture content and evaporative loss in the litter of a mixed deciduous forest was reported by Moore and Swank (1975). Weather data requirements included daily totals for sunlight and rainfall and daytime means for atmospheric temperature and relative humidity. The model contains three hydrologic compartments—atmosphere, litter, and soil. In addition to evaporation from litter to the atmosphere and liquid drainage from litter to the soil, litterfall and litter dry weight (or decomposition) are modeled. Though they are acknowledged as important at times, atmospheric condensation and upward flow of moisture from the soil are not modeled. Independent data for validating model predictions during a 7-day period in the summer and an 11-day period during the winter were collected at the Coweeta Hydrologic Laboratory in western North Carolina. Predicted moisture contents usually were within the error limits of the experimental values. Simulated evaporation agreed to within 13% of the measured evaporation but was more accurate for the first two days after rain than under drier conditions. Moisture content comparisons during an additional 80-day verification study gave similar results. A model for computing fine fuel moisture content during the diurnal cycle was developed by Van Wagner (1987) using jack pine and lodgepole pine data; it is based on the assumption that wetting and drying by vapor exchange with the atmosphere are exponential, whereas wetting due to rainfall depends on rainfall amount and yesterday's fine fuel moisture content. In a different study, Nelson (2000) used diurnal weather and moisture content data from two field sites in the eastern United States to guide the formulation of a numerical model for computing hourly changes in the moisture content and temperature of 1.27-cm-diameter ponderosa pine dowels. In the model, water moves by capillarity and combined bound

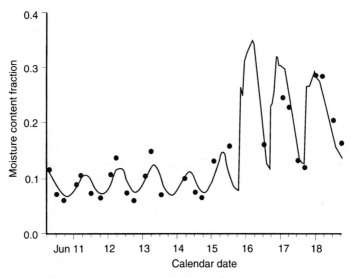

FIGURE 21   Comparison between predicted and observed moisture contents for 10-h fuel indicator sticks under field conditions in western Montana during June 1996. Rainfall occurred prior
to midnight on June 15 and 16 and at about 1800 on June 17. Solid lines represent predictions;
circles denote observations. From Nelson (2000).

water/water vapor diffusion. Either precipitation, evaporation, condensation,
or sorption determines the surface boundary condition; water takeup due to
fog is not considered. The model accepts data from remote automatic weather
stations (RAWS). Input consists of hourly readings of air temperature and relative humidity, solar radiation flux, and rainfall amount; changes in the four variables from one reading to the next are assumed to be linear. Independent observations for testing the model were compared with predicted values; predictions
generally agreed with observations to within the accuracy of the experimental
moisture content data, but rates of response appear to be slightly slower than
the observed rates (Figure 21).

IV. MOISTURE CONTENT ESTIMATION

Aside from prediction models, other means of estimating the moisture content
of live and dead fuels include weighing, correlation with Advanced Very High
Resolution Radiometer (AVHRR) data from NOAA satellites, resistance meters,
and artificial fuel sticks available with RAWS. Moisture content calculations require a sample ovendry weight. Drying in the usual convection oven is slow,
and the fuel is subject to loss of volatiles if dried too long at a high temperature.

In the literature, drying temperatures for wood and foliage range from 60 to 105°C. Perhaps the safest procedure for foliage is to dry at 100°C until a constant weight is reached. This method, of course, requires frequent weighings to prevent loss of volatiles due to prolonged drying. King and Linton (1963) concluded that materials ovendried prior to testing adsorb water about half as fast as those that have been dried only to a low moisture content. Thus experimental error may be reduced by determining ovendry weights after the measurements are made. To avoid the slowness and labor intensiveness of this method, instruments composed of a small balance and oven (or electric heater) have been developed, but this method is suitable only for small samples. Commercially available kitchen microwave ovens were tested by Norum and Fischer (1980). Wet duff (up to 230% moisture content) and 0 to 0.64-cm-diameter branchwood (up to 90%) samples weighing about 40 g were ovendried in 24 minutes or less. Live fuel moisture content can be estimated from satellites with the Normalized Difference Vegetation Index (NDVI) which operates by sensing red and near infrared radiation reflected from the surfaces of green or curing vegetation. Burgan et al. (1991) showed that the NDVI is useful for calculating overall site moisture content, whereas Piñol et al. (1998) developed a rapid method for measuring the moisture content of individual plants. Chladil and Nunez (1995) used AVHRR to follow curing of grasslands in Tasmania and concluded that the NDVI gives good results for fuel and soil moisture contents but is not a suitable standalone system for fire managers. Electrical resistance meters are useful for estimating the moisture content of wood in the hygroscopic range but are sensitive to contact resistance and presence of a moisture gradient (Skaar, 1988). Other factors mentioned by Clark and Roberts (1982) include wood species, density, and grain angle; they used a minimum of 20 samples for calibrating their meter to the species utilized in their study. The moisture content of the 10-h fuel moisture indicator sticks used in the U.S. National Fire Danger Rating System (Bradshaw et al., 1983) may be estimated with an artificial fuel moisture and fuel temperature stick that can be supplied with RAWS installations. The stick is equipped with sensors to measure its relative humidity and temperature, from which a moisture content is derived.

Most fire suppression and control organizations maintain a system with which they assess the moisture content of the live and dead fuels within their areas of jurisdiction. For operational applications, equations for estimating fuel moisture content are found in fire danger rating and fire behavior prediction systems around the world. A brief description of how moisture content is evaluated in several of the fire danger rating systems is presented in Chapter 9 in this book. In addition, operational models for prediction of dead fuel moisture content in the hygroscopic range have been discussed and compared by Viney (1991, 1992a). The reader is referred to Van Wagner (1987) and to the Forestry Canada Fire Danger Group (1992) for details of the Canadian systems, to Bradshaw et al. (1983) and Rothermel et al. (1986) for the American systems, and to Noble *et al.* (1980) and Sneeuwjagt and Peet (1985) for the Australian systems. Another aspect of fire danger involves water loss from deep soils and large fuels during extended dry periods. As before, the reader is referred to pertinent papers on the subject. Discussions of drought and drought indexes are given in key papers by Van Wagner (1987), Keetch and Byram (1968) and Alexander (1992), and Mount (1972) for the Canadian, American, and Australian systems, respectively.

NOTATION

ROMAN LETTERS

C	specific water capacity	m^{-1}, $s^2 m^{-2}$
c	constant-pressure specific heat	$J kg^{-1} K^{-1}$
D	local moisture diffusivity	$m^2 s^{-1}$
D_{av}	particle or layer average moisture diffusivity	$m^2 s^{-1}$
D_h	hydraulic diffusivity	$m^2 s^{-1}$
d	particle radius or half-thickness, layer depth	m
E	fraction of evaporable moisture	
e	base of Naperian logarithms	
G	Gibbs free energy	$J mol^{-1}$
g	gravitational acceleration	$9.8 m s^{-2}$
H	ambient air relative humidity fraction	
h	hydrostatic pressure head	m
h_m	surface emission coefficient	$m s^{-1}$
J	mass flux	$kg m^{-2} s^{-1}$
K	specific permeability	m^2
k	parameter in Eq. (47), thermal conductivity	m^{-1}, $J m^{-1} s^{-1} K^{-1}$
L	root length	m
ln	Naperian logarithm	
M	particle or layer average moisture content fraction	
M_w	molecular weight of water	$0.018 kg mol^{-1}$
M_1	M value when $\psi = -1 J kg^{-1}$	
m	local moisture content fraction	
P	fluid pressure, saturation vapor pressure	$J m^{-3}$

P_b	bound water spreading pressure	$J\,m^{-2}$
P_c	capillary pressure	$J\,m^{-3}$
p	gas phase partial pressure	$J\,m^{-3}$
Q	transpiration or evapotranspiration rate	$m\,s^{-1}$
Q_f	demand to heat dry fuel from ambient to 400°C	$kJ\,kg^{-1}$
Q_M	heat required to evaporate water at 100°C	$kJ\,kg^{-1}$
Q_T	heat of ignition	$kJ\,kg^{-1}$
q	heat flux	$J\,m^{-2}\,s^{-1}$
RWC	relative water content	
r	radial distance, flow resistance, mean radius of the gas–liquid interface	$m;\ s\,m^{-1},$ $bar\text{-}s\,m^{-1},$ $m^3\,kg^{-1}\,s^{-1};\ m$
r_1, r_2	radii of curvature of a gas–liquid interface	m
S	liquid water saturation	
s	slope of E versus $t^{1/2}$ plot	$s^{-1/2}$
T	ambient air temperature	K
t	time	s, hr
U	root sink strength	$kg\,m^{-3}\,s^{-1}$
V	specific volume	$m^3\,kg^{-1}$
v	velocity	$m\,s^{-1}$
W	plant tissue weight, duff weight consumed by fire	$kg,\ kg\,m^{-2}$
w	rate of evaporation	$kg\,m^{-3}\,s^{-1}$
x	coordinate in direction of moisture transport	m
z	vertical distance from reference elevation	m

GREEK LETTERS

α	hysteresis ratio	
β	fuel layer packing ratio	
β_n	nth root of $\beta \mathrm{Tan}\beta = Bi$	
γ	surface tension of water	$J\,m^{-2}$
Δ	finite change in associated variable	
δ	fuel layer depth	m

ε	fractional void volume	
η	kinematic viscosity	$m^2\,s^{-1}$
Θ	temperature	°C
θ	volumetric soil water content	
κ	hydraulic conductivity	$m\,s^{-1}$, kg-s m^{-1}
λ	latent heat of evaporation/condensation at T	$J\,kg^{-1}$
λ_d	differential heat of sorption	$J\,kg^{-1}$
μ	chemical potential	$J\,mol^{-1}$
ρ	mass density	$kg\,m^{-3}$
ρ_r	root density	m^{-2}
σ	particle surface-to-volume ratio	m^{-1}
τ	moisture response time	s, hr
φ	sinusoidal phase lag	hr
ϕ	phase volume fraction	
ψ	water potential	$J\,m^{-3}$, $J\,kg^{-1}$
ω	angular frequency	$\pi/12\ hr^{-1}$

DIMENSIONLESS GROUPS

Bi	mass transfer Biot number
Fo	mass transfer Fourier number

CONSTANTS

R	universal gas constant	$8.32\,J\,mol^{-1}\,K^{-1}$

SUBSCRIPTS

a	atmospheric air, dry air component of the gas phase
av	particle or layer average
b	bound water phase
c	current state
cw	cell wall
d	dry state

e	equilibrium state
f	fuel
fsp	fiber saturation point
g	gravitational potential, gas phase
h	hydraulic head
i	phase designator
in	initial state
l	leaf
m	matric potential, moisture
m(a)	apoplast matric potential
max	maximum value
o	reference state
p	pressure potential
pwb	pseudo-wet-bulb value
r	root
s	solution (or osmotic) potential, solid phase, surface
s(a)	apoplast solution (or osmotic) potential
so	soil
sat	saturated state
T	total value
t	turgid state
v	water vapor component of the gas phase
va	vacuole, vapor in ambient air
w	liquid water phase

ADDITIONAL READING

Brown, R. W. (1995). The water relations of range plants: adaptations to water deficits. *In* "Wildland Plants: Physiological Ecology and Developmental Morphology" (D. J. Bedunah and R. E. Sosebee, Eds.), pp. 291–413. Society for Range Management, Denver.

Marshall, M. J., Holmes, J. W., and Rose, C. W. (1996). "Soil Physics," 3rd ed. Cambridge University Press, Cambridge.

Smith, J. A. C., and Griffiths, H. (1993). "Water Deficits: Plant Responses from Cell to Community." BIOS Scientific Publishers, Oxford.

Waring, R. H., and Running, S. W. (1998). "Forest Ecosystems: Analysis at Multiple Scales," 2nd ed., Chapter 2. Academic Press, San Diego.

REFERENCES

Albini, F. A. (1980). "Thermochemical Properties of Flame Gases from Fine Wildland Fuels." Research Paper INT-243. USDA Forest Service, Intermountain Forest and Range Experiment Station, Ogden.

Albini, F. A., and Reinhardt, E. D. (1995). Modeling ignition and burning rate of large woody natural fuels. *Int. J. Wildland Fire* **5**, 81–91.

Alessi, S., Prunty, L., and Schuh, W. M. (1992). Infiltration simulations among five hydraulic property models. *Soil Sci. Soc. Am. J.* **56**, 675–682.

Alexander, M. E. (1992). The Keetch-Byram drought index: A corrigendum. *Bull. Am. Meteorol. Soc.* **73**, 61.

Alexander, M. E., Stocks, B. J., Wotton, B. M., and Lanoville, R. A. (1998). An example of multifaceted wildland fire research: The International Crown Fire Modelling Experiment. *In* "Proceedings, III International Conference on Forest Fire Research/14th Conference on Fire and Forest Meteorology," pp. 83–112. University of Coimbra.

Anderson, H. E. (1990a). Moisture diffusivity and response time in fine forest fuels. *Can. J. For. Res.* **20**, 315–325.

Anderson, H. E. (1990b). "Predicting Equilibrium Moisture Content of Some Foliar Forest Litter in the Northern Rocky Mountains." Research Paper INT-429. USDA Forest Service, Intermountain Research Station, Ogden.

Anderson, H. E., Schuette, R. D., and Mutch, R. W. (1978). "Timelag and Equilibrium Moisture Content of Ponderosa Pine Needles." Research Paper INT-202. USDA Forest Service, Intermountain Forest and Range Experiment Station, Ogden.

Anderson, N. T., and McCarthy, J. L. (1963). Two-parameter isotherm equation for fiber-water systems. *Ind. Eng. Chem. Process Des. Develop.* **2**, 103–105.

Babbitt, J. D. (1950). On the differential equations of diffusion. *Can. J. Res.* **A-28**, 449–474.

Barkas, W.W. (1949). "The Swelling of Wood Under Stress." Department of Scientific and Industrial Research, Forest Products Research, HSMO, London.

Blackmarr, W. H. (1971). "Equilibrium Moisture Content of Common Fine Fuels Found in Southeastern Forests." Research Paper SE-74. USDA Forest Service, Southeastern Forest Experiment Station, Asheville.

Blackmarr, W. H. (1972). "Moisture Content Influences Ignitability of Slash Pine Litter." Research Paper SE-173. USDA Forest Service, Southeastern Forest Experiment Station, Asheville.

Blackmarr, W. H., and Flanner, W. B. (1968). "Seasonal and Diurnal Variation in Moisture Content of Six Species of Pocosin Shrubs." Research Paper SE-33. USDA Forest Service, Southeastern Forest Experiment Station, Asheville.

Blackmarr, W. H., and Flanner, W. B. (1975). "Moisture Variation in Selected Pocosin Shrubs of Eastern North Carolina." Research Paper SE-124. USDA Forest Service, Southeastern Forest Experiment Station, Asheville.

Bradshaw, L. S., Deeming, J. D., Burgan, R. E., and Cohen, J. D. (1983). "The 1978 National Fire-Danger Rating System: Technical Documentation." General Technical Report INT-169. USDA Forest Service, Intermountain Forest and Range Experiment Station, Ogden.

Brown, H. P., Panshin, A. J., and Forsaith, C. C. (1949). "Textbook of Wood Technology, Vol. 1," 1st ed. McGraw-Hill, New York.

Brown, J. K., Booth, G. D., and Simmerman, D. G. (1989). Seasonal change in live fuel moisture of understory plants in western U.S. aspen. *In* "Proceedings of the Tenth Conference on Fire and Forest Meteorology" (D. C. MacIver, H. Auld, and R. Whitewood, Eds.), pp. 406–412. Forestry Canada, Environment Canada, Ottawa.

Brown, R. W. (1977). Water relations of range plants. *In* "Rangeland Plant Physiology" (R. E. Sosebee, Ed.), Range Science Series 4, pp. 97–140. Society for Range Management, Denver.

Browning, B. L. (1963). "The Chemistry of Wood." John Wiley, New York.

Burgan, R. E., Hartford, R. A., Eidenshink, J. C., and Werth, L. F. (1991). Estimation of vegetation greenness and site moisture using AVHRR data. In "Proceedings of the Eleventh Conference on Fire and Forest Meteorology" (P. L. Andrews and D. F. Potts, Eds.), pp. 17–24. Society of American Foresters, Bethesda.

Byram, G. M. (1959). Combustion of forest fuels. In "Forest Fire: Control and Use" (K. P. Davis, Ed.), pp. 61–89. McGraw-Hill, New York.

Byram, G. M. (1963). "An Analysis of the Drying Process in Forest Fuel Material." Unpublished report. USDA Forest Service, Fire Sciences Laboratory, Rocky Mountain Research Station, Missoula.

Byram, G. M., and Jemison, G. M. (1943). Solar radiation and forest fuel moisture. J. Agric. Res. 67, 149–176.

Campbell, G. S., Jungbauer, J. D., Jr., Shiozawa, S., and Hungerford, R. D. (1993). A one-parameter equation for water sorption isotherms of soils. Soil Sci. 156, 302–305.

Catchpole, E. A., Catchpole, W. R., Viney, N. R., McCaw, W. L., and Marsden-Smedley, J. B. Modelling fuel response time and equilibrium moisture content from field data. Accepted for publication, Int. J. Wildland Fire.

Chandler, C., Cheney, P., Thomas, P., Trabaud, L., and Williams, D. (1983). "Fire in Forestry: Vol. I." John Wiley, New York.

Chladil, M. A., and Nunez, M. (1995). Assessing grassland moisture and biomass in Tasmania—The application of remote sensing and empirical models for a cloudy environment. Int. J. Wildland Fire 5, 165–171.

Choong, E. T. (1965). Diffusion coefficients of softwoods by steady-state and theoretical methods. For. Prod. J. 15, 21–27.

Choong, E. T., and Skaar, C. (1972). Diffusivity and surface emissivity in wood drying. Wood Fiber 4, 80–86.

Choudhury, B. J., and Monteith, J. L. (1988). A four-layer model for the heat budget of homogeneous land surfaces. Quart. J. Roy. Meteorol. Soc. 114, 373–398.

Christensen, G. N. (1965). The rate of sorption of water vapor by thin materials. In "Humidity and Moisture: Measurement and Control in Science and Industry" (A. Wexler, Ed.), Vol. 4, pp. 279–293. Reinhold, New York.

Chrosciewicz, Z. (1986). Foliar moisture content variations in four coniferous tree species of central Alberta. Can. J. For. Res. 16, 157–162.

Clark, B., and Roberts, F. (1982). A belt weather kit accessory for measuring woody fuel moisture. Fire Manage. Notes 43, 25–26.

Cloutier, A., and Fortin, Y. (1991). Moisture content—water potential relationship of wood from saturated to dry conditions. Wood Sci. Tech. 25, 263–280.

Cloutier, A., and Fortin, Y. (1993). A model of moisture movement in wood based on water potential and the determination of the effective water conductivity. Wood Sci. Tech. 27, 95–114.

Comstock, G. L. (1970). Directional permeability of softwoods. Wood Fiber 1, 283–289.

Coté, W. A., Jr. (1967). "Wood Ultrastructure." University of Washington Press, Seattle.

Countryman, C. (1974). Moisture in living fuels affects fire behavior. Fire Manage. Notes 35, 10–13.

Cowan, I. R. (1965). Transport of water in the soil-plant-atmosphere system. J. Appl. Ecol. 2, 221–239.

Cowan, I. R., and Milthorpe, F. L. (1968). Plant factors influencing the water status of plant tissues. In "Water Deficits and Plant Growth" (T. T. Kozlowski, Ed.), Vol. 1, pp. 137–193. Academic Press, New York.

Crank, J. (1975). "The Mathematics of Diffusion," 2nd ed. Clarendon Press, Oxford.

Davis, K. P. (1959). "Forest Fire: Control and Use." McGraw-Hill, New York.

Dunlap, M. E. (1932). The drying rate of hardwood-forest leaves. J. For. 30, 421–423.

Federer, C. A. (1979). A soil-plant-atmosphere model for transpiration and availability of soil water. *Water Resources Res.* 15, 555–562.

Fernandez, M. L., and Howell, J. R. (1997). Convective drying model of southern pine. *Drying Tech.* 15, 2343–2375.

Flowers, T. J., and Yeo, A. R. (1992). "Solute Transport in Plants." Blackie Academic & Professional, Chapman & Hall, Glasgow.

Fons, W. L. (1950). Heating and ignition of small wood cylinders. *Ind. Engr. Chem.* 42, 2130–2133.

Forestry Canada Fire Danger Group. (1992). "Development and Structure of the Canadian Forest Fire Behavior Prediction System." Information Report ST-X-3. Forestry Canada, Science and Sustainable Development Directorate, Ottawa.

Fosberg, M. A. (1972). Theory of precipitation effects on dead cylindrical fuels. *For. Sci.* 18, 98–108.

Fosberg, M. A. (1975). "Heat and Water Vapor Flux in Conifer Forest Litter and Duff: A Theoretical Model." Research Paper RM-152. USDA Forest Service, Rocky Mountain Forest and Range Experiment Station, Fort Collins.

Fosberg, M. A. (1977). "Heat and Water Vapor Transport Properties in Conifer Duff and Humus." Research Paper RM-195. USDA Forest Service, Rocky Mountain Forest and Range Experiment Station, Fort Collins.

Gardner, W. R., and Ehlig, C. F. (1965). Physical aspects of the internal water relations of plant leaves. *Plant Physiol.* 40, 705–710.

Gary, H. L. (1971). Seasonal and diurnal changes in moisture contents and water deficits of Engelmann spruce needles. *Bot. Gaz.* 132, 327–332.

Gong, L., and Plumb, O. A. (1994). The effect of heterogeneity on wood drying, part 1: Model development and predictions. *Drying Tech.* 12, 1983–2001.

Hart, C. A. (1964). Principles of moisture movement in wood. *For. Prod. J.* 14, 207–214.

Hart, C. A., and Thomas, R. J. (1967). Mechanism of bordered pit aspiration as caused by capillarity. *For. Prod. J.* 17, 61–68.

Hatton, T. J., Viney, N. R., Catchpole, E. A., and de Mestre, N. J. (1988). The influence of soil moisture on *Eucalyptus* leaf litter moisture. *For. Sci.* 34, 292–301.

Helvy, J. D. (1967). Interception by eastern white pine. *Water Resources Res.* 3, 723–729.

Hillel, D. (1980a). "Fundamentals of Soil Physics." Academic Press, New York.

Hillel, D. (1980b). "Applications of Soil Physics." Academic Press, New York.

Hough, W. A. (1973). "Fuel and Weather Influence Wildfires in Sand Pine Forests." Research Paper SE-106. USDA Forest Service, Southeastern Forest Experiment Station, Asheville.

Howard, E. A., III. (1978). A simple model for estimating the moisture content of living vegetation as potential wildfire fuel. *In* "Preprint Volume, Fifth National Conference on Fire and Forest Meteorology," pp. 20–23. American Meteorological Society, Boston.

Jarvis, P. G. (1975). Water transfer in plants. *In* "Heat and Mass Transfer in the Biosphere. Part 1. Transfer Processes in the Plant Environment" (D. A. de Vries and N. H. Afgan, Eds.), pp. 369–394. Scripta Book Company, Washington, DC.

Jury, W. A., Gardner, W. R., and Gardner, W. H. (1991). "Soil Physics," 5th ed. Academic Press, San Diego.

Katchalsky, A., and Curran, P. F. (1965). "Nonequilibrium Thermodynamics in Biophysics." Harvard University Press, Cambridge.

Keetch, J. J., and Byram, G. M. (1968). "A Drought Index for Forest Fire Control." Research Paper SE-38. USDA Forest Service, Southeastern Forest Experiment Station, Asheville.

Kelsey, K. E. (1957). The sorption of water vapour by wood. *Aust. J. Appl. Sci.* 8, 42–54.

King, A. R., and Linton, M. (1963). Moisture variation in forest fuels: the rate of response to climate change. *Aust. J. Appl. Sci.* 14, 38–49.

King, N. R. (1973). The influence of water vapour on the emission spectra of flames. *Combust. Sci. Tech.* 6, 247–256.

Kozlowski, T. T., and Clausen, J. J. (1965). Changes in moisture contents and dry weights of buds and leaves of forest trees. *Bot. Gaz.* **126**, 20–26.

Kozlowski, T. T., and Pallardy, S. G. (1997). "Physiology of Woody Plants," 2nd ed. Academic Press, San Diego.

Kramer, P. J., and Boyer, J. S. (1995). "Water Relations of Plants and Soils." Academic Press, New York.

Kramer, P. J., and Kozlowski, T. T. (1979). "Physiology of Woody Plants." Academic Press, Orlando.

Kreith, F., and Sellers, W. D. (1975). General principles of natural evaporation. *In* "Heat and Mass Transfer in the Biosphere. Part I. Transfer Processes in the Plant Environment" (D. A. de Vries and N. H. Afgan, Eds.), pp. 207–227. Scripta Book Company, Washington, DC.

Lemon, E., Stewart, D. W., and Shawcroft, R. W. (1971). The sun's work in a cornfield. *Science* **174**, 371–378.

Linton, M. (1962). "Report on Moisture Variation in Forest Fuels—Prediction of Moisture Content." Commonwealth Scientific and Industrial Research Organization, Division of Physical Chemistry, Melbourne.

Little, C. H. A. (1970). Seasonal changes in carbohydrate and moisture content in needles of balsam fir (*Abies balsamea*). *Can. J. Bot.* **48**, 2021–2028.

Lopushinsky, W., and Klock, G. O. (1974). Transpiration of conifer seedlings in relation to soil water potential. *For. Sci.* **20**, 181–186.

Luke, R. H., and McArthur, A. G. (1978). "Bushfires in Australia." Australian Government Publishing Service, Canberra.

McGilvary, J. M., and Barnett, J. P. (1988). Increasing speed, accuracy, and safety of pressure chamber determinations of plant moisture stress. *Tree Planters' Notes* **39**, 3–4.

McNamara, W. S., and Hart, C. A. (1971). An analysis of interval and average diffusion coefficients for unsteady-state movement of moisture in wood. *Wood Sci.* **4**, 37–45.

Monteith, J. L. (1963a). Gas exchange in plant communities. *In* "Environmental Control of Plant Growth" (L. T. Evans, Ed.), pp. 95–112. Academic Press, New York.

Monteith, J. L. (1963b). Dew: Facts and fallacies. *In* "The Water Relations of Plants" (A. J. Rutter and F. H. Whitehead, Eds.), pp. 37–56. John Wiley, New York.

Monteith, J. L. (1980). The development and extension of Penman's evaporation formula. *In* "Applications of Soil Physics" (D. Hillel, Ed.), pp. 247–253. Academic Press, New York.

Monteith, J. L., Szeicz, G., and Waggoner, P. E. (1965). The measurement and control of stomatal resistance in the field. *J. Appl. Ecol.* **2**, 345–355.

Moore, A., and Swank, W. T. (1975). A model of water content and evaporation for hardwood leaf litter. *In* "Mineral Cycling in Southeastern Ecosystems," (F. G. Howell, J. B. Gentry, and M. H. Smith, Eds.), pp. 58–69. ERDA Symposium Series, Technical Information Center, Office of Public Affairs, U.S. Energy Research and Development Administration, Oak Ridge.

Morén, T., Salin, J.-G., and Söderström, O. (1992). Determination of the surface emission factors in wood sorption experiments. *In* "Understanding the Wood Drying Process: A Synthesis of Theory and Practice," (M. Vanek, Ed.), pp. 69–73. 3rd IUFRO International Wood Drying Conference, Boku, Vienna.

Mount, A. B. (1972). "The Derivation and Testing of a Soil Dryness Index Using Run-off Data." Bulletin No. 4. Tasmania Forestry Commission, Hobart.

Nelson, R. M., Jr. (1969). "Some Factors Affecting the Moisture Timelags of Woody Materials." Research Paper SE-44. USDA Forest Service, Southeastern Forest Experiment Station, Asheville.

Nelson, R. M., Jr. (1984). A method for describing equilibrium moisture content of forest fuels. *Can. J. For. Res.* **14**, 597–600.

Nelson, R. M., Jr. (1986a). Diffusion of bound water in wood. Part 1: The driving force. *Wood Sci. Tech.* **20**, 125–135.

Nelson, R. M., Jr. (1986b). Diffusion of bound water in wood. Part 2: A model for isothermal diffusion. *Wood Sci. Tech.* **20**, 235–251.

Nelson, R. M., Jr. (2000). Prediction of diurnal change in 10-hour fuel stick moisture content. *Can. J. For. Res.* **30**, 1071–1087.

Nissan, A. H., and Kaye, W. G. (1957). Mechanism of drying of thick porous bodies during the falling-rate period. *Nature* **180**, 142–143.

Nissan, A. H., Kaye, W. G., and Bell, J. R. (1959). Mechanism of drying thick porous bodies during the falling rate period. I. The pseudo-wet-bulb temperature. *AIChE J.* **5**, 103–110.

Noble, I. R., Bary, G. A. V., and Gill, A. M. (1980). McArthur's fire-danger meters expressed as equations. *Aust. J. Ecol.* **5**, 201–203.

Norum, R. A., and Fischer, W. C. (1980). "Determining the Moisture Content of Some Dead Forest Fuels Using a Microwave Oven." Research Note INT-277. USDA Forest Service, Intermountain Forest and Range Experiment Station, Ogden.

Olson, C. M. (1980). An evaluation of the Keetch-Byram drought index as a predictor of foliar moisture content in a chaparral community. *In* "Proceedings of the Sixth Conference on Fire and Forest Meteorology," pp. 241–245. Society of American Foresters, Bethesda.

Olson, J. S. (1963). Energy storage and the balance of producers and decomposers in ecological systems. *Ecology* **44**, 322–331.

Panshin, A. J., and de Zeeuw, C. (1980). "Textbook of Wood Technology, Vol. 1," 4th ed. McGraw-Hill, New York.

Perre, P., Fohr, J. P., and Arnaud, G. (1990). A model of drying applied to softwoods: The effect of gaseous pressure below the boiling point. *In* "Drying '89" (A. S. Mujumdar and M. Roques, Eds.), pp. 91–98. Hemisphere Publishing Corporation, New York.

Philip, J. R. (1966). Plant water relations: some physical aspects. *Ann. Rev. Plant Physiol.* **17**, 245–268.

Piñol, J., Filella, I., Ogaya, R., and Peñuelas, J. (1998). Ground-based spectroradiometric estimation of live fine fuel moisture of Mediterranean plants. *Agric. For. Meteorol.* **90**, 173–186.

Plumb, O. A., Spolek, G. A., and Olmstead, B. A. (1985). Heat and mass transfer in wood during drying. *Int. J. Heat Mass Transfer* **28**, 1669–1678.

Pompe, A., and Vines, R. G. (1966). The influence of moisture on the combustion of leaves. *Aust. For.* **30**, 231–241.

Pyne, S. J. (1984). "Introduction to Wildland Fire." John Wiley, New York.

Pyne, S. J., Andrews, P. L., and Laven, R. D. (1996). "Introduction to Wildland Fire," 2nd ed. John Wiley, New York.

Riha, S. J., and Campbell, G. S. (1985). Estimating water fluxes in Douglas-fir plantations. *Can. J. For. Res.* **15**, 701–707.

Rose, C. W. (1966). "Agricultural Physics." Pergamon Press, Oxford.

Rothermel, R. C., Wilson, R. A., Jr., Morris, G. A., and Sackett, S. S. (1986). "Modeling Moisture Content of Fine Dead Wildland Fuels: Input to the BEHAVE Fire Prediction System." Research Paper INT-359. USDA Forest Service, Intermountain Research Station, Ogden.

Rothwell, R. L., Woodard, P. M., and Samran, S. (1991). The effect of soil water on aspen litter moisture content. *In* "Proceedings of the Eleventh Conference on Fire and Forest Meteorology" (P. L. Andrews and D. F. Potts, Eds.), pp. 117–123. Society of American Foresters, Bethesda.

Running, S. W. (1978). A process oriented model for live fuel moisture. *In* "Preprint Volume, Fifth Joint Conference on Fire and Forest Meteorology", pp. 24–28. Society of American Foresters, Bethesda and American Meteorological Society, Boston.

Running, S. W. (1984a). "Documentation and Preliminary Validation of H2OTRANS and DAYTRANS, Two Models for Predicting Transpiration and Water Stress in Western Coniferous Forests." Research Paper RM-252. USDA Forest Service, Rocky Mountain Forest and Range Experiment Station, Fort Collins.

Running, S. W. (1984b). Microclimate control of forest productivity: analysis by computer simulation of annual photosynthesis/transpiration balance in different environments. *Agric. For. Meteorol.* **32**, 267–288.

Rutter, A. J. (1968). Water consumption by forests. *In* "Water Deficits and Plant Growth" (T. T. Kozlowski, Ed.), Vol. 2, pp. 23–84. Academic Press, New York.

Scholl, D. G. (1976). Soil moisture flux and evaporation determined from soil hydraulic properties in a chaparral stand. *Soil Sci. Soc. Am. J.* **40**, 14–17.

Schönherr, J. (1976). Water permeability of isolated cuticular membranes: the effect of cuticular waxes on diffusion of water. *Planta* **131**, 159–164.

Schroeder, M. J., and Buck, C. C. (1970). "Fire Weather . . . A Guide for Application of Meteorological Information to Forest Fire Control Operations." Agriculture Handbook No. 360. USDA Forest Service, Washington, DC.

Shuttleworth, W. J., and Wallace, J. S. (1985). Evaporation from sparse crops—an energy combination theory. *Quart. J. Roy. Meteorol. Soc.* **111**, 839–855.

Siau, J. F. (1995). "Wood: Influence of Moisture on Physical Properties." Department of Wood Science and Forest Products, Virginia Polytechnic Institute and State University, Blacksburg.

Siau, J. F., and Avramidis, S. (1996). The surface emission coefficient of wood. *Wood Fiber Sci.* **28**, 178–185.

Simpson, W. T. (1993a). Determination and use of moisture diffusion coefficient to characterize drying of northern red oak (*Quercus rubra*). *Wood Sci. Tech.* **27**, 409–420.

Simpson, W. T. (1993b). "Specific Gravity, Moisture Content, and Density Relationship for Wood." General Technical Report FPL-GTR-76. USDA Forest Service, Forest Products Laboratory, Madison.

Skaar, C. (1988). "Wood-Water Relations." Springer-Verlag, Berlin.

Slatyer, R. O. (1967). "Plant-Water Relationships." Academic Press, New York.

Sneeuwjagt, R. J., and Peet, G. B. (1985). "Forest Fire Behaviour Tables for Western Australia." Western Australian Department of Conservation and Land Management, Perth.

Spolek, G. A., and Plumb, O. A. (1980). A numerical model of heat and mass transport in wood during drying. *In* "Drying, 1980," pp. 84–92. Hemisphere Publishing Corporation, New York.

Spolek, G. A., and Plumb, O. A. (1981). Capillary pressure in softwoods. *Wood Sci. Tech.* **15**, 189–199.

Stamm, A. J. (1946). "Passage of Liquids, Vapors and Dissolved Materials through Softwoods." Technical Bulletin No. 929, USDA, Washington, DC.

Stamm, A. J. (1960). Combined bound-water and water-vapor diffusion into Sitka spruce. *For. Prod. J.* **10**, 644–648.

Stamm, A. J. (1964). "Wood and Cellulose Science." Ronald Press, New York.

Stamm, A. J., and Harris, E. E. (1953). "Chemical Processing of Wood." Chemical Publishing Company, New York.

Stamm, A. J., and Nelson, R. M., Jr. (1961). Comparison between measured and theoretical drying diffusion coefficients for southern pine. *For. Prod. J.* **11**, 536–543.

Stocks, B. J. (1970). "Moisture in the Forest Floor—its Distribution and Movement." Publication No. 1271. Department of Fisheries and Forestry, Canadian Forestry Service, Ottawa.

Susott, R. A. (1980). Thermal behavior of conifer needle extractives. *For. Sci.* **26**, 347–360.

Susott, R. A. (1982). Characterization of the thermal properties of forest fuels by combustible gas analysis. *For. Sci.* **28**, 404–420.

Tanner, C. B. (1968). Evaporation of water from plants and soil. *In* "Water Deficits and Plant Growth" (T. T. Kozlowski, Ed.), pp. 73–106. Academic Press, New York.

Tourula, T., and Heikinheimo, M. (1998). Modelling evapotranspiration from a barley field over the growing season. *Agric. For. Meteorol.* **91**, 237–250.

Tunstall, B. (1991). Live fuel water content. *In* "Proceedings, Conference on Bushfire Modelling and Fire Danger Rating Systems" pp. 127–136. CSIRO Division of Forestry, Yarralumla.

Urqhart, A. R. (1960). Sorption isotherms. *In* "Moisture in Textiles" (J. W. S. Hearle and R. H. Peters, Eds.), pp. 14–32. Interscience, New York.

Van Wagner, C. E. (1969a). Drying rates of some fine forest fuels. *Fire Contr. Notes* 30, 5, 7, 12.

Van Wagner, C. E. (1969b). "Combined Effect of Sun and Wind on Surface Temperature of Litter." Information Report PS-X-10. Canadian Forest Service, Petawawa Forest Experiment Station, Chalk River.

Van Wagner, C. E. (1972a). Duff consumption by fire in eastern pine stands. *Can. J. For. Res.* 2, 34–39.

Van Wagner, C. E. (1972b). "Equilibrium Moisture Contents of Some Fine Forest Fuels in Eastern Canada." Information Report PS-X-36. Canadian Forest Service, Petawawa Forest Experiment Station, Chalk River.

Van Wagner, C. E. (1979). A laboratory study of weather effects on the drying rate of jack pine litter. *Can. J. For. Res.* 9, 267–275.

Van Wagner, C. E. (1987). "Development and Structure of the Canadian Forest Fire Weather Index System." Forestry Technical Report 25. Canadian Forest Service, Ottawa.

Viney, N. R. (1991). A review of fine fuel moisture modelling. *Int. J. Wildland Fire* 1, 215–234.

Viney, N. R. (1992a). "Modelling Fine Fuel Moisture Content." PhD dissertation, Department of Mathematics, University College, University of New South Wales, Sydney.

Viney, N. R. (1992b). Moisture diffusivity in forest fuels. *Int. J. Wildland Fire* 2, 161–168.

Viney, N. R., and Catchpole, E. A. (1991). Estimating fuel moisture response times from field observations. *Int. J. Wildland Fire* 1, 211–214.

Viney, N. R., and Hatton, T. J. (1989). Assessment of existing fine fuel moisture models applied to *Eucalyptus* litter. *Aust. For.* 52, 82–93.

Viney, N. R., and Hatton, T. J. (1990). Modelling the effect of condensation on the moisture content of forest litter. *Agric. For. Meteorol.* 51, 51–62.

Wallace, J. S., Roberts, J. M., and Sivakumar, M. V. K. (1990). The estimation of transpiration from sparse dryland millet using stomatal conductance and vegetation area indices. *Agric. For. Meteorol.* 51, 35–49.

Waring, R. H., and Running, S. W. (1976). Water uptake, storage and transpiration by conifers: A physiological model. *In* "Water and Plant Life, Problems and Modern Approaches" (O. L. Lange, L. Kappen, and E.-D. Schulze, Eds.), pp. 189–202. Springer-Verlag, Berlin.

Wilson, R. A., Jr. (1990). "Reexamination of Rothermel's Fire Spread Equations in No-wind and No-slope Conditions." Research Paper INT-434. USDA Forest Service, Intermountain Research Station, Ogden.

Xanthopoulos, G., and Wakimoto, R. H. (1993). A time to ignition–temperature–moisture relationship for branches of three western conifers. *Can. J. For. Res.* 23, 253–258.

Zhang, J., and Peralta, P. N. (1999). Moisture content–water potential characteristic curves for red oak and loblolly pine. *Wood Fiber* 31, 360–369.

CHAPTER 5

Wildland Fire Spread Models

R. O. WEBER

School of Mathematics and Statistics, University College UNSW, ADFA Canberra, Australia

---

I. INTRODUCTION

The spread of individual wildland fires directly affects plant communities on a short timescale. This is distinct from fire regimes over many years which help to define vegetation mosaics. It is this fairly immediate impact which has prompted a great deal of fire modeling activity over many decades. Naturally the desire to predict the spread of wildland fires to aid in containment and management of resources during particular fire incidents has also been a strong motivational factor.

In this chapter, we will survey spread models and try to give the reader an understanding of the main ideas behind the modeling as well as an overview of the different approaches and the relative merits of the various fire spread models. The emphasis will be on the explanation of relevant physical principles and the way in which they are encapsulated into the mathematical models. In some cases, it will be possible to completely define the model; in others, only a sketch will be possible, and the reader will need to consult the references for a complete understanding.

One topic that we will not cover is the influence of firebrands on fire propagation. This is an important topic which has only been moderately researched so far (see Lee and Hellman, 1970; several USDA Forest Service Internal Research Papers by Albini and a recent ANU Ph.D. thesis still under examination) and deserves further study. Another topic of great interest is crown fires; both their initiation and their propagation. This is an active area of ongoing research (e.g., Grishin, 1984; Albini and Stocks, 1986) and was recently briefly reviewed by Albini (1996). The fire modeling that is described in the present chapter is essential background for anyone wishing to study crown fires, firebrands, etc., but these topics will not be discussed here.

II. HEAD FIRE RATE OF SPREAD (PHYSICAL PRINCIPLES AND THEIR MATHEMATICAL EMBODIMENT)

One of the first observables of a landscape fire that would occur to most people is: "How fast did it move?" By this they usually mean the speed at which the fastest section of the fireline spreads into unburnt fuels. It was considered early in this century that the principle of energy conservation, expressed mathematically, should provide the fundamental tool for modeling and predicting the rate at which the head of a landscape fire will spread. Among the earliest published papers on this topic are Fons (1946), Bruce et al. (1961), Emmons (1964), Emmons and Shen (1971), Anderson (1968, 1969) Hottel et al. (1964, 1971), McCarter and Broido (1965), McArthur (1966), Rothermel and Anderson (1966), Albini (1967), Berlad et al. (1971), Thomas (1967, 1971), van Wagner (1967), and Rothermel (1972).

Later studies include Albini (1976a, 1976b, 1981, 1985, 1986), Beer (1991a, 1991b, 1993, 1995), Catchpole (1987), Cekirge (1978), De Mestre et al. (1989), Dorrer (1984), Duarte (1997a), Frandsen and Andrews (1979), Fujii et al. (1980), Gill et al. (1995), Green et al. (1985, 1990), Grishin (1984), Konev and Sukhinin (1977), Larini et al. (1998), Pagni and Peterson (1973), Steward (1974), Telisin (1974), Weber (1989, 1990), Weber et al. (1995), and Williams (1982). Literature surveys of the history and the mathematical models themselves are available in papers by Catchpole and De Mestre (1986) and Weber (1991, also translated for the *Chinese Journal of Mechanics and Practice*).

For ecological aspects see Brown (1981, 1982), da Silva (1990), Gill (1981), Green (1983a, 1983b), Greig-Smith (1979), Klukas (1973), McArthur and

Cheney (1966), Minnich (1983), Noble and Slayter (1980), Omi and Kala-bokidis (1998), Rothermel and Deeming (1980), and Schmidt and Wakimoto (1987). For good general introductions, see Luke and McArthur (1978) and Chandler *et al.* (1983).

Fire spread is determined by the rate at which heat is transferred from the burning matter to unburnt matter. The way to describe this mathematically is with a precise statement of the rate at which the heat energy is accumulated in unburnt matter as a balance between energy inputs and energy outputs. To do this, we begin with the energy conservation expressed in a "word equation" as

$$\left\{\begin{array}{c}\text{Rate of accumulation}\\\text{of heat energy}\end{array}\right\} = \left\{\begin{array}{c}\text{Rate at which}\\\text{energy flows in}\end{array}\right\} - \left\{\begin{array}{c}\text{Rate at which}\\\text{energy flows out}\end{array}\right\}$$

This is to be applied individually to small elements of the unburnt fuel, with the fuel being considered to be a collection of many such small elements. When coupled with some ignition criteria, it then allows one to determine how quickly successive elements of fuel ignite and consequently the rate of fire spread. An algebraic relation can be obtained from this approach (Williams, 1977, 1982, 1985); namely,

$$\rho V Q_{ig} = q \qquad \text{with} \qquad Q_{ig} = \int_{T_u}^{T_i} c_p \, dT$$

This can be rearranged to give (what some people refer to as the fundamental equation for) the rate of spread

$$V = \frac{q}{\rho Q_{ig}}$$

or in words

$$\left\{\begin{array}{c}\text{Rate of}\\\text{spread}\end{array}\right\} = \frac{\{\text{Heat flux from the active combustion}[1]\}}{\{\text{Heat required for fuel ignition}\}}$$

Early purely thermal models (as described in Weber, 1991) embodied the idea of many small elements in a differential equation for the temperature. An example of this style of modeling can be found in the equations in De Mestre *et al.* (1989), which then needs to be solved by integration. In fact, the solution of these models usually proceeds by assuming a constant rate of spread for the fire and changing frame of reference to a coordinate "moving with the fire." It is in this new coordinate system that the differential equation can be integrated and boundary conditions applied to complete the determination of the arbitrary constants of integration and the rate of spread. The essential requirement for

[1] Equates with Anderson and Rothermel's "Propagating Flux"; see Section III.

accurate prediction of the rate of spread is a good description of the heat transfer processes from the fire to the fuel. This is what determines the value for the parameter A in the following example. The use of books on heat transfer, such as Gebhart (1971) or Ozisik (1973), can assist in exploring all of the options for radiative, convective, and conductive heat transfer. The fire spread papers cited in this chapter have explored many of the possibilities, with limited success when comparisons are made with experimental results.

Example

$$\rho c_p \frac{\partial T}{\partial t} = A e^{-\alpha(x-Vt)} - IS_v(T - T_a)$$

Energy Conservation Equation

$$T(x - Vt, t) = T_i$$

Ignition Interface Condition

$$T(x \rightarrow \infty, t) = T_a$$

Ambient Condition

Step 1   Introduce a moving reference frame

$$X = x - Vt:$$

Then $$\frac{\partial T}{\partial t} = -V \frac{\partial T}{\partial X}$$

Hence $$-\rho c_p V \frac{\partial T}{\partial X} = A e^{-\alpha X} - IS_v(T - T_a)$$

and $$T(X = 0, t) = T_i$$

Step 2   Integrate this differential equation to obtain the temperature profile:

$$V\rho c_p \{T(X, t) - T_a\}$$

$$= A e^{IS_v X/\rho c_p} \int_{\infty}^{X} e^{-\alpha X - (IS_v/\rho c_p)X} \, dX$$

$$= \frac{A e^{-\alpha X}}{\left(\alpha + \dfrac{IS_v}{\rho c_p} \right)}$$

Step 3 Use the condition $T(X = 0, t) = T_i$ to find the speed:

$$V = \left[\frac{A - lS_v(T_i - T_a)}{\alpha \rho c_p(T_i - T_a)} \right]$$

Notes

1. The symbol V has been used for the fire spread velocity. Often R or ROS are used in the wildland fire literature because they conjure up rate of spread. Although any definition is equally appropriate in principle, certain symbols can cause confusion when moving between research fields. As an obvious example, R is often used to denote a radius in geometry problems.

2. Simple models, such as the preceding example which can be solved exactly, can provide valuable understanding of processes and the effect of variables such as ignition temperature on measurables such as spread rate. Other similar examples, such as the lag time for fire build up, can be found in the literature (Weber, 1989). They may also form the basis for more complete simulation of a fire incident as will be shown later in this chapter, and they can also be used as the input to other models such as mesoscale meteorological models.

3. The prediction for rate of spread in this model can be seen to fit precisely with the word equation for rate of spread given at the beginning of this chapter. The heat flux from the flames, moderated by heat losses, is in the numerator, and the heat of ignition is in the denominator.

4. This sort of model only considers heat transfer aspects of fire spread. It does not model the combustion process. Rather, it assumes that the radiative output from the flame can be independently determined or estimated. More sophisticated models which include the combustion process modeled by some simplified chemical kinetic scheme, such as Grishin *et al.* (1983), Larini *et al.* (1998), and Weber *et al.* (1997), require sophisticated numerical methods for their solution and are well beyond the scope of this chapter.

III. HEAD FIRE RATE OF SPREAD: AUSTRALIA

The most widely used rate of spread models in Australia are the McArthur models for grassland fires and forest fires (McArthur, 1966; Noble *et al.*, 1980). These models make no attempt to include any physical mechanisms for fire spread. Rather, they are a purely statistical description of test fires and as such can be very successful in predicting observations in circumstances similar to the test fires. The McArthur models were developed and tested in dry grassland and forest litter in South Eastern Australia during dry winter months. Initially, the results were used to make meters, a circular slide rule which required the user to select appropriate values for degree of curing, wind velocity, and so on. The meter could then be used to obtain a fire danger index and a fire spread rate. Noble *et al.* (1980) derived mathematical relationships to represent the action of the circular slide rules. For example, the equations for the Mark 4 grassland fire danger meter are

$$F = 2 \exp[-23.6 + 5.01 \ln C_d + 0.0281 \, T_a - 0.226 \, H_r^{1/2} + 0.663 \, U_{10}^{1/2}]$$

$$V = 0.036 \, F$$

where C_d is the degree of curing in %, T_a is the air temperature in °C, H_r is the relative humidity in %, U_{10} is the wind velocity in m/s measured at a height of 10 m, F is the fire danger index, and V is the fire spread rate in m/s.

This type of model is easy to use and has been successful when used to predict fire spread in conditions similar to those during the test fires; for example, low-intensity fuel reduction burning. However, other fuel types such as heathlands and markedly different environmental conditions are beyond the range of variables for which these models were constructed.

IV. HEAD FIRE RATE OF SPREAD: UNITED STATES

The model most widely used in the United States was named after R. C. Rothermel (1972) who provided a practical implementation of the empirical model of Frandsen (1971). This used the principle of conservation of energy to write down an equation for rate of spread, but it did not distinguish between different modes of heat transfer. Rather, laboratory experiments, covering a range of environmental and fuel variation, were used to empirically determine the propagating flux q, as a function of reaction intensity I_R:

$$q = \frac{I_R}{(192 + 7.894\sigma)} \exp[(0.792 + 3.760\sigma^{1/2})(\beta + 0.1)]$$

where σ is the fine particle surface area to volume ratio in cm^{-1} and β is the fraction of fuel bed volume occupied by solid fuel (called the packing ratio and also equivalent to one minus the void volume fraction). The reaction intensity is related to the rate of fuel mass loss per unit area, dw/dt, according to

$$I_R = -H\frac{dw}{dt}$$

where H is the heat of combustion of the fuel. By substituting into the energy balance $q = \rho Q_{ig}V$ and rearranging, it is possible to obtain an expression for the spread rate

$$V = \frac{(H\, dw/dt)}{\rho Q_{ig}(192 + 7.894\sigma)}\exp[(0.792 + 3.760\sigma^{1/2})(\beta + 0.1)]$$

This equation highlights the important fuel bed parameters of ρ, σ, and β. The empirical nature is reflected in the functional forms found through laboratory experimentation.

Wind and slope effects upon the rate of fire spread are also included in the Rothermel model by multiplying the formula for the spread rate V by

$$(1 + \phi_W = \phi_S)$$

$$\phi_W = C_W U_{10}^{B_W}\left(\frac{\rho}{\rho_0}\right)^{-E_W}$$

$$\phi_S = 5.275\,\beta^{-0.3}\tan^2\theta$$

where θ is the angle of the slope measured from the horizontal, B_W, C_W, and E_W are parameters involving σ, and β_0 in the optimum packing ratio. There seems to be no obvious physical reasons for the functional forms used for ϕ_W and ϕ_S, or indeed for the way in which the spread rate formula is modified to accommodate ϕ_W and ϕ_S. It is essentially an empirical model which works well in many situations. A careful examination of the parameters (Beer, 1991a) showed that the exponent to which the wind speed is raised could exceed unity (see Table 1), meaning that for sufficiently strong winds, the model would predict that the fire spreads faster than the wind moves. This is not a failure of the model. Rather, it is a consequence of the inappropriate choice for parameters, required in cases where the model was never validated. The ad hoc solution has been to place upper limits on the values of ϕ_W.

TABLE 1   Representative Values for C_W, B_W, E_W

σ (cm^{-1})	B_W	C_W	E_W
285	3.38	1.46	0.03
154	2.43	2.52	0.13
69	1.57	3.97	0.34
40	1.17	4.77	0.46
4	n/a	n/a	n/a

Following Beer (1991a).

Up until the Yellowstone fires of 1988, it was considered that the Rothermel model contained sufficient variability to accommodate most wildland fire situations. However, there has been a concerted effort over the last decade to more completely understand the limitations of the existing Rothermel model and to make improvements which will ultimately result in a new version. Not withstanding these reservations, the Rothermel model is generally a great success. It has been incorporated into a complete operational management tool and, as such, forms an integral part of the BEHAVE software—available on the World Wide Web at www.fire.org.

V. HEAD FIRE RATE OF SPREAD: CANADA

The Canadian Forest Service have conducted measurements and field experiments over a 25-year period to compile the Canadian Forest Fire Behaviour Prediction System, now available in various book forms and electronically. This provides a systematic method for including vegetation, topography, and weather variables appropriate to Canadian ecosystems and assessing wildland fire behaviour potential. It consists of mathematical formulae developed empirically and it is usually presented in tabular form or through a computer interface. With any model such as this, we can expect the predictions to be quite accurate provided our study area is reasonably represented by one of the choices available.

VI. SMOLDERING

Smoldering combustion is essentially the slow exothermic consumption of fuels (particularly organic matter in the case of wildland fires) under limited oxygen (anaerobic) conditions. This occurs quite widely in the world, from the coal seams in tropical Kalimantan (Indonesia) which cause occasional surface fires (Goldammer, 1990), to peat bogs in cool temperate areas of the world.

Large litter layers, such as those found in the boreal forests of North America, can become quite dry and support smoldering combustion, particularly in the so-called duff layer (see Chapter 13 in this book for more on smoldering duff).

The prediction of smoldering rate of spread is in principle much easier than for flaming combustion as the heat transfer aspect is greatly simplified with conduction of heat, the only mechanism of significance (Ohlemiller, 1985, 1988; Jones, 1993; Drysdale, 1999).

However, the kinetics of smoldering combustion and hence the rate of release of heat under various circumstances is not well understood. This is particularly the case when there are small traces of minerals, such as in duff layers.

For this reason, smoldering models are usually based upon the simple energy balance as introduced earlier in this chapter, and a fixed temperature is chosen for the smoldering interface.

One method for estimating the rate at which heat is released in a smoldering front is to consider the length scale of the front, namely, the distance over which the temperature rises from ambient, T_a, to the smoldering temperature, T_{sm}. If this distance is called x and if we also assume that the conductive heat transfer has reached a steady state, then the flux of heat per unit cross-sectional area is

$$q = k \frac{T_{sm} - T_a}{x}$$

We can use this in our basic combustion front propagation equation to obtain the speed

$$V = \frac{\dfrac{k(T_{sm} - T_a)}{x}}{\rho c_p (T_i - T_a)} = \frac{k}{x \rho c_p} \cdot \left(\frac{T_{sm} - T_a}{T_i - T_a} \right)$$

Given that $T_{sm} \cong T_i$, for cellulosic materials the thermal diffusivity is $k/\rho c_p \cong 10^{-7} \, \mathrm{m^2 \, s^{-1}}$, and $x \cong 1$ mm, we can estimate

$$V \cong 10^{-4} \, \mathrm{ms^{-1}} \qquad (\text{i.e., } 0.1 \text{ mm/s})$$

The heat flux would have been

$$q \cong 0.01 \times \frac{300}{154} = 3 \times 10^4 \, \mathrm{Js^{-1} \, m^{-2}}$$

compared to the heat of ignition

$$\rho c_p (T_i - T_a) \cong 73 \times 1370 \times 300 = 3 \times 10^7 \, \mathrm{Jm^{-3}}$$

To put this into a familiar context, we note that a typical cigarette has a little less than 60 mm of tobacco. Hence, it will smolder for approximately 10 minutes if left alone in suitable circumstances.

The transition from smoldering to flaming combustion has been and continues to be an active area of research (Drysdale, 1999). It has been established beyond doubt that the main factor in the transition are limits on the supply of oxygen.

VII. WHOLE FIRE MODELING—FIRE SHAPE

So far, the fire spread models have only considered head fire rate of spread. Naturally, a whole fire consists of more than merely the fastest moving front, and the whole fire has an impact upon the landscape and particularly the vegetation. As one might expect, the intensity will vary as one traverses the fire perimeter just as the rate of spread also varies. For this reason, it has long been of interest to consider fire shape models and also to investigate mechanisms or algorithms for whole fire growth.

Some of the earliest whole fire modeling was documented by Van Wagner (1969), Kourtz and O'Regan (1971) and Anderson et al. (1982); although Curry and Fons (1938) and Peet (1967) had considerably earlier made selected observations on fire shape. Given homogeneous fuel and weather conditions and assuming a constant moderate wind, a fire growing from a point ignition source will evolve to a shape which we would describe as "elliptic." This is not to say that the fire shape is exactly elliptical (or some other similar mathematical curve) but that a reasonable description, in terms of accuracy and relative ease of calculating significant features, is an ellipse growing with time [but see also McAlpine and Wotton (1993)]. In the next section, this whole fire model will be described in detail, but first we should make it clear that there are many different approaches to whole fire modeling, and we will survey several of these in the remainder of this chapter (including cellular automata and percolatier approaches), but this is a growing (pun intended!) area of fire research for both management training and conservation or ecological purposes.

A. ELLIPSE MODEL OF FIRE GROWTH

Following Anderson et al. (1982), consider first a homogeneous fuel on level terrain with no wind, so that the conditions may be summarized as being isotropic. In this case, we expect (on average) that the fire will appear as a circular front, growing in time, and we can describe this mathematically by the parametric equations

$$x = at \cos \chi$$
$$y = at \sin \chi$$

(1)

where x and y are the coordinates in the plane of a point on the front of the fire, a is the rate of spread (uniform across the whole front in this idealised, isotropic situation), t is the time elapsed since ignition (or some other convenient temporal reference time such as the time at which the fire is first observed in which case t would increase from some nonzero initial value t_0) and χ is a parameter which can be interpreted as an angular coordinate determining the location of the front at angles between 0° and 360° from the x-axis. That these parametric equations describe a circle can be seen by rearranging Eq. (1) and using the trigonometric identity $\sin^2 \chi + \cos^2 \chi = 1$ to give a standard form for the equation of a circle in the x, y plane; namely $x^2 + y^2 = a^2t^2$. We note that this model has the radius of the circle increasing linearly with time at a rate determined by a, the rate of spread, and a, in turn, is determined by fuel type, temperature, and moisture content.

Having established the nomenclature for describing a growing circular fire in isotropic conditions, it is possible to modify the parametric equations to include the effect of a constant wind and yield an elliptical fire front. This is done by writing the coordinates x, y of any point on the fire front as

$$x = at(f \cos \chi + g)$$
$$y = at(h \sin \chi)$$

(2)

Clearly for the no wind case, $f = h = 1$ and $g = 0$; however, a windspeed of U will change the values of f, g, and h. The dependence upon wind speed, which we may indicate by writing $f(U)$, $G(U)$, and $h(U)$, needs to be determined in some other way, either empirically by fitting to the existing fire data or possibly from physical arguments. Note that Alexander (1985) was able to determine suitable values for a, f, g, and h for particular fires. Next, we note that Eq. (2) describe an ellipse with semiaxes atf and ath, and whose center is moving (as the fire grows) in the x-direction with a speed ag. This can most easily be seen by again using the trigonometric identity $\cos^2 \chi + \sin^2 \chi = 1$ and rearranging Eq. (2) to give

$$\left(\frac{x - agt}{aft}\right)^2 + \left(\frac{y}{aht}\right)^2 = 1$$

Anderson et al. (1982) derived this elliptical model for a growing fire and were then able to show that it concurred with a "modified" Huygen's principle (familiar from the ray propagation theory of light). This was a useful observation for the subsequent computer implementation of this spread model but need not concern us for the immediate purposes of this chapter.

B. ELLIPTICAL SHAPE AND WIND SPEED

The parameters in the elliptical model can be combined to give the heading, flanking, and backing rate of spread as follows:

$$a(f + g), \qquad ah, \qquad a(f - g)$$

In Figure 11 of Anderson *et al.* (1982), graphs of time series for the forward rate of spread, the lateral rate of spread, and the wind speed are all shown for a grass fire conducted by the CSIRO in Australia. This and other similar plots are not conclusive in relation to the wind speed dependence of f, g, and h. However, it is encouraging to observe that variation in the ratio of forward to lateral rate of spread, which we can write as

$$\frac{a(f + g)}{ah} + \frac{f + g}{h}$$

and plot against wind speed U (as in Figure 12 of Anderson *et al.*, 1982), can be accounted for by wind speed variation alone.

C. ELLIPTICAL SHAPE AND FIRE INTENSITY

An additional benefit of the elliptical model is that it presents a useful and reasonably simple method for extending the Byram index of head fire intensity to a measure of intensity which changes around the perimeter of the fire. The index presented by Byram (1959) is written mathematically as

$$I = H w V$$

where H is the heat of combustion of fuel, w is the mass of fuel consumed per unit area, and V is the heading rate of spread of the fire. In the elliptical model of the whole fire, we would identify $V = a(f + g)$. Hence, we could write intensity at the head, flank, and rear as

$$H w a(f + g), \qquad H w a h, \qquad H w a(f - g)$$

Alternatively, following Catchpole *et al.* (1982), we may prefer to determine an expression for the intensity at any point on the perimeter. For this purpose, one needs to introduce the angle of the normal to any point on the perimeter of the ellipse ψ. The result of this is an intensity equation[2]

$$I = H w a(g \cos \psi + \sqrt{f^2 \cos^2 \psi + h^2 \sin^2 \psi})$$

Catchpole *et al.* (1982) then demonstrate how this intensity varies around the perimeter for two cases idealized by

    a. Medium wind: $f = 2, g = 1.8, h = 1$
    b. Strong wind: $f = 4, g = 3.8, h = 1$

---

[2]Note that the angle ψ has value 0 at the head of the ellipse and values $\pi/2$ and π at the flank and rear, respectively. In fact, $-\pi \leq \psi \leq \pi$.

In addition to using the elliptical model to determine the intensity at any point on the perimeter at any moment in time, it is also possible to determine a total fire flux by integrating the intensity around the entire perimeter.

Catchpole *et al.* (1982) performed this integration for the ellipse model to find that the total fire flux is

$$2\pi H \, w \, a^2 f \, h \, t$$

This quantity is a measurement (in units of joules per second or watts) of the energy released by the total fire per unit time. Note that this quantity increases linearly with time as does the fire perimeter.

In principle, the elliptical model allows for a comprehensive analysis of the entire progression of a fire, from its ignition and will allow estimates of fire intensity to be made at all points in space and time. Using the temperature modeling in the fire plumes chapter, these fire intensity estimates could then be used to predict temperature-time exposures for vegetation and assess the ecological impact of a fire incident, with the possibility of really comparing predictions of fire impact with observed fire impact. At this stage, such a complete incident analysis has never been undertaken.

D. OTHER FIRE SPREAD ISSUES

There have been several other approaches to modeling fire spread, often motivated by a need to deal with inhomogeneous distributions of fuel and varying topography. Unlike the elliptical models, the basic ideas have come from a very local point of view of fire spread. For example, the cellular automata approach, which was first introduced by Green *et al.* (1983, 1985), considers small cells of fuel and develops a rule approach for deciding if the fire will spread to adjacent cells. This presents a very efficient and simple method for implementing into management software and provides quite reasonable fire shapes, although there has been considerable investigation into the effect the cell shape (triangular, square, hexagonal) will have on the eventual fire shape.

There have also been several probabilistic models for fire spread, which include a probability of spread from cell to cell in a given mesh. The earliest of these was by Kourtz and O'Regan (1971), using a square mesh and fuel of varying moisture content in the mesh, distributed according to a Monte Carlo method. This then required an algorithm for determining the route of the fire. This concept was also used by Catchpole *et al.* (1989), who used a Markov chain approach to allow for several different types of fuel and also by Guertin and Ball (1990). In all of these, the concepts of transition from fuel element to fuel element and also randomness are important elements. For this reason, it has become of interest to consider the percolation approach of Albinet *et al.* (1986)

(see also Stauffer and Aharony, 1992). This arose out of developments in statistical physics (De Gennes, 1976) which were directed at finding a unified approach to spread processes with a random element. Other ecological processes
(Gardner et al., 1987; Turner et al., 1989; Reed, 1999) have also been considered as being well modeled as percolation processes. On the experimental front,
Beer (1990), Beer and Enting (1990), and Duarte (1997) have conducted laboratory studies using arrays of fuel elements. They have found good agreement
between the predicted exponents from percolation theory and the clusters
found experimentally.

NOTATION

ROMAN LETTERS

A	radiation intensity from fire front	W m^{-1}
a	rate of spread in the absence of wind	m s^{-1}
B_W	wind effect parameter dimensionless	
C_d	degree of curing %	
C_W	wind effect parameter dimensionless	
c_p	specific heat at constant pressure	$\text{J kg}^{-1}\,\text{K}^{-1}$
E_W	wind effect parameter dimensionless	
F	fire danger index	m s^{-1}
f	ellipse parameter dimensionless	
g	ellipse parameter dimensionless	
H	heat of combustion of fuel	J kg^{-1}
H_r	height of wind velocity measurement	m
h	ellipse parameter dimensionless	
I	index for headfire intensity	W m^{-1}
I_R	reaction intensity	W m^{-2}
k	thermal conductivity	$\text{W m}^{-1}\,\text{K}^{-1}$
l	heat loss coefficient	$\text{J K}^{-1}\,\text{m}^{-2}$
Q_{ig}	heat of ignition	J kg^{-1}
q	heat transfer from active combustion	W m^{-2}
S_v	whole fuel bed surface area to volume ratio	m^{-1}
T	temperature	K

t	time	s
U	average wind velocity	m s^{-1}
U_{10}	wind velocity at a height of ten meters	m s^{-1}
V	rate of spread of fire front	m s^{-1}
w	mass of fuel bed per unit area	kg m^{-2}
x	fixed spatial coordinate	m
X	moving spatial coordinate	m
y	fixed spatial coordinate	m

GREEK LETTERS

α	radiation absorptivity	m^{-1}
β	packing ratio dimensionless	
θ	angle from the horizontal radians	
ρ	density	kg m^{-3}
σ	fine particle surface area to volume ratio	cm^{-1}
Φ_S	slope factor dimensionless	
Φ_W	wind factor dimensionless	
χ	angular parameter radians	
ψ	angle between normal and tangent radians	

SUBSCRIPTS

a	ambient property
ig	ignition
sm	smoldering

REFERENCES

Albinet, G., Searby, G., and Staufer, D. (1986). Fire propagation in a 2-D random medium. *J. Physique* 47, 1–7.

Albini, F. A. (1967). A Physical model for fire spread in brush. *In* "11th Symp. (Int.) on Combust.," pp. 553–560. The Combustion Institute, Pittsburgh.

Albini, F. A. (1976a). "Estimating Wildfire Behaviour and Effects." Gen. Tech. Rep. INT-30, USDA Forest Service, Ogden.

Albini, F. A. (1976b). "Computer-Based Models of Wildland Fire Behaviour: A User's Manual." Intermt. For. and Range Exp. Stn., Ogden.

Albini, F. A. (1981). A model for the wind-blown flame from a line fire. *Combust. Flame* **43**, 155–174.

Albini, F. A. (1985). A model for fire spread in wildland fuels by radiation. *Combust. Sci. Technol.* **42**, 229–258.

Albini, F. A. (1986). Wildland fire spread by radiation—A model including fuel cooling by convection. *Combust. Sci. Technol.* **45**, 101–113.

Albini, F. A. (1996). Iterative solution of the radiation transport equations governing spread of fire in wildland fuels. *Fizika Goreniya i Vzryva* **32**, 71–82.

Albini, F. A., and Stocks, B. J. (1986). Predicted and observed rates of spread of crown fires in immature Jack Pine. *Combust. Sci. Technol.* **48**, 65–76.

Alexander, M. E. (1985). Estimating the length-to-breadth ratio of elliptical forest fire patterns. *In* "Proc. 8th Conf. Forest and Fire Meteorology," pp. 287–304. Soc. Am. Foresters, Detroit.

Anderson, D. H., Catchpole, E. A., De Mestre, N. J., and Parkes, T. (1982). Modeling the spread of grass fires. *J. Aust. Math. Soc. (Ser. B)* **23**, 451–466.

Anderson, H. E. (1968). "Sundance Fire: An Analysis of Fire Phenomena." Res. Pap. INT-56, USDA Forest Service, Ogden.

Anderson, H. E. (1969). "Heat Transfer and Fire Spread." Res. Pap. INT-69, USDA Forest Service, Ogden.

Beer, T. (1990). Percolation theory and fire spread. *Combust. Sci. Tech.* **72**, 297–304.

Beer, T. (1991a). Bushfire rate of spread forecasting: Deterministic and statistical approaches to fire modeling. *J. Forecasting* **10**, 301–317.

Beer, T. (1991b). The interaction of fire and wind. *Boundary Layer Meteorol.* **54**, 287–308.

Beer, T. (1993). The speed of a fire front and its dependence on wind speed. *Int. J. Wildland Fire* **3**, 193–202.

Beer, T. (1995). Fire propagation in vertical stick arrays: The effects of wind. *Int. J. Wildland Fire* **5**, 43–49.

Beer, T., and Enting, I. G. (1990). Fire spread and percolation modeling. *Math. Comput. Modeling* **13**, 77–96.

Berlad, A. L., Rothermel, R. C., and Frandsen, W. (1971). *In* "13th Symp. (Int.) on Combust.," pp. 927–933. The Combustion Institute, Pittsburgh.

Brown, J. K. (1981). Bulk densities of nonuniform surface fuels and their application to fire modeling. *For. Sci.* **27**, 667—684.

Brown, J. K. (1982). "Fuel and Fire Behaviour Prediction in Big Sagebrush." Res. Pap. INT-290, USDA Forest Service, Ogden.

Bruce, H. D., Pong, W. Y., and Fons, W. L. (1961). The effect of density and thermal diffusivity of wood on the rate of burning of wooden cribs. Tech. Pap. No. 63, USDA Forest Service, Ogden.

Byram, G. M. (1959). Combustion of forest fuels, "Forest Fire: Control and Use" (K. P. Davis, Ed.). McGraw-Hill, New York.

Catchpole, E. A., and De Mestre, N. J. (1986). Physical models for a spreading line fire, *Aust. For.* **49**, 102–111.

Catchpole, E. A., De Mestre, N. J., and Gill, A. M. (1982). Intensity of fire at its perimeter, *Aust. For. Res.* **12**, 47–54.

Catchpole, E. A., Hatton, T. J., and Catchpole, W. R. (1989). Fire spread through nonhomogeneous fuel modeled as a Markov process, *Ecological Modeling* **48**, 101–112.

Catchpole, W. R. (1987). "Heathland Fuel and Fire Modeling." PhD thesis, University of New South Wales.

Cekirge, H. M. (1978). Propagation of fire fronts in forests. *Comput. Math. Applic.* **4**, 325–332.

Chandler, C., Cheney, N. P., Thomas, P. H., and Trabaud, L. (1983). "Fire in Forestry, Volume 1: Forest Fire Behaviour and Effects." Wiley, New York.

Curry, J. R., and Fons, W. L. (1938). Rate of spread of surface fires in the Ponderosa pine type of California. *J. Agric. Res.* 57, 239–267.

da Silva, J. M. (1990). La gestion forestiére et la sylviculture de prévention des espaces forestiers menacés par les incendies au Portugal. *Rev. For. Fr. XLII*, 337–345.

De Gennes, P. G. (1976). La percolation: Un concept unificateur. *La Recherche* 72, 919–927.

De Mestre, N. J., Catchpole, E. A., Anderson, D. H., and Rothermel, R. C. (1989). Uniform propagation of a planar fire front without wind. *Combust. Sci. Technol.* 65, 231–244.

Dorrer, G. A. (1984). A model of the spreading of a forest fire. *Heat Transfer—Sov. Res.* 16, 39–52.

Drysdale, D. D. (1999). "An Introduction to Fire Dynamics." Wiley, New York.

Duarte, J. A. M. S. (1997). Fire spread in natural fuel: Computational aspects. *Ann. Rev. Comput. Phys.* V, 1–23.

Emmons, H. (1964). Fire in the forest. *Fire Res. Abs. Rev.* 5, 163–178.

Emmons, H., and Shen, T. (1971). Fire spread in paper arrays. *In* "13th Symp. (Int.) on Combust.," pp. 917–926. The Combustion Institute, Pittsburgh.

Fons, W. L. (1946). Analysis of fire spread in light forest fuels. *J. Agric. Res.* 72, 93–121.

Frandsen, W. H. (1971). Fire spread through porous fuels from the conservation of energy. *Combust. Flame* 16, 9–16.

Frandsen, W. H., and Andrews, P. L. (1979). Fire behaviour in nonuniform fuels. Res. Pap. INT-232, USDA Forest Service, Ogden.

Fujii, N., Hasegawa, J., Pallop, L., and Sakawa, Y. A. (1980). A non-stationary model of fire spreading. *Appl. Math. Modeling* 4, 176–180.

Gardner, R. H., Milne, B. T., Turner, M. G., and O'Neill, R. V. (1987). Neutral models for the analysis of broad-scale landscape pattern. *Landscape Ecology* 1, 19–28.

Gebhart, B. (1971). "Heat Transfer." McGraw-Hill, New York.

Gill, A. M. (1981). Coping with fire. *In* "The Biology of Australian Plants" (J. S. Pate and A. J. McComb, Eds.). Univ. W.A. Press, Perth.

Gill, A. M., Burrows, N. D., and Bradstock, R. A. (1995). Fire modeling and fire weather in an Australian Desert. *CALM Sci. Suppl.* 4, 29–34.

Goldammer, J. G. (1990). "Fire in the Tropical Biota." Springer Verlag, Berlin.

Green, D. G. (1983a). Shapes of simulated fires in discrete fuels. *Ecol. Model.* 20, 21–32.

Green, D. G. (1983b). The ecological interpretation of fine resolution pollen records. *New Phytol.* 94, 459–477.

Green, D. G., Gill, A. M., and Noble, I. R. (1983). Fire shapes and the adequacy of fire spread models. *Ecol. Model.* 20, 33–45.

Green, D. G., House, A. P. N., and House, S. M. (1985). Simulating spatial patterns in forest ecosystems. *Mathematics and Computers in Simulation* 27, 191–198.

Green, D. G., Tridgell, A., and Gill, A. M. (1990). Interactive simulation of bushfires in heterogeneous fuels. *Math. Comput. Modeling* 13, 57–66.

Greig-Smith, P. (1979). Pattern in vegetation. *J. Ecol.* 67, 755–779.

Grishin, A. M. (1984). Steady-state propagation of the front of a high-level forest fire. *Sov. Phys. Dokl.* 29, 917–919.

Grishin, A. M., Gruzin, A. D., and Zverev, V. G. (1983). Mathematical modeling of the spreading of high-level forest fires. *Sov. Phys. Dokl.* 28, 328–330.

Guertin, D. P., and Ball, G. L. (1990). Using a fire spread model to determine ignition probabilities. *Proc. 1st ICFFR*, Coimbra, **B-13**, 1–5.

Hottel, H. C., Williams, G. C., and Kwentus, G. K. (1971). Fuel pre-heating in free-burning fires. *In* "13th Symp. (Int.) on Combust.," pp. 963–970. The Combustion Institute, Pittsburgh.

Hottel, H. C., Williams, G. C., and Steward, F. R. (1964). Modeling of fire-spread through a fuel bed. *In* "10th Symp. (Int.) on Combust.," pp. 997–1007. The Combustion Institute, Pittsburgh.

Jones, J. C. (1993). "Combustion Science." Millennium Books, Sydney.

Klukas, R. W. (1973). "Control Burn Activities in Everglades National Park." Proc. Annual Tall Timbers Fire Ecology Conf., Tall Timbers REs. Stn., Tallahassee, FL.

Konev, E. V., and Sukhinin, A. I. (1977). The analysis of flame spread through forest fuel. *Combust. Flame* **28**, 217–223.

Kourtz, P. H., and O'Regan, W. G. (1971). A model for a small forest fire to simulate burned and burning areas for use in a detection model. *For. Sci.* **17**, 163–169.

Larini, M., Giroud, F., Porterie, B. and Loraud, J. C. (1998). A multiphase formulation for fire propagation in heterogeneous combustible media. *Int. J. Heat Mass Transfer* **41**, 881–897.

Lee, S. L., and Hellman, J. M. (1970). Firebrand trajectory study using an empirical velocity-dependent burning law. *Combust. Flame* **15**, 165–274.

Luke, R. H., and McArthur, A. G. (1978). "Bushfires in Australia." Aust. Govt. Publishing Service, Canberra.

McAlpine, R. S., and Wotton, B. M. (1993). The use of fractal dimension to improve wildland fire perimeter predictions. *Can. J. For. Res.* **23**, 1073–1077.

McArthur, A. G. (1966). Weather and grassland fire behaviour. Leaflet No. 100, Aust. For. Timb. Bur., Canberra.

McArthur, A. G., and Cheney, N. P. (1966). The characterization of fires in relation to ecological studies. *Aust. For. Res.* **2**, 36–45.

McCarter, R. J., and Broido, A. (1965). Radiative and convective energy from wood crib fires. *Pyrodynamics* **2**, 65–85.

Minnich, R. A. (1983). Fire mosaics in Southern California and Northern Baja California. *Science* **219**, 1287–1294.

Noble, I. R., Bary, G. A. V., and Gill, A. M. (1980). McArthur's fire-danger meters expressed as equations. *Aust. J. Ecol.* **5**, 201–203.

Noble, I. R., and Slatyer, R. O. (1980). The use of vital attributes to predict successional changes in plant communities subject to recurrent disturbance. *Vegetatio* **43**, 5–21.

Ohlemiller, T. J. (1985). Smoldering combustion. *Prog. Energy Combust. Sci.* **11**, 277–282.

Ohlemiller, T. J. (1988). Smoldering combustion, Chapter 23 (Section 1). In "SPFE Handbook of Fire Protection Engineering," Vol. 1, pp. 352–359. National Fire Protection Association, Gaithersburg.

Omi, P. N., and Kalabokidis, K. D. (1998). Fuels modifications to reduce large fire probability. In "Proc. 3rd ICFFR," Vol. II, pp. 2073–2088. Luso.

Ozisik, M. N. (1973). "Radiative Transfer." Wiley, New York.

Pagni, J., and Peterson, G. (1973). Flame spread through porous fuels. In "14th Symp. (Int.) on Combust.," pp. 1099–1107. The Combustion Institute, Pittsburgh.

Peet, G. B. (1967). The shape of mild fires in jarrah forest. *Aust. For.* **31**, 121–127.

Reed, W. J. (2000). Forest fires and oil fields as percolation phenomena. *Science,* submitted.

Rothermel, R. C. (1972). A mathematical model for predicting fire spread in wildland fuels. Res. Pap. INT-115, USDA Forest Service, Ogden.

Rothermel, R. C., and Anderson, H. E. (1966). Fire spread characteristics determined in the laboratory. Res. Pap. INT-30, USDA Forest Service, Ogden.

Rothermel, R. C., and Deeming, J. E. (1980). Measuring and interpreting fire behaviour for correlation with fire effects. Gen. Tech. Rep. INT-93, USDA Forest Service, Ogden.

Schmidt, W. C., and Wakimoto, R. H. (1987). Cultural practices that can reduce fire hazards to homes in the interior west. In "Proc. Symposium and Workshop on Protecting People and Homes from Wildfire in the Interior West," pp. 131–141. USDA Forest Service, Ogden.

Stauffer, D., and Aharony, A. (1992). "Introduction to Percolation Theory." Taylor and Francis, London.

Steward, F. R. (1974). Fire spread through a fuel bed, "Heat Transfer in Fires: Thermophysics, Social Aspects, Economic Impact" (P. L. Blackshear, Ed.). Wiley, New York.

Telisin, H. P. (1974). Some aspects of the growth and spread of fires in the open. "Heat Transfers in Flames" (N. H. Afgan and J. M. Beer, Eds.). Wiley, New York.

Thomas, P. H. (1967). Some aspects of the growth and spread of fires in the open. *Forestry* **40**, 139–164.

Thomas, P. H. (1971). Rates of spread of some wind-driven fires. *Forestry* **44**, 155–175.

Turner, M. G., Gardner, R. H., Dale, V. H., and O'Neill, R. V. (1989). Predicting the spread of disturbance across heterogeneous landscapes. *Oikos* **55**, 121–129.

Van Wagner, C. E. (1967). "Calculations on Forest Fire Spread by Flame Radiation." Rep. No. 1185, Canadian Dept. Forestry, Edmonton, Alberta.

Van Wagner, C. E. (1969). A simple fire-growth model. *For. Chron.* **45**, 103–104.

Weber, R. O. (1989). Analytical models for fire spread due to radiation. *Combust. Flame* **78**, 398–408.

Weber, R. O. (1990). A model for fire propagation in arrays. *Math. Comput. Modeling* **13**, 95–102.

Weber, R. O. (1991). Modeling fire spread through fuel beds. *Prog. Energy Combust. Sci.* **17**, 67–82.

Weber, R. O., and De Mestre, N. J. (1990). Fire spread on a single ponderosa pine needle: Effect of sample orientation and external flow. *Combust. Sci. Technol.* **70**, 17–32.

Weber, R. O., Mercer, G. N., Mahon, G. J., Catchpole, W. R., and Gill, A. M. (1995). Predicting maximum inter-hummock distance for fire spread. *In* "Bushfire '95 Proceedings." Tasmania Parks and Wildlife Service.

Weber, R. O., Mercer, G. N., Sidhu, H. S., and Gray, B. F. (1997). Combustion waves for gases and solids. *Proc. Roy. Soc. (Lond)* **A453**, 1105–1118.

Williams, F. A. (1977). Mechanisms of fire spread. *In* "16th Symp. (Int.) on Combust.," pp. 1281–1294. The Combustion Institute, Pittsburgh.

Williams, F. A. (1982). Urban and wildland fire phenomenology. *Prog. Energy Combust. Sci.* **8**, 317–354.

Williams, F. A. (1985). "Combustion Theory," 2nd ed. Benjamin Cummings, Menlo Park, CA.

CHAPTER 6

Wind-Aided Fire Spread

F. E. FENDELL AND M. F. WOLFF

Space and Technology Division, TRW Space and Electronics Group,
Redondo Beach, California

I. INTRODUCTION

Fire spread in wildland encompasses very complicated phenomena, typically entailing turbulent reacting flow and intricate fuel arrangement, with processes occurring over a wide range of scales. Consequently, even simple observations and measurements are difficult, and replication is almost impossible. We present here one approach to quantifying such challenging phenomena by undertaking controlled, replicable laboratory-scale experiments which capture at least some of the important phenomena of large-scale fire spread. We then develop a semi-empirical process-based model to simulate wind-aided fire spread and test the model by using the laboratory-scale experimental setup to generate additional data for comparison. The model may subsequently be validated in large-scale wildland fires; when a model which incorporates the key features of laboratory-scale fire spread is applied to a large-scale test, attention must be concentrated on those additional phenomena that did not play a significant role in the smaller scale testing (e.g., see Chapter 8 in this book).

In general, a small number of extreme fire events is responsible for most of the total area burned [e.g., 10% of the fires in southern California account for over 90% of the area burned (Minnich, 1998)]. Thus, we focus on extreme conditions of very dry fuels and high winds which may result in exceptionally high-intensity, rapidly spreading fires *involving surface fuels*. The objective is to predict the spread of such wind-aided fires (i.e., given the current location of the firefront and given requisite information about the adjacent unburned region prior to front arrival, where will the firefront be at future times?). Traditional approaches to this problem have constrained the firefront perimeter to an elliptical shape (see Finney, 1998) or used percolation theory (Christensen *et al.*, 1993) to speculate about probabilities of future locations of the firefront. However, such approaches often delve little into physical mechanisms and may not readily permit the incorporation of observational data. Given the difference in our current state of knowledge concerning low-intensity and high-intensity fire, we do not think that it is a wise investment of effort to try to develop a tractable fire spread model that would be applicable in all situations. Therefore, our model focuses on exceptional conditions conducive to the initiation and persistence of rapid wildland fire spread (i.e., fire spread across dry, moderately sparse fuel under fairly sustained substantial wind of fairly constant direction). Such conditions are exemplified by, for example, Santa Ana-wind-driven fires in southern California chaparral (Pyne, 1982; Reid, 1994; Goldstein, 1995) and related fires arising in other Mediterranean-climate locales (Table 1; Figure 1).

For idealized scenarios, the travel of the firefront in time is straightforward (Richards, 1993). For a uniform fuel distribution on level ground in the absence of wind, an initially circular front advances with radial symmetry and remains circular (Figures 2a and 2b). For a uniform fuel distribution on level ground in the presence of a near-ground (say, at 10-m height) wind that is steady in magnitude and constant in direction, a firefront growing from a localized ignition often becomes oval in shape (Figure 2c). The head of the fire (where the propagation is with the wind) spreads relatively rapidly, the back or rear of the fire (where the propagation is against the wind) spreads relatively

TABLE 1　Sites of Wind-Aided Spread in Mediterranean Fire Regimes

Site	Typical vegetation	Typical wind	Notes
Southern Hemisphere			
Chile (central)	matorral (underbrush)	upslope	no strong coastward wind
South Africa (near Cape of Good Hope)	fynbos (small bush)	berg (foehn wind blows coastward)	4–24-year cycle of regeneration
Australia (NSW, Victoria, South Australia)	mallee (shrub-like eucalypt)	off interior desert, often followed by cold southerly burster	degraded soil, modest hills, occasional summer rain; SW Australian fires not strongly wind-aided
California			
Northern	chaparral (scrub oak; chamise, manzanita)	Mono	20–40-year cycle of regen-eration, hilly terrain
Southern		Santa Ana	
Mediterranean Basin			
Greece, Spain, France	maquis (high shrub) garrigue (low shrub)	bora (dry cold winter wind off Julian Alps to-ward Adriatic);	hilly terrain
Greece, Spain	phrygana (Greece) tomillares (Spain) (dwarf shrubs)	sirocco (dry hot South spring wind—Italy, Spain);	
Spain	matorral (underbrush)	mistral, tramontane (dry cold NW wind off Alps toward Gulf of Genoa—France, Spain)	
Italy	macchia (shrub zone)		

slowly, and the flanks (where the propagation is across the wind) progress only moderately differently from the rate for the no-wind conditions. Clearly it is the component of the wind which is locally perpendicular to the fire perimeter that establishes whether advection carries or inhibits the travel of hot product gases from just-burned fuel over the next-to-burn fuel to help preheat it. The long axis of an oval fire perimeter typically continues to increase more rapidly than the short axis. An oval perimeter frequently is narrower at the head than at the rear (where the width is the dimension perpendicular to the wind direction), and the width is often maximal near that windward site (along the bisector of the burned area) that marks the point of origin. "In uniform fuel under con-stant environmental conditions, the local rates of spread can be expected to be

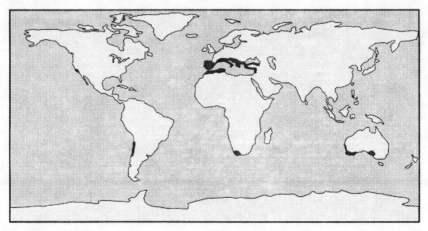

FIGURE 1   Mediterranean climate locales, in which near-coastal scrublands typically have low-fertility soils, and in which fire spread through surface vegetation is seasonally encountered, include southern California, central Chile, much of the Mediterranean basin, the Cape region of South Africa, and Australia (especially South Australia and Victoria).

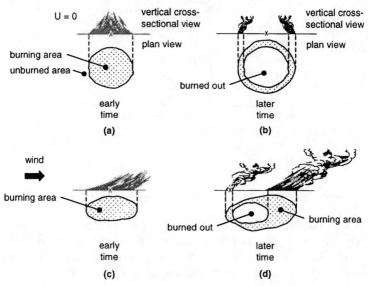

FIGURE 2   (a) At an ignition site in a uniform expanse of unburned fuel on level terrain, under calm, a firefront (which propagates slowly outward, against convectively induced advection) evolves to demarcate (b) an interior region with burned-out thin fuel, from an exterior region of unburned fuel. (c) For the corresponding event in the presence of a near-ground ambient wind sustained in magnitude and direction, the firefront evolves to an oval configuration. (d) The rate of propagation is appreciably enhanced along the portion of the perimeter where the wind aids spread and inhibited where the wind opposes spread. The enhanced spread occurs due to hot combustion-product gases preheating unburned downwind fuel.

constant for head, flanks, and back of fire. Thus the area burned over per unit time can be expected to grow linearly with time, the perimeter expected to grow at a steady rate, and the actively burning area inside the perimeter to grow in proportion to the perimeter" (Albini, 1992, p. 44).

The oval shape of the firefront perimeter arising in simplistic scenarios has been taken by many researchers to have the special configuration of an ellipse (see Finney, 1998, and references cited therein). There seems no fundamental reason for the elliptical shape to hold; in fact, the ellipse is but approximately representative of the firefront perimeter, even for burns under uniform conditions (Pyne *et al.*, 1996, pp. 58–59). Nevertheless, it is virtually unavoidable that reference to ellipses permeate the ensuing discussion of the firefront-perimeter literature. Because of the emphasis sometimes placed on elliptical configuration for the firefront in wildland vegetation, we note that there are parameter ranges for which the firefront is *not* even roughly elliptical, despite high regularity of the fuel distribution, virtual flatness of the terrain, and constancy of the wind magnitude and direction. In the hummock grasslands of arid Western Australia, the patchy fuel distribution admits fire spread only for winds above a finite significant threshold of 3–5 m/s or so (at 2 m height). Under a strong wind, backing fire spread is minimal, flanking fire spread is modest, and head fire spread from a localized ignition results in a downwind-widening-wedge shape for the burned clumps of vegetation (Burrows *et al.*, 1991; Gill *et al.*, 1995; Cheney and Sullivan, 1997). Because the burning of these spinifex plains, for which 30–50% of the ground may be covered by spinifex-type grass (with the remainder being bare arid soil), raises significant issues, we return to spinifex later.

Of course, in realistic scenarios, neither the wind, fuel loading, nor topography remains constant, and no useful general statement about how the configuration of the fire perimeter evolves seems likely to hold. The nonuniformities result in a typical, convoluted fire perimeter (in a plan view, Figure 3) having concavities (indentations called "pockets") and convexities (protrusions called "fingers") (Pyne, 1984); the possibilities are myriad. For example, buoyancy induces a wind that aids upslope spread and inhibits downslope spread; yet, if the ambient wind were blowing across the slope, then what might be considered a flank on the basis of slope might become a head on the basis of the forced convection. Such reorientation is also to be expected if the wind alters direction, as apparently is commonly encountered in southeastern Australia bush fires when sirocco-type winds, drawn off the hot desert interior prior to cold-front passage, are succeeded by cooler moister southwesterly winds (southern burster) off the ocean after cold-front passage (Pyne, 1991). Corresponding reorientation is less common in southern California because weather fronts pass to the north, especially in dry years. However, at the end of Santa Ana-wind-driven fire spreads in southern California, during which compressionally heated dry winds descending off the Colorado plateau accelerate through narrow passes in the southern Sierra Mountains to cause rapid propagation from the interior toward

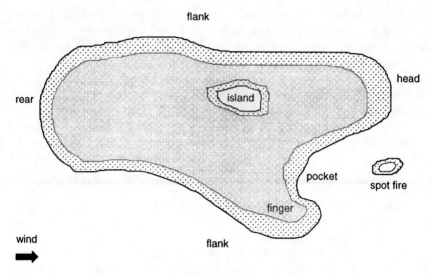

FIGURE 3 Some commonly used terminology to describe phenomena associated with wind-assisted fire spread through wildland fuel and portions of the firefront perimeter.

the coastline, the eventual return of gentler onshore marine winds may lead to a daytime reversal of the direction of spread. If so, patches of fuel (islands or "refugia") left unburned may then be consumed (Pyne, 1995). Since (as with other relevant parameters) the slope of the terrain is perennial but exceptionally rapid fire spread is dominantly strong-wind-associated, we infer that wind incrementation of spread rate can appreciably exceed slope incrementation. Accordingly, the emphasis here is on fire spread under windy conditions.

II. LABORATORY-SCALE SETUP

We regard the conducting of well-instrumented experiments to be an essential part of wildland fire modeling. In accord with Emmons (1971), we place emphasis on the control of experimental conditions to yield reproducible results. The requirements for studying fire spread in a wildland situation are appreciable: sensor distribution, maintenance, and readout; the area of surveillance and the need for periodic updating of fuel properties and quarter-hourly (if not subminute) updating of weather data; the requisite information collection, storage, processing (including interpolation and extrapolation of the input data from specific monitored sites to encompass the entire field of interest), display and distribution. Furthermore, the obvious dangers of working in the types of uncontrolled extreme wildfires of interest preclude the intentional execution of such events. Also, as mentioned previously, without adequate controls of the various variables such as fuel loading, moisture content, and wind speeds, it is difficult to test the roles of these factors. Finally, wildland fire situations are difficult to replicate. For these reasons, we have taken the approach of conducting laboratory-scale experiments to develop and test a fire spread model. We use fuel beds of regularly arranged discrete fuel elements that are well-defined, with properties that can be systematically varied.

A. TEST FACILITY

Although a test facility with a large cross-sectional area is preferable, the larger the cross section, the bigger is the blower needed to generate the wind that permits examination of flow-assisted fire spread. The cost to build and operate the facility increases with size; this can defeat the objective of extensive use of the facility for data collection for the multiparametric phenomena of interest. A list of international laboratory facilities for testing wind-aided fire spread across discrete-fuel beds is given by Pitts (1991).

We prefer a facility that blows the airstream from upwind, through the test section and over the test bed of fuel. Sucking air from the downwind side of the test section is a laboratory artifice that inhibits the entrainment of air from downwind into the hot, buoyant firefront gases (Cheney and Sullivan, 1997, pp. 80–81).

We also prefer a facility with a movable ceiling that can translate in the direction of the wind (Figure 4). Buoyant firefront gases may rise with a minimum

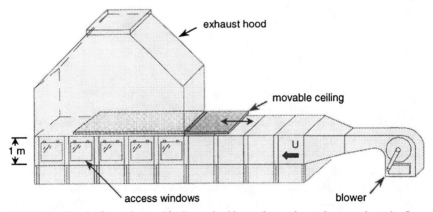

FIGURE 4 Sketch of a wind-tunnel facility with a blower that pushes ambient air through a flow-conditioning section, to produce an approximately uniform stream of low-turbulence air. The approximately 1-m-wide, 5-m-long test section has a movable ceiling that can translate in the direction of the wind.

of obstruction if the downwind edge of the ceiling is dynamically positioned just upwind of the propagating firefront. A movable ceiling maintains a constant cross-sectional area for the portion of the test section that is upwind of the propagating firefront so that the speed of the wind arriving at the firefront remains invariant with firefront advance. Mounting of a video camera at the downwind end of the ceiling furnishes feedback for positioning the ceiling during a test and also provides an excellent overview. The height for a fixed ceiling to be nonintrusive (so hot product gas is not artificially forced over the downwind, still-unburned portion of the fuel bed) is uncertain. Since the height of the visible flames during testing in our facility (to be described later) is typically a meter or two, a factor-of-ten "clearance" would require a blower capable of supplying winds in excess of, say, 5 m/s for a test section 10–20 m in height! Alternatively, if the wind from the blower is allowed to expand into a test section with a cross-sectional area much larger than that of the flow-preparation duct so that the test section is effectively without a ceiling, then the wind speed decreases along the length of the test bed. Only with a movable ceiling is a quasi steady rate of propagation in a uniform wind speed achievable, if a quasi steady rate exists for the given conditions. Data from our facility show that the rate of fire spread is significantly increased when the ceiling height is held constant at 1 m relative to results obtained for the same conditions with the movable ceiling; also, the rate of fire spread is slowed for results obtained for testing without deployment of the ceiling, so the blower-generated wind expands into a very large test section (Wolff et al., 1991, Figure 9). To our knowledge, our facility is the only one with a movable ceiling, and Table 2 indicates that fire spread rates obtained in it are less than corresponding rates measured in other facilities with fixed ceilings. We omit citation of qualitative sketches from the compilation included in Table 2.

B. INSTRUMENTATION

We concentrate exclusively on seeking the fire-propagation rate as a function of the prefire fuel-bed observables, taken as uniform in space, and of the wind, taken as constant in time. The observables include (1) the initial fuel loading, (2) the fuel height, (3) the thickness of fuel elements, (4) the fuel moisture, (5) the fuel-bed width, and (6) the wind speed.

To measure the rate of fire spread, we use type-k, chromel–alumel thermocouples inserted at regular intervals (along the centerline of the fuel bed, in the direction of the wind) from underneath so that they protrude up through the bed. The response time of the commercial thermocouples (OMEGA, Stamford, CT) is about 1 s—adequate since the typical propagation rate is observed to be below 10 cm/s.

TABLE 2   Fire Spread Rate v_f Dependence on Wind Speed U ($v_f \sim U^n$)

Researcher	n exponent, U^n	Fuel type	Scale
Thomas (1971)	0.42–0.65[a]	heather and gorse	field
Wolff et al. (1991)	0.49[b]	white pine toothpicks	lab
Fons (1946)	0.67[a]	ponderosa pine twigs	lab
Steward (1974)	0.84[a]	poplar wood shavings	lab
Steward and Tennankore (1979)	0.9–1.9[a]	birch dowels	lab
Catchpole et al. (1998)	1[b]	regular and coarse excelsior, ponderosa pine needles/sticks	lab
Gould (1991)	1[b]	kerosene/criachne grass	field
Marsden-Smedley and Catchpole (1995)	1.31[b]	buttongrass	field
Steward (1974)	1.38[a]	poplar match splints	lab
Nelson and Adkins (1988)	1.51[b]	slash-pine needles	lab
Rothermel (1972)	1.9[c]	$\sigma = 97.7 \text{ cm}^{-1}$	model (semi-empirical)
Burrows (1999a)	2.22[b]	jarrah litter	lab
Burrows (1999b)	2.67[b]	jarrah litter	field

[a]Data interpretation performed by Wolff et al. (1991).
[b]Data interpretation performed by the researchers themselves.
[c]Data interpretation performed by Gould (1991).
Note: The symbol σ denotes surface-area-to-volume ratio.

The time histories of the centerline thermocouples are used to ascertain when and where a quasi steady rate of fire spread is achieved, and what that rate is. Quasi steady propagation is achieved if the time history of a centerline thermocouple is effectively identical to those of its neighbors, with temporal offsets based on spatial separation. A typical thermocouple time history involves

1. A slow rise in temperature (interpreted as preheating);
2. A subsequent, more rapid rise (interpreted as onset of flaming); and finally
3. A gradual fall (interpreted as forced-convective cooling of the thermocouple owing to the blowing).

Thermocouples are relatively inexpensive, are easily connected to a data-acquisition system, and furnish data readily processed via automated (computer) analysis, in contrast with video or infrared-camera recordings.

Thermogravimetric analysis (TGA) of white-pine toothpicks (the basic fuel type used in our laboratory testing) indicates that 70% of the mass evolves as gas (via pyrolysis) during gradual heating from 575 to 675 K. The speed and turbulence level of the oncoming wind are determined by hot-wire anemometry. The moisture content within the fuel is determined by weighing the fuel before and after drying.

III. FIRE SPREAD MODEL

Using the previously described test facilities provides measurement of spread rate as a function of the various test parameters; the results also guide our thermo-gas-dynamic modeling of wind-aided fire spread.

A. PROPERTIES OF THE MODEL

Directly applying the conservation of mass, momentum, and energy to the fuel bed and atmosphere for tracking the progression of the firefront is very difficult. The highly disparate spatial scales, intricate geometry, and unsteady motion remain formidable obstacles. Typically, all cumulatively significant phenomena occurring on scales smaller than the computational grid scale are parameterized (phenomenologically formulated) in terms of grid-scale variables. In general, it is unknown whether a valid representation of such kind is even possible, and, even if possible, such representation is usually not known. Validation of adopted conjectures requires detailed comparison of predictions with observations in a multitude of cases, and this demanding exercise is rarely carried out.

Instead, an approximate *semiempirical* model of firefront-propagation phenomenology is sought (Beer, 1990, 1991). For such a model to be useful in practice, the inputs should require no more detail concerning the local topography, meteorology, and combustible matter than would normally be available. Our objective is to predict the position of the firefront (idealized as a mathematically thin interface) at a later time from knowledge (1) of its position at an earlier time and (2) of a *minimal* amount of (plausibly available) information about essential properties of the environment in the vicinity of the firefront. In treating the firefront as an interface, we are tentatively regarding estimation of firefront thickness and time for thin-fuel-element burnout as lower priority issues.

We concentrate here on a single firefront from a single ignition; nevertheless, the firefront, owing to the presence of firebreaks, spotting, or other circumstances, might become multiply-connected. Thus, burned-over areas should be tallied, lest repeated burning of the same fuel owing to shifting wind (or merger of two initially distinct firefronts) be erroneously permitted. Furthermore, the firefront need not be a closed curve at any time, but we deal with the closed-curve case here.

For typical southern-California-hillside scenarios, the (immobile) combustible matter consists predominantly of shrubs, such as chamise and manzanita, and small trees. There is no noteworthy organic matter in the soil and no noteworthy overstory of crowns. Thus, we focus exclusively on predicting firefront propagation through surface fuels and give a lower priority to predicting flame height. Because the time for flame to progress from one discrete fuel element to another is much less than the temporal resolution of practical interest, we may treat the distribution as an effectively continuous, equivalent mass of combustible matter per unit planform area. Accordingly, we infer the progress of the firefront to occur smoothly, not in discontinuous jumps.

Undertaking a simple approach first, we take the dependence of the firefront-propagation speed to be on the values of local parameters, not on the spatial or temporal gradients of the parameters. Nevertheless, the value of any parameter is permitted to change in space and/or time. The procedure entails a quasi steady approximation in that the flame speed holding for a given set of parameter values is instantaneously equilibrated to that set of values, whatever the antecedent set of values. That is, if we regard the instantaneous firefront as a perimeter of ignition sites, the rate of fire spread is taken instantaneously to reach the equilibrium rate of spread (into *unburned* fuel) for the local values of the parameters related to propagation.

We do not incorporate spotting, the discontinuous mode of spread associated with the lofting of firebrands. If the fuel distribution varies but moderately ahead of the current firefront position, then neglect of short-range spotting typically results in a modestly inaccurate rate of spread, since the firefront probably soon arrives at the brand-ignited site by the continuous mode of spread. Longer range spotting may permit the fire to traverse effectively fuel-free expanses that would constitute firebreaks for the continuous mode of spread. We do not address the interaction of the fire with the environment, so the nominal environmental conditions are the conditions adopted for use in the model. For example, any alteration of the wind by transit of already-burned-over expanses of fuel is taken to be a higher order effect in describing the wind arriving at the currently burning fuel. This simplification is anticipated to require modification in some circumstances. However, ignoring fire modification of the wind typically may furnish an overprediction of spread rate, so the consequences are known. We also do not explicitly incorporate the consequences of fire fighting on the firefront-propagation model (e.g., owing to air-tanker or helicopter drops of fire retardant or water on the fuel bed), except as those consequences may be formulated in terms of parameters (such as fuel-moisture or soil-moisture content), plausibly incorporated in the model for intervention-free spread. Of course, tracking the fate of an attempted backfire may be regarded as little different from tracking the firefront progress evolving from any other discrete (independent) firestart.

B. A GENERAL FORMULATION

The firefront (interface) that encloses the area with burned thin combustible matter is denoted by the "simple" smooth two-dimensional, plane-projection curve $\Gamma(s, t)$, where s denotes distance along the curve at time t (Figure 5). We choose the parametric representation such that the burned area lies to the left during a transit of the perimeter in the direction of increasing s. The direction of fire spread through the fuel bed (the movement of the interface), in the immediate vicinity of any point $x_i(s, t)$ on the interface, is perpendicular to the interface. The interface-translation speed F at each point on the interface is taken to depend on the values assigned to the state variables $u^{(j)}$ at that point on the interface at the time of interest. The number of state variables is the minimal number of properties of the fuel bed, the meteorological environment, and the topography necessary for a description (of the interface movement) that is of sufficient accuracy to meet our requirements. Thus, for $t > 0$, with subscript $i = 1$ or 2,

$$\frac{\partial x_i(s, t)}{\partial t} = F\{u^{(j)}[x_i(s, t)], v_i(s, t)\}, \qquad x_i(s, 0) \text{ given} \qquad (1)$$

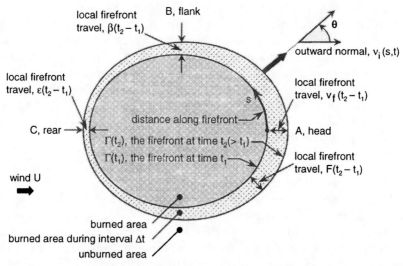

FIGURE 5   A schematic of the increase in burned area (shaded region) due to the firefront travel (stippled region) during the time interval $(t_2 - t_1)$ under a wind of magnitude U in the direction $\theta = 0$. The distance s along the firefront at time t_1 is measured from $\theta = 0$; a corresponding distance may be computed for the firefront at time t_2. The rate of spread F depends (*inter alia*) on the local unburned-thin-fuel loading and the component of wind normal to the front, so F varies with θ. The general rate F takes on the particular values v_f, β, and ε at the head, flank, and rear of the fire, respectively.

where $v_i(s, t)$ denotes the outward normal. We let $x_1(s, t)$ denote the x position at time t of the point s along the perimeter, where x is a Cartesian coordinate along the direction of the time-averaged wind at reference height above the ground; $x_2(s, t)$ denotes the corresponding y position, where the coordinate y is perpendicular to x, in accord with the right-hand convention. We regard the dependence of the functional F on $u^{(j)}$ to be known *empirically*. Only the component of the wind locally normal to the firefront is taken to aid the advance of the local fire perimeter. In terms of the functions of computer software developed by the Fire Sciences Laboratory (Missoula, MT) of the Forest Service, U.S. Department of Agriculture, for a two-dimensional simulation of fire growth through surface fuels: (1) integration of Eq. (1) corresponds to the tasks undertaken by the code FARSITE (Pyne *et al.*, 1996; Finney, 1998); and (2) the *key* functional F is furnished by the code BEHAVE (Rothermel, 1972; Andrews, 1991), which serves as a subroutine for FARSITE. Below we shall address issues concerning mathematical accuracy and real-time implementation for FARSITE, physical-formulation issues for BEHAVE, and limitations of a quasi steady, semi-empirical approach. An Australian analogue of FARSITE is *SiroFire* (Coleman and Sullivan, 1996). Were there sufficient record of the firefront position $x_i(s, t)$ during large fires, we might address the inverse approach of trying to infer the functional F from knowledge of $x_i(s, t)$, but such an undertaking is not currently practical.

This "parametric" model of firefront travel is probably not capable of accurately resolving any fine-scale detail. Thus, in the current context, we ascribe no credibility to details of behavior predicted by the model if the curvature of the boundary is significant, as would be locally the case if the length of the boundary between two points on it were comparable to the local radius of curvature. Thus, prediction of large-curvature behavior is suspect as a mathematical artifact of an approximate incorporation of the physics of the situation in the formulation. How to resolve such artifacts is perhaps best ascertained as one gains experience with the use of the model. To reiterate, the incidence of singularities in curvature [usually a consequence of negative curvature ("pockets" in Figure 3) arising on the fire perimeter] may be the "most fascinating propagation characteristics" (Sethian, 1996, p. 28) mathematically. However, their occurrence may reflect simply an incomplete incorporation of the physical laws.

C. Approaches to Solution of the Formulation

In practice, of course, a finite-difference approximation to the differential equation [Eq. (1)] is solved. A number of marker points distributed around the interface are advanced by stepping forward in time, with a smaller distance between marker points Δs and a smaller time step Δt (within limits) seemingly likely to incur smaller error because such choices permit a better approximation to the derivative. However, in this approximate formulation, a coarser resolution could conceivably furnish a prediction closer to observation. Simplistic procedures for treating the evolution equation include a first-order Euler time-stepping procedure, or a second-order, predictor-corrector procedure; the order of the numerical method also factors into the numerical error. Roberts (1989) emphasizes a distinction between (1) the advancing of the distributed marker points adopted to approximate the firefront curve at time t and (2) the advancing of line segments adopted to approximate the curve. The distinction seems somewhat blurred in that a line segment must be associated with each marker point in order to define the local normal to the curve, a necessary step to defining the local direction for advancement.

The commonly utilized computational procedure is described as "adopting Huygens' principle" (see, e.g., Anderson et al., 1982; Knight and Coleman, 1993; Richards, 1993, and the references cited therein). The procedure usually amounts to applying an elliptic stencil to describe how the fire grows at marker points and drawing a smooth curve through the most advanced points on the ellipses to obtain the updated location of the interface. The procedure seems arbitrary and not particularly convenient for convoluted perimeters and, in general, possibly highly inaccurate for the intricate nonhomogeneous situations encountered in practice. Huygens' principle is usually encountered in the context of ascertaining the evolution of an acoustic or electromagnetic wavefront. The corresponding conjecture to permit tracking the propagation of a firefront across a fuel bed is taking each point on the perimeter of a large fire to be an ignition site for a new fire, which grows with the shape and orientation of the original perimeter, the result being a larger fire of the same shape (Albini, 1984). In practice, with few exceptions, the result is a larger fire of the same shape (Albini, 1984). In practise, with few exceptions, the adopted firefront shape evolving from an ignition site is always taken to be an ellipse, with the ignition site taken as coincident with rear focus of the ellipse. The commitment to an elliptic firefront evolving from an ignition site on the firefront perimeter implies that some already-burned fuel is treated as if it may reburn—and then may burn once more, because the ellipses from neighboring ignition sites typically overlap. An ignition at the perimeter of a burned-over area is not equivalent to an ignition within an unburned area; the experience that an ignition site evolves to an oval firefront under a wind is derived from observation of spread through an entirely unburned expanse of fuel. An upshot is that, under the just-described Huygens' principle methodology for fire growth, the firefront is not always propagated normal to itself into unburned fuel without ad hoc adjustment. Finney (1998, pp. 28, 32) acknowledges that subjective spread-rate adjustment factors are introduced to bring predictions of the Huygens' principle-based FARSITE model into agreement with observations of fire growth. Adjustment is necessary because spread-rate predictions exceed the observed fire growth for all fuel models treated. It is currently unresolved and perhaps unresolvable whether the discrepancy between prediction and observation is owing to (1) shortcomings in the BEHAVE model of the spread rate for a given set of state variables (see discussion that follows), (2) adoption of the elliptical-stencil methodology to treat the firefront-evolution equation, (3) the spatial/temporal resolution of the data put into the calculation, or (4) the step size used in space and time to advance the solution—or a combination thereof. Finney (1998) suggests that finer resolution in space and time in the modeling will result in closer agreement between prediction and observation. In contrast, Cheney and Sullivan (1997) suggest that, for a heading fire, wind measured at 10 m above the ground be averaged for at least 10 minutes, preferably 15–20 minutes, for use in any currently available predictive approach, and that comparison of predictions of spread rate be made with measurements also taken over a period of 15–20 minutes because accurately accounting for shorter term variation in spread caused by gusts and lulls and spatial variations in the fuel does not seem currently achievable.

As already noted, subdividing the firefront perimeter into small line segments (each of which is then propagated in the direction of the normal to the arc into unburned fuel, much as a straightline firefront would advance in accord with the quasi steady fire spread rate for the local fire environment) may result in overlap of the propagated arcs for regions of negative curvature. A physical basis pertinent to fire spread through discrete fuel elements is needed to guide "smoothing." Sethian (1996) categorizes the tracking of a set of marker particles to approximate the tracking of a moving front as a discretized Lagrangian approach. He suggests that a volume-of-fluid technique (Chorin, 1980) is an Eulerian approach (describing where the front is in terms of a fixed spatial grid), and probably superior. Only implementation can resolve questions of accuracy and feasibility for tracking a complicated firefront geometry by use of alternate discretized forms of Eq. (1). We question arbitrarily adopting changes to the propagation-speed function F to serve mathematical convenience. Osher and Sethian (1988) and Sethian (1996) suggest (by vague analogy, as if fire propagation were basically a hyperbolic phenomenon) that the fire-propagation function F be modified to the form $(F - \varepsilon\kappa)$, where ε is a small positive constant with the units of a diffusion coefficient and κ is the local curvature of the firefront; this modified form is amenable to so-called level-set methods, which still give rise to physically implausible cusps in the firefront evolution but do avoid mathematical multivaluedness. However, the only evidence (that fire propagation is related to firefront curvature) cited by these authors is a speculation by Markstein for flame propagation in homogeneous mixtures of reactive gases. Markstein (1964, pp. 21–23) addresses instances in which the radius of flame curvature is comparable to flame-structure thickness because, under other circumstances, the role of curvature typically would be expected to be negligible (Carrier *et al.*, 1991). Markstein's speculation holds for phenomena in which classical diffusional processes in the direction of spread are essential to the mechanism of spread and in which the classical diffusional transport must be modified from planar representation to infer the correct rate of spread. Fire spread through chaparral-type surface fuels under wind aiding (diffusion-flame burning in a boundary layer of reactive vapor pyrolyzed from a condensed fuel) is not among such phenomena (see Section III.D). With respect to observations justifying the adoption of a curvature-dependent spread rate for fire propagation through surface-type fuels, Cheney and Sullivan (1997) do observe that if the head of a wind-aided grassfire remains narrow, the fire spread is slower than for a larger fire with a broad head. However, for a wind of 7 km/h, the effect of curvature on spread rate is negligible when the (say, maximum) width of the oval-configured perimeter of the wind-aided fire is 25 m; for a wind of 14 km/h, when the width is 100 m; for 21 km/h, 150 m; for very strong winds, about 200 m. There is a tendency for regions of the wind-aided-fire perimeter with large curvature to evolve to smaller curvature. Fires developing under strong hot winds increase their head-fire width quickly, and the time needed to attain the maximum rate of spread for the given fire environment (slope, loading, wind, etc.) is about 12 minutes (Cheney and Sullivan, 1997). The lesson is that significant effect of curvature on the quasi steady, one-dimensional spread rate is confined to a brief interval after ignition, and to a limited locale near the head under wind-aiding. The effect of curvature in general is insignificant for the practically interesting temporal and spatial scales of wind-aided fire spread pertinent to fire-fighting efforts. Furthermore, letting the fire-propagation function $F \to F[1 - j(\kappa)]$, where the function $j(\kappa)(< 1)$ monotonically decreases as κ decreases, may better express the limited field observations. If the wind ceased, aside from local effects at a large-curvature head, no inherent propensity toward a more circular perimeter from an established oval configuration is reported, to our knowledge; yet this is an implication of adopting the form $(F - \varepsilon\kappa)$ for vanishing F. In summary, the important geometric property derived from the flamefront configuration is the direction of the local normal vector pointing into the unburned medium. For any anomalous situation (in fire spread through a bed of fixed surface-fuel elements) in which the firefront curvature were to persist as a significant factor in firefront evolution, the simplistic formulation of Eq. (1) is probably inadequate to treat the phenomenon. If any such anomaly arises in an otherwise smooth evolution of the firefront configuration, physical judgment may well be a better basis for smoothing than any mathematical formalism, especially a formalism that creates implausible cusps and corners. With experience, an automated procedure for firefront propagation may be attained.

D. CLOSURE OF THE FORMULATION FOR FIRE
SPREAD IN CHAPARRAL-TYPE FUEL LOADING

The major challenge for implementing the semiempirical approach of Eq. (1) is to identify the dependence of the firefront-propagation-speed functional F on the environment-describing variables (for fire-spread-conducive, large-area-fire-generating conditions). We rewrite Eq. (1), which defines the firefront-perimeter velocity to be the local normal-to the-front advance rate, as ($i = 1, 2$)

$$\frac{\partial x_i}{\partial t} \equiv (V_f)_i = F(U, \theta; m, \ldots)v_i, \qquad x_i(\theta, 0) \text{ given} \qquad (2)$$

where U denotes the (time-averaged) magnitude of the wind velocity; θ denotes the angle, measured counterclockwise, from the (time-averaged) direction of the ambient wind to the local outward normal to the firefront perimeter, v_i; m is the mass loading of thin fuel burned with firefront passage; and other parameters independent of U and θ may enter the argument of F. For simplicity, we deal here with a homogeneous fuel distribution with respect to type, thickness, height, and the like with a single simple ignition site, and with flat terrain, so that a value of θ uniquely defines a point on the firefront perimeter (Figure 6). Hence, we can dispense with the introduction of the parameter s. An interesting question is whether, under Eq. (2) and for a finite wind U, there exist translating, dilating, but geometrically self-similar firefront configurations. To answer this question, we should not write any approximation to Eq. (2) which involves F depending on anything other than a scalar function of U and θ. Discussion is limited to a homogeneous fuel bed of fixed properties and to no slope.

We now composite a very limited number of results for special values of θ, to constitute a smooth functional F applicable for all values of θ. We seek to formulate F so that, for a given fuel bed, the results from Eq. (2) agree locally with one-dimensional firefront-propagation rates known to hold for that fuel bed in each of three directions: the downwind direction ($\theta = 0$), the upwind direction ($\theta = \pi$), and the cross-wind direction [($\theta = (\pi/2)$ and $(3\pi/2)$]. In fact, these "one-dimensional" rates are accessible from experiment (if executed) or modeling (if credible) but are currently at least partly unavailable. We now undertake to make plausible statements about spread in each of these three directions.

The peak rate of spread is in the downwind direction ($\theta = 0$) (i.e., near site A in Figure 5). Accurately quantifying that rate is typically regarded to have highest priority in characterizing the functional F. In the downwind direction, conductive–convective heat transfer is the *dominant* mechanism that leads to fire spread; turbulent eddy transfer probably plays no important role in that part of the firefront perimeter. According to laboratory testing for *head-on* wind-

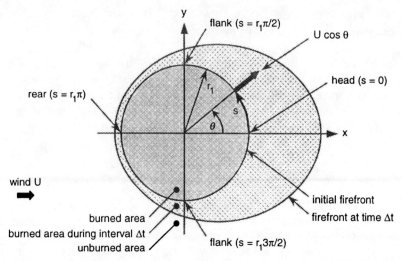

FIGURE 6 A particular example of the phenomena of Fig. 5, in which the initial perimeter of the burned area (at a time taken to be $t = 0$) is a circle of radius r_1. A wind of magnitude U arises at time $t = 0+$ and results in a head (where spread is with the wind), flanks (where spread is across the wind), and rear (where spread is against the wind). The fire spread during a time interval Δt (stippled area) results in the burned-area perimeter sketched at time Δt. The x axis, the primary ray for the angle θ, is taken to lie along the direction of the wind. If the wind changes direction to angle θ_1 with respect to the x axis, the revised Cartesian coordinates (x_1^*, x_2^*) of the point on the fire perimeter heretofore denoted (x_1, x_2) are identified by relations describing a simple rotational transformation of coordinates through angle θ_1; the x^* axis now is aligned with the wind, serves as the prime ray of the angle θ^* used in place of θ, and serves as the datum for the parametric distance s^* used in place of s.

aided fire spread across a dry uniform bed on flat terrain, $F(U, 0) \sim U^{1/2}/m^{1/2}$, where U is the speed of the head-on wind and m is (recalled to be) the dry thin-fuel loading (per unit planform area) burned during firefront passage (Wolff *et al.*, 1991); this result is shown to be consistent with the just-discussed mechanism for spread-sustaining heat transfer. Henceforth, we designate the functional $F(U, 0)$ by the symbol v_f. The expression ultimately must be modified because clearly there is a minimum loading for which fire spread does not occur, and clearly there may be a much reduced but perhaps still finite rate of spread as $U \to 0$. Nevertheless, the result is pertinent near site A. In Figure 5, we have taken the wind speed U to be constant in magnitude and direction over the spatial domain sketched, during the time interval $t_1 < t < t_2$, in accord with our simplified approach; we defer comment on the time averaging that would be required for so treating the wind encountered in a fire scenario.

At the flanks of the fire [$\theta = (\pi/2)$ or $(3\pi/2)$, e.g., near site B in Figure 5], the local normal component of the wind is zero, but the speed of propagation is significantly larger than the speed for propagation under quiescent conditions. The firefront propagation is supported by (a) turbulent eddy transfer (lateral gusts) and possibly (b) the same mechanism that permits upwind transfer ($\theta = \pi$) (i.e., buoyancy-induced influx of fresh air down in the fuel bed) and somewhat localized radiative transfer among fuel elements in that bed. Much of the combustion may be of smoldering type. Cheney and Sullivan (1997) suggest that burning at the flanks alternates between heading fire and backing fire. Henceforth, we designate $F(U, |\pi/2|)$ by the symbol $\beta(U)$.

At site C in Figure 5 (i.e., at $\theta = \pi$), there is fire spread against the wind. This upwind propagation is attributed to mechanism (b) discussed in the last paragraph. The upwind propagation is here taken to be dependent on U since the airflow through the fuel bed competes with mechanism (b). For a fuel bed composed of discrete combustible fuel elements, Steward (1974, pp. 342, 372) suggests that the rate of fire spread directly against a wind U involves a balance of (1) radiative preheating and (2) convective cooling and finite-reaction-rate effects. Henceforth, we designate $F(U, \pi)$ by the symbol $\varepsilon(U)$. As a composite for F motivated by the discussion of phenomena at the three particular directions, we tentatively suggest ($\mu \approx 2-4$, as a first try):

$$F(U, \theta) = v_f(U \cos \theta) + \beta(U \sin^\mu \theta) \qquad \text{in } |\theta| < (\pi/2) \qquad (3a)$$

$$F(U, \theta) = \beta(U \sin^\mu \theta) + \varepsilon(U \cos^2 \theta) \qquad \text{in } |\theta| > (\pi/2) \qquad (3b)$$

We expect, except possibly at very low values of U, that

$$\varepsilon(U) < \beta(U) < v_f(U) \qquad (4)$$

In words, we expect the value of the spread rate F holding at the head ($\theta = 0$) to exceed the value of the spread rate holding at the flank [$\theta = (\pi/2)$ or $(3\pi/2)$], which, in turn, exceeds the value of the spread rate holding at the back ($\theta = \pi$).

At any given time, the locus of the firefront is given by

$$\left(\frac{\partial x_1}{\partial x_2} \right)_{t \, const.} = -\cot\left(\frac{\pi}{2} - \theta \right) = -\tan \theta \qquad (5)$$

This equation is derived from trigonometric relations holding for the *tangent* to the firefront.

The firefront-tracking problem is obtained from trigonometric relations holding for the normal to the firefront, specifically,

$$\frac{\partial x_1}{\partial t} = F \cos \theta \qquad (6)$$

$$\frac{\partial x_2}{\partial t} = F \sin \theta \qquad (7)$$

with F given by Eq. (3), together with the constraint given by Eq. (5) and initial conditions giving the firefront locus at time $t = 0$: $x_1(\theta, 0)$, $x_2(\theta, 0)$. Numerically, Eqs. (6) and (7) may be used to advance the firefront perimeter over a brief interval of time Δt. From use of $x_1[\theta,(0), (\Delta t)]$ and $x_2[\theta(0), (\Delta t)]$ in Eq. (5), we obtain $\theta\{\Delta t; x_1[\theta(0), (\Delta t)], x_2[\theta(0), (\Delta t)]\}$. The cycle is then repeated: we obtain $x_1[\theta(\Delta t), 2(\Delta t)], x_2[\theta(\Delta t), 2(\Delta t)]$ from Eqs. (6) and (7), and then $\theta\{2(\Delta t); x_1[\theta(\Delta t), 2(\Delta t)], x_2[\theta(\Delta t), 2(\Delta t)]\}$ from Eq. (5). Of course, more sophisticated, iterative schemes for advancing the solution in time may also be used. A smooth interpolation for θ is needed, especially since the number of points on the firefront perimeter used to track the perimeter needs to increase as the perimeter becomes longer.

For explicitness, a possible trial form, consistent with Eqs. (3a) and (3b), is given by (for $\mu = 2$)

$$F(U, \theta) = \{\varepsilon_0 \cos^2 \theta + c_1(U \cos \theta)^{1/2}\} \\ + \{\varepsilon_0 \sin^2 \theta + aU \sin^2 \theta \exp[-bU \sin^2 \theta]\}, \qquad |\theta| < (\pi/2) \qquad (8)$$

$$F(U, \theta) = \{\varepsilon_0 \sin^2 \theta + aU \sin^2 \theta \exp[-bU \sin^2 \theta]\} \\ + \{\varepsilon_0 \cos^2 \theta \exp[-\varepsilon_1 U \cos^2 \theta]\}, \qquad |\theta| > (\pi/2) \qquad (9)$$

This form for $F(U, \theta)$ implies that we have adopted the trial forms for the windspeed behavior at the head, flanks, and back to be

$$v_f(U) = \varepsilon_0 + c_1 U^{1/2}, \qquad \beta(U) = \varepsilon_0 + aU \exp(-bU),$$
$$\varepsilon(U) = \varepsilon_0 \exp(-\varepsilon_1 U) \qquad (10)$$

respectively. The parameters c_1, ε_0, ε_1, a, and b are independent of U and θ, but in general they depend on m, the mass loading of fuel, and other parameters characterizing the fuel bed. The parameters are anticipated to be so chosen so that, except for $U \rightarrow 0$, the ordering is $v_f(U) > \beta(U) > \varepsilon(U)$, in accord with Eq. (4). Marsden-Smedley and Catchpole (1995) characterize the spread rate at the flanks and back to be 40 and 10%, respectively, of the spread rate observed at the head of fires in Tasmanian buttongrass moorlands. Incidentally, Eqs. (8) and (9) give $F = \varepsilon_0$ at all θ for $U \rightarrow 0$: a circular perimeter remains circular as it expands through a uniform bed on flat terrain in the absence of wind. The form for v_f stems from the previously mentioned experiments by Wolff et al. (1991), although $\varepsilon_0 = 0$ in those experiments; the form for β, pertinent to the flanks, takes the variation of spread rate with wind speed to be nonmonotonic; the form for ε, pertinent to the back of the fire, takes the spread rate to be monotonically decreasing with increasing wind speed.

In Eq. (10), suppose we have a fuel bed for which: (1) $\varepsilon_0 = a = 0$ [no spread in the absence of wind, or at the flanks even with wind (Burrows et al., 1991)], and (2) $v_f(U) = c_1 U^{1/2} H(U - U_{crit})$, where $H(x)$ is the Heaviside unit step function and U_{crit} denotes a finite minimal wind for the onset of spread under wind aiding (addressed in Section III.F). Then $F(U, \theta) = 0$ for $U < U_{crit}$; for $U > U_{crit}$,

$$F(U, \theta) = c_1(U \cos \theta)^{1/2} \quad \text{if } |\theta| < \cos^{-1}(U_{crit}/U)$$

$$\text{and } F(U, \theta) = 0 \quad \text{otherwise}$$

(11)

According to the model, for a localized, pointlike ignition, we obtain a burned region confined within a wedge, with vertex at the ignition site, and with each arm of the wedge inclined at an angle $\cos^{-1}(U_{crit}/U)$ from the downwind direction. The burned region is only very roughly in the configuration of a sector of a circle because the firefront advance along the bisector of the wedge (i.e., in the direction of the wind) exceeds the firefront advance near the arms of the wedge. Experimental testing is needed to check this conjecture concerning the relation between (1) the finite minimal wind speed for flame spread onset after line ignition perpendicular to the wind direction, and (2) the configuration of the burn pattern emanating from the ignition of a single element in an equivalent fuel bed.

If we now generalize to a case in which the wind direction may vary in space and time, then we return to the use of the parametric length s of the firefront. At any time, the length s is measured from the initial ray $\theta = 0$, which is taken to be aligned with the wind direction at the moment of time under discussion. For a circular perimeter of radius r_1 at time $t = 0$ (Figure 6), with the x coordinate fixing the origin of the firefront length parameter s, the initial firefront is identified by

$$x_1(s, 0) = r_1 \cos \theta(s, 0) = r_1 \cos(s/r_1), \quad x_2(s, 0) = r_1 \sin(s/r_1). \quad (12)$$

Thenceforth, from Eq. (1),

$$\frac{\partial x(s, t)}{\partial t} = [\cos \theta(s, t)]F, \quad \frac{\partial y(s, t)}{\partial t} = [\sin \theta(s, t)]F \quad (13)$$

$$\tan \theta(s, t) = -\frac{[\partial x(s, t)/\partial s]_t}{[\partial y(s, t)/\partial s]_t} \quad (14)$$

where the subscript t signifies that time t is held constant in taking the derivative with respect to the length parameter s. This particular simple scenario would arise, for example, if there were an onset, at what we designate as time $t = 0$, of a wind U in what previously had been a point-ignited spread through uniform fuel on flat terrain under nil wind.

E. Coordination of an Experimental Program

The calibration/validation of the model might be initiated against historic data, but we suspect that any such data set is too incomplete for highly informative comparison of prediction with observation. Probably, comparison with historical fires ought to be complemented with comparisons with field tests involving prior inventorying and instrumentation. These field tests need not be carried out exclusively on the very large spatial scale that may be of ultimate interest; the algorithm may be usefully tested against relatively modest-scale fires under spread-aiding conditions. Indeed, once the model gains a modicum of credibility because its predictions compare favorably with observations, the model might be advantageously used in test planning. However, it seems doubtful that field testing would ever be approved for the very high fire-danger conditions for which a validated model seems most needed. Fire can propagate through chaparral at speeds up to about 3 m/s, with intensity of 10 MW/m of perimeter (Albini, 1984), whereas the typical upper limit for prescribed burning is roughly 0.5 MW/m of perimeter (Luke and McArthur, 1978).

Because the model is semiempirical in that information on the functional dependence of the local firefront-propagation speed on the key local state variables is taken to be input, extensive experimentation is required for model development, even prior to model calibration/validation. We have emphasized the importance of experimental results pertinent to the rate of fire spread at the head of a wind-aided fire, v_f, in formulating the more general expression pertinent to the rate of advance of the entire, often-closed firefront perimeter, F. To obtain an expression for v_f, we emphasize the suitability of quasi one-dimensional fire-spread experiments in which a quasi steady rate of spread is achieved (and measured) under *uniform* (and measured) conditions of slope, spread-aiding wind, and fuel-bed properties. In this paragraph, we address some considerations in carrying out such experiments. First, in the firefront frame of reference, a quasi steady rate of spread implies that a flaming zone of finite depth is propagating at constant speed, through a bed of sufficient length in the direction of propagation; a burned-over portion of fuel bed lies upwind of the currently burning zone, and a yet-to-be-burned portion of fuel bed lies downwind. At the very least, in the direction of firefront propagation, the upwind and downwind portions of the bed should be half as thick as the burning zone. The length of run to attain quasi steady spread increases with wind speed. Results in which the streamwise expanse of the flaming zone (i.e., the flame depth) increases with the length of travel (Anderson and Rothermel, 1965) may not have attained a quasi steady rate of spread. Also, departure from one-dimensionality is inevitably encountered at the lateral edges of the fuel bed, since downwind elements are preheated from two sides in the core of the bed, but from only one side near the lateral edges of the layer. Thus, ideally, fuel beds of several different widths should be tested to establish that the spread rate taken to hold near the streamwise centerline of the fuel bed is a plausible approximation to that holding for an indefinitely wide bed (see Section IV.A). Such bed experiments are most conveniently initiated by simultaneously igniting the entire upwind edge of the fuel bed and tracking the continued progress (if any) of the firefront after the starting transient dies out and a constant rate of propagation is achieved. Further, it is preferable that experiments to obtain the rate of fire spread be conducted on the scale of physical interest. However, we repeat for emphasis: control to maintain constancy of the properties of the wind, fuel bed, and topography becomes more problematic in the field, and requirements on the amount and robustness of instrumentation increase. Because of limited control of conditions in large-scale experiments, reproduction of results is difficult to attempt, and even harder to achieve. Yet reproducibility seems essential for confidence that the key experimental parameters are clearly identified, and the observations of results are sufficiently accurate. Inevitably, the cost, difficulty, and turnaround time in experimentation increases as the scale increases, especially for spread under severe conditions; as a consequence, the desired fire spread data are most accessible on smaller scale (Carrier *et al.*, 1991; Wolff *et al.*, 1991; Weise, 1993; Catchpole *et al.*, 1998). The more limited opportunity for testing on larger scale is seemingly most advantageously used to confirm results already inferred from the smaller scale test data, or to modify the guidance drawn from the smaller scale test data owing to the role of phenomena (such as radiative transfer of heat, or even stratification of the atmosphere) that tend not to contribute prominently in smaller scale testing.

Though the rate of spread is the primary output for the intended application to flow-assisted fire spread, other quantities, such as suitably defined "flame height" and "flame tilt," may also be tabulated as functions of parameters characterizing key *prefire* properties of the environment (previously categorized as fuel-bed properties, meteorology, and topography). However, we do not emphasize relations among the spread rate, flame height, and flame tilt (e.g., Mongia *et al.*, 1998); such relations describe one unknown quantity in terms of another unknown and, by themselves, make a limited contribution. We do ascribe a significant role to analysis in assisting the selection of parameter values to be tested, and in the interpretation of data, especially if the melding of data taken on different scales to account for different contributing phenomena is entailed. For example, by modeling, we may be able to suggest tentatively, for validation or rejection by "one-dimensional fire spread testing," how to modify previously obtained, semiempirical results for the rate of spread on flat terrain, to incorporate the effect of a uniform positive (fire spread-assisting) slope in the direction of propagation. At the firefront, if we plot the vertical profile of the enthalpy (in excess of the ambient enthalpy) for the gas phase above the fuel bed, we may identify a characteristic height H for the profile. A positive slope of the terrain, δ, may imply a smaller value for H and, therefore, an enhanced characteristic heat-transfer rate from the hot gas to the portion of the fuel bed undergoing preheating to the pyrolysis-onset temperature (Fons, 1946). In fact, the enhanced heat-transfer rate owing to slope may be comparable to that for a higher spread-aiding wind speed U for a flat terrain with the same fuel bed. A simple theoretical analysis might give a strong indication in what functional form (whether an additive term or a multiplicative factor), involving which given parameters (in addition to δ, if any), to seek to summarize the data. Such a procedure ought to be more efficient than collecting data systematically for a large number of environment-describing parameters and then seeking formal correlations. On the other hand, we regard rapid fire spread modeling without close ties to testing to have little credibility.

Attempts to bypass this demanding agenda and to predict fire-growth rate across finite-element "parcels" by adopting conjectures about spread-rate "probabilities" without the aid of an extensive data base [as seen in the application of percolation theory to fire spread (Christensen et al., 1993)] seem equivalent to guessing intuitively. One might just as well dispense with the probabilistic formalism and directly guess the firefront positions at future times. Without a basis for assigning probability densities, the probabilistic approach seems an arid formalism, and one might as well simply guess the answer immediately. Validation seems especially crucial for such an approach. Furthermore, "tuning" such an approach with data seems likely to be a highly inefficient utilization of the data, relative to the utilization of the data for upgrading more physically based approaches.

F. Contribution of Modeling to Data Correlation

A plethora of functional forms have been proposed for how rapidly the rate of fire spread v_f increases with the ambient wind speed U (Table 2); the literature contains even more perplexing results concerning whether the spread rate increases or decreases with increased thin-fuel loading (e.g., Cheney, 1981; Weber, 1991; Weber and de Mestre, 1991) (Table 3). For guidance in the interpretation of our own laboratory data for wind-aided fire spread across well-defined arrays of thin wooden fuel elements (Figure 7), we undertake a simple model and add complication only as a simpler theory proves inadequate to describe the observations. As a first trial, evolved from a suggestion by Taylor (1961), we take the rate of wind-aided spread to be that consistent with the known rate of entrainment of gas into the weakly buoyant, two-dimensional plume above a line source of heat (without associated release of mass or mo-

TABLE 3   Fire Spread Rate v_f Dependence on Initial Fuel Loading m ($v_f \sim m^p$)

Researcher	p exponent, m^p	Fuel type	Scale
Wolff et al. (1991)	−0.49[a]	white pine toothpicks	lab
Cheney and Sullivan (1979)	0[a]	grass	field
Rothermel (1972)	0[b]	semi-empirical	lab; model
Burrows (1999a, 1999b)	0[a]	jarrah litter	lab; field
Marsden-Smedley and Catchpole (1995)	0[a]	buttongrass	field
Luke and McArthur (1991)	1[a]	grass	field

[a]Data interpretation performed by the researchers themselves.
[b]Data interpretation performed by Gould (1991).

mentum). If we denote the strength of the line source (or fire intensity) by the symbol $\dot{Q}$, with units of heat per length (of line source) per time, then the proposal is that the speed of the cross-wind U and the characteristic speed of the buoyant updraft in the plume, W, are related by

$$U = 2\alpha W = 2\alpha[g\dot{Q}/(\alpha \rho_\infty c_p T_\infty)]^{1/3}$$

(see, e.g., Fleeter et al., 1984). Here α denotes the dimensionless, empirical entrainment constant introduced by Taylor [$\alpha = O(0.1)$ for weakly buoyant

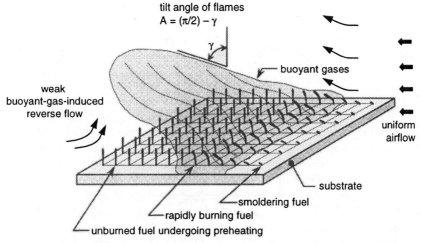

tilt angle of flames
$A = (\pi/2) - \gamma$

γ

buoyant gases

weak buoyant-gas-induced reverse flow

uniform airflow

substrate
smoldering fuel
rapidly burning fuel
unburned fuel undergoing preheating

FIGURE 7   A schematic of the phenomenology of wind-aided fire spread across a bed of identical, upright, regularly arranged, thin, wooden fuel elements, supported in holes in an inert substrate. Artificial constraints are placed on the influx of air from the downwind side of the flamefront if, in the laboratory, wind is generated by drawing from downstream, rather than by blowing from upstream.

plumes, but α might be taken somewhat larger in order to apply the model to the highly entraining base of the plume]; g denotes gravity; c_p, the characteristic specific heat capacity at constant pressure of the gas (mainly air) in the plume; ρ_∞, the ambient density near ground level; and T_∞, the ambient temperature near ground level. If 2α is almost of order unity, we expect that, for $W > U$, a fairly vertical plume arises. However, for $U > W$, the conditions for our model to hold are not present, and we expect the plume to be more blown over. The flames then are inclined from vertical and typically extend downwind over the still-unburned portion of the fuel bed. The significance of the ratio $U^3/(50\dot{Q})$ (where U is in m/s and $\dot{Q}$ is in MW/m, and we have assigned the value $\alpha \approx 0.4$, as well as plausible values to the other factors) has been discussed for over a half century in the literature on local discharges in a cross-flow. If we take the speed of propagation of a line fire to be that associated with a steadily translating line source of heat, we write $\dot{Q} = Qmv_f$, where Q denotes the effective chemical heat per mass of fuel burned, m denotes the mass (per unit planform area) of thin fuel burned during firefront passage, and v_f denotes the speed of fire spread. (For a given intensity $\dot{Q}$ of a propagating line fire, the quantities v_f and mQ are inversely related. However, the heat release rate per length of fireline, $\dot{Q}$, itself depends on fuel-preheating considerations, and thus possibly on mQ.) By substitution for $\dot{Q}$ in the preceding expression for U, after rearrangement, $v_f \sim U^3/m$—a result well at odds with our laboratory results (Wolff *et al.*, 1991). We suspect that this first-try model is inadequate because a model based on weakly buoyant representation of processes in an above-fire plume omits the diffusive processes occurring in the flaming gas-phase boundary layer and in the fuel bed. However, for those previously discussed fuel beds for which fire spread is sustained only for wind speed above some finite threshold value U_{crit} (see the end of Section III.D), we may be able to use the plume-based model to estimate U_{crit}. It *may* be that

$$U_{crit} \approx 2\alpha[g\dot{Q}_{crit}/(\alpha\rho_\infty c_p T_\infty)]^{1/3}$$

In turn, we identify $\dot{Q}_{crit}$ by requiring that the flame length of a blown-over plume be of sufficient length to span the vegetation-bare ground between a burning hummock and its downwind neighbor, and thus to achieve contact ignition of the downwind neighbor. Typically, if the center-to-center distance between neighboring hummocks is denoted s, where $s \gg d$ (d being the diameter of a hummock), then, empirically, the flame length $L = p\dot{Q}^q$, where, for L in meters and $\dot{Q}$ in kW/m, $p \approx 0.15$ and $q \approx 0.4$ are typical values (e.g., Marsden-Smedley and Catchpole, 1995). Upon setting $L = s$ to identify $\dot{Q}_{crit}$, $U_{crit} \sim s^{5/6}$. Although the accuracy of the exponent is in doubt, U_{crit} increases as the distance between hummocks increases (Burrows *et al.*, 1991) because the updraft over a fire of greater intensity must be tilted to effect flame-contact ignition of the downwind neighbor. For small separation s, a fire of weak intensity suffices to ignite the neighbor, and U_{crit} is small and may be equal to even zero: molecular heat-transfer mechanisms then may permit the flames temporarily supported by a burning element to ignite the neighbor, but such circumstances lie outside the focus of this discussion.

To relate the spread rate more accurately to wind and loading, in our next try we ignore gravitational effects. We develop a model in which the oncoming wind, heated by the gas-phase-diffusion-flame burning just above the already-pyrolyzing portion of the fuel bed, continues downwind to heat fresh unburned fuel from ambient temperature to the pyrolysis-onset temperature T_{pyr} to sustain the firefront propagation (Carrier *et al.*, 1991).

In the frame of reference of a steadily propagating firefront, we take the origin of coordinates to lie at the furthest-downwind site at which the fuel-bed surface is at the (known) pyrolysis temperature T_{pyr} (Figure 8). In such a frame, the fuel bed is translating upwind (in what we take to be the negative x direction) at a constant speed v_f. The ambient wind is flowing in the positive x direction at the constant known speed $(U - v_f) \approx U$, for cases of interest. We anticipate tentatively that the thin-fuel loading can be characterized adequately for present purposes by the overall quantity m, recalled to be the mass of thin combustible matter (per unit planform area of the bed) consumed with firefront passage. The fuel is sufficiently dry that moisture content is not a key consid-

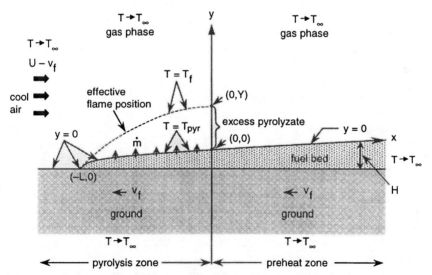

FIGURE 8   A simplistic idealization of wind-aided fire spread across a bed of discrete fuel elements, for the purpose of modeling a quasi steady spread at the rate v_f, typically (for the conditions of interest here) much slower than the speed $(U - v_f)$ of the ambient temperature oncoming air stream. The depiction is in the frame of reference of the firefront.

eration, and we tentatively ignore heat loss to inert content of the bed. Insofar as the portion of the thin fuel loading that burns with firefront passage is unknown prior to passage, we violate our dictum to relate v_f to parameters known prior to the fire passage. However, by using the total presence for the consumed loading m, we obtain a lower bound on the rate v_f (see discussion that follows). Furthermore, fires in grass and small shrub usually remove virtually all of the vegetation mass in the flaming combustion zone at the fire perimeter, and there is little if any smoldering residue after the flaming zone has passed, according to Albini (1992). More specifically, Minnich (1998, pp. 139, 147) characterizes chaparral fires as stand-replacement burns that denude most above-ground canopy owing to the high continuity and thinness of the fine fuels. There is scarce carryover fine fuels through burn cycles. On the other hand, Cheney and Sullivan (1997) suggest that head fires in Australian grasslands are somewhat less efficient and leave much of the lower fuel bed unburned. This is in marked contrast with prairie fires in the American Midwest, which typically leave very little unless the fuel is on the ground, and often not even then (P. Zedler, private communication).

We take all the heat derivable by combustion of the pyrolyzate (fuel vapor evolved from the polymeric loading m) to be released over the pyrolyzing surface of the fuel bed. "The flames of a wind-driven fire are more erect and often much taller at the front (i.e., head) than elsewhere on the perimeter" (Albini, 1992, p. 43). We ignore the fact that some of the combustible fuel vapor is not burned, or is burned (with ambient oxygen) downwind of the pyrolyzate-front position $x = 0$. Then, the conservation of energy per time per unit depth (perpendicular to the plane of Figure 9) is given by

$$v_f Q m = \rho_0 c_p T_f U Y \tag{15}$$

where $v_f Q m$ is recalled to be the intensity; Q denotes the net heat released by combustion per mass of vegetation; ρ_0 denotes the density of gas near the flame; T_f, the adiabatic flame temperature of the pyrolyzate/air diffusion flame; c_p, the specific heat capacity (at constant pressure) of the gas near the flame; and Y, the stand-off distance (at the pyrolysis-front position $x = 0$) from the surface of the fuel bed ($y = 0$) of the peak gas-phase temperature T_f. The value Y is an average, since Y varies on the integral scale of the turbulence. In Eq. (15) we are equating (1) the heat content per depth per time entering the gas phase from the pyrolyzing portion of the fuel bed and (2) the heat content (above ambient) per depth per time of the gas stream crossing the pyrolysis-front plane $x = 0$. The datum throughout this analysis is the ambient temperature, taken to be the same for the air and the bed, for convenience. We are ignoring any heating of the oncoming air stream (by a warmed substratum) upwind of the fuel-bed-burnout site, just as we are ignoring any gas-phase velocity-boundary-layer formation upwind of that burnout site.

The downward heat flux (in energy per area per time) from the gas phase

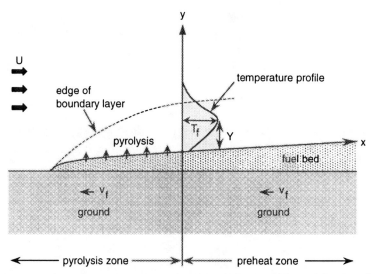

FIGURE 9   A schematic (supplementary to Fig. 8) in which it is emphasized that the diffusion-flame burning of pyrolyzate with ambient air occurs within the forced-convection boundary layer of the adopted model. The characteristic height of the maximum temperature T_f at the onset-of-pyrolysis front $x = 0$ is noted, along with a rough conjecture of the entire gas phase temperature profile at that streamwise position; the temperature approaches its ambient value at the edge of the boundary layer, and the onset-of-pyrolysis temperature at the fuel bed.

to the fuel bed over the preheating zone $\infty > x > 0$ (Figures 7 and 8) is expressed by

$$q = \frac{kT_f}{Y} f\left(\frac{x}{Y}\right) \tag{16}$$

where k denotes the thermal conductivity of the gas and the dimensionless function $f(x/Y) \to 0$ as its argument increases. This form implies that a convective-diffusive mechanism effects the preheating of fresh fuel from ambient temperature to pyrolysis temperature T_{pyr}. For later reference, were radiation (de Ris, 1979) the mechanism for preheating of fresh downwind fuel, if ε denotes the absorption coefficient of the hot gas and σ, the Stefan-Boltzmann constant, then, with Y again denoting the thickness of the hot layer,

$$q = Y\varepsilon\sigma T_f^4 g\left(\frac{x}{Y}\right) \tag{17}$$

where the dimensionless function $g(x/Y) \to 0$ as its argument increases. This behavior of $g(x/Y)$ is probably best envisioned for Y fixed and x becoming large.

For the heat balance within the fuel bed for the preheating zone, it is convenient (for this paragraph only) to reverse the sense of the coordinate axes, such that $\tilde{x}$ is positive upwind and $\tilde{y}$ is positive downward into the fuel bed;

$\tilde{x} = 0$ corresponds to that x at which the value of $f(x/Y)$ is effectively nil. Then, if subscripts $\tilde{y}$ and $\tilde{\xi}(\equiv \tilde{x}/Y)$ denote partial derivatives, and κ_b denotes the bulk thermal diffusivity of the fuel bed, the temperature (above ambient) in the preheating zone in the fuel bed is described by

$$\kappa_b T_{\tilde{y}\tilde{y}} - \frac{v_f}{Y} T_{\tilde{\xi}} = 0 \qquad (18)$$

Any flow-associated transport of heat within the fuel bed is ignored. The Laplace-transform pair is recalled to be

$$\bar{h}(s) = \int_0^\infty [\exp(-s\tilde{\xi})] h(\tilde{\xi}) \, d\tilde{\xi},$$

$$h(\tilde{\xi}) = \frac{1}{2\pi i} \int_{-i\infty+\beta}^{i\infty+\beta} [\exp(s\tilde{\xi})] \bar{h}(s) \, ds \qquad (19)$$

where the real number β is chosen to be sufficiently large that all singularities lie to the left of $\mathrm{Re}(s) = \beta$ in the complex s plane. [$\mathrm{Re}(s)$ denotes the real part of the complex variable s.] Formally applying the transform to the temperature-field equation and then solving under the condition of boundedness (for a fuel bed approximated, for this manipulation only, to be of semiinfinite depth, for simplicity of expression) yields, if $A(s)$ denotes a function of integration to be identified,

$$\bar{T}(\tilde{y}, s) = A(s)\exp\left[-\left(\frac{v_f s}{\kappa_b Y} \right)^{1/2} \tilde{y} \right] \qquad (20)$$

When q is expressed as a function of $\tilde{x}$, it is denoted $\tilde{q}$; when f is expressed as a function of $\tilde{x}/Y$, it is denoted $\tilde{f}$. Application of the boundary condition for convective-diffusive preheating gives, after use of the Laplace transform,

$$\bar{\tilde{q}}(s) = \frac{kT_f}{Y} \bar{\tilde{f}}(s) = -k_b \bar{T}_{\tilde{y}}(0, s) = k_b \left(\frac{v_f s}{\kappa_b Y} \right)^{1/2} A(s) \qquad (21)$$

Therefore,

$$\bar{T}(0, s) = A(s) = \frac{k}{k_b} T_f \left(\frac{\kappa_b}{v_f Y} \right)^{1/2} \frac{\bar{\tilde{f}}(s)}{s^{1/2}} \qquad (22)$$

At the pyrolysis front, $\tilde{\xi} = \tilde{\xi}_{pyr}$ at the interface $\tilde{y} = 0$, upon inversion of the transform,

$$T_{pyr} = \frac{k}{k_b} T_f \left(\frac{\kappa_b}{v_f Y} \right)^{1/2} N, \quad \text{where } N \equiv \frac{1}{2\pi i} \int_{-i\infty+\beta}^{i\infty+\beta} \frac{\bar{\tilde{f}}(s)}{s^{1/2}} \exp(s\tilde{\xi}_{pyr}) \, ds \qquad (23)$$

that is, N is just a dimensionless positive real number whose value depends on details (of the heat-transfer profile) that we do not specify (because we believe that such matters will remain accessible only by experiment for a long time into the future). If Eq. (16) is solved for Y and substituted in Eq. (23), we obtain, if $\kappa_b \equiv k_b/(\rho_b c_b)$,

$$\frac{v_f}{U} = N \left(\frac{k\rho_o c_p}{k_b \rho_b c_b} \right)^{1/2} \left(\frac{T_f}{T_{pyr}} \right) \left(\frac{kT_f}{QmU} \right)^{1/2} \tag{24}$$

The dependence on the square root of the ratio of the so-called conductance for the gas phase to that for the "solid phase" is conventional in such phenomena; we regard the product $k_b \rho_b c_b$ as a *composite property* of the bed and do not regard it as appropriate to attempt an approximate evaluation of any one factor in terms of other quantities that have been introduced. The appearance of the product Qm indicates that it is the net obtainable exothermicity (per unit planform area of the fuel bed) that is significant for the spread rate.

The presentation of Eq. (24) is nondimensional, but in practice no factor other than m and U is easily varied appreciably (Albini, 1992). The relation $v_f \sim (U/m)^{1/2}$ can hold *over only a limited range of parametric values*, since we expect that (1) other processes permit finite (if slow) rates of spread in the absence of wind, or against the wind; (2) sufficiently high wind could result in forced-convective extinction of thin fuel elements; and (3) nonpropagation may occur for either sufficiently sparse loading (because an upwind element may burn out without preheating its downwind neighbor to its onset-of-pyrolysis temperature) or sufficiently dense loading (because of oxygen deprivation within the bed, or radiative heat loss from thick fuel elements). In short, many finite-rate processes not accounted for in the foregoing continuum-like treatment of the fuel bed may enter, so only experiment can confirm for what (if any) thin-element loading and ambient wind speed this model holds. Of course, we would not have included this convective/diffusive-preheating development without experimental verification (Wolff et al., 1991) that $v_f \sim (U/m)^{1/2}$ holds for a toothpick-type fuel bed with $0.0 < U < 4.6$ m/s, 0.11 kg/m$^2 < m < 0.88$ kg/m^2, so $0 < (U/m) < 40$ m^3/(kg s).

Indeed, we just briefly note that the corresponding radiative-preheating development, in which we adopt Eq. (17) in place of Eq. (16) and obtain [M is a dimensionless positive real number with value related to the here-unspecified details of the profile $g(x/Y)$]

$$\frac{v_f}{U} = \left(\frac{\rho_o c_p T_f}{mQ} \right)^{3/2} \left(\frac{T_{pyr}}{T_f} \right) \frac{(\rho_b k_b c_b U)^{1/2}}{M\varepsilon\sigma T_f^3} \tag{25}$$

The brevity of this discussion stems from our not having observed this result in our testing. We believe (Carrier et al., 1991) that this result [i.e.,

$v_f \sim (U/m)^{3/2}$] *might* be observed for an exceptionally heavy fuel loading, say, $m = O(2 \text{ g/cm}^2)$.

Williams (1977) notes that the rate of quasi steady, one-dimensional spread of fire through a combustible medium is often usefully discussed by adopting a firefront structure in which there is a nonflaming ("preheat") zone, within which, continually, fresh fuel is warmed from ambient temperature to ignition temperature:

$$\rho_b v_f h_i = I, \qquad \text{where } h_i = c_b T_i \tag{26}$$

and it is recalled that the datum for temperature is the ambient temperature; I, with units of heat per area (perpendicular to the direction of propagation) per time, denotes the flux furnished to the fresh fuel by convection, conduction, and/or radiation from the flaming region (not encompassed within this examination, which is limited to a *subdomain* of the complete firefront structure). The ignition temperature T_i may be *roughly* identified with the onset-of-*vigorous*-pyrolysis temperature T_{pyr} for the vegetation of interest (Albini, 1992). Williams (1977) points out that this one-dimensional treatment is *not* readily applied to scenarios in which a fuel bed yields pyrolyzate (burned in the gas phase above the bed), as the fuel-bed surface regresses in a direction perpendicular to the direction of propagation. Indeed, Frandsen (1971) notes the inherent two-dimensionality of fire spread over a fuel bed (especially wind-aided fire spread) by emphasizing the need to supplement the heat flux (to fresh fuel) in the direction of propagation with a contribution involving the heat flux (to fresh fuel) perpendicular to the direction of propagation.

Pyne *et al.* (1996, p. 37) write: "Rothermel's fire spread model (1972) is the basis for most computer-based fire management applications in the United States, with significant use in other countries. . . . Rothermel's model was developed from a strong theoretical base in order to make its application as wide as possible. This base was provided by Frandsen (1971) who applied the conservation of energy principle to a unit volume of fuel ahead of an advancing fire in a homogeneous fuel bed. In his analysis, the fuel-reaction zone is viewed as fixed and the unit volume moves as a constant depth toward the interface. The unit volume ignites at the interface. Rate of spread is then a ratio between the heat received from the source and the heat required for ignition by the potential fuel. Frandsen's equation . . . contained heat flux terms for which the mechanisms of heat transfer were unknown. To solve the equation, it was necessary to use experimental [data]." Since Rothermel's model is widely used (Andrews, 1991), we consider the model in detail. We have our reservations about seeking universality in semiempirical models. Also, we question the citation of Frandsen's work as a basis for Rothermel's model, which is in the category of one-dimensional models typified by Eq. (26); the just-described model [Eqs. (15)–(25)] explicitly encompasses the role of heat transfer transverse to the direction of propagation, and perhaps better includes the processes thought essential by Frandsen. [Incidentally, the multidimensional treatment of firefront advance traces to work by Thomas and Simms in 1963, as Frandsen (1971) himself acknowledges.] One-dimensional models for fire spread through wildland vegetation were state-of-the-art when discussed by Emmons (1963, 1965), but multidimensional treatments of spread have now been undertaken. Further, in his one-dimensional modeling, Emmons treated the flaming zone and bed burnout, as well as the preheat zone, and adopted an explicit, radiative mechanism for the preheating of fresh fuel. We are able to discern a treatment of preheating, but no clear treatment of the rest of the firefront structure, in the Rothermel model. Indeed, Albini (1984, p. 594), in discussing the genesis of Rothermel's model, writes, "the pivotal step in the development of the model was [an] inspired conjecture." Specifically, Rothermel takes (Pyne *et al.*, 1996), in terms of the notation adopted earlier, and with ξ denoting the fraction of the product $mQ\Gamma$ that serves to preheat fresh fuel to ignition (so propagation may continue),

$$I = mQ\Gamma\xi(1 + \phi_s + \phi_w) \tag{27}$$

where the factor Γ is termed a velocity by Rothermel (though it has the units of a frequency), and ϕ_s and ϕ_w are dimensionless empirical expressions giving the increment to the fire spread rate owing to finite slope and finite wind, respectively. Although a great variety of functional forms can be fitted to a limited amount of data, we know of no analysis that motivates adopting the form of Eq. (27) for the quantity I. Whether the model is based around convection and/or diffusion and/or radiation, as the primary mechanism of preheating of fresh fuel, is not explicit. We are left uninformed about the basis of taking the effects of slope and wind on spread rate to be additive and effectively independent (Mongia *et al.*, 1998). By inspection, Eq. (27) requires that fire propagate in a situation without wind or slope, if wind or slope are to augment the spread rate—a requirement for which counterexamples are readily demonstrated (e.g., Burrows *et al.*, 1991). In fact, taking ϕ_w to be an additive contribution, universally proportional to a power of the wind speed U, precludes encompassing cases in which there is an abrupt increase of spread rate above a threshold wind, then a modest increment with further increase of wind speed (Gill *et al.*, 1995). Further, in the Rothermel model, the rate Γ is taken to be primarily an empirical function of (besides m) the (true) density of the vegetative matter, fuel-bed depth, the ratio of surface area to volume of fuel particles, fuel mineral content, and fuel moisture content; fuel availability is related empirically to moisture content only. But could not elevation of the fuel bed (ambient pressure), diffusional and advective processes within the vegetative matter and in the gaseous interstices, and chemical kinetics play a comparably important role in the rate Γ? Because the forms adopted for Γ, ϕ_s, and ϕ_w are curve fits which give v_f correctly for the available data, but for which even the simplistic, primitive motivation that went into Eqs. (15)–(25) is lacking, it would seem crucial that the model be validated against a wider data base than was used in its formulation. The model was distributed and in wide use before this validation was performed; shortcomings of the model and the need for revision have since become apparent (Albini, 1984). Perhaps the spread rate v_f, for use in forming the functional F in Eq. (1), might be best ascertained by proceeding with both testing and modeling from scratch, and regarding the Rothermel model as a historic milestone. There are likely to be several other attempts to be abandoned [perhaps including Eqs. (15)–(25)] before an adequate statement of the dependence of the spread rate v_f on prefire parameter values is in hand.

IV. PRELIMINARY TESTING OF THE MODEL

About 200 experiments on wind-aided fire spread across a well-defined, horizontally oriented array of very small diameter (1.3–4.4 mm), discrete fuel elements were carried out in the previously described, specially designed and dedicated firetunnel. This experimental setup allowed investigation of factors such as fuel-bed width, fuel arrangement, inert content, moisture, and substratum. Of course, it is the thin fuels that are burned exothermically with firefront passage. These thin fuels are quickly responsive to changes in moisture/heat and most readily combustible. Post-firefront combustion of some of the thicker fuel has relatively little bearing on the speed of the firefront advance.

A. FUEL-BED WIDTH

A curved firefront evolves from ignition along a straight line perpendicular to the direction of the oncoming wind, owing to the fact that, at a given distance downwind, those fuel elements situated near the lateral edges of a finite-width bed are less effectively preheated by the approaching, upwind firefront than the fuel elements situated in the interior of the bed (i.e., further from the edges). The retardation in the rate of fire spread is manifested further and further from the lateral edges with increasing downwind distance, so eventually a curvature characterizes the entire firefront of any finite-width fuel bed (Figure 10). We have noted that only the component of the oncoming wind that is locally normal to the firefront aids spread; once the flanks lag, they continue to lag. Inserting structures at the lateral edges of the bed cannot replicate the perfectly noncatalytic, perfectly slippery, perfectly adiabatic conditions holding for a perfectly one-dimensional fire spread; ascertaining the consequences of adding physically realizable constraints at the lateral edges (Catchpole et al., 1998, Figure 3) is problematic. Investigating the consequences on fire spread rate of increasing

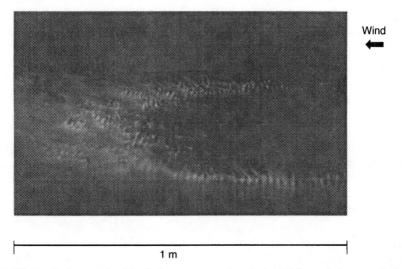

Wind

1 m

FIGURE 10 A photograph of the firefront curvature for wind-aided fire spread across a bed of vertical white-pine toothpicks. Here, the fuel loading $m = 0.11$ kg/m^2, the wind speed $U = 1.6$ m/s, and the fuel-bed width $W = 1.0$ m. The site is roughly 2 m downwind from the leading edge of the fuel bed, where the firefront was initiated along a straight line.

the bed width in the hope of extrapolating the spread rate to the value holding for an indefinitely wide bed is probably not feasible for most facilities. Furthermore, it may be important to consider that, as tests are executed for successively wider fuel beds within a test section of fixed width, some oncoming air, diverted around a "heat island" for a fire spread test with a narrower bed centered in the test-section, may be constrained to more two-dimensional streaming for a wider bed which approaches the test-section width of the facility. The upshot is that spread rate may be dependent on the width of the facility as well as on the width of the fuel bed. The information that seems accessible is the comparison of how spread rate varies with fuel loading, wind speed, and the like for a fixed bed width within a given facility.

B. FUEL ARRANGEMENT

To avoid meticulous and tedious inventorying of fuel, one hopes that the fire spread rate can be characterized by a single parameter, the fuel loading m, as long as the fuel elements are wooden, "thin" [less than $\sim 0.3 - 0.6$ cm in thickness (Cheney, 1981; Albini, 1992), so the elements remain isothermal under gradual heating], and upright. Then, recording the fuel mass before and after flaming-firefront passage is simplified, since we have noted that virtually all the fuel is consumed. Testing with well-defined fuel beds facilitates reproduction of test conditions to establish error bounds. For toothpicklike fuel elements [of density ρ_s, (exposed) height H, and characteristic cross-sectional area d^2] inserted upright into regularly distributed holes (drilled into a substrate), the fuel loading m is given by [$(1 - \phi)$ is defined as the packing ratio and $\rho_s(1 - \phi)$, the bulk density]

$$m = \rho_s H(1 - \phi) = \rho_s H(nd^2) = \rho_s H(Nd^2/s^2) \qquad (28)$$

The porosity (voidage) ϕ is varied to change the fuel loading m; more specifically, the number of fuel elements per unit area of the bed, n, is varied—where $n = N/s^2$, with N denoting the number of fuel elements (however distributed) in an area s^2, where the distance s characterizes the square-grid spacing. For heavier loading, we may insert many toothpicks in a given drilled hole; for lighter loading, we may leave many holes unoccupied. If we further fix n for a single-height constant-cross-sectional-area fuel-element type, then only small-scale variability is possible within a fuel bed of a given loading. The basic "building block" is the smallest square delineated by four drilled holes; the square (conveniently taken with one side parallel to the leading edge of the fuel bed) is meticulously repeated to comprise the entire array for the tests. For example, one can obtain an average of one toothpick in each hole by placing two toothpicks in every other hole. The presence of more than one toothpick per hole augments the possibility of shading one fuel element from some radiation owing to the presence of another. For more complicated arrays, with mixed elements, a larger basic building block is used. While [for fixed wind speed, fixed fuel-loading parameters (ρ_s, H, n, d^2), and fixed bed width] some variability of the rate of fire spread, especially for lighter loadings, is observed, *still, our preliminary testing shows that the rate of spread may be described usefully in terms of the macroscopic description m.* For example, Figure 11 shows the sensitivity of the rate of fire spread to microscale details, for the case in which there is the equivalent of two white-pine toothpicks per hole. What little sensitivity there is, is owing to a slightly faster propagation for those arrangements in which a downwind fuel element is closer to an upwind element, with both oblique and in-line considerations of importance. Incidentally, we regard the value of the fuel density ρ_s as effectively invariant for cases of interest.

However, because so simplistic a treatment of the fuel bed may be adopted inappropriately, in the rest of this subsection we dwell at length on circumstances that render the one-parameter characterization unsuitable. For example, we have yet to discuss the consequences of altering H and/or d (within the constraints of a "thin" fuel); these matters are examined in Section V. Cheney and Sullivan (1997, p. 25) write, "Fuel load by itself does not influence rate of spread of grassfires. While previous publications have reported that fuel load directly influences a fire's rate of spread, we have found that this occurs in grass-

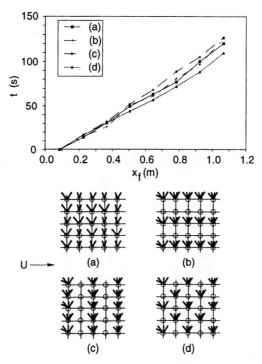

FIGURE 11   Effect of the fuel-loading pattern on the rate of firefront propagation v_f for fuel loading $m = 0.88$ kg/m², test-bed width $W = 0.55$ m, and wind speed $U = 0.7$ m/s. The quantity x_f is
the streamwise centerline position of the firefront (downwind from the leading edge of the fuel
bed), and t is the time since ignition. The hole-center-to-hole-center spacing is 1.25 cm.

lands only if changes in fuel load also reflect changes in fuel condition, in
particular changes in fuel continuity." Of course, all fuel beds of interest in a
wildland-fire context consist of discrete combustible elements; those fuel arrays
described by Cheney and Sullivan (1997) as "continuous" have relatively small
separation distance s: if oxygen is accessible, a finite (but, for our purposes,
typically very low) rate of fire spread may be sustained even in the absence of
wind. Cheney and Sullivan (1997, pp. 24–25) report that, in grasslands typified
by upright stalks (and presumably comparable values of ρ_s), the rate of spread
is largely independent of the values of m (the loading), H (the bed height), and
Nd^2 (the fineness of the stalks) but is quite sensitive to the value of s (the "continuity"); *if* all the thin fuel burned with firefront passage, so that m were identical with the initial fuel loading, then, in view of Eq. (28), these statements are
inconsistent. (See also their Fig. 4.15 on p. 32.) Cheney and Sullivan seem to
suggest that the graininess (patchiness) of a discrete-fuel distribution may be

so pronounced that the convenient single continuum descriptor m is insufficient for fire spread rate characterization, and one must revert to description of the fuel in terms of multiple constituent factors such as H and s, perhaps even taking account of the variation of properties with height within the bed. Circumstances in which such detail is essential to fuel-bed description for fire spread prediction undoubtedly arise. However, the fact that Mediterranean garrigue or Australian mallee may be so discontinuously distributed that sustained fire spread is likely only under strong surface winds does not in itself ensure that the single descriptor m is not useful for fire spread rate estimation. For an example discussed by Cheney and Sullivan, some grassland in the arid center of Australia consists of combustible dense clumps (spinifex hummocks) and bare interstices, both of roughly 1-m lateral scale, so the interstices serve as firebreaks, except under high wind that abets otherwise precluded fire spread. Perhaps the closest American analogue is sagebrush, an upright aromatic shrub with narrow grey leaves. Sagebrush grows to a height of about 0.5–3.5 m on semidesert plateau between the Rocky Mountains and Pacific Ocean. The patchy vegetation supports fire spread only under high wind. Another American analogue is the grass tussocks on the northern-Alaska tundra. Characterizing spinifex hummock grassland in terms of the single parameter m, even if in general inadequate for characterizing fire spread rate, may suffice for the special *subset* of conditions pertinent to relatively rapid fire spread. In fact, Gill *et al.* (1995, pp. 31–32, including Figure 6) state, "When the wind is above the threshold for spread, the fire may spread at the same rate as in continuous fuel"—so characterization of the surface-fuel bed in terms of m for fire spread rate estimation may be no more or no less appropriate at *any finite* rate of spread, independently of the value of s. Incidentally, Gill *et al.* (1995, pp. 32–33) discuss the frustration in using formulae for spread rate inferred from data obtained in one series of field experiments on hummock grassland to predict the fire spread rate in another series; such experience suggests that a useful contribution may be more accessible through reproducible testing under well-defined, well-controlled conditions in the laboratory (Section II). We reiterate that our primary focus is on fire spread in high-continuity, thermally thin fuels such as the dense, carpet-like, mature stands of chaparral in southern California for which graininess (coarseness of the fuel distribution) is often not as prominent a consideration.

Cheney and Sullivan (1997) find that the rate of fire spread is independent of the initial total thin-fuel loading (prior to firefront passage), but is highly sensitive to (1) the curing state of a grassland, expressed as the fraction of dead material in the sward, and (2) the drying out of the dead grasses. Specifically, the rate of spread is nil for a curing state below 50% and increases monotonically and significantly with curing state above 50%; the rate of spread is often nil for fuel-moisture content above 20% of ovendry weight and increases monotonically and significantly for lower moisture content. If we recall that the parameter m refers to the loading of thin fuel *burned during firefront passage* (Section III.D), sometimes termed the "available" thin-fuel loading, then perhaps we narrow the apparent discrepancy between field and laboratory observations with respect to the dependency of the rate of spread on the amount of fuel. We recall (Section III.F) that Albini (1992) comments that virtually all thin fuel typically is burned during firefront passage in grass and small shrub; Minnich (1998) notes that such completeness of burns holds for chaparral in southern California. Burrows *et al.* (1991, pp. 196, 198) state that fire spread in spinifex entails complete burning of all above-ground parts of the plant in the flaming zone, though they confuse the issue by also discussing flaming hummocks behind the main head fire. However, Cheney and Sullivan (1997) found that fast-spreading grass fires left appreciable thin fuel unburned in Australian field tests. Marsden-Smedley and Catchpole (1995) make the same observation concerning fire spread in Tasmanian buttongrass moorlands. Accurate prediction, over the range of fire spread conditions, of the available (as distinct from the total) thin-fuel loading is sometimes challenging. Unfortunately, the rate of fire spread may be related to a quantity which, while a convenient gross descriptor, is not always known prior to firefront passage.

Laboratory experiments with poorly reproducible fuel beds (Burrows, 1999a, 1999b) result not only in large scatter in the spread-rate data but also in no clarification of the spread-rate dependence on available fuel.

C. INERT CONTENT

The presence of noncombustible material in the fuel bed (effectively including higher moisture-content, thicker fuel that does not burn with firefront passage) may introduce heat-sink and/or wind-retardation effects. When the inert material is well-mixed with combustible polymers, an oxygen-deprivation effect may also arise, so that vigorous burning is inhibited and copious smoke may be produced. This oxygen-deprivation effect is not readily investigated with the upright-fuel-element/upright-inert-element arrangement adopted here for examination. We consider fuel beds in which finishing nails are regularly distributed (12.8 kg/m^2) (Figure 12), in the midspan only, for three tests (Figure 13) in which the discrete fuel elements are white-pine toothpicks (0.22 kg/m^2). Both upwind and downwind of the midspan, the white-pine fuel loading is identical to that of the midspan, but no nails are present. Figures 14 and 15 present the results of other tests with nails present; the inserts symbolize the loading of the midspan, with a left-leaning mark denoting a nail-filled hole in the substrate, a right-leaning mark denoting a toothpick-filled hole, and a small dot with no mark denoting an empty hole. We define R to be the ratio of the number of nails to the number of toothpicks.

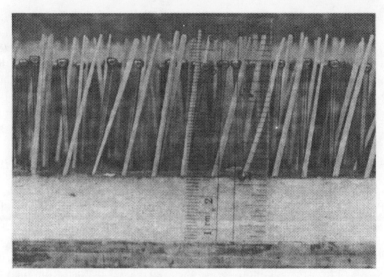

FIGURE 12   A side view of the midspan (of a pretest fuel bed) consisting of common nails interspersed regularly amid white-pine toothpicks.

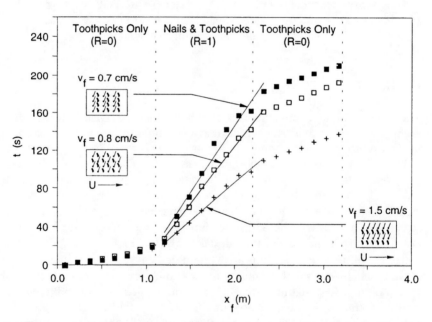

FIGURE 13   The firefront position x_f vs. time t for tests conducted with white-pine toothpicks at a fuel loading $m = 0.22$ kg/m², a wind speed $U = 2.5$ m/s, and a fuel-bed width $W = 0.55$ m. The ratio R (the number of nails divided by the number of toothpicks) is unity for the midspan of the fuel bed and zero elsewhere. The inset depicts the nail-and-toothpick arrangement in the midspan, where left-leaning lines represent nails and right-leaning lines represent toothpicks.

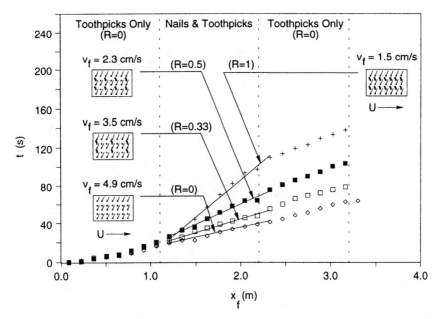

FIGURE 14 Same as Fig. 13, except the ratio R takes on the successive values, 0, 0.33, 0.50, and 1.0 (for the midspan only) for the four cases tested. In the inset, a dot without a line denotes an empty hole in the substratum.

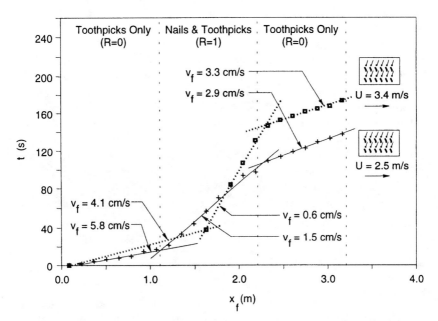

FIGURE 15 Same as Fig. 13, but with the ratio $R = 1$ for two tests with different wind speeds, $U = 2.5$ m/s and $U = 3.4$ m/s.

Figure 13 illustrates that the pattern of loading of inert material alters the fire spread rate, even for fixed total amount of combustible and inert material. The ratio R in the midspan is unity in all three tests shown, but the rates of fire spread differ by a factor of two in the midspan. The preheating capacity of a burning fuel element is strongest on a neighbor located immediately downwind; if that neighbor is an inert nail, the fire spread is slowed. Therefore, for these tests, the arrangement of the fuel has an effect on the resulting fire spread rate, and the fuel bed cannot be characterized by a single loading value, as indicated earlier for an arrangement of upright combustible elements only.

Figure 14 presents results for $R = 0, 0.33, 0.5$, and 1, with m and U held constant; v_f decreases monotonically as R increases. Increasing the amount of inert material in the bed significantly reduces the fire spread rate. The relative importance of (1) the heat-sink effect and (2) the wind-retardation drag (so the effective wind speed U is reduced), owing to the presence of the nails in the midspan, is uncertain.

In Figure 15, the spread rate v_f decreases in the midspan (where nails are present) when the wind speed U is increased from 2.5 to 3.4 m/s, whereas an increase in wind speed increases the spread rate in the nail-free regions. The observed behavior in the midsection may be evidence of a so-called finite-Damköhler-number effect in fire spread across discrete fuel elements, where the Damköhler number D_1 is defined to be the ratio of a characteristic reaction rate to a characteristic flow rate. For D_1 sufficiently small, the chemical reaction is extinguished ("chemically frozen" flow); if D_1 is sufficiently large, the chemical reaction proceeds at chemical-equilibrium rates; for intermediate values of D_1, transport rates and reaction rates are competitive, and a faster flow implies a slower rate of chemical reaction. Upwind and downwind, presumably D_1 is sufficiently large for the flow to be in chemical equilibrium; the faster flow is responsible for a faster rate of spread under rate-of-preheating-controlled considerations. In the midsection, the temperature may be reduced owing to the presence of a heat sink (the nails), and the chemical-reaction rate typically decreases exponentially as the temperature is reduced. Rather than being under preheating-mechanism control, the spread rate is under reaction-rate control, and an enhanced wind speed implies a reduced spread rate.

The presence of effectively noncombustible material seems typically not a major issue in the chaparral context of primary interest.

D. MOISTURE

The ambient moisture content of the white-pine fuel elements used in the testing is roughly 8%. The maximum amount of water which can be absorbed by the (commercial) white-pine toothpicks is roughly 30% of the dry toothpick

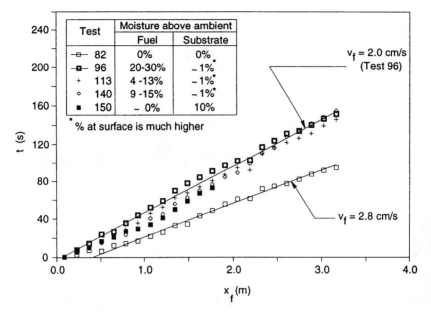

FIGURE 16 The streamwise centerline firefront position x_f as a function of the time t for tests with 0.55-m-wide beds of 4.6-cm-high white-pine toothpicks with pretest fuel loading $m =$ 0.44 kg/m^2, at a wind speed $U = 2.5$ m/s. The moisture content is expressed as a mass percent above the initial mass (where the initial mass includes ambient moisture, typically 6 to 8% of the oven dry mass). The moisture enhancement is achieved by pretest confinement in a saturated environment.

mass, whereas chamise chaparral may have a ratio of water mass to dry mass of 40 to 50% in midsummer.

The rate of fire spread is substantially reduced when additional moisture (greater than ambient conditions) is added to the toothpicks or the substrate (Wolff *et al.*, 1991, Section 3.5; Figures 16 and 17). However, as noted earlier, our primary interest is the extreme fire spread conditions associated with Santa Ana wind episodes (after a long warm dry summer) during which the relative humidity may drop to ~10% or lower for days.

E. SUBSTRATUM

The rate of fire spread observed in the laboratory testing appears to be independent of the substrate composition, at least for the clay and ceramic materials tested. The ceramic has density of 425 kg/m^3, heat capacity of 1130 J/(kg K), and thermal conductivity of 0.080 W/(m K) at 447 K and 0.223 W/(m K) at 1225 K. The respective properties of the clay are 1750 kg/m^3, 1000 J/(kg K),

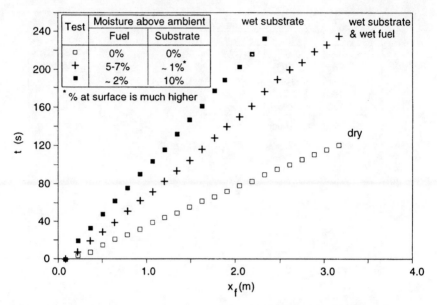

FIGURE 17 The same as Fig. 16, but for tests at a fuel loading $m = 0.22$ kg/m² and for a wind speed $U = 1.0$ m/s. The retardation of the rate of fire spread v_f owing to substrate moisture is distinguished roughly from the retardation owing to the fuel moisture content.

and 0.585 W/(m K) at room temperature. Although the square root of the conductance of the clay is about three times that of the ceramic at 1255 K, in fact the two inert materials yield about the same fire spread results. However, as just noted, retention of even residual liquid water by either material can lead to fire spread rate results distinctly altered from those obtained for the dry substrate.

V. TEST RESULTS FOR THE EFFECT OF WIND SPEED AND FUEL LOADING ON THE RATE OF FIRE SPREAD

As noted in Section III.D, and in accord with the results obtained from a diffusive-convective-preheating model in Section III.F, our laboratory testing with arrays of regularly arranged, upright, effectively identical white-pine toothpick-type discrete fuel elements gives good agreement with the formula (Figure 18)

$$v_f = C(U/m)^{1/2} \tag{29}$$

The premultiplier C is a function of H, d, inert content, water content, and bed width, but was not found to be a function of wood species within the limits of

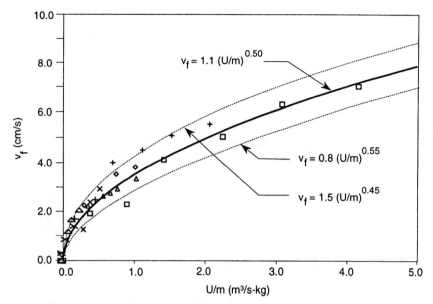

FIGURE 18   Quasi steady firefront propagation speed v_f through upright white-pine toothpicks as a function of the ratio of wind speed U to fuel loading m, from tests with $0 \leq U \leq 4.6$ m/s, 0.11 kg/m$^2 \leq m \leq 1.76$ kg/m^2, and fuel-bed width $W = 0.55$ m.

our testing. The obviously finite range of the parameters U and m for which this formula has been shown to hold was stated in Section III.F.

In Section III.F, we showed that Eq. (29) is consistent with control of fire spread rate by convective-diffusive heat transfer from hot gas to unburned fuel elements. The dominance of the heat-transfer mechanism is anticipated to hold over some range $(U/m)_\ell < (U/m) < (U/m)_u$, where $(U/m)_\ell \to 0$, and $(U/m)_u$ was not encountered, for the range of parameters tested (Figure 18). At the upper limit, the forced-convective strain out of the burning of thin elements (Fendell, 1965), and/or the inefficiency of preheating downwind elements for too sparse loading probably intrudes. At the lower limit, the vigor of buoyant ascent, and/or the inaccessibility of sufficient oxygen for completion of fuel-vapor oxidation (Fendell and Kung, 1993), and/or the inefficiency of radiative preheating of downwind elements, may intrude, so the spread rate is small or nil.

A. MIXED FUEL ELEMENTS

If, for $U = 2.5$ m/s and $m \approx 0.42$ kg/m^2, one uses 4.4-mm-diameter elements to obtain the same fuel loading achieved with 1.3-mm elements, the spacing between elements is large for the thicker elements, and fire spread is not readily

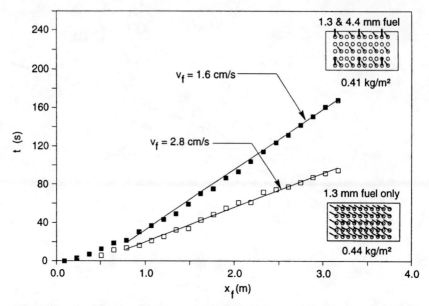

FIGURE 19   The streamwise centerline firefront position as a function of time for tests with a
0.55-m-wide bed, composed of 4.6-cm-high elements, at a wind speed $U = 2.5$ m/s. In the inset,
empty circles designate an unfilled hole in the ceramic substrate, a vertical line denotes a hole occupied by a 4.4-mm-diameter birch dowel, and an inclined line denotes a hole occupied by a 1.3-mm-
diameter white-pine toothpick.

sustained. This suggests that perhaps the thickness for elements to behave as
thin fuel is below 4.4 mm. (The range between 3 and 6 mm is marginal.) Adding roughly 25% by mass of the thinner elements between the thicker elements
facilitates sustained fire spread.

For example, Figure 19 juxtaposes a test involving 1.3-mm (27% by mass)
and 4.4-mm (73% by mass) fuels, with a test involving entirely 1.3-mm-
diameter fuel. The rate of fire spread in the mixed fuel bed is 1.6 cm/s, and in
the thin-element-only bed, 2.8 cm/s. If we adopt the test with only thin fuel as
a baseline and adopt a mass-weighted diameter of 3.6 mm for the mixed-fuel
test, then (Wolff $et\ al.$, 1991) from the generalized relation (H is recalled to denote fuel-element height in the prefire bed)

$$v_f \sim \left(\frac{U}{m}\right)^{1/2}\left(\frac{H}{d}\right)^{p}, \qquad p \approx 2/3 \qquad (30)$$

we obtain $v_f = 1.5$ cm/s—approximately what is observed. Since the multiplicative constant C is found to change significantly as specific fuel-element properties are altered (with m held constant), we must forsake some of the simplic-

ity of Eq. (29). If H and/or d vary from the nominal values used to assign C in Eq. (29), we must adopt Eq. (30).

Whereas both thick and thin fuel elements burned simultaneously in the just-discussed mixed-fuel test (Figure 19), in another test, only the thin fuel elements burned completely and a majority of the thicker fuel mass remained after firefront passage. The fuel loading was the same in both mixed-bed tests, but in the incomplete-burn test, all the thin toothpicks were moved to the columns between the columns with the thick elements. (By definition, a column consists of regularly spaced holes forming a straight line which is parallel to the oncoming-wind direction.) The wind speed was 2.5 m/s in the complete-burn test and 4.6 m/s in the incomplete-burn test. This example, in which a fire, in a bed with multidiameter fuel elements, consumed only the thinner elements and left the thicker elements virtually intact, mimics the burning of only those elements less than about 6 mm in thickness during firefront passage in wildland vegetation.

A bed of regularly arranged toothpicks is hardly a close surrogate for a chaparral stand, especially one that has experienced patch burning. Nevertheless, with respect to the pertinence of the just-described testing, in chaparral-type brush, we reiterate that most of the vegetation is consumed with firefront passage (Albini, 1992), especially for high-intensity fire (Pyne, 1984, p. 117). Furthermore, the range of height in southern California chaparral is limited to about 1.5–3 m, with 2 m being a typical characterization of mature chaparral. Also, because of the typical absence of much grazing, harvesting, "thinning," or patch burning in southern California, over large expanses the loading m increases with the passage of years, without high nonuniformity in the loading or size at a given time at a given site. Although the appellation (evergreen-shrub-type) chaparral admits the presence of multiple species, nevertheless large stands are often fairly uniform.

VI. CONCLUSIONS

On the chaparral-covered hillsides of southern California, autumnal episodes of persistent strong low-humidity warm winds, after the vegetation is desiccated during the long dry summer, can result in wind-aided fire spread over large areas. An annual critical fire season is encountered in other Mediterranean-climate locales on several continents.

To predict the evolution of the firefront that typically separates burned and unburned areas of vegetation, one could advance arc segments of the firefront perimeter. Each segment is to be advanced along its normal, according to the equilibrium rate of spread holding for the local, instantaneous conditions. This is to be carried out at the back and flanks of the perimeter, but it seems especially important to quantify accurately the advance of the head of the wind-aided fire.

In a plausible, semiempirical approach, the equilibrium rate of spread may be ascertained by reproducible, fast-turnaround laboratory testing of well-defined fuel beds in dedicated well-instrumented facilities, along with an integrated program of field burning. Simultaneously pursuing tractable, analytic, fundamentally sound models of the test phenomena may afford insight into the design of experiments and the interpretation of data.

Although the equilibrium rate of spread may not be fully achieved in the buildup phase of a firefront evolving from an ignition, for the high-spread-rate conditions of primary interest here, the fire in minutes typically achieves a size in which the equilibrium rate is an excellent approximation. Results from well-defined smaller scale experiments can be used to develop a capability to anticipate the evolution of the firefront configuration of a larger scale fire. It is this larger scale fire not suppressed in its early stage that challenges fire-management capacity at the urban-wildland interface.

However, by themselves, a limited number of minimally instrumented field events conducted in heterogeneous, possibly discontinuous, incompletely characterized fuel beds, which are burned under uncontrollable conditions, constitute isolated, perhaps unrepeatable anecdotes, not a data base to assist quantification.

NOTATION

ROMAN SYMBOLS

c_b	specific heat capacity of a fuel bed	$J\ kg^{-1}\ K^{-1}$
c_p	specific heat capacity at constant pressure of gas near the flame	$J\ kg^{-1}\ K^{-1}$
D_1	(first) Damköhler number	
d	thickness of a combustible element in a fuel bed	m
F	firefront propagation speed	$m\ s^{-1}$
g	gravity	$m\ s^{-2}$
H	height of a fuel bed	m
h_i	ignition enthalpy	$J\ kg^{-1}$
I	heat per area per time to preheat fuel (1-D model)	$J\ m^{-2}\ s^{-1}$
k	thermal conductivity of gas	$J\ m^{-1}\ s^{-1}\ K^{-1}$
k_b	thermal conductivity of a fuel bed	$J\ m^{-1}\ s^{-1}\ K^{-1}$

L	flame length	m
m	thin-fuel loading burned during firefront passage	$kg\,m^{-2}$
n	number of fuel elements per unit area in a regular arrangement of upright fuel elements	m^{-2}
Q	effective exothermicity per mass of vegetative fuel burned	$J\,kg^{-1}$
$\dot{Q}$	line-fire intensity	$J\,m^{-1}\,s^{-1}$
$\dot{Q}_{crit}$	minimal line-fire intensity for fire spread	$J\,m^{-1}\,s^{-1}$
q	heat flux from gas phase downward to the pre-heating portion of a fuel bed	$J\,m^{-2}\,s^{-1}$
R	ratio characterizing presence of inert mass to combustible mass in a fuel bed	
s	distance between fuel elements in a regular arrangement	m
s	distance along a firefront perimeter	m
T_f	adiabatic flame temperature of pyrolyzate/air diffusion flame	K
T_i	ignition temperature	K
T_{pyr}	pyrolysis-onset temperature	K
T_∞	ambient temperature	K
t	time	s
U	ambient wind speed	$m\,s^{-1}$
U_{crit}	minimal ambient wind speed for fire spread	$m\,s^{-1}$
$(V_f)_i$	rate of change of x_i with time	$m\,s^{-1}$
v_f	rate of spread at the head of a wind-aided fire	$m\,s^{-1}$
W	updraft in a buoyant plume	$m\,s^{-1}$
x_i	coordinates of a point on the firefront perimeter ($i = 1, 2$)	m
x_1	x coordinate of a point on the firefront perimeter	m
x_2	y coordinate of a point on the firefront perimeter	m
x	Cartesian coordinate parallel to the ground	m
Y	stand-off distance (at the pyrolysis front), from the surface of the fuel bed, of the peak gas-phase temperature	m
y	Cartesian coordinate normal to the ground	m

GREEK SYMBOLS

α	Taylor entrainment constant	
β	rate of spread at the flank of a wind-aided fire	m s^{-1}
Γ	a reaction rate in Rothermel's theory	s^{-1}
$\Gamma(s, t)$	two-dimensional curve delineating firefront position	
ε	rate of spread at the back of a wind-aided fire	m s^{-1}
ε	small parameter describing diffusion	$\text{m}^2 \text{ s}^{-1}$
ε	radiative-absorption coefficient	m^{-1}
θ	angle subtended by the local outward normal to the firefront perimeter and the wind direction	
κ	local curvature of the firefront perimeter	m^{-1}
κ_b	effective thermal diffusivity of a fuel bed	$\text{m}^2 \text{ s}^{-1}$
ν_i	components of local outward normal vector to the firefront perimeter ($i = 1, 2$)	
ξ	fraction of exothermicity that serves to sustain propagation, in Rothermel's theory	
ρ_s	(true) density of vegetation	kg m^{-3}
ρ_b	bulk density of vegetation	kg m^{-3}
ρ_o	density of gas near the flame	kg m^{-3}
ρ_∞	ambient gas density	kg m^{-3}
σ	Stefan-Boltzmann constant	$\text{J m}^{-2} \text{s}^{-1} \text{K}^{-4}$
ϕ	porosity (voidage)	
ϕ_s	term to account for the effect of slope on the rate of head-fire spread, in Rothermel's model	
ϕ_w	term to account for the effect of wind on the rate of head-fire spread, in Rothermel's model	

ACKNOWLEDGMENTS

We are grateful to G. F. Carrier of Harvard University for his suggestions, guidance, and encouragement throughout our work on fire dynamics. We also thank F. A. Albini of Montana State University and S. J. Pyne of Arizona State University for many helpful comments on the manuscript.

RECOMMENDED READING

Albini, F. A. (1984). Wildland fires. *Amer. Scientist* 72, 590–597.

Albini, F. A. (1992). Dynamics and modeling of vegetation fires: Observations. *In* "Fire in the Environment—The Ecological, Atmosphere, and Climatic Importance of Vegetation Fires" (P. J. Crutzen and J. G. Goldammer, Eds.), pp. 39–52. John Wiley, New York.

Carrier, G. F., Fendell, F. E., and Wolff, M. F. (1991). Wind-aided firespread across arrays of discrete fuel elements. I. Theory. *Combust. Sci. Tech.* 75, 31–51.

Glassman, I. (1996). "Combustion," 3rd ed. Academic Press, New York.

Lyons, J. W. (1985). "Fire." Scientific American, New York.

Pyne, S. J. (1984). "Introduction to Wildland Fire—Fire Management in the United States." Wiley-Interscience, New York.

Wolff, M. F., Carrier, G. F., and Fendell, F. E. (1991). Wind-aided firespread across arrays of discrete fuel elements. II. Experiment. *Combust. Sci. Tech.* 77, 261–289.

REFERENCES

Albini, F. A. (1984). Wildland fires. *Amer. Scientist* 72, 590–597.

Albini, F. A. (1992). Dynamics and modeling of vegetation fires: observations. *In* "Fire in the Environment—The Ecological, Atmosphere, and Climatic Importance of Vegetation Fires" (P. J. Crutzen and J. G. Goldammer, Eds.), pp. 39–52. John Wiley, New York.

Anderson, D. H., Catchpole, E. A., de Mestre, N. J., and Parkes, T. (1982). Modeling the spread of grass fires. *J. Austral. Math. Soc. (Ser. B)* 23, 451–466.

Anderson, H. E., and Rothermel, R. C. (1965). Influence of moisture and wind upon the characteristics of free-burning fires. *In* "Tenth Symposium (International) on Combustion," pp. 1009–1019. Combustion Institute, Pittsburgh.

Andrews, P. L. (1991). Use of the Rothermel fire spread model for fire danger rating and fire behavior prediction in the United States. *In* "Conference on Bushfire Modelling and Fire Danger Rating Systems—Proceedings" (N. P. Cheney and A. M. Gill, Eds.), pp. 1–8. CSIRO Division of Forestry, Yarralumla.

Burrows, N. D. (1999a). Fire behavior in jarrah forest fuels: 1. Laboratory experiments. *CALMScience* 3(1), 31–56.

Burrows, N. D. (1999b). Fire behavior in jarrah forest fuels: 2. Field experiments. *CALMScience* 3(1), 57–84.

Burrows, N., Ward, B., and Robinson, A. (1991). Fire behavior in spinifex fuels on the Gibson Desert Nature Reserve, Western Australia. *J. Arid Environments* 20, 189–204.

Beer, T. (1990). The Australian national bushfire model project. *Mathematical and Computer Modeling* 13(12), 49–56.

Beer, T. (1991). Bushfire-control decision support systems. *Environment International* 17, 101–110.

Carrier, G., Fendell, F., Chen, K., and Cook, S. (1991). Evaluating a simple model for laminar-flame-propagation rates. II. Spherical geometry. *Combust. Sci. Tech.* 79, 229–245.

Carrier, G. F., Fendell, F. E., and Wolff, M. F. (1991). Wind-aided firespread across arrays of discrete fuel elements. I. Theory. *Combust. Sci. Tech.* 75, 31–51.

Catchpole, W. R., Catchpole, E. A., Butler, B. W., Rothermel, R. C., Morris, G. A., and Latham, D. J. (1998). Rate of spread of free-burning fires in woody fuels in a wind tunnel. *Combust. Sci. Tech.* 131, 1–37.

Cheney, N. P. (1981). Fire behavior. *In* "Fire and the Australian Biota" (A. M. Gill, R. H. Groves, and I. R. Noble, Eds.), pp. 151–175. Australian Academy of Science, Canberra.

Cheney, P., and Sullivan, A. (1997). "Grassfires—Fuel, weather and fire behavior." CSIRO, Collingwood.

Christensen, K., Flyvbjerg, H., and Olami, Z. (1993). Self-organized critical forest-fire model: Mean-field theory and simulation in 1 to 6 dimensions. *Phys. Rev. Let.* 71, 2737–2740.

Chorin, A. J. (1980). Flame advection and propagation algorithms. *J. Comp. Phys.* 35, 1–11.

Coleman, J. R., and Sullivan, A. L. (1996). A real-time computer application for the prediction of fire spread across the Australian landscape. *Simulation* 67, 230–240.

de Ris, J. (1979). Fire radiation—a review. *In* "Seventeenth Symposium (International) on Combustion," pp. 1003–1016. Combustion Institute, Pittsburgh.

Emmons, H. W. (1963). Fire in the forest. *Fire Abstr. Revs.* 5, 163–178.

Emmons, H. W. (1965). Fundamental problems of the free burning fire. *In* "Tenth Symposium (International) on Combustion," pp. 951–964. Combustion Institute, Pittsburgh.

Emmons, H. W. (1971). Fluid mechanics and combustion. *In* "Thirteenth Symposium (International) on Combustion," pp. 1–18. Combustion Institute, Pittsburgh.

Fendell, F. E. (1965). Ignition and extinction in combustion of initially unmixed reactants. *J. Fluid Mech.* 21, 281–303.

Fendell, F. E., and Kung, E. Y. (1993). The pyrolyzation of vegetation by brief intense radiation. *J. Thermophys. Heat Transf.* 7, 510–516.

Finney, M. A. (1998). FARSITE: fire area simulator—model development and evaluation. Research Paper RMRS-RP-4. USDA Forest Service, Rocky Mountain Research Station, Fort Collins.

Fleeter, R. D., Fendell, F. E., Cohen, L. M., Gat, N., and Witte, A. B. (1984). Laboratory facility for wind-aided firespread along a fuel matrix. *Combust. Flame* 57, 289–311.

Fons, W. L. (1946). Analysis of fire spread in light forest fuels. *J. Agric. Res.* 72, 93–121.

Frandsen, W. H. (1971). Firespread through porous fuels from the conservation of energy. *Combust. Flame* 16, 99–16.

Gill, A. M., Burrows, N. D., and Bradstock, R. A. (1995). Fire modeling and fire weather in an Australian desert. *CALMScience (W. Australian J. Conservation Land Management)* 24 (Supplement 4), 29–34.

Goldstein, P. (1995). The wise man of the mountains. *Los Angeles Times Mag.* (Feb. 12), 20–23, 33–34, 36.

Gould, J. S. (1991). Validation of the Rothermel firespread model and related fuel parameters in grassland fires. *In* "Conference on Bushfire Modelling and Fire Danger Rating Systems—Proceedings" (N. P. Cheney and A. M. Gill, Eds.), pp. 51–64. CSIRO Division of Forestry, Yarralumla.

Knight, I., and Coleman, J. (1993). A fire perimeter expansion algorithm based on Huygens' wavelet propagation. *Int. J. Wildland Fire* 3, 73–84.

Luke, R. H., and McArthur, A. G. (1978). "Bushfires in Australia." Australian Government Publishing Service, Canberra.

Markstein, G. H. (1964). Perturbation analysis of stability and response of plane flame fronts. *In* "Nonsteady Flame Propagation" (G. H. Markstein, Ed.), pp. 15–74. Pergamon, New York.

Marsden-Smedley, J. B., and Catchpole, W. R. (1995). Fire behavior modelling in Tasmanian buttongrass moorlands. II. Fire behavior. *Int. J. Wildland Fire* 5, 215–228.

Minnich, R. A. (1998). Landscapes, land-use and fire policy: Where do large fires come from. *In* "Large Forest Fires" (J. M. Moreno, Ed.), pp. 133–158. Bookhuys, Leiden.

Mongia, L. M., Pagni, P. J., and Weise, D. R. (1998). "Model Comparisons with Simulated Wildfire Flame Spread Data." Western States Section Spring Meeting, Paper WSS/CI 985–68. Combustion Institute, Pittsburgh.

Nelson, R. M., and Adkins, C. W. (1988). A dimensionless correlation for the spread of wind-driven forest fires. *Can. J. For. Res.* 18, 391–397.

Osher, S., and Sethian, J. A. (1988). Fronts propagating with curvature-dependent speeds: Algorithms based on Hamilton-Jacobi formulations. *J. Comput. Phys.* 79, 12–49.

Pitts, W. M. (1991). Wind effects on fire. *Prog. Energy Combust. Sci.* **17**, 83–134.

Pyne, S. J. (1982). "Fire in America—A Cultural History of Wildland and Rural Fire." Princeton University Press, Princeton.

Pyne, S. J. (1984). "Introduction to Wildland Fire—Fire Management in the United States." Wiley-Interscience, New York.

Pyne, S. J. (1991). "Burning Bush—A Fire History of Australia." Henry Holt, New York.

Pyne, S. J. (1995). "World Fire—The Culture of Fire on Earth." Henry Holt, New York.

Pyne, S. J., Andrews, P. L., and Laven, R. D. (1996). "Introduction to Wildland Fire," 2nd ed. John Wiley, New York.

Reid, R. (1994). After the fire. *Amer. Scientist* **82** (1; Jan.–Feb.), 20–21.

Richards, G. D. (1993). The properties of elliptical wildfire growth for time-dependent fuel and meteorological conditions. *Combust. Sci. Tech.* **92**, 145–171.

Roberts, S. (1989). "A Line Element Algorithm for Curve Flow Problems in the Plane." Center for Mathematical Analyses Report CMA-R58-89. Australian National University, Canberra.

Rothermel, R. C. (1972). "A Mathematical Model for Predicting Firespread in Wildland Fuels." Research Paper INT-115. USDA Forest Service, Intermountain Forest and Range Experiment Station, Ogden.

Sethian, J. A. (1996). "Level Set Methods—Evolving Interfaces in Geometry, Fluid Mechanics, Computer Vision, and Materials Science." Cambridge University, Cambridge.

Steward, F. R. (1974). Firespread through a fuel bed. *In* "Heat Transfer in Fires: Thermophysics/ Social Aspects/Economic Impact" (P. L. Blackshear, Ed.), pp. 315–378. Scripta, Washington.

Steward, F. R., and Tennankore, K. N. (1979). "Fire Spread and Individual Particle Burning Rates For Uniform Fuel Matrices." Report, Fire Science Center. University of New Brunswick, Fredericton.

Taylor, G. I. (1961). Fire under influence of natural convection. *In* "International Symposium on the Use of Models in Fire Research" (W. G. Berl, Ed.), Publication 786, pp. 10–28. National Academy of Sciences—National Research Council, Washington, DC.

Thomas, P. H. (1971). Rates of spread of some wind-driven fires. *Forestry* **44**, 155–175.

Weber, R. O. (1991). Modeling firespread through fuel beds. *Prog. Energy Combust. Sci.* **17**, 67–82.

Weber, R. O., and de Mestre, N. J. (1991). Buoyant convection: a physical process as the basis for fire modeling. *In* "Conference on Bushfire Modelling and Fire Danger Rating Systems—Proceedings" (N. P. Cheney and A. M. Gill, Eds.), pp. 43–50. CSIRO Division of Forestry, Yarralumla.

Weise, D. R. (1993). "Modeling Wind and Slope-Induced Wildland Fire Behavior." PhD dissertation. University of California, Berkeley.

Williams, F. A. (1977). Mechanisms of firespread. *In* "Sixteenth Symposium (International) on Combustion," pp. 1281–1294. Combustion Institute, Pittsburgh.

Wolff, M. F., Carrier, G. F., and Fendell, F. E. (1991). Wind-aided firespread across arrays of discrete fuel elements. II. Experiment. *Combust. Sci. Tech.* **77**, 261–289.

CHAPTER 7

Fire Plumes

G. N. Mercer and R. O. Weber

School of Mathematics and Statistics, University College UNSW,
ADFA Canberra, Australia

I. INTRODUCTION

Ecological studies of the impact of wildland fires need to be concerned with the temperatures to which plants will be subjected. Even though many measurements of the temperatures obtained within and above wildland fires have been made, such as those by Trabaud (1979), to date no one has been able to account for the full variation of temperature with height. Indeed, sampling heights have usually been restricted to less than a few meters and only the plume region above the fire has been modeled. This may suffice for studies of crown death, such as those conducted by Van Wagner (1973), but it is insufficient for a full understanding of fire effects upon vegetation. Knowing the temperature profile above a fire is important in studying the impact of fire on vegetation. Issues such as leaf scorch, seed death, and stem death all rely on a complete knowledge of the heat exposure of the vegetation.

What is a plume? Anyone who has observed rising smoke or condensed water has seen a plume in action, and anyone considering a fire will immediately realize that the hot gases from that fire will rise and have an impact upon vegetation above the fire (as well as obviously consuming the vegetation in the fire). In broad terms, a plume is characterized by rising buoyant fluids, the buoyancy forces arising from intense localized sources of heat (and/or mass). Comprehensive textbooks on fluid mechanics, such as Yih (1969), or on heat transfer, such as Gebhardt (1971), will contain some discussion of plumes, normally in connection with natural convection above point, line, and plane sources of heat. The discussion will usually focus on the conservation of mass, momentum, and energy and the expression of these conserved quantities as mathematical equations, the Navier-Stokes equations. There are many ways of analyzing the Navier-Stokes equations, including numerical methods (similar to those employed for mesoscale meteorological modeling as in Chapter 8 in this book). Alternatively, it is sometimes possible to derive approximate solutions, which can be very convenient for application to fire situations, particularly when an estimate of the temperatures will suffice. One such approximate solution is a "similarity solution" of the Navier-Stokes equations, where the partial differential equations are simplified by the use of suitable variables, and then the equations can be simplified. The validity of using an approximate solution can be justified by comparison with careful laboratory experiments, which allow one to determine the conditions for which the theory and the phenomena are well understood. Although a valuable technique, similarity solutions will not be used in this chapter, as the mathematical level is more advanced than required for an understanding of the ecological effects of fire. Another, related approximation technique is "dimensional analysis," where the processes involved are considered in terms of the fundamental variables and the dimensions of these variables. Balancing these dimensions can then lead to valuable relationships for quantities of interest, particularly observables such as temperature rise. The most fundamental equation for modeling the ecological effects of fire, Eq. (1), can be derived from dimensional analysis. (In fact, it also arises as a similarity solution of the Navier-Stokes equations with appropriate boundary conditions.) Readers are encouraged to consult the references, such as Yih (1969), for the details of these mathematical techniques.

Early theoretical and experimental work on plumes was conducted during World War II and was inspired by the need to disperse fog from airfields, both in a still atmosphere and in the presence of appreciable wind (see Yih, 1951,

1953; Rouse *et al.*, 1952; Thomas, 1963). Out of this early theoretical and experimental work there arose an understanding that plumes arising from sources of heat only (no mass source at all) could be well described by similarity laws familiar to engineers and scientists (especially in the area of fluid mechanics). The success for both laminar and turbulent flows inspired researchers such as Thomas (1963, 1965) to attempt to use these results to model fire plumes, both in the laboratory and in the field. Van Wagner (1973, 1975) used this model of a fire plume to try to predict scorch heights in forests but only had limited success as the parameters and functional relationships did not turn out as neatly as anticipated. There could be many reasons for this, including the inappropriateness of a plume model with no mass source for modeling a fire or the extended three-dimensional nature of a fire as compared to the idealized line source for the plume model. Whatever the reasons, the landscape fire community was sufficiently "buoyed" by the results to continue trying to gain insight into temperature rise above fires by using relatively simple plume models (e.g., Mercer *et al.*, 1994; Weber *et al.*, 1995a,b).

The most significant relationship for turbulent plumes above line sources of heat (Yih, 1953) can be written as

$$\Delta T = \frac{kI^{2/3}}{z} \tag{1}$$

Here ΔT is the temperature increase above ambient at a height z above the line source of intensity I (measured in power output per unit length). The constant k is a proportionality factor, which Van Wagner (1973) determined from 13 outdoor fires in various forests. If units are used such that ΔT is measured in degrees Kelvin, z is measured in meters, and I is measured in kilowatts per meter, then the constant of proportionality, in SI units, is (M. E. Alexander, 1998):

$$k = 4.47 \ Km^{5/3}kW^{-2/3} \tag{2}$$

Naturally this value is not to be used as a "universal constant." Rather it was determined by fitting 13 data points and should only be used as a guide. Furthermore, the two-thirds power of the intensity will not be strictly accurate for the plume above a fire. It also is best used as a guide because it originally arose in a slightly different context and has been adopted for fire plume research in the spirit of elementary mathematical modeling.

A. ALTERNATIVE MODELING APPROACHES

Mathematical models of fire plumes can be constructed and presented in a variety of ways. Possibly the simplest is to adopt a relationship such as Eq. (1) and fit data sets to determine how well the temperature rise is described by such a simple relationship. This empirical approach can be statistically intensive, but it offers the advantage of a well-defined result, hopefully with confidence intervals. For the particular case of Eq. (1), Yih (1953) and Van Wagner (1973), among others, have shown it to provide a reasonable description of turbulent plumes, at least in light wind situations.

An alternative approach is to follow Morton *et al.* (1956); see also Morton (1965) and Williams (1982). This involves a consideration of the main forces affecting the dynamics of a plume and allows one to approximate the equations for conservation of mass, momentum, and energy. The approximate equations can then be solved, usually numerically, and the results expressed graphically. This method can be used to good advantage in more complex situations, such as wind-blown plumes, and will be discussed later in this chapter. The disadvantage is that the final model needs to be implemented in computer software, albeit reasonably simple software, making it a little more cumbersome to use.

The most sophisticated approach is to consider a complete (as far as this is possible) description of the fluid mechanics of a source of heat (and mass) discharging into the atmosphere. This generally involves solving a set of nonlinear, coupled, partial differential equations, the Navier-Stokes equations, and possibly a turbulence model for the fluid motion associated with a fully developed fire plume. Modern computational facilities are quite able to be used efficiently and reasonably quickly to solve problems of this nature; however, it needs to be noted that a great deal of "submodeling" is required and that the end result can really only be used by experts. By submodeling, we mean that the equations as solved have important parameters which need to be pre-determined, the boundaries of the computational domain need to be treated carefully and appropriate conditions must be imposed on fluid velocities and temperature at these boundaries, the source of heat (and mass) must be prescribed in a realistic manner, and the turbulence model must be suitable for a large range of spatial scales including any vegetation on the surface and the full size of the plume. An example of the implementation of such a model is described by Morvan *et al.* (1998), and further examples of this approach can be found in Chapter 8 in this book on coupling atmospheric and fire models. Finally, it should be mentioned that there has been considerable research into plumes in the context of building or structure fires. Some of this can be accessed from the survey paper by Baum (1999) and recent papers such as Trelles *et al.* (1999).

B. APPLICATION TO WILDLAND FIRES

Fires in the field occur over a large range of environmental conditions, with variations in topography, vegetation, and weather. A proper detailed model would need to be able to account for all of these, at least in principle. Certainly the comprehensive description of the transport of scalar quantities (such as heat) and vector quantities (such as momentum) will require a full understanding of the fluid flow as it interacts with vegetation. Fundamental studies of this have been completed; see for example the review by Raupach and Thom (1981) or the book by Kaimal and Finnigan (1994). However, the subject of airflow in vegetation has been found to be quite complex, even in the absence of any strong sources of heat and mass such as a fire. For this reason, we must consider the subject of detailed fluid mechanical computations of fire plumes and the interaction with vegetation to be in its infancy.

Consequently, for the remainder of this chapter, we will refrain from very detailed modeling of fire plumes. Rather we shall consider the observables, principally temperature and velocity, and how they are affected by the source strength (fire intensity) and environmental conditions (principally wind). Additionally, we shall consider the impact a particular fire event (rather than a fire regime) can have upon the different sorts of vegetation often encountered in a landscape mosaic.

C. Fire Plume Temperatures

Thermocouple measurements are the primary means for determining temperature rise in and around a fire. Considerable effort has been expended by many researchers (Stocks and Walker, 1968; Trabaud, 1979; Gill and Knight, 1991; Xanthopoulos, 1991) in devising a measurement system which is both robust enough to withstand repeated field work and sensitive enough to give accurate temperature readings. The resolution has relied on a compromise for thermocouple size, with the need for wires to be reasonably strong and also fairly quick to register temperature changes. With an accepted instrument and experimental method, one can then set about the arduous task of gathering temperature data in the laboratory (e.g., Xanthopoulos, 1991), or the field (e.g., Gill and Knight, 1991).

II. MODELING FIRE TEMPERATURE MAXIMA

A. Introduction

In this section we describe a model for the maximum temperature reached as a function of height as a wildland fire passes. The fuel bed, the flaming region above the fuel bed, and the fire plume are all included in the model, but separate temperature–height functions are used in each of the three regions. Continuity of the temperature and the heat flux (through the temperature gradient) across the borders of the regions can then be used to eliminate unnecessary constants. In this way, it is possible to match a constant-temperature region in the fuel bed with the region of flaming (described by an exponential function characteristic of a reaction-diffusion process) and the classical turbulent plume above the fire.

B. PLUME STUDIES

Yih (1953) calculated the temperatures reached in a turbulent plume resulting from an idealized line source of heat and derived Eq. (1). We note that Eq. (1) is only applicable directly above the stationary source, $x = 0$. Elsewhere, the generalization of this equation, also found by Yih (1953), is

$$\Delta T = \frac{kI^{2/3}}{z} \exp(-x^2/\beta^2 z^2) \tag{3}$$

where x is the horizontal distance from the source and β is an entrainment constant ($\beta \cong 0.16$ according to Lee and Emmons, 1961).

Although, strictly speaking, Yih's (1953) results apply only for a stationary line source in a quiescent atmosphere, Thomas (1963) and subsequently Van Wagner (1973) have used Eq. (1) in modeling the temperature rise above wildland fires. As one reaches a height far above a real wildland fire, it can be reasonably approximated as a line source, and it appears to move only very slowly. Thus, Thomas (1963) and Van Wagner (1973) have had some success—Van Wagner (1973) in particular with providing a first model for crown death.

A laboratory study of stationary pool fires by Kung and Stavrianidis (1982) provides one of the best experimental tests of the applicability of the dimensional analysis of Yih (1953) to fires. The experimental results show impressive agreement in the plume region. However, there is a significant region of measured temperature increases, in and around the flames, where the theoretical results are inadequate.

Our main motivation is to provide a model for the entire temperature profile for wildland fires. The two main benefits of this are

i. The ability to predict maximum external temperature rise to which vegetation would be subjected at any height,
ii. A clarification of the height above a fire at which plume theory can be reliably applied.

Item (i) depends upon the combustion characteristics of fuel types as well as the fluid mechanics of the fire, and we can only provide a partial realization of this here. Item (ii) depends solely upon the fluid mechanics and we show that our model admits an understanding of the significance of the height at which the classical plume theory becomes applicable.

Other factors which one might like to include are the movement of the fire, fire depth, wind profiles, and terrain.

C. TEMPERATURE MEASUREMENTS

The model presented earlier [see Eq. (3)] was fitted to temperature measurements made in a series of experiments in Ku-Ring-Gai National Park near Sydney in New South Wales, Australia. The measurements involved the use of a single vertical array of sheathed Type K (chromel–alumel) thermocouples to measure temperature rises above ambient in experimental fires in heathy fuels. Fires were lit only on days of very light winds. The depth of the fuel varied greatly, from 0.5 to 2 m, and the flames ranged in height from 1 to 10 m. A discussion of the utility of such measurements can be found in Gill and Knight (1991). Typical results are as shown schematically in Figure 1. A detailed report of the experiments is not included here; readers are referred to the detailed report on the experiments, as opposed to the temperature measurements, that can be found in Bradstock and Auld (1995).

It was found that many of the experiments gave a similarity of form, despite quite different fuel depths, which is very encouraging for the development of a

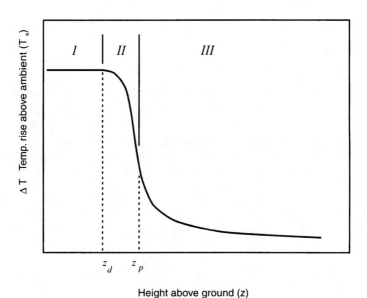

FIGURE 1   Typical form of the curve of the temperature rise above ambient versus the the height above the ground.

universal model. This was the original aim of Thomas (1963) and Van Wagner (1973), founded on the expectation that dimensional reasoning provides relationships which are scale-invariant. The key problem that arose was the inability to determine a universal value for the plume constant [k in Eq. (1)], and indeed there is no reason to expect that in such a simple model k could be a universal constant. In the present analysis, we wish to shift the emphasis away from trying to find universal constants. Rather, it is our expectation that the full temperature versus height profile can be understood with a three-region model, as described in the next two sections with the Ku-Ring-Gai fires as a case study. Furthermore, it will allow a novel comparison of fires and their potential impact upon vegetation.

D. A THREE-REGION MODEL

A typical curve of temperature rise above ambient, ΔT, versus height, z, can be divided into three regions, as shown schematically in Figure 1.

$$\text{I:} \qquad \Delta T_I = K, \qquad\qquad 0 \leq z \leq z_d \qquad\qquad (4)$$

$$\text{II:} \qquad \Delta T_{II} = K e^{-\alpha(z - z_d)^2}, \qquad z_d \leq z \leq z_p \qquad\qquad (5)$$

$$\text{III:} \qquad \Delta T_{III} = C/z, \qquad\qquad z \geq z_p. \qquad\qquad (6)$$

where K, C, α, z_d, and z_p are constants which need to be determined. In Region I, it is anticipated that the presence of combusting solid will create a constant high-temperature region which extends through a height z_d, perhaps comparable to the fuel bed depth. In Region II the flames mix with entrained air and an exponential decrease in temperature rise, following a Gaussian distribution, is assumed. Region III is the plume region and extends above a height z_p; hence the plume equation (1) is used (but with the intensity I absorbed into the constant C).

To reduce the number of constants which need to be determined, and to provide a smooth ΔT versus z curve, the temperature rise and the gradient will be matched across the boundaries between regions:

$$\Delta T_I(z = z_d) = \Delta T_{II}(z = z_d) \qquad\qquad (7)$$

$$\left. \frac{d}{dz}(\Delta T_I) \right|_{z=z_d} = \left. \frac{d}{dz}(\Delta T_{II}) \right|_{z=z_d} \qquad\qquad (8)$$

$$\Delta T_{II}(z = z_p) = \Delta T_{III}(z = z_p) \qquad\qquad (9)$$

$$\left. \frac{d}{dz}(\Delta T_{II}) \right|_{z=z_p} = \left. \frac{d}{dz}(\Delta T_{III}) \right|_{z=z_p} \qquad\qquad (10)$$

There is a little algebra which needs to be done (see Weber *et al.*, 1995b), but then these conditions determine two of the constants in terms of the other three:

$$C = K z_p e^{-\alpha(z_p - z_d)^2} \tag{11}$$

$$\alpha = \frac{1}{2 z_p (z_p - z_d)} \tag{12}$$

Therefore, the model for ΔT versus z consists of Eqs. (4)–(6) subject to Eqs. (7)–(10). The following three remaining parameters need to be found:

 i. K, the maximum temperature reached anywhere. It may be possible to estimate this from a combustion calculation, assuming a certain proportion of total heat generated is lost to the atmosphere.
 ii. z_d, perhaps related to fuel bed depth or zone of persistent flame.
 iii. z_p, perhaps related to the height of the flames in the zone of flame flickering.

It is valuable to have these guiding roles for K, z_d, and z_p when one comes to fit wildland fire data. A detailed survey of ΔT versus z data from wildland fires is required to fully justify these guiding roles for K, z_d, and z_p. In this context, one should note the extreme paucity of published data which uses thermo-couple array or other temperature measurement means. The most detailed studies known to the authors are Tunstall *et al.* (1976), Van Wagner (1975), and Williamson and Black (1981), none of which, in their current form, can be compared with our three-zone model.

E. Fitting Experimental Data

To determine the parameters in the three-region model, seven of the experimental fires are now examined in detail. The three model parameters were first determined approximately by simply viewing the ΔT versus z plots of the experimental data. This instantly provided a smooth curve which gave a good fit to the data. Further refinement using a simple least-squares routine to minimize the error was then done. This usually provided a small improvement to the original fit. In Figures 2 and 3, we present the results of curve fitting from two of the experimental fires. For the fire presented in Figure 2, the fuel bed was quite deep and the flames quite high; hence, we chose $K = 700°C$, $z_d = 1.5$ m, and $z_p = 2.15$ m. For the fire presented in Figure 3, the fuel bed was much shallower, although the flames were of a similar size; hence, we chose $K = 790°C$, $z_d = 0.25$ m, and $z_p = 2.00$ m.

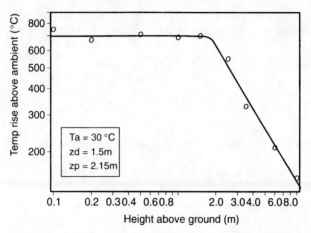

FIGURE 2   Log-log plot of the temperature rise for a deep fuel bed.

The ability of the curves to fit the experimental data with a minimum of fuss is most impressive. It should be stressed that the model is not yet predictive. Indeed detailed measurements in a given fuel type would be required to calibrate the model prior to using it in a predictive sense. Environmental factors such as wind, humidity, and fuel moisture would also need to be taken into account. Despite this, even prior to any calibration, the results provide an insight into where the fire plume region begins. Namely, for the heath communities in Ku-Ring-Gai, NSW, it would seem that 2 m is the minimum height at which plume theory can be successfully applied.

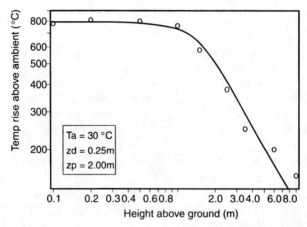

FIGURE 3   Log-log plot of the temperature rise for a shallow fuel bed.

The three regions identified here could match those identified by McCaffrey (1979) for fires burning natural gas above ceramic plates in the laboratory (viz., a continuous-flame region, an intermittent flame region, and a plume region). In wildland fires, the fuel is solid—unlike the fuel in McCaffrey's fires—so it is possible that our regions, especially Region I, may be related to fuel-bed characteristics such as depth. Region I is effectively missing in shallow fuels, such as litter (Stott 1986, author's unpublished data), but readily identified in deeper fuels, such as shrublands (present work and also Trabaud, 1979) and tall grasses (Tunstall et al., 1976). Therefore, the situation described here is a more general one.

F. TIME DEPENDENCE—
TEMPERATURE RISE AND FALL

Not only is the temperature profile in the plume above a fire of interest, but the temperature–time history is also of critical importance (particularly in regard to vegetation survival). Assuming that a wildland fire is moving at a rate of spread U and that it approximates a line source at position $x = Ut$, then we may recast Eq. (3) as

$$\Delta T = \frac{A}{z} \exp(-t^2/\hat{\beta}^2 z^2) \tag{13}$$

where $\hat{\beta} = \beta/U$, $A = kI^{2/3}$, and $t = 0$ at the time of maximum temperature. Thus, t is negative prior to the arrival of the fire and positive after the fire has passed. This equation gives the temperature rise as the fire approaches but is not valid for the temperature fall.

To determine the temperature fall, note that, from simple Newtonian theory, the cooling rate of a hot object is related to its temperature elevation above ambient by

$$dT/dt = -\gamma(T - T_a) \tag{14}$$

where γ is a constant for a given fuel. Integration of (14) yields the following expression for the ΔT as a function of time:

$$\Delta T = B \exp(-\gamma t) \tag{15}$$

with the obvious condition that B is equal to the ΔT_{max} and thus $B = A/z$ necessarily, where A/z was defined for the temperature rise. This should apply exactly to fires where the combustion residence time is reasonably short. From a fluid-mechanical analysis of the cooling phase, assuming Newtonian cooling, a value for γ of order 0.1 is obtained. This is much greater than values fitted from experimental data (see discussion that follows) and is thus equivalent to much faster cooling than occurs in practice. The answer to this lies in the fact that real wildland fires leave a trail of partially combusted fuel, from smoldering ash through coals to actively flaming logs; hence, the combustion residence time is not necessarily short. Therefore, γ obtained from a least-squares fit to data from such fires will be an "effective" cooling.

In addition to the experimental fires in Ku-Ring-Gai Chase National Park, near Sydney, data were also fitted using experiments at the CSIRO Kapalga Experimental Station in Kakadu National Park, Northern Territory. The Ku-Ring-Gai fuels were shrublands, varying in depth from 0.5 to about 2 m (Weber *et al.*, 1995b), while the Kapalga vegetation consisted of mainly long dry grass and intermittent shrubs and eucalypts to about 15 m maximum height (Moore *et al.*, 1995). Data from seven Ku-Ring-Gai fires and 20 Kakadu fires, obtained over a 3-year period, were analyzed. Here we choose to only use results at the 3.5- and 6.0-m level above ground since the $1/z$ plume theory is more likely to apply and the moving fire is more like a line source when viewed from that height. Figures 4 and 5 show the results of fitting Eqs. (13) and (15) to one of the Ku-Ring-Gai fires at 3.5 and 6.0 m above ground level, respectively. Both curves were constrained to pass through ΔT_{max}. Note the rapid rise into maximum (within 60 s) after the fire's onset, followed by a slow fall in temperature over more than 300 s following the maximum. However, at large positive values of time, temperatures tend to an elevated level above those predicted by the Newtonian cooling model. This is reflected in the fitted values for the cooling parameter γ

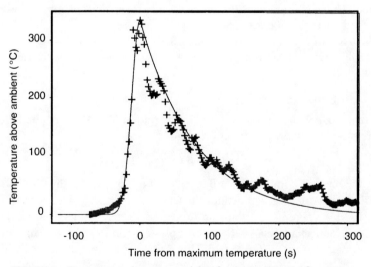

FIGURE 4   Temperature–time curve and data for the Ku-Ring-Gai fire at 3.5 m.

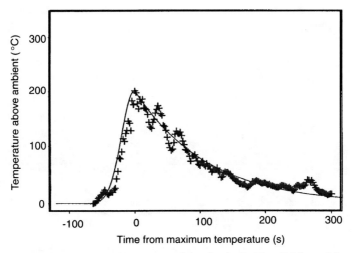

FIGURE 5   Temperature–time curve and data for the Ku-Ring-Gai fire at 6.0 m.

which were 0.013 and 0.009 at the 3.5- and 6.0-m levels, respectively. These are an order of magnitude smaller than the theoretical Newtonian value of about 0.1. The fluctuations in ΔT which occur during cool-down probably correspond to small-scale flaring of fuel elements and also to slow fluid-mechanical phenomena connected with air entrainment. Figures 6 and 7 depict temperatures against time for a Kakadu fire, for 3.5 m and 6.0 m above ground level. The

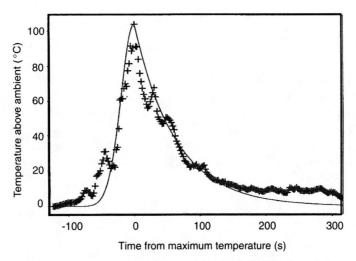

FIGURE 6   Temperature–time curve and data for the Kakadu fire at 3.5 m.

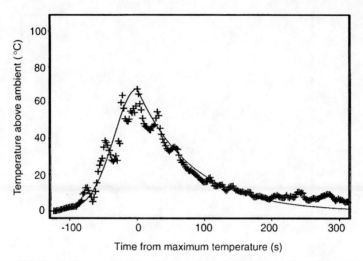

FIGURE 7   Temperature–time curve and data for the Kakadu fire at 6.0 m.

scaling is quite different from that of the Ku-Ring-Gai fire, yet the same qualitative behavior is evident. Again, the model *underpredicts* the decline because it
is unable to accommodate prolonged smoldering or persistent combustion after
the firefront has passed. The fitted value of γ was 0.016 at 3.5 m and 0.013 at
6.0 m. In both fires, the fit around the model is very good.

G. TOTAL TIME-ABOVE-TEMPERATURE

See also Chapter 14 in this book by Dickinson and Johnson. The simple approach to predicting thermal death of plant materials in fires would be to take
the length of exposure above certain temperatures and compare the results with
those from constant temperature exposure in a furnace or water bath. Given experimental temperature data or fitted curves such as those in Figures 4–7, the
total time for which a given value of ΔT is exceeded may be read directly off the
graph. However, with a little algebra, we may express time above ΔT as a function of ΔT as follows: From Eq. (13) the time above ΔT during which the temperature is rising is

$$t_r = \beta z (\ln(A/z\Delta T))^{1/2} \tag{16}$$

From Eq. (15), the time above ΔT during which the temperature is falling is

$$t_f = (1/\gamma) \ln(A/z\Delta T) \tag{17}$$

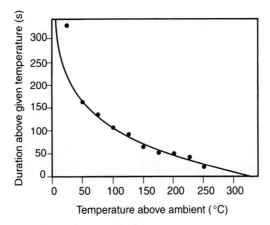

FIGURE 8   Duration of the fire above a given temperature for the Ku-Ring-Gai fire at 3.5 m.

Therefore, the total time above ΔT is

$$t_{tot} = \beta z (\ln(A/z\Delta T))^{1/2} + (1/\gamma)\ln(A/z\Delta T) \tag{18}$$

Fitted values of A, β, and γ were inserted into Eq. (18), and the resulting curves are shown in Figures 8–11.

It is also possible to obtain an estimate of time above ΔT directly from the thermocouple data. This is fraught with problems as it requires an interpolation between data points. However, in order to evaluate our method of fitting temperature–time curves, it seemed appropriate to compare the two methods

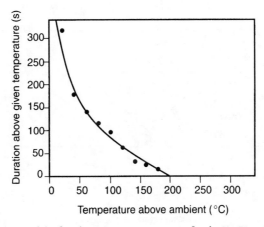

FIGURE 9   Duration of the fire above a given temperature for the Ku-Ring-Gai fire at 6.0 m.

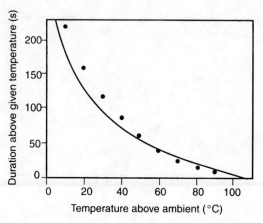

FIGURE 10   Duration of the fire above a given temperature for the Kakadu fire at 3.5 m.

of finding time above ΔT. Hence the "data" points on Figures 8–11 exhibit the same trend as the curve from Eq. (18).

H.   USING TIME ABOVE ΔT TO ESTIMATE DEATH

Laboratory measurements of temperature–time exposures which cause leaf death usually come from bathing the sample in a constant temperature and measuring the time till death (see Chapter 14 in this book for more on this topic). This is not a true representation of the temperature exposure in a fire, due to many factors including the variability in thermal environment associ-

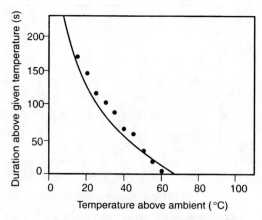

FIGURE 11   Duration of the fire above a given temperature for the Kakadu fire at 6.0 m.

ated with wildland fires and the different thermal properties of water and air but, for obvious reasons, is the convenient experiment to perform. It provides curves like those in Martin *et al.* (1969), characterized by the equation

$$\ln t_d = a - bT \qquad (19)$$

where t_d is the time to death at an exposure temperature T.

A way in which these laboratory curves might be used together with our temperature–time relationship [given by Eq. (18)] to predict leaf death and other fire effects follows. It is the heat flux and the ability of the vegetation to dissipate heat that governs the temperature rise of a sample. However, the flux is difficult (if not impossible) to estimate for a given fire, at a given height and time, even with temperature information. Hence, in the absence of detailed understanding of the fluid mechanics, we are forced into considering only the temperature information available to us.

We first notice that the lethal time at a constant temperature, T_{const}, will be more than the lethal time at a varying temperature, $T_{var}(t)$, where the minimum is always equal to or greater than T_{const}. It is then clear that we can perform a direct comparison of the time above a given temperature curve from a fire with the death curves found in the laboratory. This provides a bound on the effects of the fire on vegetation. Namely, if the time–temperature curve is ever at a higher temperature than the death curve, then the vegetation being considered will perish. However, if the death curve is always above the time–temperature curve, we cannot be certain of the fate of the vegetation. These possibilities are shown together in Figure 12.

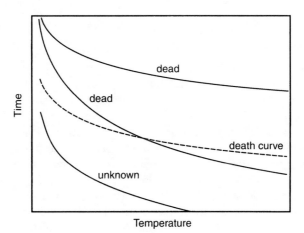

FIGURE 12   A typical death curve and some of the possibilities for the time–temperature curves and the effect on vegetation.

A better method to determine the fate of the vegetation, particularly in this uncertain zone, consists of the following. Divide the time–temperature curve into discrete temperature ranges and determine the time spent within a particular range. The ratio of this calculated time to the laboratory-measured time to death at a representative time in the same range is then determined. This allows for both the heating and cooling phases that the vegetation is subjected to. If we assume additivity of these exposures, then the sum of these ratios will give an indication of the likelihood of death. Mathematically, this can be expressed as a "death number," D,

$$D = \sum_i \frac{t(T_{i-1} < T < T_i)}{t_d(T_i^*)} \tag{20}$$

where $t(T_{i-1} < T < T_i)$ is the time spent in the temperature range (T_{i-1}, T_i) and $t_d(T_i^*)$ is the laboratory measured time to death of a representative temperature (T_i^*) in the range (T_{i-1}, T_i). If the sum of these ratios, the death number (D), is greater than 1, then death is likely, and if the sum is less than 1, then the vegetation is likely to survive. Of course, for values near 1, the outcome is still uncertain. The larger the number, the closer together the discrete temperature ranges considered; hence, the narrower the temperature range, the more accurate this method will be. In the limit as the width of the temperature range tends to zero, Eq. (20) can be rewritten

$$D = -\int_{\tau = T_a}^{T_{max}} \left(\frac{1}{t_d(\tau)} \frac{dt(\tau)}{d\tau} \right) d\tau \tag{21}$$

which, using the model for the external temperature [Eq. (18)] and the model of Martin *et al.* (1969) for the time to death [Eq. (19)], gives the death number as

$$D = \left(\frac{\beta z}{2} + \frac{1}{\gamma} \right) e^{-a} \int_{\tau = T_a}^{T_{max}} \frac{e^{b\tau}}{\tau - T_a} d\tau \tag{22}$$

What is needed are the laboratory-measured time to death of the vegetation (leaves, fruit, stem) for as many different temperatures as possible to determine $t_d(T)$ and the external temperature profile either from experiments or the model detailed previously. The fires considered here both have a maximum temperature well above 60°C; hence, leaf scorch will occur at the heights in question. The effects of fire exposure on the fruits and stems could be determined using the preceding method. Unfortunately, at present, the data needed to use Eq. (20) or to fit to the model of Martin *et al.* (1969) [Eq. (19)] to find $t_d(T)$ are not available in the literature. Mercer *et al.* (1994) have used the model outlined here as the external temperature input to their model for the temperature exposure of seeds in woody fruits. Judd and Ashton (1991) have conducted experimental heating of seeds to consider survival, and Bradstock *et al.* (1994) have examined banksia seed survival during wildland fires. All this work is really the preliminary stage of these investigations, and there is a need for mathematical modeling and experimental studies to be conducted together to provide reliable methods for predicting the full impact of wildland fires on vegetation.

III. PLUMES ABOVE FIRES IN A CROSS WIND

A. INTRODUCTION

Here a more detailed two-dimensional model of the plume above a line fire is presented by considering the balancing of mass, momentum, and energy associated with the fire and a cross wind. The case where the fire influences the wind profile, which is possible for intense fires, is not considered here. For the sake of this model, the wind profile is an input parameter. This has advantages because different wind profiles such as over grasslands or in tree canopies are easily incorporated. Of interest here is how a cross wind affects the plume above a line fire in both tilting the plume and with increased entrainment into the plume. For a comprehensive approach where the wind profile and fire are coupled, see Chapter 8 in this book.

The previous model was based on the assumption that the wind was light enough that it had little impact on the plume above the fire. Van Wagner (1973) derived correlations for the scorch height of vegetation above a fire in a light wind. This work was based on the assumption that the plume is simply tilted without being distorted and did not take into account the increased entrainment into the plume due to the cross wind. Nevertheless, it was used to try to predict the scorch height for vegetation above wind-blown wildland fires and is an interesting example of a relatively simple correlation model. A model of the wind-blown flame from a line fire was presented by Albini (1981a) who replaced entrainment to the plume velocity (this is the no-wind vertical plume assumption for entrainment) by accretion proportional to the cross-flow velocity. This work concentrated only on the flame and was not extended to the plume above the fire. Fleeter *et al.* (1984) considered Gaussian distributions for the velocity and temperature above the fire and analyzed the case when the cross-wind velocity is comparable to the entrainment velocity.

There have been numerous detailed (but similar) models developed to describe buoyant plumes from point heat sources (often referred to as axisymmetric plumes) in a cross flow (Schatzmann, 1979; Davidson, 1986a,b; Krishnamurthy and Hall, 1987, to name but a few). In contrast, buoyant plumes above a line heat source in a cross flow have had considerably less attention. The model used here is developed in a similar manner to these point heat source (axisymmetric) plume models which are often used to model the discharge from stacks but can also be used to model the plume above a pool fire. The models consist of a system of coupled ordinary differential equations derived from the conservation of mass, momentum, and energy of the plume. There are generally two types of distribution of velocity and temperature (or density) used in these models: top-hat distributions where the velocity and temperature are considered to be constant across the plume and have the ambient level outside the plume and Gaussian where the velocity and temperature are taken to have Gaussian distributions across the plume. It has been shown by Davidson (1986a) that the differences between the top-hat and Gaussian approaches for the point heat source (axisymmetric) plume are small over the range of parameters of practical interest. The top-hat approach, which requires the solution of a simpler set of equations, will be used here. If more detail than the mean plume behavior (obtained from the top-hat model) is required, then it can be determined by fitting Gaussian distributions from the top-hat results (Davidson, 1986a).

B. MATHEMATICAL FORMULATION

The situation modeled here is an idealization of a blown line fire plume as shown in Figure 13. The plume is considered to be initially rising vertically with velocity w_i, half width b_i, temperature T_i, and density ρ_i. These initial conditions are taken to be far enough above the flames so that the detailed structure of the flame is not required. Values for these initial conditions must be obtained from measurements or from flame models. The ambient atmosphere is at T_a and has density ρ_a and a height-dependent wind velocity $U_a(z)$. Mass entrainment

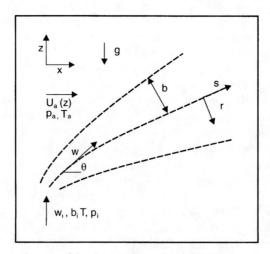

FIGURE 13   A schematic of the plume and cross wind showing the nomenclature used.

of the ambient air into the plume is represented by the linear sum of the horizontal and vertical components. That is, for the plume that is at an angle θ to the flow, the entrainment speed, v_e, is given by

$$v_e = \alpha(w - U_a \cos \theta) + \beta U_a \sin \theta \tag{23}$$

where w is the velocity of the plume. For the case of the axisymmetric plume, values for α and β are documented with $\alpha \approx 0.1$ and $\beta \approx 0.5$ (Krishnamurthy and Hall, 1987). For the line plume, there are less data available in the literature for the value of these parameters, and further measurements are needed to generate reliable estimates for them. Throughout this paper, the value of α will be taken as 0.16 as in Lee and Emmons (1961), and β will have the value 0.5.

In a similar manner to that described in Davidson (1986a) for the axisymmetric plume, the governing equations are derived from balancing the fluxes of momentum and energy. Conservation of mass gives

$$\frac{d(\rho_p b w)}{ds} = \rho_a v_e \tag{24}$$

where s is the arc length along the plume centerline and ρ_p and b are the density and half width, respectively. Due to the top-hat assumption, the density (temperature) and velocity are assumed constant across the width. Conservation of s-momentum gives

$$\frac{d(\rho_p b w^2)}{ds} = \rho_a v_e U_a \cos \theta - b(\rho_p - \rho_a)g \sin \theta \tag{25}$$

Conservation of r-momentum (r is the direction normal to s) gives an equation for the trajectory angle

$$\frac{d\theta}{ds} = -\frac{\rho_a v_e U_a \sin \theta + b(\rho_p - \rho_a)g \cos \theta}{\rho_p b w^2} \tag{26}$$

The conservation of thermal energy gives

$$\frac{d(\rho_p b w T_p)}{ds} = \rho_a v_e T_a \tag{27}$$

where T_p is the plume temperature. Rearranging (27) by using (24) gives

$$\frac{d(\rho_p b w \Delta T)}{ds} = -\rho_p b w \sin \theta \frac{dT_a}{dz} \tag{28}$$

where $\Delta T = T_p - T_a$ is the difference in the temperature of the plume above the ambient air and dT_a/dz is the measure of the temperature stratification in the ambient atmosphere which, for the purposes of this model, can be taken as zero; that is, here we are considering a neutral atmosphere. This is easily altered if a stratified or unstable atmosphere is to be considered. The ideal gas law gives

$$\rho_p = \rho_a \frac{T_a}{T_p} \tag{29}$$

and the plume trajectory is defined by

$$\frac{dx}{ds} = \cos \theta \tag{30}$$

$$\frac{dz}{ds} = \sin \theta \tag{31}$$

The model then consists of solving the seven equations (24)–(26) and (28)–(31) for the seven unknowns b, w, T_p, ρ_p, θ, x, and z as functions of distance along the plume trajectory subject to the initial conditions

$$b = b_i, \qquad w = w_i, \qquad T_p = T_i, \qquad \rho_p = \rho_i,$$

$$\theta = \pi/2, \qquad x = 0, \qquad z = 0 \tag{32}$$

Introducing the change of variables

$$y_1 = \rho_p b w, \qquad y_2 = \rho_p b w^2, \qquad y_3 = \theta,$$

$$y_4 = \rho_p b w \Delta T, \qquad y_5 = x, \qquad y_6 = z \tag{33}$$

and using Eq. (29), the system of equations can be written

$$\frac{dy_1}{ds} = \rho_a v_e \tag{34}$$

$$\frac{dy_2}{ds} = \rho_a v_e U_a \cos y_3 + \frac{y_1 y_4}{y_2} \frac{g \sin y_3}{T_a} \tag{35}$$

$$\frac{dy_3}{ds} = -\frac{\rho_a v_e U_a \sin y_3}{y_2} + \frac{y_1 y_4}{y_2^2} \frac{g \cos y_3}{T_a} \tag{36}$$

$$\frac{dy_4}{ds} = 0 \tag{37}$$

$$\frac{dy_5}{ds} = \cos y_3 \tag{38}$$

$$\frac{dy_6}{ds} = \sin y_3 \tag{39}$$

In this form, it is a straightforward matter to solve this system of six coupled first-order ordinary differential equations subject to the initial conditions given by Eq. (32) in the appropriate new variables given by Eq. (33). Here the solutions were determined using the NAG (1991) routine D02BBF which integrates a system of first-order differential equations using a Runge-Kutta-Merson method.

To obtain the simplified version for the no-wind case, it is a simple matter of setting $U_a = 0$ and $y_3 = \theta = \pi/2$ in the preceding equations and noting that Eqs. (36) and (38) are no longer relevant because there is no plume angle or horizontal distance to consider. The arc length variable s also then equates to the height above the source that is $z = s$, and Eq. (39) is also not needed.

C. WIND PROFILE

For the equations of the model to be solved, the wind profile $U_a(z)$ must be specified. One of the strengths of this model is that any horizontal wind profile can be used and hence a variety of conditions can be modeled. For example, the plumes above grassland fires or fires above and below canopies, all of which have different velocity profiles, can each be modeled. In this chapter, no attempt is made to model the impact of the fire upon the prevailing wind field. Near the fire, this is unrealistic because there can be substantial effects upon the wind field due to the fire. However, a much more detailed model that takes into account all of the fluid mechanics and possibly also the combustion properties of the fire would be needed to accurately predict the effect of the fire on the velocity profile (the fluid mechanics is included in the discussion in Chapter 8 in this book).

As a first (somewhat unrealistic) approximation, we consider the wind to be constant with height. This is the situation assumed by Van Wagner (1973) and Fleeter et al. (1984). By using a similarity analysis, it has been shown for the uniform wind case that the plume is self-similar (Raupach, 1990); hence, away from the source, the midline trajectory is a straight line as a function of downwind distance, that is $z \propto x$. This agrees with the experimental work of Ramaprian and Haniu (1989) for two-dimensional plumes in a cross-flow. Shown in Figures 14a and 14b are plots of the plume midline trajectory and width for the initial conditions $b_i = 0.2$ m, $w_i = 1.0$ m/s, $T_i = 900$ K with $T_a = 300$ K for $U_a = 1, 3$ m/s, respectively. Clearly, away from the source, the trajectory is a straight line as expected from theory and experimental results. The effect of increasing the wind strength is also evident. With increasing wind strength, the plume is blown over more and hence temperatures at a given height above the heat (fire) are lower as the velocity increases. Therefore, for a fire with a given fixed heat release rate that is independent of the wind strength, vegetation above the fire is exposed to higher temperatures than in a higher wind case. Of course, this does not take into account the effect the wind can have on the heat release rate of the fire.

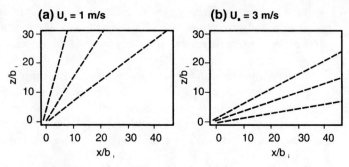

FIGURE 14   Plume trajectory and width for a constant wind.

Two different and more realistic wind profiles are more relevant to the present model. The first is a logarithmic profile as used by Albini (1983), namely,

$$U_a(z) = U_H \frac{\ln\left(\dfrac{z - 0.64V}{0.13V}\right)}{\ln\left(\dfrac{H - 0.64V}{0.13V}\right)} \tag{40}$$

where V is the vegetation height and U_H is the velocity measured at a height H. This velocity profile is applicable above a vegetation cover of trees or shrubs. More applicable to grassland or bare-ground is a power law profile

$$U_a(z) = U_H\left(\frac{z + z_0}{H}\right)^n \tag{41}$$

where n is usually taken as 1/7 (Albini, 1981b, 1983). Here z_0 is the base of the plume and z is measured from this point. Figures 15a and 15b are plots of the plume midline and width for the initial conditions as for Figure 14 but with a

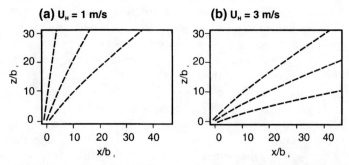

FIGURE 15   Plume trajectory and width for a power law wind profile with U_H given at $H = 10$ m and the plume starting at $z_0 = 1$ m.

power law velocity profile with $U_H = 1, 3$ m/s at $H = 10$ m. The differences between this more realistic profile and the uniform velocity profile considered earlier are evident comparing Figures 14 and 15. The plume for the power law velocity profile is more upright close to the ground as is expected from the lower wind velocities in this region when compared to the uniform velocity case. Far enough above the height where the initial conditions are given, the midline trajectory is approximately linear with downwind distance since the wind velocity is only slowly varying far enough above the source. The major difference between the results for the two different wind profiles is the height to which the plume extends. With a more realistic power law velocity profile, the plume is substantially higher than with a uniform velocity profile. This results in a significant difference when modeling the impact of the fire on vegetation above it and should therefore be taken into account in any such models.

D. TEMPERATURE–TIME PROFILES

Before the effect of the fire plume on vegetation can be determined, the temperature–time profiles at various heights above the fire must be calculated. These temperature–time profiles can then be used as inputs into models for the effect of temperature exposures on vegetation (Mercer *et al.*, 1994). In the present model, the fire is assumed stationary; hence, temperature versus downwind distance profiles can be determined. To convert from this to a temperature–time profile, an estimate must made of the rate of spread of the fire. The distance from the stationary fire can be related to time exposure due to a moving fire. In many circumstances, the rate of spread of fire is small compared to the wind speed; hence, the fact that the fire is moving has little effect on the results presented. In the case when the rate of spread is not small compared to the wind speed, the velocity profile should be altered so that it is relative to the moving fire; that is, the frame of reference should be with the moving fire. In the no-wind case, this is similar to having a wind in the opposite direction to that of the moving fire.

As described in Davidson (1986a), Gaussian distributions can be fitted to the top-hat results of this model to give the distribution within the plume as

$$T_G = T_a + \frac{N}{\lambda^2}(T_p - T_a)e^{-(r^2/\lambda^2 b^2)} \tag{42}$$

where r is the distance normal to the centerline trajectory, N is the plume edge criterion which is usually taken to be 2, and λ^2 is the spreading ratio of mass and heat to momentum which here is taken to be 1 (Davidson, 1986a). Figures 16a and 16b show the temperature versus downwind distance for the

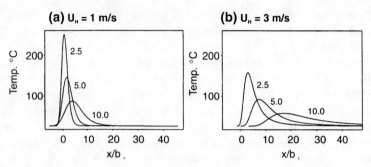

FIGURE 16 Temperature vs. nondimensionalized downwind distance (x/b_i) for the plumes given in Figure 15 at nondimensional heights of $z/b_i = 2.5$, 5.0, and 10.0.

plumes given in Figures 15a and 15b at three nondimensional heights $z/b_i = 2.5$, 5.0, 10.0. With increasing wind strength, the maximum temperature reached at any height decreases, and the profiles become more skewed as the plumes are bent over farther. For the lightest wind considered, $U_H = 1$ m/s, the plume is not far from symmetrical (the no-wind case), whereas for $U_H = 3$ m/s, there is considerable skewing of the plume downwind.

E. DISCUSSION

This model can be used to predict the temperature–time curves above line fires and hence impact on vegetation above the fire. Certain inputs are needed before the model can successfully be used. The ambient wind velocity profile is needed. This can either be from velocity profiles outlined earlier (or similar analytic profiles) or from measured or predicted velocity profiles depending on the terrain and vegetation involved. Estimates for the initial plume conditions b_i, w_i, and T_i are also required. These can be obtained from either measurements of experimental fires or from models such as the one by Albini (1981a). These will depend on the fire and hence on many parameters such as the fuel type, the loading and moisture content, the wind velocity, the ambient temperature, and the relative humidity to name but a few. Indeed the initial conditions can be related to the heat of the fire

$$Q = \int_{-b_i}^{b_i} C_p w_i \rho_i (T_i - T_a)\, dx \tag{43}$$

which simplifies to

$$Q = 2 C_p w_i \rho_i \Delta T_i b_i \tag{44}$$

where $\Delta T_i = T_i - T_a$ is the difference in temperature of the plume from the air. So if an estimate for the heat release rate of the fire is known, then only two of w_i, T_i, or b_i are needed; the third can be estimated from Eq. (44). The most difficult of these initial conditions to determine is w_i, the initial velocity in the plume. The initial plume and temperature are easier quantities to measure experimentally or theoretically from fire models.

The present relatively simple model can be used as an aid in predicting fire impact upon vegetation in many situations. A common reason for using prescribed fire in some parts of the world are fuel reduction burns used to minimize the intensity of wildfires. These have an impact on the surviving vegetation that needs to be estimated before the burns are conducted. Typically fuel reduction burns are conducted in light winds as an aid to their control. They therefore have a larger impact on the vegetation above the fire than a fire of similar intensity would in a stronger wind where the plume is bent farther and does not impinge on the vegetation to such a degree. Of course, the effect of the wind on the fire should also be taken into account. This model can then be used to determine temperature impacts such as the height of scorch of the leaves and potential death of canopy stored seed [in conjunction with models like the one in Mercer et al. (1994)].

Since this is only a plume model, no attempt has been made to model the flame of the fire. Models such as the one presented in Albini (1981a) should be used to obtain details of the flaming zone. The present model takes no account of small-scale phenomena such as flare ups; hence, its use is confined to broad areas rather than small areas such as an individual fuel element. Phenomena such as smoldering of fuel after the main fire has passed and the cooling of vegetation and the ground, all of which can possibly increase temperature exposure of the vegetation above the fire, are not considered.

On a larger scale than that considered here, this model can also be incorporated into a model for spotting ahead of a line fire. In this case, the assumption that the atmosphere is neutral is not justifiable; hence, the dT_a/dz term in Eq. (28) should not be taken as zero. An estimate for this temperature stratification in the atmosphere would be needed which would depend upon the atmospheric conditions prevailing at the time. The velocity within the plume has been calculated, and this can be used to determine the height that burning brands will reach before they fall back and possibly ignite a fire downwind of the main fire front. The model presented here differs from previous models of firebrand trajectories in two ways. First, the present model includes the entrainment into the plume due to the cross wind. This is in contrast to existing models for firebrand trajectories (Lee and Hellman, 1969, 1970; Albini, 1979, 1981b, 1983) which use the no-wind assumption that the entrainment velocity is proportional to the vertical velocity. This will overpredict the height to which the firebrand will rise since the plume rises higher in the no-wind case than in the case with wind present. It is more realistic to include the entrainment as presented in this model in determining the firebrand's actual trajectory. Second, in the present work, a line fire is considered as opposed to previous work by Lee and Hellman (1969) and Albini (1979, 1981b) who considered a spot fire (axisymmetric source) and Albini (1983) who considered a buoyant line thermal as a right elliptical cylinder. The line source is appropriate where the spotting distance is relatively short so that, on the scale of the distance the firebrand has traveled, the fire appears as a line source. For spotting a large distance from the fire, a point source is accurate enough because at a large distance from the fire structure (spot fire or line fire) is less relevant.

NOTATION

ROMAN LETTERS

b	plume half width	m
C	constant	m K
C_p	specific heat	$\text{J kg}^{-1}\text{K}^{-1}$
g	gravitational acceleration	m s^{-2}
H	height of wind velocity measurement	m
I	line fire intensity	kW m^{-1}
k	constant	$\text{K m}^{5/3}\text{W}^{-2/3}$
K	constant	K
N	plume edge criterion dimensionless	
n	power law parameter dimensionless	
Q	heat release	W
r	coordinate normal to s	m
s	distance along plume	m
ΔT	temperature rise above ambient	K
T	temperature	K
$U_a(z)$	ambient wind velocity	m s^{-1}
V	vegetation height	m
v_e	entrainment speed	m s^{-1}
w	plume velocity	m s^{-1}
x	horizontal distance	m
z	vertical distance from plume base	m

z_0	height of the plume base	m
z_d	height related to the fuel bed depth	m
z_p	height related to the start of the plume	m

GREEK LETTERS

α	entrainment coefficient	m^{-2}
β	entrainment coefficient dimensionless	
Δ	change from ambient	
λ^2	spreading ratio of mass and heat versus momentum	
θ	plume trajectory angle from the horizontal	
ρ	density	$kg\ m^{-3}$

SUBSCRIPTS

a	ambient property
i	initial property
p	plume property
G	Gaussian

REFERENCES

Albini, F. A. (1979). "Spot Fire Distance from Burning Trees—A Predictive Model." United States Department of Agriculture, Forest Service, General Technical Report INT 56. Intermountain Forest and Range Experimental Station, Ogden.

Albini, F. A. (1981a). A model for the wind-blown flame from a line fire. *Combustion and Flame* 43, 155–174.

Albini, F. A. (1981b). "Spot Fire Distance from Isolated Sources—Extensions of a Predictive Model." USDA Forest Service Research Note INT 309. Intermountain Forest and Range Experimental Station, Ogden.

Albini, F. A. (1983). Transport of fire-brands by line thermals. *Combust. Sci. Tech.* 32, 277–288.

Alexander, M. E. (1998). "Crown Fire Thresholds in Exotic Pine Plantations of Australia." PhD thesis. Australian National University.

Baum, H. R. (1999). "Large Eddy Simulations of Fires—From Concepts to Computations." Society of Fire Protection Engineers, 1999 Arthur B. Guise Medal Lecture.

Bradstock, R. A., and Auld, T. D. (1995). Soil temperatures during experimental bushfires in relation to fire intensity: Consequences for legume germination and fire management in southeastern Australia. *J. Appl. Ecol.* 32, 76–84.

Bradstock, R. A., Gill, A. M., Hastings, S. M., and Moore, P. H. R. (1994). Survival of serotinous seedbanks during bushfires: Comparative studies of Hakea species from south-eastern Australia. *Aust. J. Ecol.* 19, 276–282.

Davidson, G. A. (1986a). Gaussian versus top-hat profile assumptions in integral plume models. *Atmospheric Environment* 20, 471–478.

Davidson, G. A. (1986b). A discussion of Schatzmann's integral plume model from a control volume viewpoint. *J. Climate Appl. Meteorol.* 25, 858–867.

Fleeter, R. D., Fendell, F. E., Cohen, L. M., Gat, N., and White, A. B. (1984). Laboratory facility for wind-aided firespread along a fuel matrix. *Combustion and Flame* 57, 289–311.

Gebhardt, B. (1971). "Heat Transfer." McGraw-Hill, New York.

Gill, A. M., and Knight, I. K. (1991). Fire Measurement. *In* Conference on Bushfire Modeling and Fire Danger Rating Systems (N. P. Cheney and A. M. Gill, Eds.). pp. 137–146. CSIRO Division of Forestry, Canberra.

Judd T. S., and Ashton D. H. (1991). Fruit clustering in the Myrtaceae: Seed survival in capsules subjected to experimental heating. *Aust. J. Bot.* 39, 241–245.

Kaimal, J. C., and Finnigan, J. J. (1994). "Atmospheric Boundary Layer Flows: Their Structure and Measurement." Oxford University Press, Oxford, UK.

Krishnamurthy, R., and Hall, J. G. (1987). Numerical and approximate solutions for plume rise. *Atmospheric Environment* 21, 2083–2089.

Kung, H-C., and Stavrianidis, P. (1982). Buoyant plumes of large-scale pool fires. *In* "19th Symposium on combustion," pp. 905–912. The Combustion Institute (Pittsburgh).

Lee, S-L., and Emmons, H. W. (1961). A Study of natural convection above a line fire. *J. Fluid Mechanics* 11, 353–369.

Lee, S-L., and Hellman, J. M. (1969). Study of fire-brand trajectories in a turbulent swirling natural convection plume. *Combustion and Flame* 13, 645–655.

Lee, S-L., and Hellman, J. M. (1970). Fire-brand trajectory study using an empirical velocity-dependent burning law. *Combustion and Flame* 15, 265–274.

Martin, R. E., Cushwa, C. T., and Miller, R. L. (1969). Fire as a physical factor in wildland management. *In* "Proc. 9th Ann. Tall Timb. Fire Ecol. Conf.," pp. 271–288.

McCaffrey, B. J. (1979). "Purely Buoyant Diffusion Flames: Some Experimental Results." U.S. National Bur. Standards Rep. No. NBSIR 79-1910.

Mercer, G. N., Gill, A. M., and Weber, R. O. (1994). A time dependent model of the fire impact of seeds in woody fruits. *Aust. J. Bot.* 42, 71–81.

Moore, P. H. R., Gill, A. M., and Kohnert, R. (1995). Quantifying bushfires for ecology using two electronic devices and biological indicators. *CALM Science Supplement* 4, 83–88.

Morton, B. R. (1965) Modeling fire plumes. *In* "Tenth Symposium (International) on Combustion," pp. 973–982. The Combustion Institute (Pittsburgh).

Morton, B. R., Taylor, G. I., and Turner, J. S. (1956). Turbulent gravitational convection from maintained and instantaneous sources. *Proc. Roy. Soc. (Lond.)* A234, 1–15.

Morvan, D., Porterie, B., Larini, M., and Loraud, J. C. (1998). Numerical simulation of turbulent diffusion flame in cross flow. *Combust. Sci. Tech.* 140, 93–122.

NAG (1991). "Numerical Algorithms Group Fortran Library," Mark 15, Vol. 2. NAG, Oxford, UK.

Ramaprian, B. R., and Haniu, H. (1989). Measurements in two-dimensional plumes in cross flow. *J. Fluids Eng.* 111, 130–138.

Raupach, M. R. (1990). Similarity analysis of the interactions of bushfire plumes with ambient winds. *Math. Comput. Modeling* 13, 113–121.

Raupach, M. R., and Thom, A. (1981). Turbulence in and above plant canopies. *Ann. Rev. Fluid Mech.* 13, 97–129.

Rouse, H., Yih, C. S., and Humphreys, H. W. (1952). Gravitational convection from a boundary source. *Tellus* 4, 201–210

Schatzmann, M. (1979). An integral model of plume rise. *Atmospheric Environment* 13, 721–731.

Stocks, B., and Walker, J. (1968). Thermocouple errors in forest fire research. *Fire Technol.* **4**, 59–62.

Stott, P. (1986). The spatial pattern of dry season fires in the savanna forests of Thailand. *J. Biogeog.* **13**, 345–358.

Thomas, P. H. (1963). The size of flames from natural fires. *In* "9th Symp. (Int.) on Combustion," pp. 844–859. The Combustion Institute (Pittsburgh).

Thomas, P. H. (1965). Buoyant diffusion flames: Some measurements of air entrainment, heat transfer and flame merging. *In* 10th Symp. (Int.) on Combustion," pp. 983–996. The Combustion Institute (Pittsburgh).

Trabaud, L. (1979). Etude du comportement du feu dans la garrigue de chêne kermes à partir des températures et des vitesses de propagation. *Ann. Sci. Forest.* **36**, 13–38.

Trelles, J., McGrattan, K. B., and Baum, H. R. (1999). Smoke transport by sheared winds. *Combustion Theory and Modeling* **3**, 323–341.

Tunstall, B. R., Walker, J., and Gill, A. M. (1976). Temperature distribution around synthetic trees during grass fires. *For. Sci.* **22**, 269–276.

Van Wagner, C. E. (1973). Height of crown scorch in forest fires. *Can. J. Forest Res.* **3**, 373–378.

Van Wagner, C. E. (1975). Convection temperatures above low intensity forest fires. *Can. For. Service Bi-monthly Res. Notes* **31**, 21 and 26.

Weber, R. O., Gill, A. M., Lyons, P. R. A., and Mercer, G. N. (1995a). Time dependence of temperature above wildland fires. *CALM Sci. Suppl.* **4**, 17–22.

Weber, R. O., Gill, A. M., Lyons, P. R. A., Moore, P. H. A., Bradstock, R. A., and Mercer, G. N. (1995b). Modeling wildland fire temperatures. *CALM Sci. Suppl.* **4**, 23–26.

Williams, F. A. (1982). Urban and wildland fire phenomenology. *Prog. Energy Combust. Sci.* **8**, 317–354.

Williamson, G. B., and Black, E. M. (1981). High temperature of forest fires under pines as a selective advantage over oaks. *Nature* **293**, 643–644.

Xanthopoulos, G. (1991). "A Model for the Prediction of Crowning in Forest Fires." PhD thesis. University of Montana.

Yih, C. S. (1951). Free convection due to a point source of heat. *In* "Proc. 1st U.S. Nat. Congre. Appl. Mech.," pp. 941–947.

Yih, C. S. (1953). Free convection due to boundary sources. *In* "Fluid Models in Geophysics," Proc. First Symposium on the use of Models in Geophysics, pp. 117–133. U.S. Gov't Printing Office, Washington.

Yih, C. S. (1969). "Fluid Mechanics." McGraw-Hill, New York.

CHAPTER 8

Coupling Atmospheric and Fire Models

MARY ANN JENKINS
Department of Earth and Atmospheric Science, York University, Toronto, Ontario, Canada

TERRY CLARK AND JANICE COEN
National Center for Atmospheric Research, Mesoscale and Microscale Meteorology, Boulder, Colorado

I. INTRODUCTION

The interactions of forest fires and air flow are highly nonlinear. The heat and moisture supplied through the burning of ground and canopy fuel during a forest fire create extreme levels of buoyancy forcing. The horizontal gradients of buoyancy produce vortices or fire whirls of tornado strength, which in turn affect the nature of the fire spread through advection of hot gases and burning material, while fire vortices enhance mixing of air with the flame which leads to higher flame temperatures, increased combustion efficiency, and greater intensity. Winds at the fire scale can be either strongly modified or even solely produced by the fire, depending on the level of atmosphere–fire coupling. This coupling or feedback occurs over spatial scales from tens of meters at the flame front to kilometers on the scale of the total burn area.

In this chapter, we attempt to describe a fairly recent and major advance in the modeling of wildfires: the coupling of a cloud-resolving numerical prediction model with a simple fire-spread and wildfire behavior model, so that the atmosphere–fire is treated as a single, dynamical system. With this modeling approach, it is possible to simulate the small-scale atmosphere–fire interactions and feedbacks that are important to wildfire behavior, especially severe wildfire behavior, and the possible impacts of evolving, larger scale atmospheric forcing on the fire and vice versa.

A "small" fire is more or less a surface fire, in steady state, that grows in size but not in intensity. Small fires spread through more or less constant heat transfer by radiation with additional heat transfer by convection, or winds, at the head of the fire. Small fires are governed by the properties of the fuel bed in which the fires burn and the air movements within and a few meters above the fuel bed. A small fire becomes a "large" fire when the fire's intensity becomes sufficiently great to produce a convection column stronger than the ambient wind field. A fire that multiplies its rate of energy output many times in a short period and burns with an intensity far out of proportion to apparent burning conditions is called a "blowup" fire.

Large fires do not conform to fire behavior expected of the far more frequent, low-intensity surface fire. Large fires are said to "make their own weather" because they can noticeably alter the temperature, humidity, and wind fields in their vicinity. A large fire exhibiting extreme behavior can put up a convection column to a height of 7000 m or more. Large fires and blowup fires have higher fire-line intensities (defined as a heat flux or the rate of fire line spread times the heat per unit area generated by the available fuel), where high fire-line intensities are of two forms: a high rate of spread when the fire is wind driven or high rate of heat release when the fire is convection dominated. Large fires have different modes of propagation and distribution patterns than the ordinary surface fire. Their fire fronts can have significantly higher spread rates (up to 61 m/min as opposed to 6 m/min for surface fires with steady state headfire spread rates; McRae *et al.*, 1989) and can accelerate under the influences of convection. Convective processes in the column can produce extreme fire behavior such as crown fires or large-scale spotting caused by firebrands lofted ahead of the fire by wind. In the aftermath, the burn pattern may show complex boundaries and inexplicably unburned islands. It is the task of fire researchers to explain such behavior and patterns, and ultimately to predict them.

In some cases, a large fire becomes a conflagration, a fire that takes on storm characteristics with associated strong, mesocyclonic rotation. Massive fire whirls up to 400 m wide and several thousand meters in height can develop on the outer perimeter. Whirls on the edge of the burn can be powerful enough to rip out standing trees and send them up into the convection column. They can travel for considerable distances (4 km), removing slash pieces and organic material in their path, exposing the bare mineral soil underneath (McRae and Stocks, 1987; McRae and Flannigan, 1990). Large pieces of slash (4 m long, 20 cm in diameter) can be driven into the ground (45 cm into a sandy soil), much like spears. There is postfire evidence of multiple whirls within these burns.

Although relatively rare, blowup fires, large fires, and conflagrations are responsible for the major loss of life in forest fires (Chandler *et al.*, 1983), and in the western United States, 1% of the largest fires account for 80–96% of the area burned (Strauss *et al.*, 1989). Large-scale forest fires result in billions of dollars in financial losses annually in the United States alone. They can locally disrupt complete ecosystems and can, in cases of extremely large fires, eject significant amounts of smoke high into the atmosphere with possible long-range air quality and long-term global meteorological effects.

Each blowup fire or conflagration raises the question of how to recognize the conditions causing extreme fire behavior and how to predict them. The transition from a small fire to a large fire is typically sudden, 15 minutes or so, and at present, the causes for the sudden transition from an ordinary, steady state to large or blowup fire are not known. Wildfire behavior models developed by the forestry community have traditionally been used to predict fire spread rate and heat release for a prescribed set of fuel, slope, and wind conditions and are designed primarily to assess the behavior and risk of surface fires. These models are not capable of assessing the complex behavior and risk of fires that have progressed past the small stage.

In the following sections, the basics of coupled fire–atmosphere numerical prediction, an approach that is capable of simulating the complex behavior of large and small fires, are presented. The vorticity dynamics that play an important role in determining fire behavior are described. To define what is meant by fire–atmosphere coupling, a nondimensional Froude number, the ratio between the kinetic energy of the flow in a fire and the potential energy provided by the fire, is presented and its importance to fire dynamics and extreme fire behavior is discussed. The elements of a fire model and of a three-dimensional numerical atmospheric prediction model, and how these models are coupled to produce a physically based dynamical fire–atmosphere model appropriate for wildfire modeling, are described. The sources of error in a numerical approach to coupled fire–atmosphere prediction, the problems of resolution, and the technical and physical difficulties to overcome are discussed. Recent numerical modeling studies of wildfire behavior using a coupled atmosphere–fire model are presented, and how infrared video technology is used to observe and analyze important physical features and properties of fire-scale convection from a safe distance is described. Finally, we discuss where the coupled fire–atmosphere modeling approach is, where it is likely to go in the near future, and what it should become.

II. VORTICITY DYNAMICS IN A FIRE

When the fire's intensity becomes great enough so that the flow responds strongly to the heating supplied by the fire, the fire can force its own fire-scale circulations. Convective eddies are seen as horizontal rolls or vertical vortices. In a convectively dominated fire, intense vertical rotors can develop, with wind speeds up to 100 m s^{-1} (Corlett, 1974; King, 1964). To understand the convective processes that come into play, we must understand the dynamics of the fluid and how these rotating rolls or vortices develop.

Vertical shear in the wind (wind speed changing with height) provides a source of rotation about a horizontal axis. To understand why, look at the situation depicted in Figure 1. Imagine a paddle wheel placed in the wind field, with the wind coming from the left. Since the wind hitting the bottom of the paddle wheel is stronger than the wind hitting the top, the wheel spins in a counterclockwise direction (Figure 1a). Similarly, parcels of air in the shear wind field rotate because the bottoms of the parcels are moving faster than the tops. This spin of individual air parcels or particles is called vorticity. When the winds begin to interact with a strong updraft, as shown in Figure 1b, the rotation about the horizontal axis is tilted, becoming rotation about a vertical axis. Similarly, the rotation of individual air parcels in the fluid is also tilted into the vertical producing vertical vorticity.

Strong horizontal gradients in buoyancy near the surface along the leading edge of the fire front produce especially strong horizontal rotation or vorticity. When the horizontal vorticity is tilted into the vertical by the fire updraft, very intense vertical vorticity can develop where the large buoyancy gradients and horizontal gradients of the vertical wind coexist.

Vertical vorticity is also produced by the solenoidal effect. In the absence of other forces, fluid parcels or particles will not spin when density and pressure fields are constant or when they change but in the same direction. But when

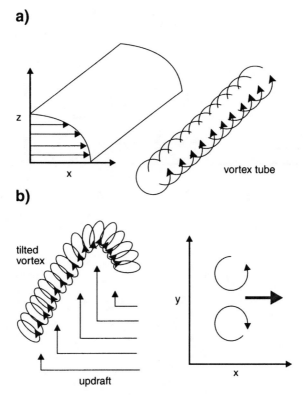

FIGURE 1   Spin about a horizontal axis, illustrated by the vortex tube, is present when wind speeds change (in this case decrease) with height (a). When the vertically sheared winds are deflected by a strong updraft (b), the axis of spin is tilted into the vertical. For the conditions depicted in (a) and (b), the resulting upflow has a counterclockwise spin to the north of the updraft core and a clockwise spin to the south.

horizontal gradients of pressure and density occur and are not parallel, fluid parcels experience opposing forces over their lengths, that combine to exert a torque, causing them to rotate, producing vertical vorticity.

To illustrate this, consider what happens to a small rectangular parcel or bar of fluid in the situation depicted in Figure 2. First imagine a pressure gradient in the fluid, perpendicular to the parcel and uniform along its length. In this case, the parcel simply moves through the fluid from high to low pressure without rotating. Now imagine that the fluid density is not constant but is larger at one end of the parcel than the other. The forces $\vec{F}_{y_1}$ and $\vec{F}_{y_2}$ acting on either end of the parcel are approximated by

$$\vec{F}_{y_1} = -\frac{1}{\rho_1}\frac{\Delta p}{\Delta y}, \qquad \vec{F}_{y_2} = -\frac{1}{\rho_2}\frac{\Delta p}{\Delta y}$$

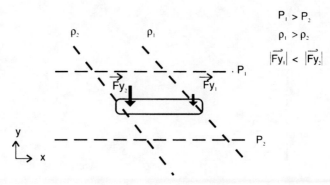

FIGURE 2 Production of vertical vorticity by the solenoidal effect. See text for explanation.

Since

$$\rho_1 > \rho_2, \qquad \frac{1}{\rho_1} < \frac{1}{\rho_2}$$

then

$$|\vec{F}_{y_1}| < |\vec{F}_{y_2}|$$

The torque applied by the forces, $\vec{F}_{y_1}$ and $\vec{F}_{y_2}$, causes the fluid bar to rotate counterclockwise. Here ρ is fluid density, and Δp is the difference in fluid pressure over distance Δy, where y is the horizontal north–south direction.

Once tilted into the vertical or generated by the solenoidal effect, vertical vorticity can be intensified by the processes of horizontal convergence and vertical stretching. Imagine an air parcel enclosed by a surface. When additional air flows horizontally through this surface into the parcel, we say that mass converges into the parcel or that the parcel experiences horizontal mass convergence. More air enters the air parcel ($u + \Delta u$ or $v + \Delta v$, where u and v are wind components in the horizontal east–west or x and north–south or y directions, respectively, and Δu, Δv, are changes in u, v) than leaves it (u or v). If no adjustment in its physical parameters follows, the parcel experiences compression or, what is the same thing, an internal pressure increase. The parcel expands to relieve this outward-directed pressure force. But the horizontal mass influx prevents horizontal expansion to a great degree, so the parcel expands or stretches vertically.

Vertical stretching and horizontal convergence also occur in the following way. Imagine a vertically spinning air parcel as a short, wide vortex tube. The vortex tube will stretch vertically if the top rises faster than the base. The difference in rise is $\Delta w / \Delta z$, where Δz is distance in the vertical and Δw is the change in vertical velocity w over Δz. When $\Delta w / \Delta z$ is positive, there is more air leaving the top of the air parcel ($w + \Delta w$) than entering the bottom (w). This is called divergence, and the tube is being stretched vertically. But the parcel's mass must be conserved. Conservation of mass therefore requires horizontal convergence, which shrinks the tube's radius, narrowing the diameter of the spinning air parcel.

A fundamental law of physics requires that an air parcel's angular momentum about its vertical axis, which is proportional to its momentum (mass times velocity) and to its distance from the axis, be conserved. As the air parcel's distance from the central axis of rotation decreases, its velocity must increase. This is the principle of conservation of angular momentum and can be thought of as the "ice skater effect." The air parcel begins to spin faster, just as a figure skater spins faster when she pulls her arms into her body.

Horizontal convergence and vertical stretching narrow the diameter of the spinning air parcel, further increasing its rotation as the angular momentum of the air parcel, now rotating over a smaller distance, is conserved (i.e., the ice skater effect). Vertical vorticity therefore is intensified.

Vorticity can also increase or decrease at any given locality just by air with different vorticity blowing into its region. Local vorticity increases when replaced by air of greater vorticity and decreases when replaced by air of lesser vorticity. This process is called advection, and it works in both horizontal directions and in the vertical direction. The term advection is normally reserved for horizontal transport, whereas vertical advection is known as convection.

Finally, turbulent or frictional drag in a fluid can affect vorticity. Turbulent or frictional drag can sometimes increase rotation and, very importantly, sometimes reduce or even eliminate rotation in a fluid. Friction slows wind near any solid boundary such as the ground, and in this case friction always acts in the opposite direction from the wind. Frictional drag against the ground reduces rotation or causes spin-down regardless of wind direction.

A vortex is air moving around a central axis about as fast as it moves toward and along the axis. The development and intensification of a vertical vortex is a combination of multiple processes working together: the generation of vertical vorticity by the solenoidal effect and tilting, the dynamic-pipe effect, advection of vorticity in the vertical, and convergence and vertical stretching (Appendices II and III).

Figure 1 illustrates the generation of vertical vorticity by tilting. Once the horizontal vorticity in Figure 1a is tilted into the vertical (Figure 1b), the local vorticity owing to this tilting will be positive to the north of the updraft core and negative to the south. As a result, a counterrotating vortex pair is established with counterclockwise rotation to the north and clockwise rotation to the south of the initial updraft.

Along the rotating column of each vortex, the pressure field is in balance with the strongly curved wind field. The inwardly directed force acting on air parcels as a result of the reduced pressure at the center of the column is countered by the outwardly directed centrifugal force resulting from the parcels' rotation about the center. In this condition, called cyclostrophic balance, air moves easily around and along the axis of the vortex, but radical motions toward or away from the axis of rotation are strongly suppressed. The rotating column acts as a "dynamic pipe"; it is like a hose of a vacuum cleaner, except that instead of being channeled by the wall of the hose, the airflow in the vortex is constrained by its own swirling motion.

The low pressure in the center of the vortex provides a vertically directed pressure gradient force (the centrifugal pump effect described in Appendix III). Since air is accelerated from high to low pressure, an updraft develops from below, or a downdraft develops from above, in the rotating core. When the vertical pressure gradient force directs air parcels upward, the air flowing along the vortex's axis is sucked in through its lower end, and as air parcels converge into the base of the pipe, they turn and accelerate upward. In doing so, they are stretched vertically (as in Figure 3). Stretching narrows the diameter of the vortex, further increasing the angular speed of its winds and the vorticity of its air

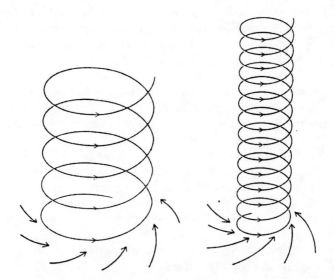

FIGURE 3 Convergence and stretching intensify the vertical rotation initiated by tilting and extend it upward. Once rotation is vertical, a balance between the inwardly directed pressure gradient force and the outwardly directed centrifugal force develops. The centrifugal force suppresses flow toward or away from the axis of rotation (the "dynamic pipe" effect), while the low pressure at the center of rotation provides a vertically directed force. When the force is directed upward, the winds converge at the lower end of the rotating updraft. The enhanced convergence compresses the column and stretches it upward. Because the column is narrower, it must spin faster to conserve its angular momentum.

parcels, as the angular momentum of the air, now rotating over a smaller distance, is conserved (i.e., the ice skater effect).

Thus, for an upwardly directed pressure gradient force, the dynamic pipe intensifies the rotation at its lower end, which in turn extends the pipe downward as the centrifugal force gets strong enough to establish cyclostrophic balance. Owing to inertia, inflow that gets sucked in through the pipe's lower end can actually overshoot its equilibrium radius, conserving its angular momentum and picking up speed as it approaches the center of the core before turning sharply to spiral upward. As a result, the vortex is narrower at the bottom, and the highest wind speeds are found in a small ring at the base of the vortex.

Frictional drag against the ground upsets the cyclostrophic balance of the vortex by slowing the wind and reducing the centrifugal force to zero. But the inward radial pressure-gradient force does not vanish there. As a result, strong inflow close to the ground, driven by the unbalanced inward pressure gradient force, transports fluid parcels closer to the axis than would have otherwise been possible without friction, whereupon the flow turns vertical. The resulting wind speeds can be significantly greater than those of the circulation in the vortex above the surface friction layer.

Note that while vertical progression due to the dynamic pipe effect can take place, the vortex can also build upward or downward through vertical advection of vorticity and intensify by convergence and stretching. When vorticity and radial inflow are constant with height, high angular momentum air approaches the axis from below and aloft simultaneously. The vortex in this case forms nearly independently of height. In a fire, however, buoyancy forcing supplied through the burning of surface fuel and the generation of near-surface vertical vorticity by the solenoidal effect and tilting likely provide that the vertical vorticity and radial inflow are maximized near the ground. Apparently due to vertical advection of vorticity and horizontal convergence and vertical stretching, the fire vortex forms from the ground upward. It is also observed that fire vortices can be short-lived. One possible explanation for this is that the pressure gradient forcing can boost updraft strength at one level, quickly destroy it at another. A detailed examination of where and what processes dominate to build and destroy fire vortices has yet to be done.

It is also possible that, as a result of powerful updrafts in the cores of vortices, compensating downdrafts form between the updraft cells, which, if strong enough, can separate and move the centers of convection to the right and left of the original updraft depicted in Figure 1b.

The vortex dynamics just presented are purposely idealized or simplified to convey the essence of the problem. A seemingly simple complication arising in reality is that rotating vortices are observed to sometimes break apart, as discussed in the next paragraph. Exactly how vortices restructure themselves in such cases is unknown and something that fire modeling may eventually resolve. Neither is the observed flow field necessarily a one-cell vortex characterized by an updraft at the core of the vortex. Powerful vortices can also develop an internal multiple vortex structure that has a downdraft in the central core surrounded by two or more subsidiary vortices with updrafts along their vertical axes. To learn more about the complications of the internal structure of laboratory vortices, we refer the interested reader to Church and Snow (1979) and Gall (1982).

Vertical rotation within fires can result from at least two similar but different mechanisms. Adopting the terminology from Clark *et al.* (1999), the first mechanism is enhanced vortex tilting. In this case, the tilting of horizontal vorticity is believed to result from a local region of buoyancy that acts strongly enough to break or disconnect the tilted vortex, leaving one or two rotating, standing hot fire whirls or plumes, which transport heat almost vertically into the atmosphere.

The second mechanism is the "turbulent burst"- or "horseshoe"- or "hairpin"-shaped vortex. The turbulent burst is a concept developed in the study of boundary layer turbulence (e.g., Moin and Kim, 1982; Kim and Moin, 1986). The boundary layer is defined as the portion of the atmosphere in which the flow field is strongly influenced by interaction with the surface of the Earth. Again vortex tilting occurs, but in a manner that results in an unbroken vortex that moves both vertically and horizontally; as the vortices rise and come closer together, their combined motion results in the vortex tilting forward at a relatively sharp angle giving a hairpin shape. When the horizontal motion is strong enough, the rotors shoot or burst forward over unburned fuel, sometimes for considerable distances. It is believed that this type of vortex tilting led to the hairpin vortex observed by Radke *et al.* (2000). Observations captured a flame shooting horizontally as far as 100 m in only 2 s (Figure 4), and the fire fighters wisely stayed clear of this particular fire behavior at the time of the observations.

For a more in-depth description, a mathematical treatment of vorticity is given in Appendix I, the development of vertically rotating convective cells is described in Appendix II, and the development of an updraft in a rotating convective cell is described in Appendix III.

III. COUPLING BETWEEN ATMOSPHERE AND FIRE

As discussed in previous sections, winds at the fire scale can be either strongly modified or even solely produced by the fire, depending on the level of atmosphere–fire coupling. This coupling or feedback occurs over spatial scales from tens of meters at the flame front to kilometers on the scale of the total burn area. Therefore, the knowledge of the interaction of the fire with the atmosphere is

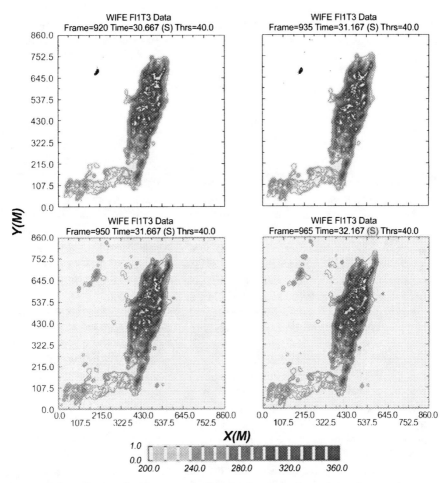

FIGURE 4 Airborne infrared imagery for fire spread analysis shows "hairpin" vortex (approximate location 753 m in y direction and 540 m in x direction) bursting forward at 70 m in 2 s. (After Radke *et al.*, 2000.)

needed to understand and predict fire behavior correctly. It is first necessary to define this feedback or coupling between the atmosphere and the fire.

In grass fires, it has been observed that the importance of wind speed diminishes after a critical speed is reached. Above certain speeds, further increases in wind speed do not stimulate additional increases in rate of spread or fire line intensity. To interpret this transition with increasing wind speed, Byram (1973) suggested a convective Froude number, F_r. The Froude number is defined as the ratio of the kinetic energy of the air and potential energy provided by the fire. The measure of kinetic energy is the relative speed of the air passing over the fire, $U_0 = U - v_f$, where U is the ambient wind speed and v_f is the fire spread (i.e., rate of spread at the head of a fire). The measure of the potential energy used for the fire is the fire-line width in the mean wind direction times the buoyancy of the heated air. One possible formulation for F_{r_i} (Clark *et al.*, 1996a) is

$$F_{r_i}^2 \approx \frac{U_0^2}{g\dfrac{<\Delta T>}{T}W_f}$$

Here T is temperature and g is acceleration due to gravity. The $<\Delta T>$ is the average or mean value of the temperature anomaly ΔT over the region of intense heating and through W_f, the width of this region. Buoyancy is defined as $g(<\Delta T>/T)$. The Froude number is a nondimensional number. The values of U_0, W_f, T, and $<\Delta T>$ can therefore differ without the combination $U_0^2/[g(<\Delta T>/T)W_f]$ being different. Providing other governing parameters are also similar, we can expect the same behavior for flow with the same Froude number (i.e., all such flows are then dynamically similar).

An increasing F_{r_i} means that for increasing ambient wind speeds the fire deposits less heat into parcels of air as they move over the fire, until at some ambient wind speed the effect can be neglected. The process can be compared to a freight train of partly filled grain cars moving under a loading chute (Chandler *et al.*, 1983). If the train moves too slowly, the cars fill up and are unable to carry off as much grain as is being fed to them. Once the train reaches a speed at which the cars do not become filled, it is able to carry away all the grain, and an increase in speed will not change its ability to do so. To illustrate large F_{r_i}, simply run your hand over a burning match; when performed at comfortably high values of F_{r_i}, the interaction between hand and fire is negligible.

F_{r_i} can be used to gauge the level and type of coupling between the fire and atmosphere. The smaller the $F_{r_i}^2$, the larger the effect of buoyancy forcing on the acceleration of the flow. When $F_{r_i}^2 \approx 1$, the flow enters a regime where both fire and dynamics are important. When $F_{r_i}^2 < 1$, the flow responds strongly to the heating supplied by the fire, and the fire regime is plume-driven or convection dominated. If fire convection is to control the larger-scale dynamics, values of $F_{r_i}^2$ should be small. In high-wind-speed cases, as in wind-driven fires, $F_{r_i}^2 > 1$, and the flow is not dominated by the heating supplied by the fire. For typical values of $U_0 = U - v_f \approx 5 \text{ m s}^{-1}$, $T \approx 300 \text{ K}$, $\Delta T \approx 50 \text{ K}$, and $W_f \approx 50 \text{ m}$, the $F_c^2 \approx 0.3$. This implies that even for small, stable fires in weak or moderate winds, the effect of buoyancy forcing on the acceleration of the flow is significant, and the wind and fire are essential parts of the same system and cannot be separated. Also implied is that small F_c^2 is a necessary but not sufficient condition for severe coupled fire–atmosphere behavior.

IV. THE ELEMENTS OF FIRE MODELING

Regardless of the level of sophistication, current forest fire models require and provide, in one form or another, certain basic information. This includes initial total mass and dry mass of fuels (mass per unit area); initial moist/dry mass ratio of fuels; the rate at which fuel, once dry, burns (mass per unit area per unit time); the amount of energy provided per mass of dry fuel burnt; and the ratio of latent (moist) to sensible heat of this energy. The fire model outputs are sensible and moisture (latent) heat from the burning fuels (energy per second). Certain additions may be necessary: a percentage of the sensible heat provided by the fire dries the fuel; canopy ignition can occur when sensible heat flux from ground fuels exceeds a threshold value (energy per mass); to emulate smoldering, a random component can be added to the burn rates; and surface fuel is ignited by contact, where ignition contact follows the predicted fire spread rate and direction.

The ability to predict the rate of fire spread is recognized as the most important single requirement for successful fire suppression. Current models of forest fire spread assume that radiative heat transfer and convection, or wind, are the mechanisms that transmit heat from the flaming zone to adjacent fuels. Radiative heat transfer is increased by fuel changes that convert more of the reaction intensity into effective heat flux. Convection transfer is strengthened by increasing winds that can also increase both convective and radiant heat transfer. The important effects of terrain on fire propagation speed are also included.

There are many problems with current wildfire and fire spread rate models, but the major drawback is that, to apply them successfully, information on fuel, topography, and winds at the local, fire-scale level is required, information that is presently not available to fire managers and fighters. And because convection is a key mechanism for fire propagation after the incipient stage, knowledge of the small-scale atmospheric circulations in and around a wildfire is primary. In most applications, however, single values of wind speed and direction are applied over a large area and long time period. Therefore, microscale processes that affect the fire spread rate and involve the variable winds within the fire and adjacent fuel are not treated, whereas the formulae currently used to link the fire spread rate to the ambient winds ignore any feedback between the small-scale air motions and the fire. The solution to these deficiencies is to couple fire behavior and fire-spread-rate models to a dynamical atmospheric model.

V. MODELING THE ATMOSPHERE

To model the atmospheric circulation in and around a wildfire properly, a prognostic three-dimensional numerical model based on fundamental fluid dynamics and capable of resolving convective-scale motions is required. A model of fundamental fluid dynamics consists of the Navier-Stokes equations of conservation of momentum, conservation of mass, and conservation of energy. Three dimensions are essential as fire dynamics are inherently three-dimensional due to advection, the solenoidal effect, vortex tilting, horizontal convergence, and turbulence or frictional drag.

The Navier-Stokes equations are smooth analytical functions, or differential equations, which in the case of conservation of momentum describes the instantaneous relation between position, velocity, and acceleration throughout the motion. A numerical model discretizes these differential equations to work at a finite number of regularly spaced locations called grid points. Each grid point holds the average value for a volume of surrounding air. This volume is called a grid cell or grid volume. To determine the functions at a future time, the equations are rearranged into forecast equations, and the functions are forecast for one-time step, utilizing present (and sometimes past) conditions to determine future conditions. Each operation in the analytical equations is expressed as a finite difference between grid variables. Computational time is spent for each grid point in the domain, and the computation is repeated for a succession of time steps in order to reach a forecast of possibly several hours duration. In a Cartesian coordinate system, the sizes of the grid cells in three directions are Δx (east–west), Δy (north–south), and Δz (vertical).

There are several sources of error in a numerical forecast. One is round-off error. The computer is able to store only a limited amount of information; as a result, real numbers are only approximated. Truncation is another source of error. An analytical variable is represented at a grid point as a function of its values at other grid points, the result being an infinite sum of terms, each of greater power of Δx or Δt (i.e., a Taylor's series expansion). For practical reasons, the computer retains the terms of the lowest power, the series is truncated, and higher power terms (which slightly improve accuracy) are neglected. Another error source is numerical instability, when the numerical solution rapidly diverges from the correct solution. Truncation is one cause, but numerical instability can also occur if the wind speeds are large, the grid size small, and the time step Δt not small enough. A necessary requirement for stability in the x direction with ambient wind speed U is

$$\Delta t \leq \frac{\Delta x}{U}$$

with similar requirements in the y and z directions.

A numerical model of the atmosphere–fire system must use, from an atmospheric scientist's perspective, a grid mesh with small Δx, Δy, and Δz values; currently typical sizes of grid cells in the domain of the fire can be tens of meters. Therefore, to preserve numerical stability, the time step must be also be reduced. Typical time-step durations Δt are on the order of fractions of a second in the fire domain. A small time step has consequences for the total run time of the calculation and, since error is accumulated by the many more time steps required to reach a particular time, for the accuracy of the numerical solution.

The errors in initial conditions also affect the long-range predictability. The Navier-Stokes equations are nonlinear; as a result, the system is very sensitive to initial conditions. Such sensitivity means that a substantially different forecast can result from only slightly different initial conditions and that the forecast always becomes inaccurate with time. To minimize this error, initial conditions are specified as accurately as possible, and only short-term forecasts are considered trustworthy.

And lastly and most importantly, the physics of the fire–atmosphere system must be faithfully represented by the model formulation. The final source of error is then how accurately subgrid scale atmospheric and combustion processes are currently determined in terms of grid-scale (and therefore model forecasted) parameters. This is a central problem in numerical forecasting, and it is discussed further in the next section. Ideally the model framework should also be amenable to adding chemistry for pollution or smoke dispersal studies.

Because the computational demands of coupled atmosphere–fire modeling are great, additional features that diminish computational load while providing an accurate forecast are necessary.

One such feature is a vertically stretched, terrain-following coordinate system. In this system, the lower grid levels follow the contoured surface, while the vertically stretched coordinate allows a variable vertical resolution to the solutions of the equations near the surface and in regions of rough terrain.

Another such feature is two-way interactive grid nesting, also called grid refinement or multilevel grid nesting. Diverse scales of motion influence a wildfire's behavior, from the scale of the regional weather pattern (1000 km) down to the local and fire-generated flow (tens of meters and less), and multiple levels of interactive-grid refinement allow the numerical model to run over such a large range of scales. A large outer domain with large-sized grid cells and time steps, and initialized with coarse resolution meteorological data from, say, General Circulation Forecast Models, determine satisfactorily the largest scale atmospheric circulation. Inside or nested in this domain are successively smaller and smaller domains, each containing topography at the grid resolution of that domain, while fields are matched between the boundaries of the coarser grid and finer grid domains. Eventually the meteorology is forecast in the smallest domain, where fine-scale grid cells are used to resolve the fire's convective-scale motions.

A further refinement is an adaptive grid mesh in which the finite difference grid is refined in selected regions that may be changed in time to follow evolving features such as the moving fire front.

As in Section II, the description of a three-dimensional numerical atmospheric prediction model presented is purposely simplified to convey the essence of the approach. Although there are many more aspects involved, they are beyond the scope of this text and are therefore not included.

VI. THE COUPLED FIRE– ATMOSPHERE MODELING APPROACH

We have seen that even for small, stable fires coupling between the fire and the atmosphere can be significant. Temperature, humidity, and wind fields are noticeably altered by the fire, which in turn alter the fire. Once temperature and humidity fields are altered, pressure gradient and buoyancy forces come into play, and the velocity field changes. Velocity field changes affect fire behavior and fire spread rates, which can advance the fire front, igniting more fuel, releasing sensible and latent heat, and repeating the process.

The coupling between the fire and the atmosphere occurs mainly through the fluxes of sensible and latent heat from the ground fuel and canopy burns. Radiation is also important, both in preheating the fuels ahead of the fire and shortening their ignition time and to the atmospheric circulation in and above a wildfire. To couple current fire and numerical atmospheric prediction models, radiation and sensible heat and moisture fluxes associated with the fire are included as subgrid-scale sources of heat in the atmospheric model's conservation of energy equation. Wind is also modified by the forest, especially the vertical wind profile, and the effect of surface and forest canopy drag on the velocity field is a subgrid-scale momentum source in the atmospheric model's conservation of momentum equation.

The heat fluxes and radiation from the fire are absorbed by the air, and a method for incorporating this effect in the atmospheric model is required. A very simple approximation is that the amount of radiation absorbed into the atmosphere decays exponentially with height, where the depth of extinction may range from 50 to 100 m in the vertical. In reality, the interaction between radiation and the atmosphere is much more complex than this and is not well understood. For example, an accurate treatment of flame front radiation onto unburned fuel and resulting fire spread strongly depends on a good knowledge of the vortex dynamics at the fire line. To understand the impact of radiation on fire spread, rotors produced by the mechanism of enhanced vortex tilting might be approximated as a radiating vertical wall of flame, whereas the forward bursts of flames in hairpin vortices cannot be reasonably approximated in this way. To date, this type of interaction has not been taken into account.

Yet, however they are handled, the effects of convection, turbulent mixing and frictional drag, and black body radiation occurring on scales not resolved by the numerical atmospheric model must be represented accurately.

Because current wildfire formulae are mainly algebraic and based on statistical or empirical ideas and do not describe the microscale processes that affect the fire spread rate and involve the variable winds within the fire and adjacent fuel realistically, coupling the fire model with the atmospheric model cannot completely accomplish the task of including the feedback between the small-scale air motions and the fire. The solution is to develop a physically based, dynamic wildfire model. Yet, even if a fully physically based wildfire model were available to couple with a fully physically based atmospheric model, the technical and conceptual difficulties to be overcome are tremendous.

Within fires, there are small vortical structures on the millimeter scale that tightly bend the flame fronts, up to vortex structures on length scales of many meters. The time scales range from milliseconds in submillimeter eddies to the puffing frequencies on the order of seconds. In addition to these fire-convective length and time scales are diffusive length and time scales, from the submillimeter and submillisecond range required for resolution of chemical species diffusion gradients in flame zones, to the tens of meters and thousands of seconds required to resolve the length and time scales of fire/object interaction. More than five orders of magnitude in temporal and spatial resolution are required to simulate directly all the relevant scales within the fire. Coupling a numerical fire simulation with a numerical atmospheric model adds a further five or more orders of magnitude in resolution when attempting to incorporate atmospheric forcing by meteorological conditions on the regional weather scale.

Given the present level of computational resources, no numerical simulation tool is capable of simultaneously resolving all required length and time scales associated with forest fire phenomena. The problem is easily illustrated by considering the following: A numerical model basically divides three-dimensional space into grid cells with volume $\sim(\Delta x)^3$, and the computations will locate a moving fluid particle within one of these cubes. If greater resolution and accuracy are desired, a finer mesh with smaller grid cells is used. To improve the accuracy tenfold for instance, the size of the grid cell must be $\Delta x/10$ instead of Δx. There are $10^3 = 1000$ such small cubes in a single large one, all of which must be allocated some place in the memory of the computer. This means that the computations which can trace a parcel's location within 100 m in 1 s will run for almost 20 minutes to pinpoint the same trajectory within 10 m. By this time, of course, the model forecast can be meaningless, the parcel having in reality long since left the computed position. Alternatively, to find the answer within 1 s, some way must be found to perform the same computations 1000 times faster, which cannot be achieved by technological means only. Today, speed increases tenfold, not a thousandfold, from one generation of computers to the next. (The current observation is that computer speed and memory capacity double about every 18 months.) This problem is not specific to coupled fire–atmosphere modeling; in every area of science, there are a great many computations that cannot be performed now or in any foreseeable future.

Basically there are two approaches to dealing with the numerical modeling of phenomena operating on a large range of scales. One is starting with the smallest length scales and resolving all the physical processes up to the largest length scales that can be computed. The other is starting with the largest length scales and resolving downscale with as many grid cells that can be computed. Although the first approach is preferable from a scientific perspective, the largest length scales within a fire model that currently can be numerically modeled are on the order of centimeters. Even with the tremendous growth in both speed and memory for computing hardware, this approach will likely be limited to length scales significantly smaller than the resolution of the smallest grid used by coupled atmosphere–fire models. The second approach uses a technique known in the atmospheric science community as subgrid-scale parameterization. Physical mechanisms with length and time scales below the minimum that can be resolved by numerical models are expressed either empirically or mathematically in terms of observed variables or model forecasted parameters in an ad hoc attempt to represent the physical phenomena as realistically as possible. Subgrid-scale parameterization is not specific to atmospheric numerical modeling; this is also the technique used by combustion scientists to model turbulence in the flow, combustion processes, soot formation, and thermal radiation of combustion products in a fire, along with the turbulent eddy viscosity or friction near solid surfaces. Attempts are made to develop transport models that connect fire propagation rates to the full conservation equations of energy, momentum, species concentrations, mass, and turbulence and that represent the essence of a combination of many microscale processes without resolving each process in complete detail (Linn, 1997). In both atmospheric and fire combustion numerical modeling research, subgrid-scale parameterizations have proven to represent successfully many unresolved physical processes in the system to the degree that is needed, and the development of and improvements to parameterizations are ongoing efforts.

A. THE IMPORTANCE OF IGNITION TRACING ON FIRE–ATMOSPHERE COUPLING

One major technical difficulty with a coupled fire–atmosphere model is ignition tracing. In a numerical model, the fuel is necessarily divided into grids. The simplest way to advance the fire through the fuel grid mesh is to ignite entire fuel cells on contact. For example, four tracers are assigned to a grid cell once the ground fuel for that particular cell is ignited. Based on the prescribed spread rate, the tracers move parallel with and against, and normal right and left of, the tracer-advecting wind. Once a ground tracer reaches an unburned grid box, that fuel cell is ignited, and four new tracers are initiated at the point of contact. Tracers should not move into burned out or burning grid boxes. One of the immediate problems with this approach is that ignition tracers do not necessarily maintain or move to increase the fire perimeter within a fuel box, and negative distances can be covered. This happens when the local fire convection changes the wind direction, and an ignition tracer that was moving with the wind suddenly finds itself in an opposing wind condition. With this simple approach to ignition parameterization, it is difficult to prevent the model ignition tracer moving back into burned or burning regions within a fuel grid cell. Obviously this simple ignition parameterization fails to maintain or move the ignition pattern to increase the fire perimeter within a fuel box in a realistic fashion.

Another result of this simple approach to ignition is that the problems of grid mesh size are exacerbated. Model resolution is always an issue, but it is especially so for the fire–atmosphere system where model resolution has an extremely important influence on heat transport away from the fire. When the entire fuel cell is allowed to ignite on contact and the model resolution is too coarse, the model fluid becomes inefficient at transporting heat, artificially changing the heat balance.

When the fire begins, there is initially no convection to remove heat efficiently, causing very high air temperatures to build near the surface. Once the convection is well established, the temperature equilibrates to significantly lower values as the convective heat transport balances the heat flux from the fire. The time the fire takes to establish this balance in a numerical simulation using a 20-m grid resolution is approximately 1–2 minutes. After convective adjustment, the atmosphere transports heat away from the fire such that the convection is in balance with its heat source. The sensible heat flux into the atmosphere is calculated as

$$F_s = c_p <\rho w' T'>$$

where c_p is the specific heat capacity at constant pressure p, ρ is air density, and w' and T' are the respective vertical velocity and temperature perturbations or fluctuations from the mean. The $< >$ denotes the average or mean value over the region of intense heating. When using a 20-m grid resolution, realistic numerical model values for the maximum vertical velocity anomaly Δw are approximately $20-30$ m s^{-1}, and for the maximum temperature anomaly ΔT are approximately $40-80$ K over the region of intense heating. A rough estimate of F_s supplied by these model perturbations is, therefore,

$$F_s \approx c_p <\rho \Delta w \Delta T> \approx 0.8 \text{ to } 2.4 \text{ MW m}^{-2} \tag{1}$$

which easily accounts for the ~ 0.8 MW m^{-2} of sensible heat flux typically emitted by the fire. ΔT is limited to values as low as $40-80$ K in a forest fire due to the efficient heat transport by convection, even though the flames may have radiant temperatures of $\sim 800-1600$ K.

To establish a horizontal fire resolution when the entire fuel cell ignites on contact, take $\Delta x_{fuel} \leq v_f \tau$, where τ is the time it takes the fire to burn the small fuel, and v_f is the fire spread rate. A fire line typically moves 10–25 m in time τ. For $v_f \approx 0.5$ m s^{-1} and $\tau \approx 50$ s, $\Delta x_{fuel} \leq 25$ m should provide a continuous fire behavior. If Δx_{fuel} is too large, the heat flux from the fire is discontinuous, which forces air to always readjust to the fire as it moves through the fuel grid mesh. Because the convective adjustment time is comparable to the time required for the small fuel to burn, namely about 100 s, the effects of the heat flux from a model with coarse fuel resolution will not average out with time. A $\Delta x_{fuel} = 10$ m horizontal resolution for the fire is generally sufficient to avoid discontinuities in the numerical model solutions when the entire fuel cell ignites on contact. This relatively small grid cell necessarily increases the computational load.

If horizontal and vertical resolutions in the atmospheric model are substantially reduced, then the heat flux is input into the model atmosphere in an unrealistic fashion, and the model fluid becomes inefficient at transporting heat. The numerical results are much higher temperature fluctuations and a much smaller convective Froude number. An almost immediate model blowup fire with uncontrollable fire spread rates in excess of 50 m s^{-1} can occur. The ignition parameterization and resolution must not dictate the values of F_s and F_{r}, and thereby the overall fire behavior.

An effective tracer ignition parameterization is instead where the subgrid-scale ignition pattern moves through each fuel cell at the fire spread rate directed by the fire-scale winds. What is needed and what has been recently developed is an improved tracer ignition parameterization in which the coordinates of ignition tracers define a particular burn area within the cell. These coordinates are time dependent and allow fires of arbitrary orientation and shape to move through a mesh of fuel grids at fire spread rates, with minimal adverse effects on the model that result when the entire fuel cell is allowed to ignite on contact. This type of ignition parameterization provides spatially smooth predictions of heat and moisture fluxes from the fire and can perform at higher grid resolutions. Larger grid mesh values can be used with this ignition approach, although it is still necessary to consider the influence of model resolution on heat transport away from the fire, and to resolving the fine-scale convective processes that maintain or move the ignition pattern to increase the fire perimeter within a fuel box in a realistic fashion.

Figure 5 shows an example of this type of tracer ignition for a spot fire expanding outward under zero wind conditions. The plot shows a region of 21-by-21 fuel cells with the solid circular contour outlining the fire line. The insert shows the regularly shaped fire line constructed from the shapes provided by four ignition tracers within each fuel cell nearest the fire line. Most of the fuel cells inside the fire line are fully ignited except for those near the fire line which are only partially ignited.

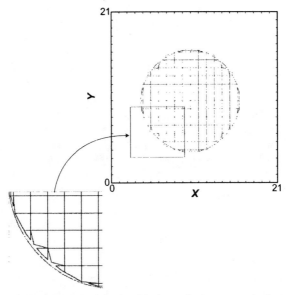

FIGURE 5   Circular-shaped fire line simulated for a spot fire in zero wind. The circular solid contour represents the fire line, and the interior rectangles represent the individual burnt or burning fuel cells. See text for explanation.

VII. IDEALIZED STUDIES OF WILDFIRE BEHAVIOR

Despite the difficulties to overcome, efforts at numerical coupled atmosphere–fire modeling are currently helping to improve our understanding of wildfire behavior and will in the future predict it. In the following we describe the numerical results from model simulations of Clark *et al.* (1996a,b) and Coen *et al.* (1998). Their simulations produced a number of realistic features of natural fires. These include the bowing of fire lines into conical shapes, the formation of fire whirls that touch down within the fire line and break it apart, microscale vortices in the fire line that strongly accelerate the local winds and fire spread rate, how negative vertical wind shear can contribute to extreme fire behavior, and the deflection along the ridge of a line fire propagating up a relatively sharp slope. We also describe the numerical results from model simulations by Bossert *et al.* (2000) and Linn *et al.* (2000). Their simulations demonstrate the sensitivity of fire behavior to different fuel characteristics and configuration. The dependent fuel parameters studied were fuel load, moisture, and type, and a two-storey fuel canopy. We discuss the use of the infrared video camera, a remote sensing device that provides high spatial and temporal observational data of wildfire winds and heat fluxes.

A. FIRE-SCALE CONVECTION
DRAWING THE FIRE LINE

The perimeter of a moving fire line normally does not spread evenly, but with a bias. The center spreads slightly more rapidly than the ends; a line of fire narrows and forms a head. In moderate ambient winds, prescribed burns ignited in a line almost always develop a parabolic shape to the fire front. This feature occurred in Clark *et al.* (1996a,b), in a series of idealized numerical experiments designed to test the effect of different constant background winds on the evolution of a short ($\approx$ 400-m-long) fire line.

In weak ambient winds ($<$ 3 m s^{-1}), the model results show a chaotic and broken fire front. Under these conditions, air feeding the hot convection column comes directly from below. The vorticity dynamics described in Section VII.B (and Appendix II) produce vertical vortices embedded in the fire line, and the erratic fire-scale, fire-generated winds break up the fire line.

Under moderate ambient wind conditions ($\approx$ 3 to 5 m s^{-1}), numerical simulations produce a stable fire line, where moderate winds move the mean surface effects of convection ahead of the fire line. The convection cell aloft remains just downstream of the fire, drawing the fire with it, and its near-surface convergence pattern induces a smoothly, continually evolving, parabolic shape to the fire line. Figure 6a shows the evolution of a fire line in a constant background or ambient wind of 3 m s^{-1}. Figure 6b shows that a slightly faster constant ambient wind speed of 5 m s^{-1} accentuates this effect. This convective feedback is a linear fire-scale phenomenon, in that smaller scale interactions between developed flow do occur, but the overall effect remains.

Once a fire line is long enough, it cannot sustain a single convective updraft column. The essentially two-dimensional updraft along a linear fire line experiences along-line roll instabilities that can cause the line to break into multiple convective updrafts or towers. Providing conditions similar to those required to form the local parabolic shape exist, multiple protrusions or fire heads form.

An example of this (Figure 7) occurred in the 1985 Onion fire in Owens Valley of California (C. George, personal communication). The fire line developed protrusions or fingers that were spaced about 1 km apart and that appeared only with an ambient wind. When the mean wind died down, the fuel between the fingers burned out, forming a more linear and stalled smoldering fire line. With the return of the wind, fire spread resumed with fingers reforming. Numerical simulations with a coupled atmosphere–fire model reproduce this phenomenon. Modeling results show one convective finger developing along an initially 400-m fire length and two convective fingers along an 800-m fire line.

These simulations demonstrate a useful aspect of a numerically coupled atmosphere–fire model: the ability to examine the effect of a single variable on the fire behavior. In these experiments only the background wind was changed.

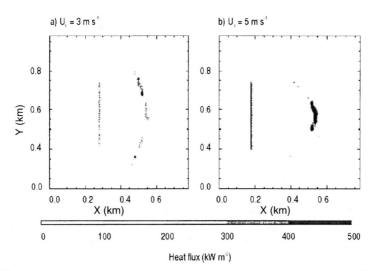

FIGURE 6 Numerical simulations of a fire line (heat fluxes greater than 0.1 MW m^{-2}) in constant background winds of (a) $U_0 = 3$ m s^{-1} and (b) $U_0 = 5$ m s^{-1}. (After Clark *et al.*, 1996a.) The bottom left to right axis is the *x* axis, the vertical axis on the left is the *y* axis, and units of *x*, *y* are km. The straight line parallel to the *y* axis is the initial fire line.

FIGURE 7 The 1985 Onion sage brush fire in Owens Valley, California. See text for description. (Courtesy of C. George, USDA Forest Service, Intermountain Research Station, Missoula, Montana.)

Other ambient meteorological conditions were set to neutral. No topography, Coriolis force (deflection of moving air by rotation of the Earth), or diurnal effects were included. The fuel was homogeneous so that any scale selection or breakup in fields and fire line were due to the dynamic and thermodynamic interactions between the fire model and the meteorology. Different fuel properties had been suggested as being responsible for the fingering fire behavior shown in Figure 7. The simulations show that the ambient wind strength, not the properties of the fuel complex, was responsible for the differences in fire line behavior.

B. FIRE-SCALE VORTICES DISRUPT THE FIRE LINE

In prevailing winds, the leeward side of the convection column is a preferential spot for the development of tornadic fire whirls and, under the right wind conditions, these fire-scale vortices can disrupt the fire line. Clark *et al.* (1996b) set up numerical experiments to test the effect on a short (400-m-long) fire line of the change in ambient wind speed with height. Two ambient wind profiles were used (Figure 8), one a constant westerly wind (3 m s^{-1}), the other a hyper-

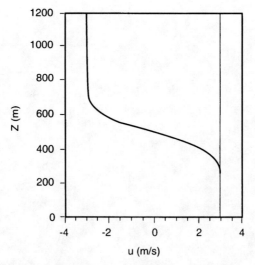

FIGURE 8   Two idealized wind profiles used to demonstrate fire behavior caused by coupling between convective motions and ambient vertical winds. See text for explanation.

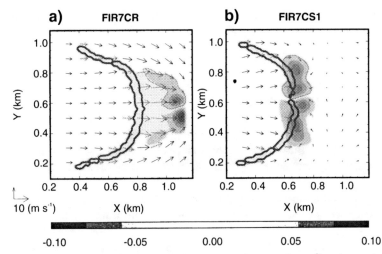

a) FIR7CR   **b)** FIR7CS1

FIGURE 9   Fire lines (solid outlines of heat fluxes greater than 0.1 MW m^{-2}) and vertical vorticity (shading) are shown for (a) the constant wind and (b) the hyperbolic tangent wind profiles in Figure 8. Vectors denote local winds at 15 m above ground level. (After Clark et al., 1996b.)

bolic tangent profile westerly (3 m s^{-1}) at the surface, decreasing with height to 500 m above the ground, and then reversing direction to an easterly wind (-3 m s^{-1}) aloft.

Figure 9 shows the near-surface vorticity patterns and fire line ignitions for the two experiments. The constant ambient wind profile case develops a parabolic fire line shape as described in Section VII.A, whereas the hyperbolic tangent profile case shows a fire line separated at its center.

Both cases produced a counterclockwise rotating vortex ahead of the fire line and south of the updraft core and a clockwise rotating vortex ahead of the fire line and north of the updraft core. The rotating vortices are the result of tilting the horizontal shear in the convection into the vertical by the fire's central updraft core, the process that is depicted in Figure 1.

In the constant ambient wind profile, the vortices, once developed, touch down, but well in front of the fire line. In the hyperbolic wind profile, the vortices develop in front of the fire line, but eventually the winds above 500 m advect the vortices back into the fire line, where they touch down in the fire. The counterrotating winds are sufficient to overcome the ambient winds, causing reversed flow and breaking of the fire line. The mathematical explanation for this development is given in Appendix II. Although the effect of vorticity dynamics on the fire line propagation is obvious, there is no evidence from the numerical results that the events in the second wind profile case weakened the fire more than the first.

C. MICROSCALE VORTICES AFFECTING FIRE SPREAD

The smallest scale vorticity at the fire front is important to the dynamics of fire spread. Tilting into the vertical of the small-scale negative shear (wind speeds decreasing with height) close to the surface provides increased winds in the direction of the fire spread. This is the situation depicted in Figure 1 and can result in a marked increase in the rate of fire spread (as illustrated by the right-pointing arrow in Figure 1b). It is simple for a numerical model to determine when and where in the fire domain maximum fire spread rates are. The strongest fire spread rates occurred for the second wind profile simulation (i.e., Figure 8 and Figure 9b), and the sequence of events was tracked by the numerical model.

The vorticity advected back into the fire and behind the fire line when the easterly wind aloft interacted with the fire; as a result, chaotic and amplified fire behavior set in at about 16 minutes into the simulation. Figure 10 shows a time sequence of vertical velocity after the amplified behavior set in. A fire-induced downdraft in the rear flow produced near-surface vertical shear that moved through the fire front, adjacent to an existing vertical rotor. Vertical tilting of this shear by local updrafts intensified the rotation rate of vertical rotor leading to anomalously large fire spread rate v_f values. The vertical wind increased to well over 5 m s^{-1} at 15 m above ground level. Model wind speeds are necessarily averaged over a large grid volume (20 m on a side in this case). Unlike model winds, tornadic whirls at 10–50 m in diameter in real fires obtain updrafts and horizontal velocities of 50–100 m s^{-1}.

D. THE EFFECT OF ATMOSPHERIC STABILITY AND WIND CHANGING WITH HEIGHT ON A FIRE PROPAGATING OVER A SMALL HILL

Observations, laboratory experiments (Weiss and Biging, 1996), and computer models (Linn, 1997; Coen *et al.*, 1998) show that when fire spreads on sloped terrain, it spreads faster uphill on steeper slopes. This is due to a combination of dynamic effects, the relative importance of each still being debated (Baines, 1990). One effect is that the fire's convection increases momentum and wind velocities near the ground. Another involves the fluid dynamics of flow over

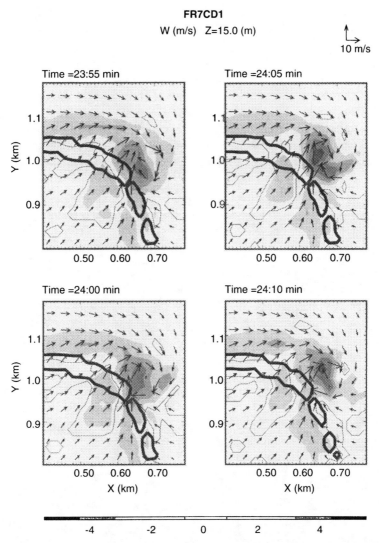

FIGURE 10 The vertical velocity (units m s^{-1}) from a numerical simulation of an initially 400-km-long fire line in a hyperbolic tangent wind profile where $U_0 = \pm 3$ m s^{-1}. See text for explanation. Four times are shown and frames are 5 s apart. Solid outlines are for heat fluxes greater than 0.1 MW m^{-2}. Vectors denote local winds at 15 m above ground level. (After Clark *et al.*, 1996b.)

hills, which leads to accelerations and decelerations in wind velocities even without the presence of a fire. A third effect is attributed to radiation and convection preheating and drying the fuel ahead of the fire (Rothermel, 1972) and to flames being brought closer to the fuel and propagating the fire by contact.

Figure 11 presents the results of an idealized simulation by Coen *et al.* (1998) of a fire line propagating over a small Gaussian-shaped ridge with a relatively sharp slope (height 200 m and half-width 300 m) and extending north–south over the entire model domain. The hyperbolic tangent wind profile of Figure 8 was again used, but with stable atmospheric conditions below 500 m and un-

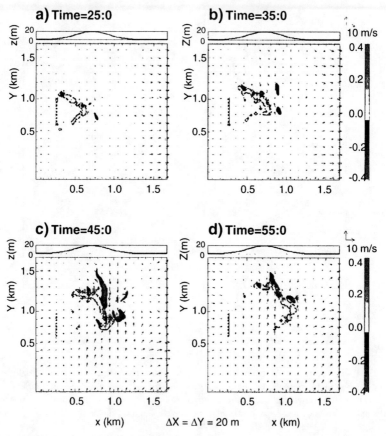

FIGURE 11   A near-surface horizontal cross section of the vertical vorticity (shaded) associated with a fire line (solid outline of heat fluxes greater than 0.1 MW m^{-2}) as it flows over a small ridge at 25, 35, 45, and 55 minutes into the simulation. Ambient atmospheric conditions are a 3 m s^{-1} low-level, westerly (from the left), stable flow. The vectors indicate the winds at the 15 m height. The west–east profile of the hill is shown above the main figure.

stable above.[1] These simulations demonstrated the unusual interactions that occur between a fire and its atmospheric environment when spreading over mountainous terrain for the chosen atmospheric conditions. Figure 11 shows that, as the fire moved up the slope, the right flank weakened and died, and then, instead of continuing to propagate down the other side, the left flank veered sharply along the ridge line. Since the flow driving the fire forward was symmetrical and cannot be the source of asymmetry, this fire behavior suggests that an instability amplified some small asymmetric disturbance.

After bowing forward into the expected bow-shaped fire line (Figure 11, time = 25 min), other interesting features appeared in this simulation. These included a large fire-scale rotation just in the lee of the ridge (Figure 11, time = 45 min), into which flowed a string of vertical vortices with alternating sign that formed initially at the rear edge of the active (left) flank of the fire.

The atmospheric situation for the simulation shown in Figure 11 is depicting a type of air flow known as nocturnal drainage flow. At night, air cools. Cold air is heavier than warm and can flow downslope with speeds of 3 to 20 m s^{-1}. These flows are characterized by a shallow, stable layer of cool air spreading quickly near the surface under a layer of less stably stratified air that is either still or traveling in the opposite direction. Figure 11 demonstrates how complex the interactions between the fire and the atmosphere can become even for common atmospheric flows like low-humidity, nocturnal drainage flows. The results of this simulation also raise a crucial question: the possible impossibility of predicting fire behavior when a sudden amplification of small, transient instabilities in the flow leads to such deviant behavior.

E. SENSITIVITY OF FIRE BEHAVIOR TO FUEL CHARACTERISTICS AND CONFIGURATION

A method for creating a more general operational atmosphere–fire model is to use a physics, as opposed to empirically, based fire transport and behavior model to represent the processes of forest fire combustion and spread rate. One such model is the FIRETEC (Linn, 1997; Linn and Harlow, 1998). The FIRETEC is based on conservation of mass, momentum, species, and energy. It includes a

---

[1] Whether an atmosphere at a location is stable or not depends on the resulting motion of a parcel of air that is displaced vertically from its initial position. If the parcel moves further away from its initial position, then the atmosphere is unstable and convection occurs; if the parcel returns toward its initial position, then the atmosphere is stable; and if the parcel does not move once displaced, then the atmosphere is neutral. The magnitude of the parcel's vertical acceleration is determined primarily by the temperature or density difference between the air parcel and its environment at a given location, time, and pressure level.

representation of fuel pyrolysis, turbulent transport of combustion products, and radiative preheating of fuel due to the approaching flame front.

The FIRETEC has been coupled with the HIGRAD, an atmospheric dynamics model developed to handle the sharp temperature and flow gradients encountered in the vicinity of a wildfire (Reisner *et al.*, 1998, 2000). Using FIRETEC/HIGRAD, Bossert *et al.* (2000) simulated a small portion of a 16,000-acre fire in steep terrain (Calabasas fire, October 21 to 22, 1996, Corral Canyon, CA). FIRETEC allows a fully three-dimensional characterization of fuel, and in this study the dependent fuel parameters were fuel load, moisture, and type. The results show how the fire responded to four different fuel types. Even though the fuel load and moisture contents were variable among the four fuel types, the numerical results were not dramatically different in three of the four cases. In contrast, one fuel type, which had the lightest fuel load and highest moisture content, produced radically different fire behavior. The fire never achieved sufficiently high enough temperatures to sustain itself up the sloped terrain and died after 8 minutes of simulation. This result shows that the FIRETEC model is sensitive to the specification of fuel load and moisture. Bossert *et al.* (2000) also produced one simulation showing steeper topography overwhelming any detailed specification of fuels, resulting in an equivalent of a blowup fire. Convective heating of a steeper portion of the slope by hot gases from the fire's plume caused the entire portion of the slope to be engulfed by fire after only one additional minute of simulated time.

FIRETEC was also used to consider multistorey fuel canopies involving ground and crown fuels. Linn *et al.* (2000) show FIRETEC/HIGRAD simulations of fire sensitivity to live moisture in the canopy fuel; as expected and in agreement with observations, high moisture in the canopy fuel and the fire was confined to surface; low moisture in the canopy fuel and the fire crowned. These results demonstrate the potential of using a physics-based fire behavior model coupled with an atmospheric prediction model for exploring such phenomena as crowning. Linn *et al.* (2000) also simulated a grass fire in which FIRETEC/HIGRAD caused irregular fuel consumption that produced a nonhomogeneous fire front and burn pattern. The simulated burn pattern (Figure 12) is similar to patterns sometimes observed in real fires: streaks of burned and unburned fuel, perpendicular to the fire front and parallel to the ambient wind.

VIII. INFRARED OBSERVATIONS OF FIRES

There is no substitute for direct observations of fire dynamics and behavior. Unfortunately severe fires are almost inaccessible to planned observations, and forestry researchers are required to build their models on sometimes shaky ob-

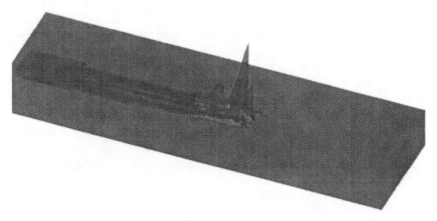

FIGURE 12   Side view image of a three-dimensional simulation of a tall-grass fire in 4 m s^{-1} ambient winds blowing perpendicular to the fire front. (After Linn *et al.*, 2000.)

servational foundations. Theoretical descriptions of convection patterns and behavior in fires still rely heavily on laboratory and computer simulations. This is changing. Advances in understanding come from the application of remote sensing instrumentation. It is now possible to probe fires from a safe distance with Doppler radar and lidar, and most recently, infrared or IR video.

Although data describing the fire spread rate and some qualitative aspects of wildfire behavior exist, none has revealed the very small convective time and spatial scales in the fire. The intense vortices of various strengths and sizes in fires have been observed both experimentally (Church *et al.*, 1980) and in model simulations (Heilman, 1992; Heilman and Fast, 1992; Clark *et al.*, 1996a,b). As discussed in the introduction, observations in one case study (McRae and Flannigan, 1990) show that fire vortices can rip out and loft standing trees. A pool fire simulation produced counterrotating vortices to characterize the structure of rising smoke (McGrattan *et al.*, 1996). Similar vortices were observed in fires using lidar and radar where the scale of the observed vortices was on the order of 100 m (Banta *et al.*, 1992).

The fire convection is rarely more than a few meters, lasting rarely more than a split second. It is necessary to observe fire-scale features on time scales of 0.05–0.1 s and spatial scales of 1–3 m. The long dwell time, the low scanning rate, and the spatial resolution of the Doppler radar ($\sim$10 μm) and lidar ($\sim$3 mm) do not provide wind speed measurements on such small scales (Banta *et al.*, 1992).

However, current IR camera capability provides high-frequency (1/30 s) and high resolution (0.05–0.16 m) radiant temperature images that are being used to derive wind fields to provide clues to motions on scales of the order of meters and time scales of fractions of a second. Infrared video cameras have a temperature measurement accuracy of better than $\pm 2°C$ at temperatures up to $2000°C$ and can resolve temperature differences of less than $0.1°C$. The IR camera views a shallow depth into the flame front and variabilities in the distribution of hot soot particles provide the information necessary to derive two-dimensional images of flow. Analysis of these images provides estimates of velocities in the vertical and in one horizontal direction. Measurements in crown fires in the Canadian Northwest Territories suggest updrafts of 10 to 30 m s^{-1}, downdrafts of -10 to -20 m s^{-1}, and horizontal motions of 5 to 15 m s^{-1} throughout the fire (Clark *et al.*, 1999).

IR temperatures and vertical winds are used to estimate sensible heat fluxes from the fire. Following Eq. (1), the sensible heat flux is

$$F_s \approx c_p <\rho w(T_{IR} - \overline{T}_{IR})> \tag{2}$$

where w is the vertical velocity at the location of IR temperature measurement T_{IR}, and $\overline{T}_{IR}$ is the time and area average of T_{IR}. Heat fluxes calculated according to Eq. (2) then can be compared with expected values based on fuel loading and observed burnout times to corroborate the image flow analysis.

It is possible to infer from an IR image flow analysis a number of physical mechanisms contributing to the fire spread rate. IR camera imagery appears to represent such fine-scale fire structures as hairpin vortices or turbulent bursts and the tilting of horizontal vortices leading to counter rotating convective towers where the estimated vertical vorticity is $4–10$ s^{-1} (Clark *et al.*, 1999). These fire structures determined by IR video measurements are being compared to and used to corroborate computer simulations such as those described in the previous section.

IX. CONCLUSIONS AND FUTURE WORK

In this chapter, we describe the coupling of an atmospheric prediction model with an empirical fire spread model so that the fire–atmosphere is treated as a single, dynamic system. This is a fairly recent and major advance in the modeling of wildfires. Unlike the wildfire models that have traditionally been used to assess the risk, behavior, and spread of fires, this model is capable of simulating the convective-scale fire–atmosphere circulations, including the small-scale vortex dynamics, that play an important role in influencing a wildfire's overall behavior, as well as fire spread rates. And unlike traditional wildfire models, the coupled fire–atmosphere modeling approach can include the possible impacts of evolving large-scale atmospheric forcing on the fire and vice versa.

There are two major problems to overcome before this approach can be used to predict forest fire behavior. First, we must learn to faithfully represent the physics of the fire–atmosphere system to the degree that is needed. Second, we must develop observational systems and techniques to provide the detailed observations of fire structures that are required to improve and validate the numerical results of the fire–atmosphere model. Real-time simulations of wildfire events may be possible in the near future using newer computer technology, provided also that topography, initial weather data, and fuel data can be gathered quickly for use in the model, and new visualization techniques be developed to display results in real time.

We have discussed a few numerical modeling results from publications (Clark *et al.*, 1996a,b; Coen *et al.*, 1998; Bossert *et al.*, 2000; Linn *et al.*, 2000) directed at improving our understanding of wildfire behavior. To gather the observations to validate numerical model results from simulations like these, we have briefly described the measurement technique of infrared video imagery. The current analysis of IR video imagery is capable of estimating two-dimensional wind fields and heat fluxes from the fire.

One weakness of the coupled fire–atmosphere modeling approach is the currently necessary employment of empirical formulae for forest fire models. The accuracy of ignition spread is dependent on the fire spread rate model, while the heat and moisture fluxes absorbed by the air are dependent on the accuracy of the fire model. Eventually forest fire models and fire spread-rate formulae must be replaced with more physics-based treatments of microscale processes within the fire and adjacent fuel. This is an extremely difficult exercise. Forest fires have complicated chemistry, radiation, and combustion properties. The interactions between forest fires and air flow are highly nonlinear. Diverse scales of motion influence a wildfire, and at almost every scale the physics require fully dimensional temporal and spatial dynamics.

The ultimate goal is to develop a fully comprehensive coupled atmosphere–fire model of forecast potential in the operational mode. This is not possible at the present time or in the near future. The reasons are that many physical processes in the fire–atmosphere system remain poorly understood and that computational demands of such a modeling approach far exceed the capability of present-day computer technology. The most important use today of physically based fire behavior and transport models is to identify conditions where the current operational empirically based forest fire models are not appropriate. Although unsuitable for faster-than-real-time applications, coupling a fully physically based treatment of the fire combustion and transport with an atmospheric prediction model has the advantage of being able to examine complex wildfire behavior under any set of conditions. Existing fire behavior and fire spread models are faster and easier to use than physically based models and are quite adequate for predicting fire behavior, but only under certain circumstances.

Physically based models will be used to extend existing operational models and to develop new operational models that are suitable for a wider range of conditions.

In the meantime, present capabilities are sufficient to model components of the fire–atmosphere system with a considerable degree of sophistication, as the combustion and atmospheric science research communities show. Given the recent remarkable advances in computer technology, significant development will be made as many model refinements are added.

Finally, we emphasize that convection is the dominant mechanism of fire propagation after the incipient stage of a fire and, despite these objections to the current state of numerical fire modeling, knowledge of the small-scale atmospheric circulations in and around a wildfire is the single most important and elemental improvement that can be made to current wildfire modeling.

APPENDIX I. CIRCULATION AND VORTICITY

Circulation is defined as

$$C \equiv \oint \vec{V} \cdot \vec{dl}$$

where the line integral of fluid velocity $\vec{V}$ is performed over a closed contour, dl is an element length of the contour, and A is the area enclosed by the contour. The line integral is performed in a *counterclockwise* or *positive* direction. Circulation is a scalar quantity and is the macroscopic measure of rotation for a finite area of fluid.

The vorticity vector $\vec{\Omega}$ is defined as

$$\vec{\Omega} = \nabla \times \vec{V}$$

the *curl* of the velocity vector field [&*obar*{V}|ucarrhd|.&] Here $\vec{V} = u\hat{i} + v\hat{j} + w\hat{k}$ is the three-dimensional velocity field and

$$\nabla = \frac{\partial}{\partial x} + \frac{\partial}{\partial y} + \frac{\partial}{\partial z}$$

is the three-dimensional del operator, where $\hat{i}$, $\hat{j}$, and $\hat{k}$ are the unit vectors directed along the $x, y,$ and z Cartesian axes. We are interested in rotation about a horizontal axis or the *vertical* component of vorticity. Therefore we take

$$\hat{k} \cdot \vec{\Omega} = \hat{k} \cdot \nabla \times \vec{V} = \frac{\partial v}{\partial x} - \frac{\partial u}{\partial y}$$

The quantity $\zeta = \hat{k} \cdot \nabla \times \vec{V} = \partial v/\partial x - \partial u/\partial y$ is called *relative* vorticity (i.e., *relative* to the ground).

While circulation is a macroscopic measure of rotation of the fluid, vorticity is a *microscopic* measure of rotation or "spin" of individual particles in a fluid. Relative vorticity is defined as

$$\zeta = \frac{\partial v}{\partial x} - \frac{\partial u}{\partial y} = \lim_{A \to 0} \frac{\oint \vec{V} \cdot d\vec{l}}{A}$$

or the circulation for an infinitesimally small horizontal area A $(A \to 0)$.

To get a physical feeling for vorticity, we use natural coordinates and consider the flow profile in Figure I.1. The natural coordinate system is defined by the orthogonal set of unit vectors $\hat{t}, \hat{n}, \hat{k}$, where $\hat{t}$ is parallel to the horizontal velocity at each point, $\hat{n}$ is normal to the horizontal velocity and directed so that it is positive to the *left* of the flow direction, and $\hat{k}$ is directed vertically upward. $s(x, y, t)$ is the curve followed by a parcel moving in the horizontal plane, and the horizontal velocity is $\vec{V} = V\hat{t}$, where the horizontal speed V is a nonnegative scalar defined by $V \equiv ds/dt$.

The circulation about the circuit in Figure I.1 is

$$C = V[\delta s + d(\delta s)] - \left[V + \frac{\partial V}{\partial n}\delta n \right] \delta s$$

where the negative sign $(-)$ specifies flow direction. Also $\delta n \sin(\delta\beta) \approx \delta n\, \delta\beta = d(\delta s)$ for small angle $\delta\beta$. So

$$C = V\,\delta s + V\,d(\delta s) - V\,\delta s - \frac{\partial V}{\partial n}\delta n\,\delta s$$

$$= V\,\delta n\,\delta\beta - \frac{\partial V}{\partial n}\delta n\,\delta s$$

$$= \left(V\frac{\delta\beta}{\delta s} - \frac{\partial V}{\partial n} \right)\delta n\,\delta s$$

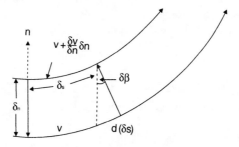

FIGURE I.1   Circulation for an infinitesimal loop in the natural coordinate system.

(a)   **(b)**

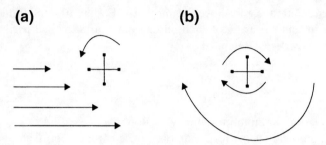

FIGURE I.2   Two types of two-dimensional flow: (a) the rate of change of wind speed normal to the direction of the flow, and (b) the turning of wind following the motion of the air parcel. In (a), a pinwheel in the flow spins cyclonically. In (b), a pinwheel in the flow spins anticyclonically.

But $R\,\delta s = \delta\beta$ where $R \equiv$ radius of curvature following the parcel motion so that

$$C = \left(\frac{V}{R} - \frac{\partial V}{\partial n}\right)\delta n\,\delta s$$

Since the relative vorticity is defined as

$$\zeta = \lim_{A\to 0}\frac{C}{A}$$

we have

$$\zeta = \lim_{\delta n\,\delta s\to 0}\frac{C}{\delta n\,\delta s} = \frac{V}{R} - \frac{\partial V}{\partial n}$$

where V/R is *curvature* vorticity and $\partial V/\partial n$ is *shear* vorticity. Curvature vorticity is the turning of the wind along the parcel motion. Shear vorticity is the rate of change of wind speed normal to the direction of the flow. Figure I.2 illustrates both shear (a) and curvature (b) vorticity. When a small paddle wheel is placed in the flow in Figure I.2a, it turns in the counterclockwise or cyclonic direction. When a small paddle wheel is placed in the flow in Figure I.2b, it turns in the clockwise or anticyclonic direction. Vorticity ζ is *positive* for cyclonic motion and *negative* for anticyclonic motion.

APPENDIX II. DEVELOPMENT OF VERTICAL ROTATION IN A FRICTIONLESS FLUID

The material and concepts discussed in Section II of this chapter are based on the following mathematical representation of vorticity dynamics. In a nonrotating reference frame (i.e., the Coriolis force due to the Earth's rotation is neglected), conservation of momentum is expressed as

$$\frac{d\vec{V}}{dt} = \frac{\partial \vec{V}}{\partial t} + (\vec{V} \cdot \nabla)\vec{V} = -\frac{1}{\rho}\nabla p - g\hat{k}$$

in a frictionless (or inviscid) fluid. Here $\vec{V} = u\hat{i} + v\hat{j} + w\hat{k}$ is the three-dimensional velocity field and

$$\nabla = \frac{\partial}{\partial x} + \frac{\partial}{\partial y} + \frac{\partial}{\partial z}$$

is the three-dimensional del operator, where $\hat{i}$, $\hat{j}$, and $\hat{k}$ are the unit vectors directed along the Cartesian x, y, and z axes. t is time, ρ is density, p is pressure, and g is the acceleration due to gravity. The force due to friction is neglected. The vector identity

$$(\vec{V} \cdot \nabla)\vec{V} = \nabla\left(\frac{\vec{V} \cdot \vec{V}}{2}\right) - \vec{V} \times \nabla \times \vec{V}$$

is used to rewrite the conservation of momentum equation as

$$\frac{\partial \vec{V}}{\partial t} = -\nabla\left(\frac{\vec{V} \cdot \vec{V}}{2}\right) + \vec{V} \times \vec{\Omega} - \frac{\nabla p}{\rho} - g\hat{k} \tag{II.A}$$

where $\vec{\Omega} = \nabla \times \vec{V}$. Taking the curl ($\nabla \times$) of Eq. (II.A), and recalling that the curl of the gradient vanishes, gives

$$\frac{\partial \vec{\Omega}}{\partial t} = \nabla \times (\vec{V} \times \vec{\Omega}) - \nabla \times \frac{\nabla p}{\rho} \tag{II.B}$$

Let $\zeta = \hat{k} \times \vec{\Omega} = \hat{k} \cdot \nabla \times \vec{V}$, the vertical component of the vorticity (Appendix I). Taking $\hat{k} \cdot$ of Eq. (II.B) gives

$$\frac{\partial \zeta}{\partial t} = \hat{k} \cdot \nabla \times (\vec{V} \times \vec{\Omega}) - \hat{k} \cdot \nabla\left(\frac{1}{\rho}\right) \times \nabla p \tag{II.C}$$

the vertical vorticity equation. After some algebraic manipulation, it is possible to rewrite Eq. (II.C) as

$$\begin{aligned}\frac{\partial \zeta}{\partial t} = &-\vec{V}_h \cdot \nabla_h \zeta - w\frac{\partial \zeta}{\partial z} - \zeta\nabla_h \cdot \vec{V}_h \\ &+ \hat{k} \cdot \left(\frac{\partial \vec{V}_h}{\partial z} \times \nabla_h w\right) + \frac{1}{\rho^2}\left(\frac{\partial \rho}{\partial x}\frac{\partial p}{\partial y} - \frac{\partial \rho}{\partial y}\frac{\partial p}{\partial x}\right)\end{aligned} \tag{II.D}$$

where $\vec{V}_h = u\hat{i} + v\hat{k}$ is the two-dimensional horizontal wind and ∇_h is the horizontal gradient in the *x-y* plane. Each term in Eq. (II.D) has a physical interpretation.

Term $\partial\zeta/\partial t$ is the *local* time change of vertical vorticity. Rather than attach our coordinate system to moving fluid parcels, we determine the rate of change of ζ for the fluid at a fixed point or location on the Earth's surface. Each term on the right-hand side of Eq. (II.D) contributes to the local rate of change of ζ.

Term $-\vec{V}_h \cdot \nabla_h\zeta$ is the advection of vorticity by the horizontal wind. For example, wind blowing from a region of positive vorticity toward a region of negative vorticity contributes positively to the local vorticity change. Term $-w(\partial\zeta/\partial z)$ is the advection of vorticity by the vertical wind, and its effect is similar to horizontal advection of vorticity except that it redistributes vorticity ζ in the vertical direction.

Term $-\zeta\nabla_h \cdot \vec{V}_h$ represents changing vorticity due to horizontal divergence. It is the fluid analog of the change in angular velocity resulting from a change in the moment of inertia of a solid body when angular momentum is conserved. If there is positive horizontal divergence, the area enclosed by a chain of fluid parcels increases with time, and if the circulation is to be conserved, the average vorticity of the enclosed fluid must decrease. This term is called the divergence or stretching term. If it is assumed that the three-dimensional flow is non-divergent, $\nabla \cdot \vec{V} = 0$, then

$$-\frac{\partial w}{\partial z} = \frac{\partial u}{\partial x} + \frac{\partial v}{\partial z}$$

Substituting this into the divergence term gives

$$-\zeta\nabla_h \cdot \vec{V}_h = \zeta\frac{\partial w}{\partial z}$$

Vertical motion is responsible for the "stretching" effect. When vertical motion narrows the diameter of the area enclosed by the chain of fluid parcels (horizontal convergence), vorticity increases.

Term $\hat{k} \cdot (\partial\vec{V}_h/\partial z \times \nabla_h w)$ represents the generation of vertical vorticity by the tilting of horizontally oriented components of vorticity into the vertical by a nonuniform vertical motion field. This term is called the twisting or tilting term and is illustrated in Figure 1 and described in Section II.

Term $-\hat{k} \cdot \nabla(1/p) \times \nabla p = (1/\rho^2)[(\partial\rho/\partial x)(\partial p/\partial y) - (\partial\rho/\partial y)(\partial p/\partial x)]$ is called the *solenoidal* term. It is illustrated in Figure 2 and described in Section II. To illustrate possible production of vertical vorticity by the solenoidal term, consider what happens to a small rectangular parcel or bar of fluid in the situation depicted in Figure 2. Imagine a pressure gradient in the fluid, perpendicular to the parcel and uniform along its length; in this case, the parcel moves through the fluid from high to low pressure. Now imagine that the fluid density is not constant, but is larger at one end of the parcel than the other. The forces $\vec{F}_{y_1}$ and $\vec{F}_{y_2}$ can be written

$$\vec{F}_{y_1} = -\frac{1}{\rho_1}\frac{\partial p}{\partial y}\hat{j}, \qquad \vec{F}_{y_2} = -\frac{1}{\rho_2}\frac{\partial p}{\partial y}\hat{j}$$

Since

$$\rho_1 > \rho_2, \qquad \frac{1}{\rho_1} < \frac{1}{\rho_2}$$

then

$$|\vec{F}_{y_1}| < |\vec{F}_{y_2}|$$

and the torque applied by the pressure gradient forces, $\vec{F}_{y_1}$ and $\vec{F}_{y_2}$, causes the fluid bar to rotate counterclockwise. Thus when the gradient of pressure ∇p and the gradient of density $\nabla \rho$ are not parallel, the parcel experiences a torque that causes it to rotate, producing vorticity.

The advection terms distribute existing ζ, while the stretching or divergence term "spins up" or "spins down" existing vorticity. Only the tilting and sole-noidal terms *generate* vertical vorticity ζ in Eq. (II.D) in an inviscid fluid. And although friction in the fluid is not included in this discussion, *turbulent* or *frictional drag* can be both a source and sink of local rotation.

A. VERTICALLY ROTATING CONVECTIVE CELLS IN A MEAN WIND FIELD

Imagine a flow field as in Figure II.1 consisting of a single convective updraft embedded in a background flow $\overline{U}$ which is a function of height z only and westerly in the positive $\hat{i}$ direction. Let

$$\vec{\Omega} = \frac{d\overline{U}}{dz}\hat{j} + \vec{\Omega}'(x, y, z, t), \qquad \vec{V} = \overline{U}(z)\hat{i} + \vec{V}'(x, y, z, t),$$

$$p = \overline{p}(z) + p'(x, y, z, t), \qquad \text{and} \qquad \rho = \overline{\rho}(z)$$

where it is assumed for simplicity that density does not vary in the horizontal and is a function of height only. The overbar ($^-$) denotes the background or base state, and the prime ($'$) denotes fluctuations due to convection or perturbations from the background or base state. Substituting $\vec{\Omega}$, $\vec{V}$, ρ, and p into the right-hand side of Eq. (II.D) gives

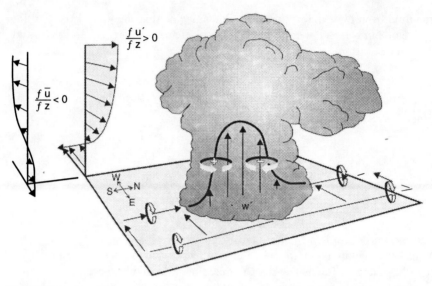

FIGURE II.1 Development of rotation in a single convective updraft (shown by arrows) along a fire front embedded in a background with mean wind $\bar{u}$ decreasing with height (shown by wind arrows in the upper left corner). Heavy solid lines show vortex lines with sense of rotation shown by circulation arrows. Plus and minus signs indicate cyclonic and anticyclonic rotation caused by vortex tube tilting.

$$\frac{\partial \zeta'}{\partial t} = -\vec{V}_h' \cdot \nabla_h \zeta' - \bar{U}\frac{\partial \zeta'}{\partial x} - w'\frac{\partial \zeta'}{\partial z} - \zeta' \nabla_h \cdot \vec{V}_h'$$

$$+ \frac{\partial w'}{\partial y}\frac{\partial \bar{U}}{\partial z} + \hat{k} \cdot \left(\frac{\partial \vec{V}_h'}{\partial z} \times \nabla_h w' \right) \tag{II.E}$$

For this flow field, there is no generation of vertical vorticity by the solenoidal effect, and in Eq. (II.E) vertical vorticity is produced only by tilting the horizontal vorticity into the vertical:

$$\frac{\partial w'}{\partial y}\frac{\partial \bar{U}}{\partial z} + \hat{k} \cdot \left(\frac{\partial \vec{V}_h'}{\partial z} \times \nabla_h w' \right) = \frac{\partial w'}{\partial y}\frac{\partial \bar{U}}{\partial z} + \frac{\partial w'}{\partial y}\frac{\partial u'}{\partial z} - \frac{\partial w'}{\partial x}\frac{\partial v'}{\partial z}$$

The terms on the right-hand side of this expression are, respectively, the tilting by the perturbed differential vertical motion field ($\partial w'/\partial y$, $\partial w'/\partial x$) of linear horizontal vorticity ($\partial \bar{U}/\partial z$) and perturbed horizontal vorticity ($\partial u'/\partial z$, $\partial v'/\partial z$).

For forced convection, such as in a wildfire, the perturbation portion of the dependent variables can be as large or larger than the mean or basic state portion. When perturbation variables and products of perturbation variables dominate, the effect by the tilting of horizontal vorticity of the base state, $\partial \bar{U}/\partial z$, is negligible compared to the tilting of horizontal shear in the convective flow, $\partial u'/\partial z$ and $\partial v'/\partial z$.

Say fire convection produces vertical wind shears where $\partial u'/\partial z > 0$. This happens for east winds $(u' < 0)$ stronger at the surface, decreasing with height, or west winds $(u' > 0)$ weaker at the surface, increasing with height. This positive vertical wind shear of u' coupled with $\partial w'/\partial y > 0$ south of the updraft core and $\partial w'/\partial y < 0$ north of the updraft core produces—once tilting of horizontal vorticity has occurred—two centers of vertical vorticity. The one south of the updraft core has counterclockwise or positive rotation; the one north has clockwise or negative rotation. This is similar, but opposite to, the situation depicted in Figure 1(c) (Section II). If the perturbation product $(\partial w'/\partial y)(\partial u'/\partial z)$ is the dominate tilting term in Eq. (II.E), then $\partial \zeta'/\partial t$ is increased south and decreased north of the updraft core.

Once ζ' exists, both advection of the perturbed vorticity by the basic state flow, $-\overline{U}(\partial \zeta'/\partial x)$, and by the perturbed horizontal wind, $-\vec{V}_h' \cdot \nabla_h \zeta'$, can contribute to the local time change of perturbation vorticity owing to the air motion. If a background wind $\overline{U}$ blows from a region of positive perturbation vorticity $(\zeta' > 0)$ toward a region negative or no perturbation vorticity $(\zeta' \leq 0)$, the advection of perturbed vorticity by the basic state flow contributes positively to the local perturbation change, and $\partial \zeta'/\partial t$ in Eq. (II.E) is increased. For the situation depicted in Figure II.1, the mean wind $\overline{u}$ will advect fire-generated vorticity ζ' east of the fire at low levels, and west of the fire at upper levels.

And once ζ' exists, the stretching term $-\zeta' \nabla_h \cdot \vec{V}_h'$ can either increase or decrease ζ' locally for horizontal winds converging or diverging, respectively, while the vertical advection term $-w'(\partial \zeta'/\partial z)$ can redistribute vorticity in the vertical. For example, upper-level vorticity can be advected to the surface by convective downdrafts.

Density does vary in all three directions, and the solenoidal effect in Eqs. (II.D) and (II.E) generates perturbed vertical vorticity that also is advected by the background wind and convectively perturbed flow and is increased or decreased by horizontal convergence and stretching. And finally, the fluid is not frictionless. Although friction and turbulence can increase local rotation, they are also crucial for the destruction of local rotation.

APPENDIX III. GENERATION OF VERTICAL MOTION IN ROTATING CONVECTIVE CELLS

Even rotating convective cells with relatively low buoyancy can produce low-level updrafts that are as strong or even stronger than their high-buoyancy counterparts. The reason for this is that the strength of vertical motion in rotating convective cells is not only dependent on buoyancy forcing. Acceleration in the core of a rotating convective cell is principally governed by buoyancy and the vertical gradient of perturbation pressure.

The total buoyancy of an air parcel is due to thermal (sensible, latent, and radiational) heating and the effect of condensate loading, while vertically directed pressure gradient forces are attributed to the so-called "centrifugal pump" effect. Section II and Appendix II describe how cyclonic and anticyclonic vertically rotating vortices can develop in a fire. As soon as a vertical vortex is established, vertical motion in the center of the vortex develops. To understand the generation of the vertical motion, take the $\nabla \cdot$ of Eq. (II.A) to yield

$$\nabla^2\left(\frac{p}{\rho}\right) = -\nabla^2\frac{(\vec{V}\cdot\vec{V})}{2} + \nabla\cdot(\vec{V}\times\vec{\Omega}) \tag{III.A}$$

where $\nabla \cdot \vec{V} = 0$ is assumed. Terms $-\nabla^2[(\vec{V}\cdot\vec{V})/2]$ and $\nabla(\vec{V}\times\vec{\Omega})$ represent dynamical forcing, and we are interested in the effect of dynamical forcing on the pressure field in a vertical vortex. We use cylindrical coordinates (r, λ, z) centered on the axis of rotation (Figure III.1) to investigate the pressure variation in a vortex. Here λ is azimuthal angle, $\hat{j}_\lambda$ is the unit vector in the azimuthal direction (with positive counterclockwise), z is vertical height, r is the radial distance from the axis of rotation, and $\hat{i}_\lambda$ is the unit vector in the r direction.

For pure rotation about a vertical z axis, the horizontal velocity field and vertical component of vorticity in cylindrical coordinates are, respectively,

$$\vec{V} = v_\lambda\hat{j} \qquad \text{and} \qquad \hat{k}\cdot\vec{\Omega} = \zeta = \frac{1}{r}\frac{\partial}{\partial r}(rv_\lambda)$$

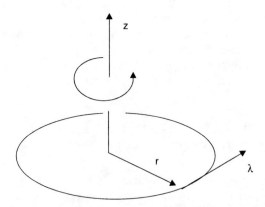

FIGURE III.1   The cylindrical coordinate system.

Therefore,

$$\hat{V} \times \vec{\Omega} = \frac{v_\lambda}{r} \frac{\partial}{\partial r}(r_\lambda v_\lambda)\hat{i}_\lambda$$

Assuming that the vertical scale is much larger than the radial scale, the Laplacian ∇^2 in cylindrical coordinates is approximately

$$\nabla^2 \approx \frac{1}{r} \frac{\partial}{\partial r}\left(r\frac{\partial}{\partial r}\right)$$

Substituting into Eq. (III.A), the dynamical component of the pressure perturbation in the vortex is expressed as

$$\frac{1}{r} \frac{\partial}{\partial r}\left(\frac{r}{\rho_0} \frac{\partial p_{dyn}}{\partial r}\right) \approx -\frac{1}{r} \frac{\partial}{\partial r}\left[r\frac{\partial(v_\lambda^2/2)}{\partial r}\right] + \frac{1}{r} \frac{\partial}{\partial r}\left[v_\lambda\frac{\partial}{\partial r}(rv_\lambda)\right]$$

$$\approx \frac{1}{r} \frac{\partial v_\lambda^2}{\partial r} \qquad\qquad\text{(III.B)}$$

where $\rho = \rho_0$ and is considered constant in the radial direction. Integrating Eq. (III.B) with respect to r gives

$$\frac{1}{\rho_0} \frac{\partial p_{dyn}}{\partial r} \approx \frac{v_\lambda^2}{r}$$

an equation for cyclostrophic balance, where the pressure gradient force $(1/\rho_0)(\partial p_{dyn}/\partial r)$ is always directed *inward*, toward the center of rotation, while the centrifugal force v_λ^2/r is always directed *outward*. The *low* pressure at the center of (either clockwise or counterclockwise) rotation provides a pressure gradient force to balance the outward centrifugal force.

This *dynamically* generated low pressure at in the center of the vortex produces a *vertically* directed pressure gradient force. Since air is accelerated from high to low pressure, the result is the development of an updraft from below, or a downdraft from above, in the rotating core. This is the centrifugal pump effect.

NOTATION

ROMAN LETTERS

c_p	specific heat capacity of dry air at constant p	J kg^{-1} K^{-1}
g	magnitude of acceleration due to gravity	m s^{-2}

p	pressure	Pa
s	distance along a parcel trajectory	m
t	time	s
u, v, w	wind components in x (eastward), y (north-ward), and z (vertical) directions	m s^{-1}
v_f	rate of spread at the head of a wind-aided fire	m s^{-1}
x, y, z	distances in the east–west, north–south, and vertical directions; Cartesian coordinates parallel to, and perpendicular to, the ground	m
C	fluid circulation	m^2 s^{-1}
$\vec{F}$	a force	kg m s^{-2}
F_s	sensible heat flux	W m^{-2}
F_{r_i}	convective Froude number	
T	absolute temperature	K
T_{IR}	infrared T measurement	K
U	ambient air wind speed	m s^{-1}
U_0	relative speed of air passing over the fire	m s^{-1}
V	speed in natural coordinates	m s^{-1}
$\vec{V}$	fluid velocity	m s^{-1}
W_f	width of region of intense heating	m
τ	burn time of small fuel	s
Δt	time step	s
Δw	w anomaly over region of intense heating;	m s^{-1}
	change in w over Δz	m s^{-1}
Δx_{fuel}	fuel grid cell size in horizontal direction	m
$\Delta x, \Delta y, \Delta z$	distances or grid cell sizes in x, y, and z directions	m
ΔT	T anomaly over region of intense heating	K
ζ	vertical component of relative vorticity	s^{-1}
ρ	air density	kg m^{-3}
$\vec{\Omega}$	vorticity vector	s^{-1}
∇	three dimensional gradient or del operator	m^{-1}
$\hat{i}, \hat{j}, \hat{k}$	unit vectors along x, y, and z axes	

CYLINDRICAL COORDINATES

$\hat{t}, \hat{n}$ unit vectors normal and parallel to fluid parcel trajectory

β angular direction of the wind or fluid flow

R radius of curvature following the parcel motion

λ, r asimuthal angle, radial distance from axis

$\hat{j}_\lambda, \hat{i}_\lambda$ unit vectors in the asimuthal and r directions

u_λ velocity component in the asimuthal direction

SPECIAL NOTATION

--   a background or base state; time and area average

< >   horizontal average or mean value

′   fluctuations or perturbations from a background or base state; fluctuations or perturbations from the horizontal average

^   unit vector

REFERENCES

Baines, P. (1990). Physical mechanisms for the propagation of surface fires. *Math. Comput. Modeling,* 13, 83–94.

Banta, R. M., Olivier, L. D., Holloway, E. T., Kropfli, R. A., Bartram, B. W., Cupp, R. E., and Post, M. J. (1992). Smoke column observations from two forest fires using Doppler lidar and Doppler radar. *J. Appl. Meteorol.* 31, 1328–1349.

Bossert, J. E., Linn, R. R., Reisner, J. M., Winterkamp, J. L., Dennison, P., and Roberts, D. (2000). Coupled atmosphere–fire behavior model sensitivity to spatial fuels characterization. Preprints of "3rd Symp. Fire and Forest Meteorology," Amer. Meteor. Soc. 80th Annual Meeting, Los Angeles, CA, pp. 21–26.

Byram, G. M. (1973). Combustion of forest fuels. "Forest Fire, Control and Use" (A. A. Brown and K. P. Davis, Eds. 2nd ed., pp. 155–182. McGraw-Hill, New York.

Chandler, C., Cheney, P., Thomas, P., Trabaud, L., and Williams, D. (1983). "Fire in Forestry Forest Fire Behavior and Effects." John Wiley and Sons, New York.

Church, C. R., and Snow, J. T. (1979). The dynamics of natural tornadoes as inferred from laboratory simulations. *J. Rech. Atmos.* 12, 111–133.

Church, C. R., Snow, J. T., and Dessens, J. (1980). Intense atmospheric vortices associated with a 100 MW fire. *Bull. Amer. Meteor. Soc.* 61, 682–694.

Clark, T. L., Radke, L., Coen, J. L., and Middleton, D. (1999). Analysis of small-scale convective dynamics in a crown fire using infrared video camera imagery. *J. Appl. Meteorol.* 38, 1401–1420.

Clark, T. L., Jenkins, M. A., Coen, J., and Packham, D. R. (1996a). A coupled atmospheric–fire model: Role of the convective Froude number and dynamic fingering at the fire line. *Int. J. Wildland Fire* 6(4), 177–190.

Clark, T. L., Jenkins, M. A., Coen, J., and Packham, D. R. (1996b). A coupled atmospheric–fire model: Convective feedback on fire line dynamics. *J. Appl. Meteorol.* **35**, 875–901.

Coen, J. L., Clark, T. L., and Hall, W. D. (1998). Simulations of the effect of terrain on fire behavior: Experiments using a coupled atmosphere–fire model. Preprints of "2nd Symp. Fire and Forest Meteorology," Amer. Meteor. Soc., Phoenix, AZ, pp. 87–90.

Corlett, R. C. (1974). Fire violence and modeling (Chapter 6). *In* "Heat Transfer in Fires: Thermophysics, Social Impacts, Economic Impact." Scripta Book Company, Hemisphere Publishing Corporation, Washington, DC.

Gall, R. L. (1982). Internal dynamics of tornado-like vortices. *J. Atmos. Sci.* **39**, 2721–2736.

Heilman, W. E. (1992). Atmospheric simulations of extreme surface heating episodes on simple hills. *Int. J. Wildland Fire* **2**, 99–114.

Heilman, W. E., and Fast, J. D. (1992). Simulations of horizontal roll vortex development above lines of extreme surface heating. *Int. J. Wildland Fire* **2**, 55–68.

Kim, J., and Moin, P. (1986). The structure of the vorticity field in turbulent channel flow. Part 2. Study of ensemble-averaged fields. *J. Fluid Mech.* **162**, 339–363.

King, A. R. (1964). Characteristics of a fire-induced tornado. *Australian Meteorological Magazine* No. 44, pp. 1–9.

Linn, R. R. (1997). "A Transport Model for Prediction of Wildfire Behavior." Los Alamos National Laboratory thesis LA-13334-T, New Mexico State University.

Linn, R. R., and Harlow, F. H. (1998). FIRETEC: A transport description of wildfire behavior. Preprints of "2nd Symp. Fire and Forest Meteorology," Amer. Meteor. Soc. 78th Annual Meeting, Phoenix, AZ, pp. 14–19.

Linn, R. R., Bossert, J. E., Harlow, F., Reisner, J. M., and Smith, S. (2000). Studying complex wildfire behavior using FIRETEC. Preprints of "3rd Symp. Fire and Forest Meteorology," Amer. Meteor. Soc. 80th Annual Meeting, Los Angeles, CA, pp. 15–20.

McGrattan, K., Baum, H. R., and Rehm, R. G. (1996). Numerical simulation of smoke plumes from large oil fires. *Atmos. Environ.* **30**(24), 4125–4136.

McRae, D. J., and Stocks, B. J. (1987). Large-scale convection burning in Ontario. "Proc. 9th Conf. on Fire and Forest Meteorology," San Diego, pp. 23–29.

McRae, D. J., Stocks, B. J., and Ogilvie, C. J. (1989). Fire acceleration on large-scale convection burns. "Proc. 10th Conf. on Fire and Forest Meteorology," Ottawa, pp. 101–107.

McRae, D. J., and Flannigan, M. D. (1990). Development of large vortices on prescribed fires. *Can. J. For. Res.* **20**, 1878–1887.

Moin, P., and Kim, J. (1982). Numerical investigation of turbulent channel flow. *J. Fluid Mech.* **118**, 341–377.

Radke, L. F., Clark, T. L., Coen, J. L., Walther, C., Lockwood, R., Riggan, P. J., Brass, J., and Higgins, R. (2000). The WildFire Experiment: Observations with Airborne Remote Sensors. *Can. J. Remote Sensing* **26**(5), 406–417.

Reisner, J. M., Bossert, J. E., and Winterkamp, J. L. (1998). Numerical simulations of two wildfire events using a combined modeling system (HI-GRAD/BEHAVE). Preprints of "2nd Symp. Forest and Fire Meteorology," Amer. Meteor. Soc. 78th Annual Meeting, Phoenix, AZ, pp. 6–13.

Reisner, J. M., Swynne, S., Margolin, L., and Linn, R. R. (2000). Coupled-atmosphere fire modeling using the method of averaging. *Mon. Wea. Review,* **128**, 3683–3691.

Rothermel, R. C. (1972). "A Mathematical Model for Predicting Fire Spread in Wildland Fuels." Research Paper INT-115. USDA Forest Service, Intermountain Forest and Range Experiment Station, Ogden, UT.

Strauss, D., Bednar, L., and Mees, R. (1989). Do one percent of forest fires cause ninety-nine percent of the damage? *For. Sci.* **35**(2), 319–328.

Weiss, D. R., and Biging, G. S. (1996). Effects of wind velocity and slope on flame properties. *Can. J. For. Res.* **26**, 1849–1858.

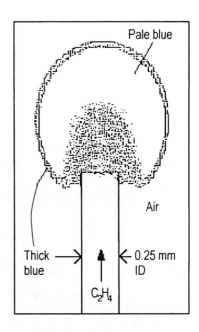

CHAPTER 2, FIGURE 3 An ethylene-air microdiffusion flame established on a 0.25-mm-diameter hypodermic needle (fuel flow rate at the burner port is 2 m/s). From Ban *et al.* (1994). *J. Heat Transfer* **116**, 331.

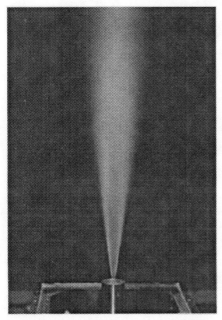

CHAPTER 2, FIGURE 4 A hydrogen-air jet diffusion flame established on a 1.43-mm-diameter stainless-steel burner nozzle (hydrogen flow rate at the burner port is 500 m/s) (Takahashi *et al.*, 1996).

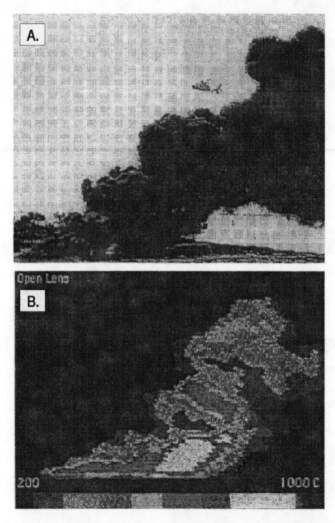

CHAPTER 2, FIGURE 9   (A) A diesel oil pool fire using a 15-m square shape open-top container (the test was conducted by the Building and Fire Research Laboratory at the National Institute of Standards and Technology, at the U.S. Coast Guard Fire and Safety Test Detachment in Mobile, AL). (B) An infrared image of the flame in Figure 9A.

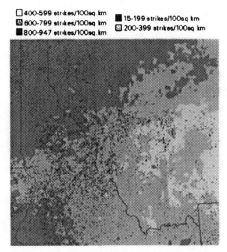

CHAPTER 11, FIGURE 1 Fire locations (black points) with lightning density (flashes/ 100 km²) as background.

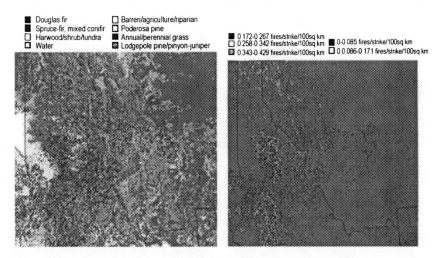

CHAPTER 11, FIGURE 2 Fire locations (black points) with fuel type as background.

CHAPTER 11, FIGURE 3 Fire locations (black points) with ignition efficiency as background.

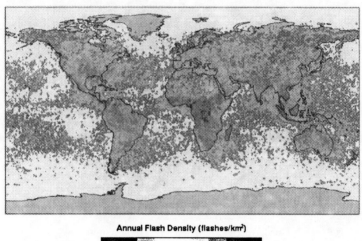

Annual Flash Density (flashes/km²)

0.1 0.2 0.5 1.0 2.0 5.0 10. 20. 50. 100.

CHAPTER 11, FIGURE 18 Worldwide lightning map as observed by the Optical Transient Detector in 1997. Courtesy of Global Hydrology and Climate Center and NASA/MSFC.

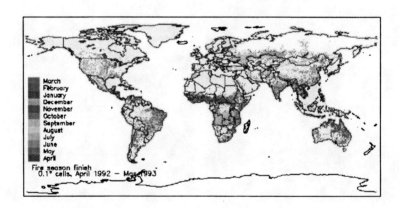

CHAPTER 11, FIGURE 19 World Fires, April 1992–May 1993 (courtesy of Dwyer *et al.*, 1999).

CHAPTER 9

Surface Energy Budget and Fuel Moisture

KENNETH E. KUNKEL

Illinois State Water Survey, Champaign, Illinois

I. INTRODUCTION

The effect of weather on fuel moisture has traditionally been estimated using statistically based empirical relationships that are a function of temperature and relative humidity (Simard, 1968; Van Wagner, 1987). These relationships have provided a useful approach to estimate the danger of forest fires. However, because they are not based on fundamental physical principles, it may not always be appropriate to use them as a tool to study fuel-drying processes. There is no assurance that these relationships will hold in a general sense. In fact, the differences in systems of Canada, the United States, and other countries suggests that any particular set of these relationships may not be generally applicable to the other regions. In Chapter 4 in this book, a detailed description of the principles governing moisture within fuels is provided. This chapter builds on that discussion by describing the broadly applicable physical principles of evaporation as it relates to fuel drying. It is hoped that this information can be used by the ecological researcher to more accurately describe the fuel conditions that are encountered in field studies and to identify the important processes and mechanisms responsible for those fuel conditions.

The moisture content of forest fuels is dependent on a number of factors. This chapter will examine the meteorological factors that influence variations in fuel moisture content. Specifically, the meteorological factors that affect evaporation rates are discussed in detail. A number of methods for estimating evaporation rates are described, with a primary emphasis on the widely used Penman-Monteith formulation. This formulation is used to describe the functional dependence of evaporation rates on important meteorological variables. Also, the state of the atmosphere in close proximity to the Earth's surface and its importance to the environmental conditions experienced by fuels are discussed. Finally, the chapter discusses state-of-the-art models that are used in atmospheric circulation models to describe the complex interactions between the land surface and the atmosphere. Such models may be of use to the ecological researcher in describing evaporative drying of fuels.

In living fuels, foliage and twigs play an important role in forest fire behavior and can exhibit significant fluctuations in moisture content (Chandler et al., 1983). The moisture content is determined by the relative balance between evaporative losses and supply of water from the root system. Water is lost through transpiration, which is controlled by meteorological factors and the degree of stomatal opening. Water uptake by the roots is controlled by soil characteristics, soil moisture content, and the vertical distribution of roots. When soil moisture is high, water loss is approximately balanced on a daily time scale by water uptake. When soil moisture is deficient, root uptake of water is restricted and may not balance transpiration losses, leading to gradual decreases in fuel moisture content. The diurnal cycle of transpiration is of large amplitude, peaking in early afternoon, and this leads to changes in fuel moisture content during the day (Chandler *et al.*, 1983).

The moisture content of dead fuels is controlled by different processes (Pyne, 1984), although a similar set of meteorological factors provide the atmospheric control on these processes. Moisture cannot be replenished by root uptake. Moisture replenishment can occur by absorption through direct contact with liquid water. Moisture exchange in the gaseous phase depends on the fuel's hygroscopic properties and is driven by the difference in water vapor pressure between the fuel particle interior and the atmosphere in immediate contact with the fuel. Fine fuels, such as grass, leaves, needles, and ground litter, can rapidly change moisture content over time scales of hours. Medium (sticks, branches) and coarse (logs) fuels change moisture more slowly, on time scales of weeks to months for medium fuels and months to years for coarse fuels (Brown and Davis, 1973). For these fuels, the persistence of meteorological conditions is an important factor in determining fire risk.

II. EVAPOTRANSPIRATION PROCESSES AND THE METEOROLOGICAL CONTROLLING FACTORS

The evaporation of water from the soil or from the surface of living or dead fuels or the transpiration of water through leaves will be referred to herein by the common term "evapotranspiration" (ET). Fundamentally, ET can occur when the water vapor pressure of the air layer immediately adjacent to the surface containing liquid water is lower than the water vapor pressure of the liquid water. If the adjacent air layer and liquid water surface are in thermal equilibrium (often the case), then ET can occur when the water vapor pressure of the adjacent air layer is below its saturated value. The saturation water vapor pressure is a function of temperature.

The latent heat of vaporization of water is high, with a value of $2454\,\mathrm{J\,g^{-1}}$ at 20°C, varying slightly with temperature. The rate of ET is often limited by availability of energy. In a closed system, an unsaturated near-surface air layer will be moistened and cooled by ET and quickly reach saturation, stopping the process of ET. In the atmosphere, ET is sustained principally by two processes. The first process is the absorption of electromagnetic radiation at the surface, which raises the temperature of the liquid water and the adjacent air. Thus, the value of saturation of the adjacent air is increased, in essence increasing the capacity of the air for further evaporation. The second process is ventilation of the surface–air interface, which replaces the moistened air near the interface with drier and/or warmer air. Several meteorological variables are of particular importance in determining the magnitude of these two processes. These are shortwave and longwave radiation, temperature, atmospheric water vapor content, and vertical mixing.

A. RADIATION

The primary source of energy for ET is electromagnetic radiation from the sun which is concentrated principally in the visible and near-infrared (shortwave) portion of the spectrum, at wavelengths less than 3×10^{-6} m. The amount of radiation that is available for ET is determined by several processes, as described later.

The amount of incoming solar radiation at the top of the atmosphere varies with latitude and season based on the Earth–sun geometry. Solar radiation is attenuated as it passes through the atmosphere by direct absorption and scattering by gases, particulates, and, most importantly, clouds. Globally averaged, about 54% of the solar energy at the top of the atmosphere reaches the Earth's surface (Salby, 1996). At the surface, some of the solar radiation is reflected back to space. The reflectivity of the surface varies considerably with surface type and is represented by a parameter called the albedo, which is the fractional part of the radiation that is reflected. Except for snow-covered surfaces, the albedo of most surfaces is less than 0.3. For example, Betts and Ball (1997), in the BOREAS experiment (Sellers *et al.*, 1997), measured summertime albedos of 0.20 over grass, 0.15 over an aspen forest, and 0.083 over a coniferous forest.

The rate of emission of radiation by matter is governed by the Stefan-Boltzmann law,

$$I = \varepsilon \sigma T^4 \tag{1}$$

where I = energy flux (W m^{-2}), ε = emissivity, σ = Stefan-Boltzmann constant (5.67×10^{-8} W m^{-2} K^{-4}), and T = temperature of the matter (K). At the range of temperatures of the atmosphere and the earth's surface, this emission is in the infrared (longwave) portion of the spectrum, mostly at wavelengths $>3 \times 10^{-6}$ m. Most land surface types have high emissivities, with forests in the range of 0.97–0.99 (Oke, 1978). Although the cloud-free atmosphere is nearly transparent at visible wavelengths, it is semiopaque at infrared wavelengths and absorbs and reemits a significant amount of infrared radiation toward the surface. The magnitude of infrared radiation emitted by the cloud-free atmosphere is largely determined by the temperature and water vapor content in the lowest portion of the atmosphere. However, clouds have a much higher emissivity than the cloud-free atmosphere. Thus, the amount and type of cloud cover is an important factor in determining the magnitude of atmospheric in-

frared radiation. The upward longwave radiation, I_s, above the Earth's surface
is the sum of the emission by the Earth's surface, governed by Eq. (1), and the
reflection of downward longwave radiation from the atmosphere, I_a. Usually,
the absorptivity (1 − reflectivity) of the Earth's surface is assumed to be equal
to the emissivity. Thus,

$$I_s = \varepsilon_s \sigma T_s^4 + (1 - \varepsilon_s)I_a \qquad (2)$$

where ε_s = emissivity of the surface and T_s = temperature of the surface (K).

A portion of the electromagnetic radiation that reaches the Earth's surface is
transformed into nonradiative forms of energy, including latent heat through
the ET process, heat storage in the soil and biomass, and direct heating of the
atmosphere through conduction. The amount of radiation available for transformation into these other forms of energy is called the net radiation (R_n) and
is defined as

$$R_n = S_d(1 - a) + I_a - I_s \qquad (3)$$

where S_d = downward solar radiation at the Earth's surface (W m^{-2}) and a =
albedo. Figure 1 shows the diurnal evolution of R_n on a typical sunny summer day in the central United States, along with the individual components in
Eq. (3), as measured by a SURFRAD station (see discussion that follows).
There is a strong diurnal dependence to R_n, with the magnitude peaking near
midday and dropping to negative values at night, primarily in response to the

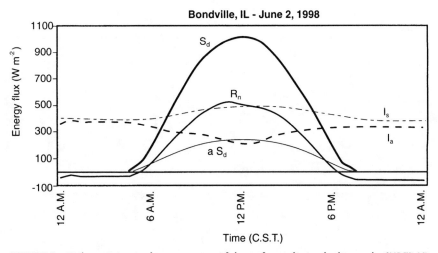

FIGURE 1   Daily variation in the components of the surface radiation budget at the SURFRAD
site at Bondville, Illinois, on June 2, 1998. Components are R_n (medium solid), S_d (thick solid), aS_d
(thin solid), I_s (thin dashed), and I_a (thick dashed).

variation of incoming solar radiation. Infrared radiation (I_s, I_a) also varies through the day, although with a lesser amplitude than solar radiation.

The daily accumulated radiative energy available to be transformed into other forms of energy, found by integrating R_n, is positive in Figure 1. The magnitude of the daily integrated R_n varies with latitude, the seasonal cycle, and meteorological conditions (particularly cloud cover). However, the climatology of R_n is not well specified because the measurement of R_n is technically difficult; data have typically been obtained only for short duration field experiments. Even in carefully controlled experiments, commercially available sensors have been found to exhibit biases (Hodges and Smith, 1997). To address the need for climatic observations of radiation, the Surface Radiation Budget Network (SURF-RAD) was established in the United States in 1993 by the U.S. Department of Commerce's National Oceanic and Atmospheric Administration. The SURF-RAD is providing accurate, long-term, continuous measurements of the surface energy budget at six sites in climatologically diverse regions of the United States. Figure 2 shows the annual cycle in 1997 of the daily average values of the radiation budget components at Goodwin Creek, Mississippi. All components exhibit a pronounced seasonal cycle, with values peaking in the summer. Daily R_n peaks at about 13 MJ m^{-2} in July and falls to a minimum of less than 2 MJ m^{-2} in December. Daily integrated values of I_a and I_s are much higher than S_d because infrared radiation occurs at night as well as during day. However, these largely cancel, and the net infrared radiation $(I_a - I_s)$ is smaller than the net shortwave radiation throughout the year.

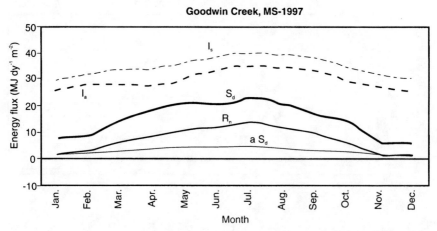

FIGURE 2 Annual cycle of the daily average values of the components of the surface energy budget at Goodwin Creek, Mississippi, during 1997. Components are R_n (medium solid), S_d (thick solid), aS_d (thin solid), I_s (thin dashed), and I_a (thick dashed).

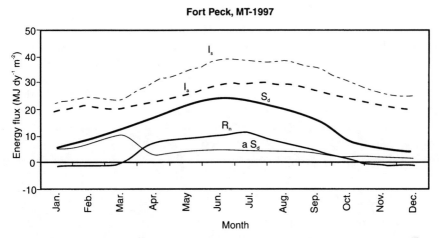

FIGURE 3 Annual cycle of the daily mean values of the surface energy budget components during 1997 at Fort Peck, Montana. Components are R_n (medium solid), S_d (thick solid), aS_d (thin solid), I_s (thin dashed), and I_a (thick dashed).

Figure 3 shows the radiation components for a more northern location: Ft. Peck, Montana. The seasonal cycle is qualitatively similar, but there are some significant differences. Incoming solar radiation peaks in summer at about 24 MJ m^{-2}, slightly higher than in Mississippi. This is a consequence of longer days and lower cloud cover at Ft. Peck, which offsets the higher solar elevation angles at Goodwin Creek. The amplitude of the annual cycle is higher at the more northern Montana location. The rather high values of reflected solar radiation (aS_d) in January, February, and March, reaching a peak of about 10 MJ m^{-2} in March, are a result of snow cover and the high albedo associated with snow cover. The peak value of R_n in summer is lower in Montana than in Mississippi. A comparison of Figures 2 and 3 indicates that this is primarily a result of lower values of incoming longwave radiation at the Montana site, probably due to lower water vapor content and lower air temperatures. During the cold season, net radiation is negative at Ft. Peck due to high values of reflected shortwave radiation.

There are no comparable long-term SURFRAD sites over a forest canopy. However, short-term experiments indicate that the radiation budget over a forest canopy is qualitatively similar to what is shown in Figures 2 and 3. For example, McCaughey *et al.* (1997) measured the longwave components of the surface energy budget over a jack pine (*Pinus banksiana*) forest in Saskatchewan, Canada, during the warm season. Net longwave radiation was very similar to what is seen in Figure 3 for Fort Peck, Montana.

At ground level within a forest canopy, net radiation is reduced sharply below that at the top of the canopy because of shielding by the canopy. The magnitude of the reduction in R_n depends upon tree density, tree architecture, and the season. J. M. Chen et al. (1997) measured net radiation in a relatively sparse aspen (Populus tremuloides) forest. Net radiation at ground level was about half of the above-canopy value before leaf emergence and about a quarter of the above-canopy value after leaf emergence. Baldocchi and Vogel (1996) measured net radiation in a deciduous forest and a boreal jack pine (Pinus banksiana) forest. Net radiation at the forest floor in the deciduous forest was less than 10% of the value above the canopy. Net radiation at the floor of the jack pine forest was 10–15% of the above-canopy level. Black and Kelliher (1989) measured net radiation in a Douglas fir (Pseudotsuga menziesii) stand and found that below-canopy net radiation was 13–16% of the above-canopy levels. These selected examples suggest that the available energy for evapotranspiration at the forest floor is usually a small fraction of what is available at the top of the canopy.

B. Atmospheric Water Vapor Content

The saturation value of atmospheric water vapor content is a highly nonlinear function of temperature. This dependence can be expressed as (Buck, 1981)

$$es = A \exp[BT/(C + T)] \tag{4}$$

where es = saturation value of water vapor pressure (hPa), T = temperature (°C), and A, B, and C are constants. Over water, the values of these constants are $A = 6.1121$, $B = 17.502$, and $C = 240.97$. Over ice, the values are $A = 6.1151$, $B = 22.452$, and $C = 272.55$. A graphical representation of Eq. (4) is shown in Figure 4, over water and over ice. The value of es approximately doubles for every 10°C increase in temperature. At temperatures below 0°C, there are small differences in es over water and over ice.

Evaporation rates in the air layer adjacent to a flat water surface are proportional to the difference $(es - e)$, where e = actual water vapor pressure, when water is freely available and the water surface is in thermal equilibrium with the adjacent air layer. This difference is referred to as the water vapor pressure deficit. Thus, evaporation rates are highly dependent on temperature because of the relationship of es to T.

Observations of atmospheric water vapor content are usually expressed in terms of dew point temperature (T_d, °C) or relative humidity (RH, %). The actual water vapor pressure (e) is related to T_d by

$$e = A \exp[BT_d/(C + T_d)] \tag{5}$$

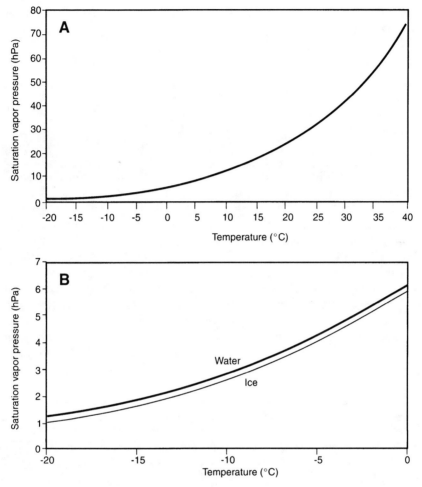

FIGURE 4   Dependence of the saturation water vapor pressure (hPa) on temperature (°C) (A) over water for $T > -20°C$ and (B) over water and ice for $T < 0°C$.

Relative humidity is defined in terms of the mixing ratio (v), which is the ratio of the mass of water vapor to the mass of dry air in a volume. The definition is

$$\text{RH} (T) = 100\% \ v/vs \ (T) \tag{6}$$

where $vs \ (T)$ is the saturation mixing ratio at the air temperature T. The relationship of vapor pressure to mixing ratio is (List, 1949)

$$v = 0.62197 \ e/(P - e) \tag{7}$$

where P = air pressure (hPa). Relative humidity can then be expressed in terms of vapor pressure as

$$RH = 100\% \frac{e(P - es)}{es(P - e)} \qquad (8)$$

Water vapor content can also be defined as the ratio of the mass of water vapor to the mass of moist air in a volume, known as the specific humidity (q). This can be expressed in terms of the mixing ratio as

$$q = v/(1 + v) \qquad (9)$$

Relative humidity is strongly dependent on temperature through its dependence on es. In most situations, there is little diurnal variation in e but large variations in RH. RH usually reaches a maximum in the early morning around sunrise when temperatures are at a minimum and falls to a minimum in mid afternoon around the time of the daily maximum temperature. This behavior is illustrated by the daily average profile for International Falls, Minnesota, in July (Figure 5). There is relatively little variation in water vapor pressure, ranging from 14.6 hPa in the early morning to 16.4 hPa in late morning, a total percentage change of 11%. By contrast, relative humidity varies from near 90% in the early morning to near 50% in late afternoon.

Some measures of ET use relative humidity as one indicator of drying rates. By itself, RH may be a deceptive indicator because of its dependence on temperature. Figure 6 shows the dependence of the water vapor pressure deficit on

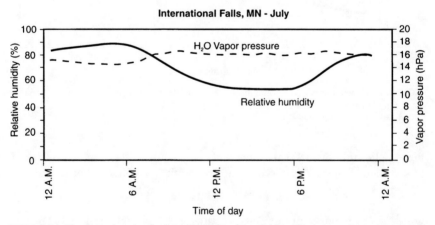

FIGURE 5   The diurnal cycle of relative humidity (%, solid line) and atmospheric water vapor pressure (hPa, dashed line) during July at International Falls, Minnesota. These are averages of hourly observations for the period 1961–1990.

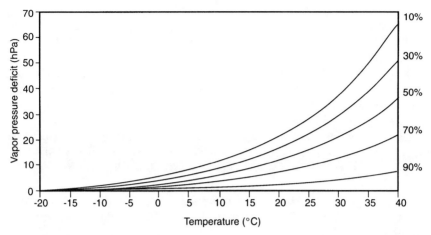

FIGURE 6 The dependence of the water vapor pressure deficit (hPa) on temperature (°C) for relative humidity values of 10, 30, 50, 70, and 90%.

temperature for selected values of relative humidity. The water vapor pressure deficit, and thus the ET rates, increases exponentially with temperature, following Eq. (4), for constant RH.

C. TEMPERATURE

The dependence of evaporation on temperature is primarily through the dependence of es on temperature. Temperature may also have indirect effects because longwave radiation increases with temperature. However, as illustrated in Figures 2 and 3, net longwave radiation $(I_a - I_s)$ is only weakly dependent on season (temperature).

D. VERTICAL MIXING (WIND AND STABILITY)

Evaporation moistens and cools the air near the surface, which has a negative feedback on evaporation rates. Mixing of air near the surface has the net effect of transporting moist air upward and replacing it with drier and/or warmer air from aloft, maintaining evaporation. One factor affecting the rate of mixing is wind speed. Evaporation rates tend to increase with increasing wind speed, other factors being equal. A second factor is the vertical air density gradient which can enhance or suppress mixing.

1. Mixing Processes

Mixing in the lowest few meters of the atmosphere is driven primarily by the interaction of the surface with the atmosphere. The Earth's surface is aerodynamically rough and interacts with moving air. This interaction results in the transfer of momentum between the surface and the atmosphere. The wind speed near the surface is reduced by this momentum transfer. Momentum transfer occurs through two processes. The first is friction; air in direct contact with surface elements is slowed by frictional forces. Vertical mixing processes transfer this slow moving air stream upward away from the surface to a region of faster moving air streams and vice versa. This exchange slows the faster moving air and accelerates the slower moving air. The second process is through the pressure force; high-pressure builds up on the windward side of surface elements that are exposed to the wind which then slows the air upwind of the element. This process does not involve exchange of air and momentum is transferred without direct contact with the surface.

Mixing occurs either by molecular diffusion in laminar flow or by turbulence. Momentum transfer to the surface results in a vertical gradient in wind speed. This gradient is often sufficiently large that laminar flow is unstable, and turbulent motion results. Turbulent mixing is much more efficient than molecular diffusion. The atmospheric transport of momentum usually occurs at much higher rates than provided by molecular diffusion because the atmosphere near the surface is generally turbulent, except immediately adjacent to the surface. In a shallow layer (of the order of 1 cm thickness) near surface objects, proximity to the surface prevents turbulent motion and the wind flow is laminar; the transfer of momentum, heat, and water vapor occurs by molecular diffusion in this shallow, laminar-flow layer.

2. Atmospheric Stability

The second factor affecting the rate of mixing, in addition to the interaction of the wind with the surface, is the stability of the atmosphere. Because pressure decreases with height, the temperature of a rising parcel of air will decrease with height at a rate known as the "adiabatic lapse rate," as governed by the gas law, whose value is $9.8°C\ km^{-1}$. If temperature decreases with height at a slower rate than adiabatic, then a rising parcel will find itself at a lower temperature (higher density) than the surrounding environment and will tend to sink back to its original level. This suppresses vertical mixing and the atmosphere is referred to as "stable." If the temperature decreases with height more rapidly than adiabatic, then a rising parcel of air will find itself at a higher temperature (lower density) than the surrounding environment and will accelerate upward. This enhances vertical mixing and the atmosphere is referred to as "unstable." It has been convenient within the science of meteorology to define stability conditions through use of a quantity known as the potential temperature (θ) which is defined as

$$\theta(z) = T(z)\left[\frac{P_0}{P(z)}\right]^{0.287} \tag{10}$$

where z = height, $T(z)$ has units of K, and P_0 is the pressure of a reference level, usually taken to be 1000 hPa. In essence, the potential temperature is a value that has been adjusted for the changes in temperature that occur due to pressure changes when a parcel rises or sinks in the atmosphere. When expressed in terms of θ, the criteria for stability become

$$\frac{\partial \theta}{\partial z} < 0 \qquad \text{unstable}$$

$$\frac{\partial \theta}{\partial z} = 0 \qquad \text{neutral} \tag{11}$$

$$\frac{\partial \theta}{\partial z} > 0 \qquad \text{stable}$$

In a stable atmosphere, the character of the wind flow, whether turbulent or laminar, is determined by a balance between the tendency for vertically sheared flow to break down into turbulence and the suppression of vertical motions by stable buoyancy. This characteristic is often expressed in a parameter known as the gradient Richardson number (Ri), defined as

$$\text{Ri} = \frac{g}{T}\frac{\partial \theta}{\partial z}\bigg/\left(\frac{\partial U}{\partial z}\right)^2 \tag{12}$$

where g = acceleration of gravity, and U = horizontal wind speed. When Ri is greater than about 0.25, turbulent motion ceases, and the flow becomes laminar. In this case, exchange of momentum, heat, and water vapor is markedly reduced. In the opposite case, when the potential temperature decreases with height, the atmosphere is buoyantly unstable, and convection transfers momentum and enhances the rate of turbulent mixing.

The lowest layer of the atmosphere, where turbulence driven by momentum and heat exchange at the surface is present, is often referred to as the planetary boundary layer (PBL). During the daytime, when incoming solar radiation creates an unstable temperature profile, the depth of the PBL is typically 1–2 km (Panofsky and Dutton, 1984). At night, under a stable temperature profile, the PBL depth is often less than 100 m, particularly when the wind speed is low. The lowest part of the PBL is referred to as the surface layer. The surface layer is not precisely defined; within this layer, vertical variations of momentum, heat, and moisture fluxes are small (<10%) and, as a first approximation, can be assumed to be constant with height. As a rule of thumb, the surface layer occupies the lowest 10% of the PBL (Panofsky and Dutton, 1984).

3. Vertical Wind Profile

The turbulent transport of momentum between the surface and the atmosphere creates a vertical gradient in wind speed. Within the surface layer, the wind speed typically increases with the logarithm of height, as documented by numerous field experiments. Mathematically, the wind speed (U) dependence on height (z) can be expressed as (Panofsky and Dutton, 1984)

$$U(z) = \frac{u_*}{k}\{\ln[(z - d)/z_0] - \psi_m\} \tag{13}$$

where u_* = the friction velocity (m s^{-1}), d = displacement height (m), z_0 = roughness length (m), k = von Karman constant (= 0.4), and ψ_m is a correction factor for stability effects. The friction velocity is a measure of the rate of momentum transport. The displacement height is a measure of the shielding of the Earth's surface by a vegetative canopy. For short grass and short crops, d is small or negligible. However, a dense forest canopy is very effective at shielding the free atmospheric flow from the surface; from the viewpoint of the atmosphere, the effective surface is much higher than the actual ground level. Typically, in dense crops or forests, the displacement height is 60–80% of the average height of the vegetation (Thom, 1975).

The roughness length is a measure of the efficiency of the interactions between the free atmospheric flow and the surface. For a very smooth, flat surface, such as a sandy desert, the interaction between the free atmospheric flow and the surface is minimal, and the corresponding roughness length of that surface is also small. By contrast, surfaces that have many undulations which provide opportunities for airflow to penetrate between the undulations and be slowed by interactions with these undulations are much more efficient at transferring momentum and the corresponding roughness length is much larger. Forests typically have rather large roughness lengths because of the tall trees that comprise the forest and the many openings that allow air to penetrate into the forest canopy. Interestingly, very dense forests where air cannot penetrate readily down into the canopy can have a smaller roughness length than a more open forest where the wind flow can more easily penetrate deeper into the canopy. In general, the roughness length is a function of the height of the surface roughness elements, their density, and their flexibility. The roughness length in general will increase with increasing height of the surface elements. The roughness length will also increase with the increasing density of the surface elements up to a point above which the roughness length will decrease as the density becomes very high. Also the roughness length generally decreases with increasing flexibility of vegetation, since the pressure force mechanism for momentum transport will be more effective for rigid objects than for flexible objects. As roughness increases, turbulence and vertical mixing rates increase. Garratt (1977) and Oke (1978) provide typical values of z_0 for a wide range of surfaces. Forests have values usually in the range of 1–6 m. By contrast, the roughness height of short grass is of the order of 1 cm.

The logarithmic wind profile applies to the free atmosphere above the canopy. Within the canopy, the vertical wind speed gradient is dependent on the structure and density of the canopy (Fritschen, 1985). Denser canopies are characterized by lower wind speeds than more open canopies. The gradient within the canopy has often been parameterized with an exponential function (Cionco, 1965):

$$U(z) = U(h) \exp[\beta(1 - z/h)] \tag{14}$$

where h = height of the canopy, and β is a parameter called the wind velocity attenuation coefficient. For forest canopies, β is typically in the range of 2–5 (Cionco, 1978). These values of β indicate that wind speeds at the floor of a forest are a small fraction of values at the top of a forest, thus reducing ventilation and ET rates.

Water vapor is to a first approximation a passive element in the surface mixing process. When momentum is transported through friction, in which slow-moving air molecules near surface objects are exchanged with faster moving air, water vapor molecules are also exchanged. However, the transport of momentum by the pressure force process does not result in exchange of air; therefore, the transport of water vapor is less efficient than momentum transport.

III. ESTIMATION OF POTENTIAL EVAPOTRANSPIRATION RATES

A. PENMAN–MONTEITH METHOD

1. General Formulation

The energy balance at the air-surface interface in nonprecipitating conditions is usually written as

$$R_n \cong LE + H + G \tag{15}$$

where L = latent heat of evaporation ($J\ kg^{-1}$), E = water vapor flux ($kg\ m^{-2}\ s^{-1}$), H = sensible heat flux ($W\ m^{-2}$), and G = soil heat flux ($W\ m^{-2}$), as illustrated in Figure 7. Measurements of the energy budget and application

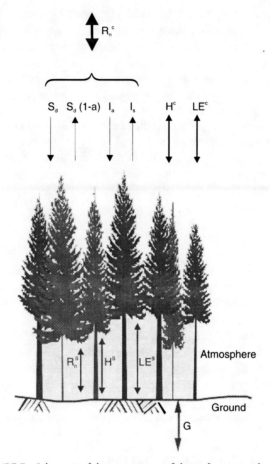

FIGURE 7   Schematic of the components of the surface energy budget.

of this formula have focused both on the interface between the top of the canopy and the free atmosphere (denoted by superscript c) and on the forest floor–atmosphere interface (denoted by the superscript s). This expression ignores photosynthesis, heat storage in plant tissue and the canopy volume, and metabolic activity, all of which tend to be smaller than the experimental error ($<10\%$ of R_n). Although this approximation is usually adequate and will be used in this chapter, it is important to point out that heat storage in plant tissue and the canopy volume can be significant in tall forests, particularly when net radiation is low. For example, Denmead and Bradley (1985) made estimates of this storage term for a mature Ponderosa pine (*Pinus ponderosa*) forest and found that it usually represented $>10\%$ of the net radiation, nonnegligible if high accuracy is required.

The product LE is referred to as the latent heat flux. The latent heat flux is usually directed upward from the surface to the atmosphere. However, at night, condensation can occur, in which case the flow of moisture is toward the surface; however, the values of condensation are usually much lower than evaporation. Sensible heat flux (H) is usually directed upward into the atmosphere during the day and downward toward the canopy at night. Likewise, the soil heat flux is usually directed downward into the ground during the day and toward the surface at night. The exact partitioning of radiative energy into these components is dependent on meteorological factors, the characteristics of the soils, and the characteristics of the fuel.

The sensible and latent heat fluxes are often parameterized in terms of the near-surface gradients of temperature and vapor pressure as follows (Bonan, 1996):

$$H = \rho c_p (\theta_s - \theta_a)/r_a \tag{16}$$

and

$$LE = \frac{\rho c_p [es_s - e]}{\gamma (r_a + r_s)} \tag{17}$$

where θ_a = air potential temperature, θ_s = potential temperature of surface, es_s = saturation water vapor pressure at the surface, $\gamma = c_p P/(0.622L)$, ρ = air density (kg m^{-3}), c_p = heat capacity of air at constant pressure (J kg^{-1} K^{-1}), r_a = aerodynamic resistance (s m^{-1}), r_s = stomatal or surface resistance (s m^{-1}), and P = pressure (hPa). T_a and e are measured at some standard height. For the practical problem of estimating fluxes, these formulae are difficult to apply because T_s and es_s are not easily measured and thus not typically available to make such estimates. Thus, many methods have been developed to estimate fluxes from more readily available measurements. In particular, numerous formulae have been developed to estimate "potential" evapotranspiration (PET) rates (i.e., that which would occur if soil water were not limited and meteorological factors only limit the rate of ET). One of the most successful and widely used of these is the Penman formula (Penman, 1948):

$$PET = \frac{\Delta(R_n - G) - \rho c_p(es - e)/r_a}{\Delta + \gamma} \tag{18}$$

This equation was modified slightly by Monteith (1965, 1981) to produce the following formula for actual evapotranspiration (AET):

$$AET = \frac{\Delta(R_n - G) + \rho c_p(es - e)/r_a}{\Delta + \gamma(1 + r_s/r_a)} \tag{19}$$

where $\Delta = d(es)/dT$ (hPa K^{-1}). These formulae are quite accurate for a wide range of meteorological conditions because they incorporate all the primary meteorological factors and are based on fundamental energy conservation principles. The first term in the numerator, called the radiation term, accounts for the direct absorption of radiative energy at the air–surface interface. The second term in the numerator, the advection term, accounts for additional evaporation due to ventilation of the surface (the variable r_a incorporates the wind speed; see the discussion that follows). This ventilation rate is primarily a function of the water vapor pressure deficit of the free atmosphere and the rate of turbulent mixing, which is embodied in the variable r_a. It is important to note that the water vapor pressure deficit used in Eqs. (18) and (19) is that at a standard measurement height, not at the air–surface interface. These often are quite different.

2. Aerodynamic Resistance

The rate of vertical diffusion of water vapor between the surface and the free atmosphere is parameterized by r_s and r_a, using an analogy to electrical circuits in Ohm's law (Chapter 4 in this book) where the evaporation rate (current) is determined by the water vapor pressure deficit (voltage difference) divided by the inverse of the diffusion rate (resistance). The aerodynamic resistance (r_a) is inversely proportional to the rate of vertical mixing of air near the surface. In general, it is a function of the wind speed, the roughness of the surface, and the thermal stability of the near-surface atmosphere. The vertical profiles of temperature and water vapor content follow a logarithmic dependence similar to Eq. (12); r_a can be derived from these logarithmic forms and expressed as

$$r_a(z) = \frac{1}{k^2 U(z)}[\ln\{(z - d)/z_0\} - \psi_m][\ln\{(z - d)/z_0\} - \psi_{m,h,w}] \qquad (20)$$

where $\psi_{m,h,w}$ is the stability correction for wind (m), temperature (h), or water vapor (w), depending on whether r_a is used to estimate momentum, sensible heat, or latent heat fluxes, respectively. This expression for r_a is based on a fundamental understanding of the nature of the turbulent near-surface atmosphere, an understanding that is supported by numerous field experiments.

Strictly speaking, the general form for r_a [Eq. (20)] is applicable to fluxes of water vapor, heat, and momentum. However, as noted earlier, the efficiency of water vapor and heat transport is less than that of momentum near the surface because the pressure force does not act to transport water vapor and heat. In essence, the resistance for water vapor and heat transport is larger than for momentum transport. This is sometimes referred to as the "excess" resistance. Experimental evidence indicates that the efficiency of water vapor transport is similar to that of heat transport and these are usually assumed to be equal. Experimental observations (Chamberlain, 1968; Garratt and Hicks, 1973) have shown that the magnitude of this excess resistance is dependent on the flexibility of surface elements. Surface elements that are rigid, such as fallen logs, tree trunks, etc., are efficient at exchanging momentum by the pressure force, and, in this case, the difference between the efficiency of momentum and water vapor/heat transport is large. By contrast, flexible surface elements, such as the leaves of trees and grasses, are not very efficient at transporting momentum through the pressure force because they bend in the wind. In this case, the efficiencies for momentum and water vapor/heat transport are more nearly equivalent. A variety of expressions have been developed to parameterize this effect. Perhaps the most common and simplest approach is to assign a roughness length for water vapor and heat transport that is smaller than that for momentum. The values of z_0 given previously are for momentum. For water vapor or heat, a typical value is $z_{0w} = z_0/7$, where z_{0w} is the roughness height for water vapor or heat (Rowntree, 1991; Shuttleworth, 1991). However, experimental evidence suggests that z_{0w} is not a constant for a particular surface and, in fact, varies with environmental conditions (Blyth and Dolman, 1995; Hignett, 1994; Sun and Mahrt, 1995). Based on work by Zilitinkevich (1970), the relationship of z_{0w} to z_0 can be expressed as

$$\ln\frac{z_0}{z_{0w}} = 0.13\left(\frac{u_* z_0}{\nu}\right)^{0.45} \tag{21}$$

where ν = kinematic viscosity of air with a value of about $1.5 \times 10^{-5}\ \mathrm{m^2\ s^{-1}}$. The quantity $(u_* z_0/\nu)$ is the roughness Reynolds number, which can be interpreted as the Reynolds number of the smallest turbulent eddy. In this relationship, the ratio of z_0 to z_{0w} increases with increasing roughness and with increasing wind speed (through u_*).

The excess resistance can be incorporated in Eq. (20) as follows:

$$r_a(z) = \frac{1}{k^2 U(z)}[\ln\{(z-d)/z_0\} - \psi_m][\ln\{(z-d)/z_{0w}\} - \psi_{h,w}] \tag{22}$$

A number of expressions have been derived for ψ_m and $\psi_{h,w}$, based on field experiments. These are often expressed as a function of a parameter known as the Monin-Obukhov length (L_{mo}), defined as

$$L_{mo} = -\frac{u_*^3 \rho c_p T}{kgH} \tag{23}$$

where L_{mo} has the units of length and is similar to the gradient Richardson number (Ri) in expressing the relative magnitude of wind shear forces to buoyancy forces. When the absolute magnitude of sensible heat flux (H) is large compared to the momentum flux, the absolute value of L_{mo} is small, and vice versa. The sign of L_{mo} is indicative of the stability (negative for unstable conditions, positive for stable conditions). Commonly used expressions for the wind (ψ_m)

and temperature/water vapor ($\psi_{h,w}$) profile correction factors are (Paulson, 1970)

$$\psi_m[(z-d)/L_{mo}] = \ln\left[\left(\frac{1+x^2}{2}\right)\left(\frac{1+x}{2}\right)^2\right]$$

$$-2\tan^{-1}x + \frac{\pi}{2} \qquad \text{for } L_{mo} < 0 \qquad (24)$$

$$\psi_m[(z-d)/L_{mo}] = -5\frac{(z-d)}{L_{mo}} \qquad \text{for } L_{mo} > 0 \qquad (25)$$

where $x = [1 - 16(z-d)/L_{mo}]^{0.25}$ and

$$\psi_{h,w}[(z-d)/L_{mo}] = 2\ln\left[\frac{1}{2}\{1 + (1 - 16(z-d)/L_{mo})^{0.5}\}\right] \qquad \text{for } L_{mo} < 0 \qquad (26)$$

$$\psi_{h,w}[(z-d)/L_{mo}] = -5(d-z)/L_{mo} \qquad \text{for } L_{mo} > 0 \qquad (27)$$

3. Stomatal or Surface Resistance

The variable r_s is the resistance of the surface to the transport of water vapor. When $r_s > 0$, the surface limits ET below that of the potential rate. For dead fuels, this can be interpreted as inversely proportional to the rate of diffusion of water to the surface of the fuel. When the diffusion rate is low, it is this factor that controls ET rates, rather than the availability of energy and the rate of turbulent mixing. For leaves, r_s is the stomatal resistance. The stomatal resistance reflects the physiological control on transpiration by the opening and closing of stomates in the leaf which is part of the process of photosynthesis. A number of meteorological factors influence the magnitude of r_s including solar radiation, soil moisture, water vapor pressure deficit, and temperature. Because it is dependent on solar radiation, r_s exhibits a pronounced and characteristic diurnal dependence as stomates open and close in response to increases and decreases in solar radiation. Stomates also respond to deficient soil moisture by partially closing to conserve water. When soil moisture is not limiting, minimum values of r_s are achieved in the middle of the day and are a function of the plant type. Crops such as corn, soybean, and wheat typically have minimum values around 50 s m^{-1}, whereas forests have higher minimum values, typically around 100 s m^{-1} or greater (Rowntree, 1991).

Some recent experiments have provided a number of observations of r_s applicable to forests. Blanken et al. (1997) found that daytime mean values of stomatal resistance over an aspen forest varied widely, from about 50 to >200 s m^{-1}, with a typical daytime mean value of around 100 s m^{-1}. Chang et al. (1997) measured stomatal resistance values of 100–150 s m^{-1} over a poplar

(*Populus trichocarpa* X *P. tacamahaca*) forest. Baldocchi *et al.* (1997) measured stomatal resistance over a boreal jack pine forest in Saskatchewan, Canada, and found rather high values, generally in excess of 200 s m^{-1} and often exceeding 500 s m^{-1}. They also obtained measurements over a well-watered deciduous forest. Values of stomatal resistance were in the range of 150 – 400 s m^{-1}, somewhat lower than the jack pine forest. They argued that the high values of stomatal resistance in the jack pine forest reflect an adaptation of the plant to an environment of relatively low water availability and limits on decomposition and nutrient cycling in the boreal region where the jack pine measurements were taken. There is also experimental evidence that stomatal resistance increases when the above-canopy vapor pressure deficit exceeds approximately 10 hPa; above this threshold, stomatal resistance varies in such a way that transpiration remains relatively constant. This has been observed over an aspen forest (Hogg *et al.*, 1997), in tropical rainforests (Meinzer *et al.*, 1993; Granier *et al.*, 1996), in temperate forests and woodlands (Lopushinsky, 1986; Price and Black, 1989; Goulden and Field, 1995), and in a Holm-oak savanna (Infante *et al.*, 1997). This may reflect a physiological response of the plant to maintain leaf water potentials above the point where plant damage would occur (Tyree and Sperry, 1988; Sperry and Pockman, 1993; Goulden and Field, 1995; Goulden *et al.*, 1997).

4. Application of Penman–Monteith Formulae

The practical application of Eqs. (18) and (19) to estimate PET and AET is sometimes difficult because the required meteorological variables are often not available. For example, the routine hourly observations taken by National Meteorological Services at airports do not include any components of the radiation budget. By contrast, a number of specialized networks that have been established by universities, state agencies, and other organizations do include measurements of downward solar radiation and soil heat flux. However, even many of these networks do not measure net radiation.

For this reason, practical application of Eq. (17) often requires estimation of radiation components. Meyers and Dale (1983) developed a method to estimate downward solar radiation from the cloud cover observations that are taken at airports. Petersen *et al.* (1995) applied this method to develop an historical solar radiation climatology for the midwestern United States. The following methods can be used to estimate R_n if hourly solar radiation observations are available or if hourly cloud observations are available to estimate solar radiation using the method of Meyers and Dale (1983) and Petersen *et al.* (1995).

A variety of investigators have found that simple regressions between R_n and S_d provide reasonably accurate estimates of R_n with r^2 values greater than 0.9 (Shaw, 1956; Monteith and Szeicz, 1961; Fritschen, 1967; Nielsen *et al.*, 1981; Zhong *et al.*, 1990; Pinker and Corio, 1984; Pinker *et al.*, 1985; Kustas *et al.*,

1994; Kaminsky and Dubayah, 1997). However, these studies have produced widely varying regression coefficients. This suggests that the exact relationship between R_n and S_d may be rather stable for a particular location but will vary from location to location depending on other factors, probably including the latitude, temperature and humidity climatology, and land-use. Even at a specific location, the relationship will likely vary with season. Use of this approach is problematic because it may be uncertain which study's regression coefficients should be adopted.

A second method avoids some of these uncertainties by making independent estimates of the longwave components of the radiation budget. In this method, the net radiation R_n is calculated following Weiss (1983) by

$$R_n = S_d(1 - a) + F_c(I_a - I_s) \tag{28}$$

where F_c is a function of cloud cover. I_a is given by Brutsaert (1975) as

$$I_a = 1.24(e/T_a)^{0.143}(\sigma T_a^4) \tag{29}$$

where the units are hPa (e) and K (T_a). I_s is estimated by

$$I_s = 0.98\sigma T_a^4 + (1 - 0.98)I_a \tag{30}$$

F_c is given by

$$F_c = 0.4 + 0.6\, C_T \tag{31}$$

where $C_T =$ cloud transmission. Soil heat flux G is estimated as

$$G = 0.15\, R_n \tag{32}$$

This method is probably more accurate than the use of linear regressions between S_d and R_n but does require additional data (water vapor pressure, temperature, cloud cover). Cloud cover data may not be readily available for non-National Weather Service sites. Also, both I_s and I_a are estimated using air temperature at a standard height even though infrared emissions are determined by the surface temperature (for I_s) and the air temperature above the site (for I_a). This further reduces the accuracy.

Another approach to the estimation of PET and AET using the Penman method has been developed by Doorenbos and Pruitt (1977). This publication provides convenient tables and procedures, particularly for estimating the radiation term.

B. RADIATION MODELS

Empirical studies have indicated that the left- and right-hand terms of the numerator in Eqs. (18) and (19) are correlated. There is a physical basis for this finding since increasing radiation (left-hand side) will tend to increase the temperature and thus the atmospheric water vapor pressure deficit (right-hand side). In humid climates, the left-hand term is typically much larger than the right-hand term. This has led to a class of models to estimate PET from radiation. Representative of this class is the Priestly–Taylor (1972) equation

$$PET = \alpha \frac{\Delta(R_n - G)}{\Delta + \gamma} \tag{33}$$

where α is an empirical constant to which they assigned a value of 1.26, based on analysis of field data. Other studies have been performed to estimate α, and most of them have found values near to that of Priestly and Taylor (1972) for situations where stomatal resistance is low. However, values of α are often somewhat lower for forests because of higher values of stomatal resistance. For example, Blanken et al. (1997) measured a value of 1.22 over a hazelnut understory in an aspen forest. However, the average value of α above the forest canopy was only 0.91 with an upper limiting value of 1.11, indicating that stomatal control of forest transpiration substantially reduced actual evaporation below the potential value that would be calculated using a standard value of $\alpha = 1.26$.

Although application of this model requires measurements of radiation, it does not require measurements of humidity which is difficult to measure with high accuracy. This is the main attraction of this class of models. It also does not require estimates of the aerodynamic resistance and the associated uncertainties in estimating surface parameters such as z_0 and d.

C. PAN EVAPORATION

Pan evaporation is measured at scattered stations around the world and provides a relatively straightforward approach to the estimation of PET. Its primary attraction is that it is a direct measure of evaporation and integrates the effects of all the meteorological controlling factors. The standard Class A pan is 121 cm in diameter and 25.5 cm deep. Data are obtained by measuring the depth of water in the pan and subtracting from the previous day's depth. Periodically, water is added to the pan. This is a simple and straightforward measurement. However, pan evaporation data must be used with care because, in general, they provide an overestimate of water loss from vegetated surfaces due primarily to two factors. First, the water volume and the pan itself absorb a greater fraction of incoming solar radiation (has a lower albedo) than vegetation. This raises the temperature of the water above the ambient air temperature and causes excessive rates of evaporation. Second, the Class A pan is raised slightly above the land surface which enhances the turbulent transport of water vapor from the water surface to the atmosphere. These factors cause pan evaporation to be typically 20–40% higher than PET.

Correction factors, called pan coefficients, have been developed for the United States by Farnsworth and Thompson (1982) and Farnsworth *et al.* (1982). When applied to the pan evaporation data, these factors provide reasonably accurate estimates of PET. For the United States, these factors are typically in the range of 0.65–0.8, the lowest values applying to the sunny regions of the southwest and the highest values to the northern Great Lakes and west coast. Doorenbos and Pruitt (1977) also provide procedures to estimate pan coefficients, given the general climatic regime of an area and the location of the pan.

D. TEMPERATURE MODELS

There is a correlation between temperature and radiation and also between temperature and water vapor pressure deficit. This has been the underlying foundation for development of models that require only temperature. One such model is the Blaney and Criddle (1950) equation which provides monthly values of PET in mm:

$$PET = C_u D(0.46\, T_m + 8) \tag{34}$$

where C_u = an empirically determined consumptive use coefficient, D = monthly mean percentage of annual daytime hours of the year, and T_m = mean monthly temperature in °C. Values of C_u have been determined for irrigated crops and typically range from 0.6 to 0.9. This method has been applied widely for irrigation applications in the western United States.

Thornthwaite (1948) developed a measure of PET that was used to classify climatic regimes, defined as

$$PET_m = 1.6\left[\frac{10\, T_m}{s}\right]^a \tag{35}$$

where

$$s = \sum_{m=1}^{12} (T_m/5)^{1.514} \tag{36}$$

and

$$a = 6.75 \times 10^{-7}\, s^3 - 7.71 \times 10^{-5}\, s^2 + 1.79 \times 10^{-2}\, s + 0.49 \tag{37}$$

where PET_m is the value of PET in cm for month m and T_m is the mean temperature (°C) for month m. This formula is reasonably accurate in the humid eastern United States but underestimates PET in arid regions.

The exact relationship between PET and temperature depends on general climatic conditions, particularly arid vs. humid climates, and any temperature model must be tailored to these conditions to minimize uncertainties and errors. Even with care, such models will likely produce estimates with significant errors when meteorological conditions are climatically unusual for a location. The real attraction of such models is that temperature data are widely available. However, their use is justifiable only when data are limited to temperature and when accuracy requirements are not high.

IV. FUNCTIONAL DEPENDENCE OF PET AND AET

The Penman-Monteith equation is used here to illustrate the dependence of PET and AET on selected meteorological variables. Figure 8 illustrates the dependence on temperature for selected values of dewpoint temperature and selected nominal values for R_n, U, P, and r_s. AET increases rather rapidly with temperature, primarily reflecting the increase in es with temperature. However, there is also a contribution from the partitioning of net radiative energy between sensible and latent forms of energy. As temperature increases, an increasing fraction of net radiation is partitioned into latent heat energy. AET decreases with increasing T_d, reflecting decreases in the saturation vapor pressure deficit.

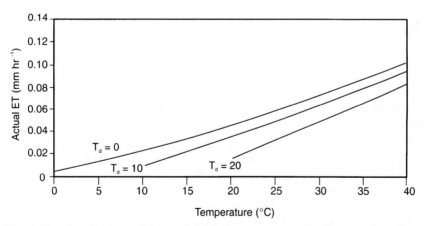

FIGURE 8   The dependence of the actual evapotranspiration on temperature for values of dewpoint temperature of 0°C, 10°C, and 20°C. Nominal values of R_n = 400 W m^{-2}, U = 5 m s^{-1}, P = 1000 hPa, and r_s = 100 s m^{-1} were used to make these calculations with Eq. (19).

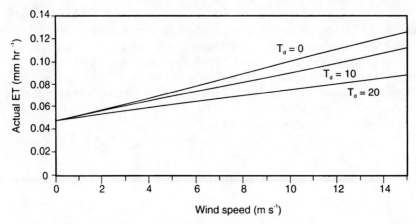

FIGURE 9   The dependence of the actual evapotranspiration on wind speed for values of dew-point temperature of 0°C, 10°C, and 20°C. Nominal values of $R_n = 400$ W m^{-2}, $T = 30$°C, $P = 1000$ hPa, and $r_s = 100$ s m^{-1} were selected for these calculations using Eq. (19).

Figure 9 illustrates the dependence on wind speed for selected values of T_d and nominal values for R_n, T, P, and r_s. There is a linear dependence on wind speed. The sensitivity to wind speed increases with decreasing T_d (increasing water vapor pressure deficit). For these nominal values, the radiation term alone is about 0.047 mm hr^{-1} (y intercept). The advection term can be comparable to, or even exceed, the radiation term for moderate wind speeds and low values of T_d.

The dependence of AET and PET on temperature for selected values of r_s is illustrated in Figure 10. The curve for $r_s = 0$ indicates PET. As r_s increases, AET decreases below PET. At higher values of r_s, the rate of change in AET with increasing temperature is rather small. This illustrates that, as the surface resistance increases, AET rates become less dependent on atmospheric forcing and more dependent on the characteristics of the fuel that determine the surface resistance.

Figures 8–10 illustrate that AET is a sensitive function of temperature, atmospheric water vapor content, wind speed, and stomatal or surface resistance. The use of the Penman-Monteith equation [Eq. (19)], which incorporates all of these variables, provides a suitable framework for accurate estimation of AET. By contrast, simpler approaches, such as the radiation or temperature models, do not incorporate all these variables and may result in large errors under certain conditions.

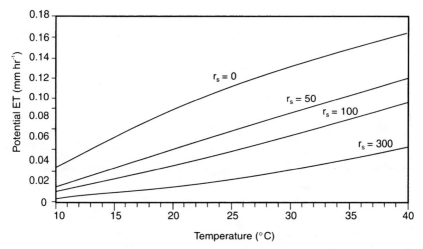

FIGURE 10   The dependence of actual evapotranspiration on temperature for values of $r_s = 0$, 50 s m^{-1}, 100 s m^{-1}, and 300 s m^{-1}. Nominal values of $R_n = 400$ W m^{-2}, $T_d = 10°C$, $U = 5$ m s^{-1}, and $P = 1000$ hPa were selected for these calculations using Eq. (19). The curve for $r_s = 0$ is the potential evapotranspiration.

V. CHARACTERISTICS OF PET

The drying of forest fuels is related to the relative magnitudes of AET and of the amount and frequency of precipitation events. Long-term measurements of AET are not widely available. However, AET is approximately equal to PET when soil moisture levels are high. Even under deficient soil moisture levels, AET usually varies proportionally with PET. Therefore, an analysis of PET, which can be estimated from the standard meteorological observations available at many locations, provides insight into factors that affect forest fire risk.

A. SEASONAL CYCLE

The seasonal variation of PET is a key factor in determining the risk of forest fires. Of particular importance is the relative magnitude of PET compared to precipitation. When PET exceeds precipitation, soil moisture can be depleted, and both live and dead fuels can lose moisture content, increasing the risk of fires. Figure 11 shows the seasonal cycle of PET at International Falls, Minnesota, based on the Penman-Monteith formula using the methodology described earlier to estimate R_n. The climate at this location can be classified as humid.

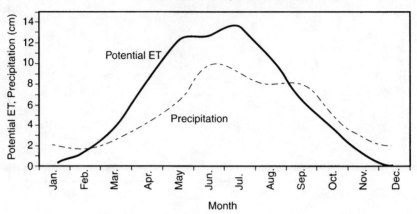

FIGURE 11   Annual cycle of average monthly potential evapotranspiration (solid) and total precipitation (dashed) for International Falls, Minnesota, based on 1961–1990 data.

Values of PET are quite low during the colder half of the year, rising rapidly during the spring and reaching a peak in mid summer. Precipitation exceeds PET during much of the colder half of the year, but during the late spring and summer months PET exceeds precipitation, on average. As a result, there is normally a depletion of soil moisture during the summer months. Often, the highest fire danger occurs in mid to late summer in this area, when soil moisture has reached its climatological minimum. The drying of dead fuels is not as sensitive to total precipitation as it is to the number of days with precipitation. This variable remains high through the summer months with about one day in three receiving some precipitation.

Variations in climatic conditions from year to year can be significant. Figure 12 shows monthly PET and precipitation at International Falls during 1980. During a very dry spring, PET exceeded precipitation substantially, resulting in early depletion of soil moisture. This can enhance the danger of spring fires before vegetation has broken dormancy when it is more flammable.

B. ELEVATION DEPENDENCE

In mountainous regions, PET decreases with increasing elevation, primarily because of the dependance of temperature on elevation. During summer, temperature decreases on average by about 6°C km^{-1} in much of the western United States and Canada. Even though actual water vapor pressure also decreases with height, the saturation vapor pressure usually decreases more rapidly, leading

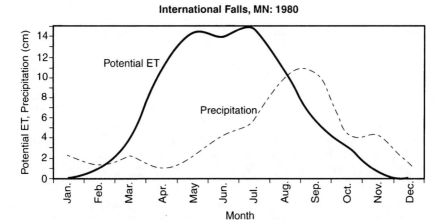

FIGURE 12 Monthly potential evapotranspiration (solid) and total precipitation (dashed) for International Falls, Minnesota, during 1980.

to a decrease in water vapor pressure deficit with increasing height. In addition, mountainous areas usually experience more cloudiness, reducing the net radiation.

An example is shown in Figure 13, which compares PET (estimated from pan evaporation data) at Flagstaff, Arizona (el. 2110 m) and Phoenix, Arizona

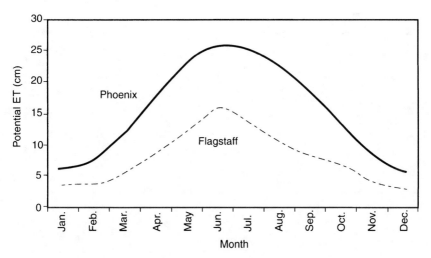

FIGURE 13 Annual cycle of monthly average potential evapotranspiration (cm) for Flagstaff, Arizona (light) and Phoenix, Arizona (dark).

(el. 323 m). The differences are due to both lower temperatures and more cloudiness at Flagstaff compared to Phoenix.

C. TOPOGRAPHIC EFFECTS

Topographical effects can cause significant local variations in the surface energy budget, specifically incoming solar radiation. South-facing slopes will receive substantially greater amounts than north-facing slopes. These differences are dependent on the latitude and the time of year. Oke (1978) presents estimates of direct-beam solar radiation for a latitude of 40° at the time of the equinoxes and the solstices. He shows that differences between south-facing and north-facing slopes are smallest at the time of the summer solstice and largest at the winter solstice. Figure 14 shows an estimate of the ratio of solar radiation on a 30° south-facing slope to the radiation on a horizontal surface as a function of time of year at a latitude of 40°N, based on Jordan and Liu (1977). The ratio is highest at the winter solstice and is greater than 1 through most of the year. However, the ratio is actually less than 1 around the time of the summer solstice, indicating that a horizontal surface receives slightly more solar radiation than a 30° south-facing slope.

Measurements by Aisenshtat (1966) in the Turkestan Mountains (41°N) during September also illustrate the effect of slope. In this case, R_n was nearly three times as great on a 33° south-facing slope as a 31° north-facing slope. Evapotransporation was nearly twice as large on the south-facing slope.

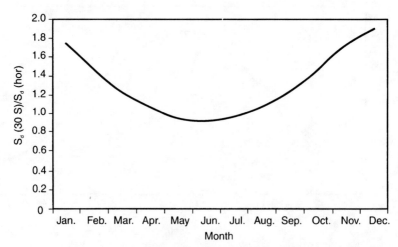

FIGURE 14   Annual cycle of the ratio of monthly average solar radiation on a 30° south-facing surface to solar radiation on a horizontal surface at a latitude of 40°N.

It is obvious that slope can have a large effect on fuel moisture content and fire risk. Fuels on south-facing slopes will, in general, tend to have lower moisture contents than those on north-facing slopes.

D. UNDERSTORY ET

Within a forest canopy, both radiation and ventilation are reduced by canopy shielding, as described earlier. Observations indicate that the magnitude of understory ET varies widely depending on the density of the understory, the type of understory vegetation, and soil moisture. Black and Kelliher (1989) reviewed field experiments and found that the ratio of understory ET to total ET varied from <0.05 to >0.50. Denmead and Bradley (1985) obtained measurements in a 16-m-high mature Ponderosa pine forest and found that understory ET was <20% of total ET. Furthermore, they found that much of the understory ET occurred from intermittent gusts of wind that ventilated the canopy. Baldocchi and Meyers (1991) measured ET at the floor of a deciduous forest and reported on the average diurnal dependence during spring and fall. Their daily average values peaked at about 25 W m^{-2}, or only about 0.04 mm hr^{-1}. Black and Kelliher (1989) reported that the ratio of understory to total ET often increases as soil moisture decreases, perhaps reflecting increases in stomatal resistance of the overstory trees as soil moisture decreases.

VI. NEAR-SURFACE ENVIRONMENT

Chapter 4 in this book describes the relationship of the moisture content of dead fuels to relative humidity and temperature. As illustrated in Figures 4–13, the equilibrium moisture content is strongly dependent on the environmental conditions experienced by the fuel. Routine meteorological observations are taken at standard heights. For example, in the United States the standard height for measurements of maximum and minimum temperature in the National Weather Services's Cooperative Observer Network is 1.5 m. However, the actual meteorological conditions at the fuel surface are often quite different because large vertical gradients are a common feature of the near-surface atmosphere, due to the momentum, heat, and water vapor exchange that occurs at the air–surface interface. These gradients can be particularly large for ground-based fuels. A critical factor in determining the magnitude of the vertical gradients is the nature of diffusion and mixing. Within the laminar sublayer, the mixing rates are determined by molecular diffusion. Because this is quite slow, the vertical gradients are largest in this laminar sublayer. Above the laminar sublayer, mixing rates increase rapidly because the atmosphere is turbulent. Immediately above

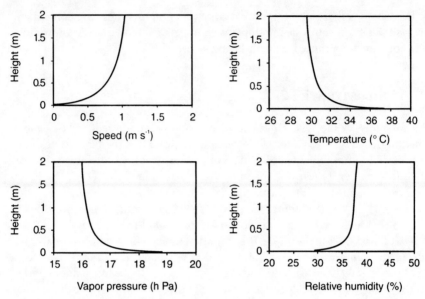

FIGURE 15 Vertical profiles of wind, temperature, water vapor pressure, and relative humidity for $H = 300$ W m^{-2} and $LE = 150$ W m^{-2}.

the laminar sublayer, the size of turbulent eddies is restricted by proximity to the surface. The dominant size of eddies becomes larger as the distance from the surface increases. Thus, turbulent mixing rates tend to increase with height. As a consequence, the vertical gradients gradually decrease with height.

Above the laminar sublayer, the logarithmic profile that applies to the height dependence of wind speed is also observed for temperature and water vapor content. The exact dependence is a function of the magnitude of the fluxes of heat, momentum, and water vapor, as well as the characteristics of the surface, particularly the roughness height. Figure 15 shows the vertical profiles of temperature, humidity, and wind for typical mid-summer conditions over a surface that is not shielded by a forest canopy. These profiles have been calculated using empirically determined formulae that include the effects of stability (Panofsky and Dutton, 1984). As can be seen, the vertical gradients are large near the surface, gradually decreasing as height increases. Although both temperature and water vapor content increase as the surface is approached, the net effect is a decrease in relative humidity because of the exponential dependence of saturation water vapor pressure on temperature.

Figure 15 does not show what happens across the laminar sublayer. For temperature, observational evidence shows that the gradients can be very large. This is not surprising since exchange of heat occurs through molecular diffusion. Figure 16 shows an example of the diurnal dependence of the temperatures of

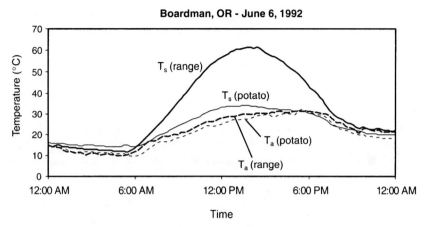

FIGURE 16 Temperature (°C) of the ground surface and the atmosphere at a height of 2 m on June 6, 1992, for a semiarid grassland (dark) and for a potato field (light) in eastern Oregon.

the surface and of the atmosphere at a height of 2 m for a semiarid grass rangeland. The temperature difference peaks at solar noon with values in excess of 30°C. This is probably typical of ground fuels. Figure 16 also shows the temperature difference for a nearby irrigated potato field. By contrast with the dry rangeland, the temperature difference over the potato field never exceeds 10°C. This contrast reflects two factors that influence this gradient. First, the rangeland is very sparsely vegetated, and momentum transport occurs largely by interactions with the solid, nonflexible ground surface rather than with the sparse grass. In the potato field, the atmospheric interactions are primarily with the flexible plants. Thus, the "excess" resistance to heat transport is higher for the rangeland than for the potato field, supporting a larger temperature gradient. The second factor is the aridity of the surface and the partitioning of net radiation into latent and sensible heat fluxes. The rangeland was quite dry, and most of the available energy was partitioned into sensible heat flux. In the irrigated potato field, latent heat flux was higher than sensible heat flux. Thus, evaporative cooling of the surface was significant in the potato field, reducing the temperature gradient.

Following Eq. (16), the surface potential temperature can be expressed as a function of the sensible heat flux, stability, wind speed, and surface roughness length, similar to Zeng and Dickinson (1998), as

$$\theta_s = \theta(z) + \frac{Hr_a(z)}{\rho c_p} \tag{38}$$

where $r_a(z)$ takes the form of Eq. (22).

Measurement of the surface temperature is relatively straightforward using small in situ probes or infrared thermometry. However, measurement of the surface water vapor content is much more difficult and such measurements have not been reported. It seems likely that the trends indicated in Figure 15 would continue across the laminar sublayer; that is, there is likely to be a substantial decrease in relative humidity at the fuel surface corresponding to the increase in temperature shown in Figure 16. Thus, ground-based fuels that are exposed to the sun experience a considerably lower relative humidity during daytime than what would be indicated by measurements at a standard measurement height. The surface water vapor content can be expressed in a similar manner to Eq. (36) as

$$e_s = e(z) + \frac{LEr_a(z)}{\rho c_p} \tag{39}$$

VII. MODELS OF LAND–SURFACE INTERACTIONS

The preceding sections illustrate some of the complexities involved in describing mathematically the interactions of the atmosphere with the land surface and how these interactions affect drying of fuels. A great deal of effort has been devoted within the atmospheric sciences community to the development of mathematical models and associated computer software packages that can be used to describe the temporal behavior of surface fluxes of heat, momentum, and water vapor and their dependence on meteorological conditions and land surface conditions. Some of these have been developed into community models that are available to interested researchers. A brief description of their capabilities, structure, and limitations is given here. In recent years, there has been a systematic effort to improve these models through intercomparisons among these models and comparisons with field studies (T. H. Chen *et al.*, 1997). This has led to significant model improvements.

The discussion here will focus on two models, the Biosphere-Atmosphere Transfer Scheme (BATS) (Dickinson *et al.*, 1981, 1986, 1993) and the Land Surface Model (LSM) (Bonan, 1996). Both of these are community models available through the National Center for Atmospheric Research (NCAR). Most other models share many of the features of these two models. These models are quite complex, which precludes a comprehensive description. The modeling of one of the simpler cases, bare ground, will be described. This case can be applied to the problem of modeling the environmental conditions (temperature and relative humidity) experienced by a duff layer, including the diurnal dependence of those conditions.

In the absence of snow cover, the ground surface energy budget can be written as follows by combining Eqs. (3) and (15):

$$S_d(1 - a) + I_a - I_s - H - LE - G = 0 \qquad (40)$$

Several of these terms are directly or indirectly a function of the surface temperature, T_s. To specify the surface fluxes, a value of T_s must be found that satisfies Eq. (40). The upward longwave radiation, I_s, is parameterized by Eq. (2). The sensible heat flux, H, is parameterized by Eq. (16). The expression for latent heat flux [Eq. (17)] does not explicitly contain T_s but is parameterized in the BATS model as

$$LE = L\rho b[q_s(T_s) - q]/r_a \qquad (41)$$

where $q_s(T_s)$ is the saturated specific humidity at the surface temperature and b is a soil wetness factor which ranges from 1 when the soil is wet to 0 when it is dry. This expression for LE is a function of T_s. To determine the ground heat flux at the surface, $G(0)$, it is necessary to solve for the heat flow in the soil. This is described next. Once this is solved, G can be parameterized as follows:

$$G(0) = \frac{2K_{g1}(T_s - T_{g1})}{\Delta z_1} \qquad (42)$$

where T_1 and Δz_1 are the temperature and thickness, respectively, of the first soil layer and K_{g1} is the thermal conductivity $(\text{W m}^{-1} \text{ K}^{-1})$ of the first soil layer.

If S_d and I_a are specified (e.g., they are assumed not to be a function of T_s), the surface temperature can be solved using the preceding parameterizations for I_s, H, LE, and G as functions of T_s. Since these forms contain higher order polynomials, exponential functions, and other complex functions, it is not possible to find an analytical solution. In the LSM (Bonan, 1996), T_s is found by iteratively solving the equation

$$S_d(1 - a) + I_a - I_s - H - LE - G - \left(\frac{\partial I_s}{\partial T_s} + \frac{\partial H}{\partial T_s} + \frac{\partial LE}{\partial T_s} + \frac{\partial G}{\partial T_s}\right)\Delta T_s = 0 \qquad (43)$$

where $\Delta T_s = T_s^{n+1} - T_s^n$ and the superscript n indicates the iteration. The time step for solution of this equation is for the order of 1 h or less.

The soil temperature profile is determined by discretizing the soil column into multiple layers. The soil heat flux at depth z is

$$G(z) = K_g(z)\frac{\delta T_g}{\delta z} \qquad (44)$$

One-dimensional energy conservation requires that

$$\rho_g(z)c_g(z)\frac{\delta T_g}{\delta t} = \frac{\partial}{\partial z}\left[K_g(z)\frac{\partial T_g}{\partial z}\right] \tag{45}$$

where $\rho_g(z)$ and $c_g(z)$ are the soil density and soil heat capacity, respectively, at depth z. In the LSM, six layers of geometrically increasing thicknesses (10, 20, 40, 80, 160, and 320 cm) are used. Equation (45) is solved by numerical methods for T_g with the boundary conditions that $G(z) = 0$ at the bottom of the soil column.

The volumetric soil water content affects the density and heat capacity of the soil and the latent heat flux, as expressed in the parameter b [Eq. (41)]. Thus, accurate estimation of the surface temperature requires calculation of soil water. The soil water budget must account for the movement of water vertically and for ET and precipitation at the soil–atmosphere interface. In the LSM, the change in soil water content due to the movement of water is based on the Richards equation:

$$\frac{\partial V_s}{\partial t} = \frac{\partial}{\partial z}\left[K_H\left(\frac{\partial V_s}{\partial z}\frac{\partial \Psi_s}{\partial V_s} + 1\right)\right] \tag{46}$$

where V_s is the volumetric soil water content ($mm^3\ mm^{-3}$), K_H is the hydraulic conductivity ($mm\ s^{-1}$), and Ψ_s is the soil matrix potential (mm). This equation is combined with the accounting for evaporation and precipitation and numerically implemented for the six-layer soil column to solve for volumetric soil water.

This system of equations [Eqs. (40)–(46)] can be used to simulate the diurnal dependence of surface temperature and humidity, given a knowledge of external meteorological forcing. It is important to specify properly key parameters such as $\rho_g(z)$, $c_g(z)$, $k_g(z)$, and $K_H(z)$, since these can be quite different for organic layers compared to mineral soils. The multilayer modeling of the soil column in LSM and many other models provides a convenient framework for modeling the commonplace situation of an organic layer overlying a mineral soil.

The modeling of the surface energy budget is much more complex when a vegetative canopy is present and will not be described in detail here. In general terms, models such as BATS and LSM break this system into four major components: the plant, the atmosphere above the canopy, the atmosphere within the canopy, and the ground surface. They then explicitly account for interactions between the vegetation and the within-canopy air volume, between the within-canopy air volume and the ground surface, and between the within-canopy air volume and the atmosphere above the canopy. Radiation exchange can occur between the atmosphere and the plants, between the atmosphere and the ground surface (through gaps in the canopy overstory), and between the plants and the ground surface. Exchange of heat and water vapor occurs between the plants and the within-canopy air volume, between the within-canopy air volume and the ground surface, and between the within-canopy air volume and the atmosphere above the canopy. Each of the processes is parameterized, and a set of state variables for each component of the system is calculated.

In addition to the modeling of radiative and sensible fluxes, another important process that is parameterized by these models is the interception of precipitation by the canopy. This can be very important in forest canopies where the surface area of the vegetation is very large. Rain will coat the vegetation with a film of water before water will fall to the ground. This water can then evaporate and return to the atmosphere. This process typically reduces the amount of rain reaching the ground surface by 10–50%. The BATS and LSM models explicitly track the storage of water on vegetative surfaces. The maximum value of this storage is assumed to be a linear function of the stem and leaf area of the vegetative canopy.

Pressing research problems, particularly the need to accurately model future climate change, are motivating work to improve these models. The ecological researcher can benefit from these advances by applying current and future land models to simulate the environmental conditions experienced by fuels.

VIII. REMOTE SENSING OF THE SURFACE ENERGY BUDGET

There often exists substantial spatial variability in the surface energy budget, arising from spatial variations in a number of conditions, including precipitation distribution, soil water holding characteristics, and vegetation characteristics. Thus, point measurements of the surface energy budget may be representative only of a relatively small area surrounding a measurement site. Also, comprehensive measurements of the components of the surface energy budget are taken at very few sites because of the cost and difficulty. For these reasons, there has been considerable research investigating the use of remote sensors on airborne or satellite platforms to make such measurements.

One widely used technique relies on the parameterization of H expressed in Eq. (16). In this relationship, H is proportional to the difference in temperature between the surface and the overlying atmosphere. Thermal infrared measurements are routinely obtained by several satellites, providing an estimate of the surface radiative temperature (T_s). The air temperature (T_a) can be estimated from the surface observing network. The aerodynamic resistance (r_a) must be estimated, requiring values for the roughness height and wind speed. Thus, some knowledge of the characteristics of the surface is necessary.

The latent heat flux (LE) is often estimated as a residual of the terms in Eq. (15). This requires an estimate of the radiative terms in Eq. (3). Values of S_d and albedo can be retrieved from geostationary satellites (Tarpley, 1979; Gautier et al., 1980; Diak and Gautier, 1983) using methods in which cloudiness is the major factor modulating solar radiation at the surface. The infrared radiation terms (I_a and I_s) can be estimated from thermal radiation satellite measurements. The soil heat flux (G) is usually small compared to H and can be estimated with sufficient accuracy using a combination of the measurement of T_s and a model, such as Eq. (42).

Although the basic approach (Jackson et al., 1977; Seguin and Itier, 1983) for estimating H has changed little, there have been a number of variations, resulting in a rather large number of models (Olioso et al., 1999; Schmugge and Becker, 1991). These models have been somewhat successful in providing large-scale estimates of H and LE (Lagouarde, 1991).

The accuracy of remote sensing estimates of surface fluxes using thermal IR measurements is limited by the precision of the measurement of T_s, which is of the order of 2–3 K. In many situations, the temperature difference $(T_s - T_a)$ is of the same order of magnitude. In these situations, the estimates of the surface fluxes are not highly accurate. Typically, under the best circumstances, the sensible heat flux can be estimated to within ± 30 W m^{-2} (Seguin et al., 1999). A complication arises when the surface is partially covered with vegetation. Often, the vegetative elements and the bare ground have very different radiative temperatures. Intercomparison of some of these models suggests that, with partial vegetative cover, separate accounting for the fluxes from the bare ground and from the vegetative elements provides superior results (Zhan et al., 1996). Estimates of vegetative cover using the Normalized-Difference Vegetation Index (NDVI) can provide more accurate estimates of the components of the surface energy budget (Boegh et al., 1999; Inoue and Moran, 1997).

There has also been a significant amount of work investigating the use of remote sensors to estimate surface soil moisture. One approach relies on passive microwave radiometry. The surface brightness temperature in the microwave portion of the spectrum is a highly sensitive function of volumetric soil moisture. Aircraft measurements using microwave radiometers have achieved accuracies of $\pm 3\%$ in measurement of the volumetric soil water (Jackson et al., 1995). One limitation of passive microwave approaches is that satellite sensors can achieve only rather crude resolution (10–30 km). Another proposed approach utilizes active microwave sensors which can obtain multifrequency, multipolarization data at rather high spatial resolution. This approach shows promise of achieving accurate estimates of soil moisture at high resolution (Bindlish and Barros, 2000).

Despite some of the limitations and uncertainties, remote sensing of the surface energy budget holds potential as a tool to assess the variability of fuel moisture on regional scales (1–100 km).

IX. FIRE WEATHER RATING SYSTEMS

Fire weather rating systems incorporate measures of drying that reflect the physical basis for ET described in this chapter. To illustrate this, the United States and Canadian rating systems are briefly discussed.

A. UNITED STATES

The National Fire Rating System of the United States is based in part on indices of the fuel moisture of both dead and live fuels. The relationships among environmental conditions and fuel moisture are empirically based. The drying of dead fuels under constant conditions is assumed to follow an exponential decay curve. Dead fuels are stratified into four categories of time response: 1, 10, 100, and 1000 h. A key concept in this system is the equilibrium moisture content (EMC) which is the "moisture content dead fuels would obtain if left in a steady state environment long enough to obtain equilibrium (no net moisture exchange)" (Bradshaw *et al.*, 1983). The EMC is calculated from temperature and relative humidity using regression equations developed by Simard (1968). These equations are

EMC =

$$
\begin{cases}
0.03299 + 0.281073\ \text{RH} & -\ 0.00058\ \text{RH}\ T & \text{RH} < 10 \\
2.22749 + 0.160107\ \text{RH} & -\ 0.0148\ T & 10 < \text{RH} < 50 \\
21.06060 + 0.005565\ \text{RH}^2 - 0.0035\ \text{RH}\ T - 0.483199\ \text{RH} & & \text{RH} > 50
\end{cases}
$$

$$(47)$$

Figure 17 shows this relationship as a function of RH for various values of temperature, indicating that the EMC is highly sensitive to RH and slightly sensitive to temperature.

An important consideration in the application of the preceding formula is that the relative humidity and temperature values are those at the surface of the dead fuel, not at the normal observational height of instruments. As noted in the previous section, these values can be quite different, particularly during sunny daytime conditions. To account for the near-surface gradient, the rating system uses adjustment factors that are applied to the observed temperature and relative humidity; these are a function of sky cover conditions and are shown in Table 1 (Bradshaw *et al.*, 1983). As this table shows, the adjustment for temperature and relative humidity are large under sunny conditions but quite consistent with observations, as illustrated in Figures 15 and 16. These adjusted factors are used in the EMC regression equations. These corrections are applied to mid-afternoon observations when relative humidity reaches its minimum value and fuel moisture content would be at a minimum.

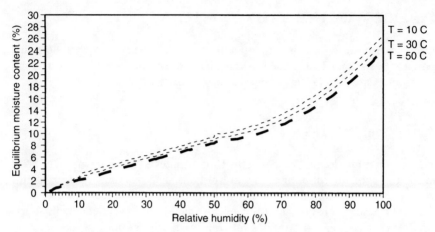

FIGURE 17 Equilibrium moisture content (%) of dead fuels as a function of relative humidity for values of temperature = 10, 30, and 50°C as calculated in the U.S. National Fire Rating System.

For 1- and 10-h fuels, mid-afternoon observations are used to estimate fuel moisture content. This reflects the substantial variation of RH within the daily cycle (Figure 5) and the need to incorporate this into the rating system for fast response fuels. To compute fuel moisture content for 100-h fuels, a weighted 24-h average EMC value is used; this is calculated from the day's maximum and minimum temperature and maximum and minimum relative humidities. For 1000-h fuels, EMC is averaged over 7-day periods.

Two types of classes are used for live fuels: grass forbs and woody shrubs. The system's developers found that the 1000-h fuel moisture function responded to meteorological conditions in a similar fashion to that expected for live fuel. The system thus applies this function to estimate live fuel moisture. Of course, plants are adapted to their environment. Those adapted to drier conditions tend to use moisture more slowly than those adapted to wetter conditions. Thus, the fire danger rating system uses parameters that are a function of climatic zone, as

TABLE 1 Temperature and Relative Humidity Adjustment Factors for Calculation of Equilibrium Moisture Content

Fractional sky cover	Temperature adjustment (°F)	Relative humidity adjustment
0.0–0.1	+25	× 0.75
0.1–0.5	+19	× 0.83
0.5–0.9	+15	× 0.91
0.9–1.0	+5	× 1.0

determined by the Thornthwaite climate classification scheme. In this system, there are four climate classes: arid and semiarid, subhumid, humid, and wet.

B. Canada

The Canadian Forest Fire Weather Index System (Van Wagner, 1987) uses three separate indices for three distinct dead fuel classes:

1. Fine Fuel Moisture Code (FFMC), which represents the moisture content of litter and other fine fuels, with a time lag of less than one day.
2. Duff Moisture Code (DMC), which represents the moisture content of loosely compacted organic matter, with a time lag of 12 days.
3. Drought Code (DC), representing a deep layer of compact organic matter with a time lag of 52 days.

There are a number of basic similarities between the Canadian system and the U.S. system, although some of the functional relationships are different. In the Canadian system, an equilibrium moisture content is calculated which is a function of relative humidity and temperature. This is used in the computation of the FFMC. The functional dependence exhibits hysteresis; specifically, the equilibrium moisture content is lower during wetting than during drying. Fuel moisture content approaches the equilibrium value in an exponential manner, much as in the U.S. system. The rate of change is a function of wind speed, relatively humidity, temperature, and (in the case of the DMC and DC) the day length (which is a surrogate for radiation). Thus, all the major meteorological factors that affect ET are incorporated in an empirical manner.

The DMC uses a constant equilibrium moisture content of 20%. The rate of drying is a function of temperature, relative humidity, and an empirical day length factor. This function does not include any wind speed dependence. This is a reflection of the control on ET by surface resistance (r_s). When r_s is large, the aerodynamic resistance, which incorporates the wind speed dependence, is relatively less important. Evaporative drying for the DC index is based on an estimate of potential evaportransporation which is calculated from an empirical function of noon time temperature and a day length factor.

NOTATION

Roman Letters

a	albedo	dimensionless
A, B, C	constants used in formula relating es to T	

b	soil wetness factor (0–1)	
c_p	heat capacity of air at constant pressure	$m^2 s^{-2} K^{-1}$
c_g	heat capacity of the soil	$J\,kg^{-1}\,K^{-1}$
C_T	cloud radiation transmission	dimensionless
C_u	empirically-determined consumptive use coefficient	
d	displacement height	m
D	monthly mean percentage of annual daytime hours of the year	%
e	atmospheric water vapor pressure	hPa
es	saturation value of atmospheric water vapor pressure	hPa
E	water vapor flux	$kg\,m^{-2}\,s^{-1}$
EMC	equilibrium moisture content	%
ET	evapotranspiration	$kg\,m^{-2}\,s^{-1}$, $W\,m^{-2}$, or $mm\,h^{-1}$
F_c	function relating net infrared radiation to cloud cover	dimensionless
G	soil heat flux	$W\,m^{-2}$
h	height of the vegetative canopy	m
H	sensible heat flux	$W\,m^{-2}$
I	infrared radiation	$W\,m^{-2}$
k	von Karman constant (= 0.4)	dimensionless
K_g	thermal conductivity of the soil	$Wm^{-1}\,K^{-1}$
K_H	soil hydraulic conductivity	$mm\,s^{-1}$
L	latent heat of evaporation	$J\,kg^{-1}$
L_{mo}	Monin-Obukhov length	m
P	air pressure	hPa
P_o	reference level pressure	hPa
PET	potential evapotranspiration	$W\,m^{-2}$ or $mm\,h^{-1}$
PET_m	accumulated value of PET for month m	cm
q	specific humidity	$kg\,H_2O/kg$ moist air

r	resistance	s m^{-1}
R_n	net radiation	W m^{-2}
RH	relative humidity	%
Ri	gradient Richardson number	dimensionless
S_d	downward solar radiation at the earth's surface	W m^{-2}
T	temperature	°C
T_d	dew point temperature	°C
T_m	mean monthly temperature	°C
u_*	friction velocity	m s^{-1}
$U(z)$	wind speed at height z	m s^{-1}
v	atmospheric water vapor mixing ratio	kg H$_2$O/ kg dry air
vs	saturation water vapor mixing ratio	kg H$_2$O/ kg dry air
V_s	volumetric soil water content	mm^3 mm^{-3}
z	height above ground surface or depth in soil	m
z_0	roughness length for momentum transport	m
z_{ow}	roughness length for water vapor transport	m

GREEK LETTERS

α	empirical constant used in Priestley-Taylor formula for PET	dimensionless
β	wind velocity attenuation coefficient	
Δ	des/dT	hPa K^{-1}
ε	emissivity	dimensionless
γ	$c_p P/(0.622L)$	hPa K^{-1}
υ	kinematic viscosity of air	m^2 s^{-1}
Ψ_s	soil matrix potential	mm
ψ	vertical profile adjustment factor for temperature, wind, and water vapor	dimensionless
ρ	air density	kg m^{-3}
σ	Stefan Boltzmann constant (5.67×10^{-8} Wm^{-2} K^{-4})	W m^{-2} K^{-4}
θ	potential temperature	K

SUBSCRIPTS

a	air or aerodynamic
g	soil
h	temperature
m	wind
s	surface of soil or vegetative canopy
w	water vapor

SUGGESTED READING LIST

Campbell, G. S., and Norman, J. M. (1998). "An Introduction to Environmental Physics." Springer-Verlag, New York.

Panofsky, H. A., and Dutton, J. A. (1984). "Atmospheric Turbulence: Models and Methods for Engineering Applications." John Wiley and Sons, New York.

Rosenberg, N. J. (1974). "Microclimate: The Biological Environment." John Wiley and Sons, New York.

Schmugge. T. J., and Andre, J.-C., Eds. (1991). "Land Surface Evaporation: Measurement and Parameterization." Springer-Verlag, New York.

REFERENCES

Aisenshtat, B. A. (1966). Investigations of the heat budget of central Asia. *In* "Sowremennye Problemy Klimatologii" (M. I. Budyko, Ed.), pp. 83–129. Meteorol. Gidrol, Leningrad.

Baldocchi, D. O., and Meyers, T. P. (1991). Trace gas exchange at the floor of a deciduous forest: I. Evaporation and CO_2 efflux. *J. Geophys. Res.* 96, 7271–7285.

Baldocchi, D. D., and Vogel, C. A. (1996). Energy and CO_2 flux densities above and below temperate and broad leafed forest and a boreal pine forest. *Tree Physiol.* 16, 5–16.

Baldocchi, D. D., Vogel, C. A., and Hall, B. (1997). Seasonal variation of energy and water vapor exchange rates above and below and boreal jack pine forest canopy. *J. Geophys. Res.* 102, 28,939–28,951.

Betts, A. K., and Ball, J. H. (1997). Albedo over the boreal forest. *J. Geophys. Res.* 102, 28,901–28,909.

Bindlish, R., and Barros, A. P. (2000). Multifrequency soil moisture inversion from SAR measurements with the use of IEM. *Remote Sens. Environ.* 71, 67–88.

Black, T. A., and Kelliher, F. M. (1989). Processes controlling understory evaporation. *Phil. Trans. Roy. Soc. (Lond)* B. 324, 207–231.

Blaney, H. F., and Criddle, W. O. (1950). "Determining Water Requirements in Irrigated Data." U.S. Department of Agriculture Soil Conservation Service Technical Paper No. 96.

Blanken, P. D., Black, T. A., Yang, P. C., Neumann, H. H., Nesic, Z., Staebler, R., den Hartog, G., Novak, M. D., and Lee, X. (1997). Energy balance and canopy conductance of a boreal aspen forest: Partitioning overstory and understory components. *J. Geophys. Res.* 102, 28,915–28,927.

Blyth, E. M., and Dolman, A. J. (1995). The roughness length for heat of sparse vegetation. *J. Appl. Meteorol.* 34, 583–585.

Boegh, E., Soegaard, H., Hanan, N., Kabat, P., and Lesch, L. (1999). A remote sensing study of the NDVI-T_s relationship and the transpiration from sparse vegetation in the Sahel based on high-resolution data. *Remote Sens. Environ.* 69, 224–240.

Bonan, G. B. (1996). "A Land Surface Model (LSM Version 1.0) for Ecological Hydrological, and Atmospheric Studies Technical Description and User's Guide." NCAR Technical Note NCAR/TN-417+STR. National Center for Atmospheric Research, Boulder.

Bradshaw, L. S., Deeming, J. E., Burgan, R. E., and Cohen, J. D. (1983). "The 1978 National Fire Danger Rating System: Technical Documentation." General Technical Report INT-169, U.S. Department of Agriculture, Forest Service, Intermountain Forest and Range Experiment Station, Ogden.

Brown, A. A., and Davis, K. P. (1973). "Forest Fire: Control and Use." McGraw-Hill, New York.

Brutsaert, W. (1975). On a derivable formula for long-wave radiation from clear skies. *Water Resour. Res.* 11, 742–744.

Buck, A. L. (1981). New equations for computing vapor pressure and enhancement factors. *J. Appl. Meteorol.* 20, 1527–1532.

Chamberlain, A. C. (1968). Transport of gases to and from surfaces with bluff and wave-like roughness elements. *Quart. J. Roy. Meteorol. Soc.* 94, 318–332.

Chandler, C., Cheney, P., Thomas, P., Trabaud, L., and Williams, D. (1983). "Fire in Forestry: Volume One. Forest Fire Behavior Effects." John Wiley and Sons, New York

Chang, H., Simmonds, L. P., Morison, J. I. L., and Payne, D. (1997). Estimation of transpiration by single trees: Comparison of sap flow measurements with a combination equation. *Agric. For. Meteorol.* 87, 155–169.

Chen, J. M., Rich, P. M., Gower, S. T., Norman, J. M., and Plummer, S. (1997). Leaf area index of boreal forests: Theory, techniques, and measurements. *J. Geophys. Res.* 102, 29,429–29,443.

Chen, T. H., Henderson-Sellers, A. Milly, P. C. D., Pitman, A. J., Beljaars, A. C. M., Polcher, J., Abramopoulus, F., Boone, A., Chang, S., Chen, F., Dai, Y., Desbouough, E. E., Dickinson, R. E., Dumenil, L., Ek, M., Garratt, J. R., Gedneny, N., Gusew, Y. M., Kim, J., Koster, R., Kowalczyk, E. A., Loval, K., Lean, J., Lettenmaier, D., Liang, X., Makfouf, J.-F., Mengelkamp, H.-T., Mitchell, K., Nanonova, O. N., Noilhan, J., Robock, A., Rozenzweig, C., Schaake, J., Schlosser, C. A., Schulz, J.-P., Shao, Y., Shimakin, A. B., Verseghy, D. L., Wetzel, P., Wood, E. F., Xue, Y., Yang, Z.-L., and Zeng, Q. (1997). Cabauw experimental results from the Project for Intercomparison for Land-Surface Parameterization Schemes. *J. Climate* 10, 1194–1215.

Cionco, R. M. (1965). A mathematical model for air flow in a vegetative canopy. *J. Appl. Meteorol.* 4, 517.

Cionco, R. M. (1978). Analysis of canopy index values for various canopy densities. *Boundary-Layer Meteorol.* 15, 81–93.

Denmead, O. T., and Bradley, E. F. (1985). Flux-gradient relationships in a forest canopy. *In* "The Forest-Atmosphere Interaction" (B. A. Hutchison and B. B. Hicks, Eds.), pp. 543–561. D. Riedel, Dordrecht.

Diak, G., and Gautier, C. (1983). Improvements to a simple model for estimating insolation from GOES data. *J. Clim. Appl. Meteorol.* 22, 505–508.

Dickinson, R. E., Henderson-Sellers, A., and Kennedy, P. J. (1993). "Biosphere-Atmosphere Transfer Scheme (BATS) Version 1e as Coupled to the NCAR Community Climate Model." NCAR Technical Note NCAR/TN-387+STR. National Center for Atmospheric Research, Boulder.

Dickinson, R. E., Henderson-Sellers, A., Kennedy, P. J., and Wilson, M. F. (1986). "Biosphere-Atmosphere Transfer Scheme (BATS) for the NCAR Community Climate Model." NCAR Technical Note/TN-275+STR. National Center for Atmospheric Research, Boulder.

Dickinson, R. E., Jaeger, J., Washington, W. M., and Wolske, R. (1981). "Boundary Subroutine for the NCAR Global Climate Model." NCAR Technical Note/TN-173+1A. National Center for Atmospheric Research, Boulder.

Doorenbos, J., and Pruitt, W. O. (1977). "Crop Water Requirements." F.A.O. Irrigation and Drainage Paper 24. F. A. O., Rome.

Farnsworth, R. K., and Thompson, E. S. (1982). "Mean Monthly, Seasonal, and Annual Pan Evaporation for the United States." NOAA Technical Report NWS 34. U.S. Department of Commerce, Washington, DC.

Farnsworth, R. K., Thompson, E. S., and Peck, E. L. (1982). "Evaporation Atlas for the Contiguous 48 United States." NOAA Technical Report NWS 33. U.S. Department of Commerce, Washington, DC.

Fritschen, L. J. (1967). Net and solar-radiation relations over irrigated field crops. Agric. Meteorol. 4, 55–62.

Fritchen, L. J. (1985). Characterization of boundary condtions affecting forest environmental phenomena. In "The Forest-Atmosphere Interaction" (B.A. Hutchison and B. B. Hicks, Eds.), pp. 3–23. D. Riedel, Dordrecht.

Garratt, J. R., and Hicks, B. B. (1973). Momentum, heat and water vapour transfer to and from natural and artificial surfaces. Quart. J. Roy. Meteorol. Soc. 99, 680–687.

Garratt, J. R. (1977). "Aerodynamic Roughness and Mean Monthly Surface Stress over Australia." CSIRO Aust. Div. Atmos. Phys. Tech. Paper No. 29. CSIRO, Melbourne.

Gautier, C., Diak, G., and Masse, S. (1980). A simple physical model to estimate incident solar radiation at the surface from GOES satellite data. J. Appl. Meteorol. 19, 1005–1012.

Goulden, M. L., Daube, B. C., Fan, S.-M. Sutton, D. J., Bazzaz, A., Munger, J. W., and Wofsy, S. C. (1997). Physiological responses of a black spruce forest to weather. J. Geophys. Res. 102, 28,987–28,996.

Goulden, M. L., and Field, C. B. (1995). Three methods for monitoring the gas exchange of individual tree canopies: Ventilated-chamber, sap-flow and Penman-Montieth measurements on evergreen oaks. Funct. Ecol. 8, 125–135.

Granier, A., Huc, R., and Barigah, S. T. (1996). Transpiration of natural rain forest and its dependence on climatic factors. Agric. For. Meteorol. 78, 19–29.

Hignett, P. (1994). Roughness lengths for temperature and momentum over heterogeneous terrain. Boundary Layer Meteorol. 68, 225–236.

Hodges, G. B., and Smith, E. A. (1997). Intercalibration, objective analysis, intercomparison and synthesis of BOREAS surface net radiation measurements. J. Geophys. Res. 102, 28,885–28,900.

Hogg, E. H., Black, T. A., den Hartog, G., Neumann, H. H., Zimmermann, R., Hurdle, P., Blanken, P. D., Nesic, Z., Yang, P. C., Staebler, R. M., McDonald, K. C., and Oren, R. (1997). A comparison of sap flow and eddy fluxes of water vapor from a boreal deciduous forest. J. Geophys. Res. 102, 28,929–28,937.

Infante, J. M., Rambal, S., and Joffre, R. (1997). Modeling transpiration in Holm-oak savanna: Scaling up from the leaf to the tree scale. Agric. For. Meteorol. 87, 273–289.

Inoue, Y., and Moran, M. S. (1997). A simplified method for remote sensing of daily canpy transpiration—A case study with direct measurements of canopy transpiration in soybean canopies. Int. J. Remote Sens. 18, 139–152.

Jackson, R. D., Reginato, R. J., and Idso, S. B. (1977). Wheat canopy temperature: A practical tool for evaluating water requirements. Water Resour. Res. 13, 651–656.

Jackson, T. J., LeVine, D. M., Swift, C. T., Schmugge, T. J., and Schiebe, F. R. (1995). Large area mapping of moisture using the ESTAR passive microwave radiometer in Washita '92. Remote Sens. Environ. 53, 27–37.

Jordan, R. C., and Liu, B. Y. H. (Eds.). (1977). "Applications of Solar Energy for Heating and Cooling of Buildings." American Society of Heating, Refrigerating, and Air-Conditioning Engineers, Inc., New York.

Kaminsky, K. Z., and Dubayah, R. (1997). Estimation of surface net radiation in the boreal forest and northern prairie from shortwave flux measurements. J. Geophys. Res. 102, 29,707–29,716.

Kustas, W. P., Pinker, R. T., Schmugge, T. J., and Humes, K. S. (1994). Daytime net radiation estimated for a semiarid rangeland basin from remotely sensed data. *Agric. For. Meteorol.* **71**, 337–357.

Lagouarde, J.-P. (1991). Use of NOAA AVHRR date combined with an agrometeorological model for evaporation mapping. *Int. J. Remote Sens.* **12**, 1853–1864.

List, R. J. (1949). "Smithsonian Meteorological Tables," 6th ed. Smithsonian Institution Press, Washington, DC.

Lopushinsky, W. (1986). Seasonal and diurnal trends of heat pulse velocity in Douglas-fir and ponderosa pine. *Can. J. For. Res.* **16**, 814–821.

McCaughey, J. H. , Lafleur, P. M., Joiner, D. W., Bartlett, P. A., Costello, A. M., Jelinski, D. E., and Ryan, M. G. (1997). Magnitudes and seasonal patterns of energy, water, and carbon exchanges at a boreal young jack pine forest in the BOREAS northern study area. *J. Geophys. Res.* **102**, 28,997–29,007.

Meinzer, F. C., Goldstein, G., Holbrook, N. M., Jackson, P., and Cavelier, J. (1993). Stomatal and environmental control of transpiration in a lowland tropical forest tree. *Plant Cell Environ.* **16**, 429–436.

Meyers, T. P., and Dale, R. F. (1983). Predicting daily insolation with hourly cloud height and coverage. *J. Clim. Appl. Meteorol.* **22**, 537–545.

Monteith, J. L. (1965). Evaporation and the environment. *Symp. Soc. Exp. Biol.* **19**, 205–234.

Monteith, J. L. (1981). Evaporation and surface temperature. *Quart. J. Roy. Meteorol. Soc.* **107**, 1–27.

Monteith, J. L., and Szeicz, G. (1961). The radiation balance of bare soil and vegetation. *Quart. J. Roy. Meteorol. Soc.* **87**, 159–170.

Nielsen, L. B., Prahm, L. P., Berkowicz, R., and Conradsen, K. (1981). Net incoming radiation estimated from hourly global radiation and/or cloud observations. *J. Climatol.* **1**, 255–272.

Oke, T. R. (1978). "Boundary Layer Climates." Methuen, London.

Olioso, A., Chauki, H., Courault, D., and Wigneron, J.-P. (1999). Estimation of evapotranspiration and photosynthesis by assimilation of remote sensing data into SVAT models. *Remote Sens. Environ.* **68**, 341–356.

Panofsky, H. A., and Dutton, J. A. (1984). "Atmospheric Turbulence: Models and Methods for Engineering Applications." John Wiley and Sons, New York.

Paulson, C.A. (1970). The mathematical representation of wind speed and temperature profiles in the unstable atmospheric surface layer. *J. Appl. Meteorol.* **9**, 857–861.

Penman, H. L. (1948). Natural evaporation from open water, bare soil, and grass. *Proc. Roy. Soc. (Lond) A.* **193**, 120–145.

Petersen, M. S., Lamb, P. J., and Kunkel, K. E. (1995). Implementation of a semi-physical model for examining solar radiation in the Midwest. *J. Appl. Meteorol.* **34**, 1905–1915.

Pinker, R. T., and Corio, L. A. (1984). Surface radiation budget from satellites. *Mon. Wea. Rev.* **112**, 209–215.

Pinker, R. T., Corio, L. A., and Tarpley, J. D. (1985). The relationship between planetary and surface net radiation. *J. Clim. Appl. Meteorol.* **24**, 1272–1268.

Price, D. T., and Black, T. A. (1989). Estimation of forest transpiration and CO_2 uptake using the Penman-Montieth equation and a physiological photosynthesis model. *In* "Proceedings of Workshop on Estimation of Areal Evapotranspiration," pp. 213–228. IAHS Pub. 177, Vancouver.

Priestley, C. H. B., and Taylor, R. J. (1972). On the assessment of surface heat flux and evaporation using large scale parameters. *Mon. Wea. Rev.* **100**, 81–92.

Pyne, S. J. (1984). "Introduction to Wildland Fire: Fire Management in the United States." John Wiley and Sons, New York.

Rowntree, P. R. (1991). Atmospheric parameterization schemes for evaporation over land: Basic concepts and climate modeling aspects. *In* "Land Surface Evaporation: Measurement and Parameterization" (T. J. Schmugge and J.-C. André, Eds.), pp. 5–29. Springer-Verlag, New York.

Salby, M. L. (1996). "Fundamentals of Atmospheric Physics." Academic Press, San Diego.

Schmugge, T. J., and Becker, F. (1991). Remote sensing observations for the monitoring of land–surface fluxes and water budgets. In "Land Surface Evaporation: Measurement and Parameterization" (T. J. Schumgge and J.-C. Andre, Eds.), pp. 337–347. Springer-Verlag, New York.

Seguin, B., and Itier, B. (1983). Using midday surface temperature to estimate daily evaporation from satellite thermal IR data. Int. J. Remote Sens. 4, 371–383.

Seguin, B., Becker, F., Phulpin, T., Gu, X. F., Guyot, G., Kerr, Y., King, C., Lagouarde, J. P., Ottle, C., Stoll, M. P., Tabbagh, A., and Vidal, A. (1999). IRSUTE: A minisatellite project for land surface heat flux estimation from field to regional scale. Remote Sens. Environ. 68, 357–369.

Sellers, P. J., Hall, F. G., Kelly, R. D., Black, A., Baldocchi, D., Berry, J., Ryan, M., Ranson, K. J., Crill, P. M., Lettenmaier, D. P., Margolis, H., Cihlar, J., Newcomer, J., Fitzjarrald, D., Jarvis, P. G., Gower, S. T., Halliwell, D., Williams, D., Goodison, B., Wickland, D. E., and Guertin, F. E. (1997). BOREAS in 1997: Experiment overview, scientific results, and future directions. J. Geophys. Res. 102, 28,731–28,769.

Shaw, R. H. (1956). A comparison of solar radiation and net radiation. Bull. Amer. Meteorol. Soc. 37, 205–206.

Shuttleworth, W. J. (1991). Evaporation models in hydrology. In "Land Surface Evaporation: Measurement and Parameterization" (T. J. Schmugge and J.-C. André, Eds.), pp. 93–120. Springer-Verlag, New York.

Simard, A. J. (1968). "The Moisture Content of Forest Fuels—I." Inf. Rep. FF-X-14. Forest Fire Research Institute, Ottawa.

Sperry, J. S., and Pockman, W. T. (1993). Limitation of transpiration by hydraulic conductance and xylem cavitation in Betula occidentalis. Plant Cell Environ. 16, 279–287.

Sun, J., and Mahrt, L. (1995). Determination of surface fluxes from the surface radiative temperature. J. Atmos. Sci. 33, 1110–1117.

Tarpley, J. D. (1979). Estimating incident solar radiation at the surface from geostationary satellite data. J. Appl. Meteorol. 18, 1172–1181.

Thom, A. S. (1975). Momentum, mass and heat exchange of plant communities. In "Vegetation and the Atmosphere, Vol. 1. Principles" (J. L. Monteith, Ed.), pp. 57–109. Academic Press, New York.

Thornthwaite, C. W. (1948). An approach toward a rational classification of climate. Geogr. Rev. 38, 55–94.

Tyree, M. T., and Sperry, J. S. (1988). Do woody plants operate near the point of catastrophic xylem dysfunction caused by dynamic water stress? Plant Physiol. 88, 574–580.

Van Wagner, C. E. (1987). "Development and Structure of the Canadian Forest Fire Weather Index System." Forestry Technical Report 35. Canadian Forestry Service, Ottawa.

Weiss, A. (1983). A quantitative approach to the Pruitt and Doorenbos version of the Penman equation. Irrig. Sci. 4, 267–275.

Zeng, X., and Dickinson, R. E. (1998). Effect of surface sublayer on surface skin temperature and fluxes. J. Climate 11, 537–550.

Zhan, X., Kustas, W. P., and Humes, K. S. (1996). An intercomparison study on models of sensible heat flux over partial canopy surfaces with remotely sensed surface temperature. Remote Sens. Environ. 58, 242–256.

Zhong, M., Weill, A., and Taconet, O. (1990). Estimation of net radiation and surface heat fluxes using NOAA-7 satellite infrared data during fair-weather cloudy conditions of MESOGERS-84 experiment. Boundary Layer Meteorol. 33, 353–370.

Zilitinkevich, S. S. (1970). "Dynamics of the Atmospheric Boundary Layer." Leningrad Gidrometeor, Leningrad.

CHAPTER 10

Climate, Weather, and Area Burned

M. D. FLANNIGAN

Canadian Forest Service, Edmonton, Alberta, Canada

B. M. WOTTON

Canadian Forest Service, Sault Ste. Marie, Ontario, Canada

I. Introduction
II. Weather and Area Burned—Synoptic Surface Features
III. Weather and Area Burned—Upper Air Features
    A. Upper Air—Circulation
    B. Vertical Structure of the Atmosphere
IV. Teleconnections
V. Future Warming and Area Burned
VI. Summary
    References

I. INTRODUCTION

Forest fires are strongly linked to weather and climate (Flannigan and Harrington, 1988; Johnson, 1992; Swetnam, 1993). Fire has been an integral ecological process since the arrival of vegetation on the landscape. For the purposes of this chapter, we will define weather as short-term processes that result in variations in the atmospheric conditions ranging from minutes to a fire season. Processes that influence the atmosphere over time periods longer than a fire season will be defined as climate. There are several factors that control the climate and weather at any one location. These factors include variations in solar radiation due to latitude, distribution of continents and oceans, atmospheric pressure and wind systems, ocean currents, major terrain features, proximity to water bodies, and local features including topography (see Trewartha and Horn, 1980, for more details). As climate varies, the corresponding weather variables can vary in magnitude and direction.

The objective of this chapter is to highlight the connection between climate/ weather and the area burned by forest fires. We have used examples primarily from Canada or North America to illustrate our points. This chapter is divided into sections that describe the relationships between surface weather and area burned, upper air features and area burned, and teleconnections and area burned. We also try to identify key knowledge gaps in the fire and weather/ climate relationship. This chapter closes with a short discussion on how global change might influence forest fire activity and area burned in the 21st century.

II. WEATHER AND AREA BURNED— SYNOPTIC SURFACE FEATURES

The day-to-day weather can dramatically influence fire behavior and area burned. In fact, much of the area burned in a region during any given year occurs on just a few days of severe fire weather (Nimchuk, 1983; Harvey *et al.*, 1986). This has led to many studies over various spatial and temporal scales that try to relate the weather to fire. These studies can be broken into two broad categories, namely, case studies and what we will call area burned studies for the purpose of this paper. Case studies examine the conditions associated with an individual fire or an outbreak of fires and usually cover a short period of time and a relatively small area (approximately the fire area). Area-burned studies relate the weather to the area burned from numerous fires over many years. These area-burned studies have spatial domain ranges from hundreds of square kilometers to continental scale and a temporal range of 10 years to about 100 years. The temporal scale is often limited by the availability of area-burned data and meteorological data. This section will address both case studies and area-burned studies. In some studies the area burned is related to indexes such as components of fire danger rating systems or drought that are derived from meteorological data.

There have been numerous case studies that address the weather associated with a fire or an outbreak of fires. Most case studies have documented the weather prior to and during major fire runs (Stocks and Walker, 1973; Stocks, 1975; Quintilio *et al.*, 1977; Flannigan and Harrington, 1987; Hirsch and Flannigan, 1990). These studies have shown that fires spread rapidly when the fuels were dry and the weather conditions were warm to hot, dry, and windy. Turner (1970) studied the effect of hours of sunshine on fire season severity. The synoptic weather types associated with critical fire weather were studied by Schroeder (1964). These studies by themselves have limited application because of the narrow scope in terms of temporal and spatial scales used. However, they are of value in terms of identifying the most likely meteorological predictors of area burned that can be used in studies with a larger time and space domain (area-burned studies).

The most important elements of surface weather with respect to area burned are cold frontal passage, dry spells, and low relative humidities. These last two elements influence fuel moisture and the associated fire danger components which were also found to be important in terms of area-burned activity. Brotak and Reifsnyder (1977a) studied the synoptic weather conditions associated with large wildland fires (over 2000 ha) over the eastern half of the United States for the 1963–1973 period. They found that nearly 80% of these fires were associated with a cold front, either prior to or following passage of a dry cold front (Figure 1). It is important to note that a surface wind shift from southwest to northwest occurs with the passage of a cold front in the northern hemisphere. This is important because the flank of a fire with a southwest wind, which would typically be an ellipse in a SW-NE direction, would now become the head of the fire with a northwest wind which can cause rapid and significant growth in the fire. Haines *et al.* (1983) used over 2500 wildfires from the northeastern United States to examine the capability of the U.S. National Fire-Danger Rating System (NFDRS) to predict four measures of fire occurrence: the probability of a fire day, the number of fires per fire day, the number of fires per day, and the probability of a large-fire day. Good fits were found for the probability of a fire day and for the number of fires per fire day, whereas significantly poorer results were obtained for number of fires per day and the probability of a large fire day.

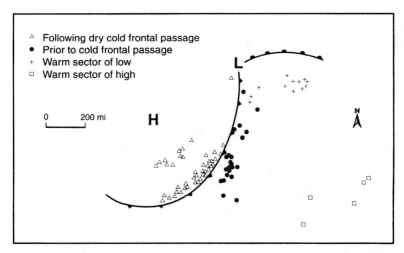

FIGURE 1   Idealized surface map showing relative position of major fires over eastern United States (from Brotak and Reifsnyder, 1977a).

Harrington *et al.* (1983) related the monthly provincial area burned in Canada to components of the Canadian Fire Weather Index (FWI) System for 1953–1980. They used mean and extreme values of components of the FWI System calculated at 41 stations across Canada. Results showed that the monthly means and extreme maximum values of the Duff Moisture Code (DMC—a model of the moisture content in the upper portion of the organic layers of the forest floor) and the Daily Severity Rating (DSR) were the best predictors of area burned. In western Canada, with the exception of the Yukon and Northwest Territories, the explained variance (r^2) averaged 33%. In the Territories and eastern Canada, the explained variance averaged 12%. Using the same data set, Flannigan and Harrington (1988) studied the relation between meteorological variables and monthly area burned by wildfire from May to August 1953–1980 for nine provincial-sized regions in Canada. They found that severe fire months were independent of rainfall amount but significantly dependent on rainfall frequency, temperature, and relative humidity. Results were similar to those obtained by Harrington *et al.* (1983) except that the meteorological variables did better in the Northwest Territories and eastern Canada than the Canadian FWI System (Van Wagner, 1987). The most important predictors were long sequences of days with less than 1.5 mm of rain and long sequences of days with relative humidity below 60%. These long sequences were assumed to be associated with blocking highs in the upper atmosphere.

The long-term average of area burned across a landscape is determined by a complex set of variables including the size of the sample area, the period under consideration, the extent of the forest, the topography, fragmentation of the landscape (rivers, lakes, roads, agricultural land), fuel characteristics, season, latitude, fire suppression policies and priorities, fire control, organizational size and efficiency, fire site accessibility, ignitions (people and lightning), simultaneous fires, and the weather. Thus results that explain 30% of the variance over large portions of the boreal forest are considered statistically significant.

III. WEATHER AND AREA BURNED—
UPPER AIR FEATURES

How do upper air features influence forest fires at the Earth's surface? There are two ways that the upper atmosphere relates to forest fire activity. First, the weather observed at the Earth's surface is a strong indicator of what is happening in the three-dimensional atmosphere. Temperature, precipitation, wind, cloudiness, and atmospheric pressure depend largely on the horizontal and vertical state of the atmosphere. The strength, location, and movement of surface highs and lows and associated warm/cold fronts are functions of the three-dimensional atmosphere. Thus surface weather conditions at the fire site are greatly influenced by upper air features. Second, many large forest fires are not constrained to the surface in that these fires may have convection columns that extend many kilometers into the atmosphere. The interaction between the convection column and the vertical structure of the atmosphere can have a significant impact on fire behavior and fire growth (see Chapter 8 in this book).

There are two aspects of this section of relating upper air features to area burned. The first aspect uses the atmospheric circulation at a specified pressure level above the Earth's surface, typically the 500-mb level, and relates patterns in the atmospheric flow to area burned by wildfires. The term *long wave* is used with respect to the atmospheric circulation to denote northward or southward displacements in the major belt of westerlies characterized by large wavelength and significant amplitude, also known as planetary or Rossby waves. Typically, three to five long waves can be found encircling the northern hemisphere at any given time. The second aspect relates to the vertical structure of the atmosphere and relates this structure to area burned by fire. Vertical profiles of temperature, atmospheric moisture, and wind speed are commonly used in studies relating atmospheric structure with area burned.

A. Upper Air—Circulation

Upper atmospheric ridging is critical in terms of area burned. Newark (1975) was among the first to notice that forest fire occurrence in northwestern Ontario during the summer of 1974 was related to 500-mb-long wave ridging. Nimchuk (1983) related two episodes of catastrophic burning during the Alberta 1981 fire season to the breakdown of the upper ridge over Alberta. These episodes, which lasted 8 days, accounted for about 1,000,000 ha burned. The breakdown of these upper ridges is often accompanied by increased lightning activity as upper disturbances (short waves) move along the west side of the ridge. Additionally, as the ridge breaks down, strong and gusty surface winds are common. The ridge breakdown is preceded by warm and dry conditions associated with the upper ridge, which dries the forest fuels. These upper ridges can persist for weeks to months (Daley, 1991). Brotak and Reifsnyder (1977a) also studied the upper air conditions associated with 52 major wildland fires (area burned 2000 ha or more) in the eastern United States during the 1963–1973 period. They found that the vast majority of the fires were associated with the eastern portion of a small but intense short wave trough at 500 mb (Figure 2). Despite the difference in geographical location, the studies by Brotak and Reifsnyder (1977a) and Nimchuk (1983) may be discussing the same situation, although the emphasis changes from trough to ridge breakdown from the former to the latter. Cold fronts are often associated with the breakdown of

356

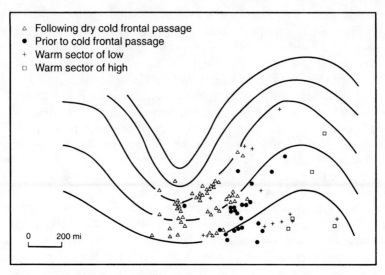

FIGURE 2 Idealized 500-mb map showing relative position of major fires over eastern United States (from Brotak and Reifsnyder, 1977a).

the ridge or the passing of a short-wave trough, which are also important in terms of major wildland fires (Brotak and Reifsnyder, 1977a).

Some researchers have used height anomalies in the upper atmosphere to address the relationship between long waves and area burned. Flannigan and Harrington (1988) found that the monthly 700-mb height anomaly for the forested regions of their provincial areas was the predictor that was selected the most when relating meteorological variables to monthly provincial area burned in Canada 1953–1980. The 700-mb height anomaly is calculated by subtracting the climatological average 700-mb height, typically a 30-year average, from the observed 700-mb height. Positive anomalies mean that the observed heights are higher than average, and, conversely, negative anomalies mean observed heights are lower than average. The height of the pressure level is a function of the air column's temperature, with lower heights generally related to lower temperatures and higher heights to warmer temperatures in the total air column from the surface to the specified pressure level. More recently, some studies have related height anomalies and blocking ridges to teleconnection patterns between North America and the surrounding oceans. Johnson and Wowchuk (1993) found that midtropospheric positive height anomalies (blocking ridges) were related to large-fire years in the southern Canadian Rocky Mountains whereas negative anomalies were related to small-fire years. They observed that these blocking ridges associated with the large-fire years were teleconnected (spatially and temporally correlated with respect to 500-mb heights) to upper-level troughs in the North Pacific and eastern North America, which is the positive mode of the Pacific North America (PNA) pattern (see teleconnection section). Skinner *et al.* (1999) found that 500-mb height anomalies were well correlated with seasonal area burned over various large regions of Canada. They also found a structure similar to the PNA pattern for the extreme fire seasons in western and west-central Canada. Current research suggests that blocking frequency is related to the wave number (the number of long waves in the westerlies—typically three to five) with blocking ridges being more frequent with higher wave numbers (Weeks *et al.*, 1997).

B. VERTICAL STRUCTURE OF THE ATMOSPHERE

Fires can be divided into two groups: convection-column driven and wind driven (Nelson and Adkins, 1988; Nelson, 1993). Fires in which the convection column is well developed are more sensitive to the atmospheric lapse rate because these columns extend farther up into the atmosphere. The lapse rate is the rate of temperature change with height above ground. Typically, temperature decreases with height above ground in the lower atmosphere. Inversions denote those situations where temperature increases with increasing altitude. Wind-driven fires are influenced by strong winds at the surface and typically have significant vertical wind shear. Wind shear is the rapid change of speed and/or direction in any direction; typically there is a significant increase of wind speed with height above the ground as friction reduces the wind speed near the Earth's surface. Under such conditions, the convection column is sheared and, as such, does not play a significant role in fire spread and behavior. Additionally, there is a dynamic interaction between the fire and the atmosphere which can lead to some erratic fire behavior (see Chapter 8 in this book) such as fire whirls, horizontal roll vortices, spotting, and the development of pyrocumulus including thunderstorms generated by the fire (Stocks and Flannigan, 1987).

The vertical structure of the atmosphere influences fire behavior. Dry and unstable air enhances the growth of forest fires in two ways: first, in the absence of strong winds, it promotes a well-developed convection column, which may produce spotting and other erratic fire behavior like fire whirls; second, when wind speeds are strong near the Earth's surface, the instability allows these high winds to be mixed down to the surface, promoting fire spread and erratic fire behavior like horizontal roll vortices (Haines, 1982). An unstable lower atmosphere was a common feature of the large fires that Brotak and Reifsnyder (1977b) studied.

To address the role of the vertical structure of the atmosphere on fire activity, Haines (1988) developed a lower atmosphere severity index (LASI) for wildland fires to account for the influence of temperature stability and moisture content in the lower atmosphere. Unstable air is a layer of air that is characterized by a vertical temperature gradient such that air parcels displaced upward will accelerate upward and away from their original altitude. Haines found that only 6% of all fire season days fell into the high-index class for the western United States. However, 45% of days with large and/or erratic wildfire occurred during those high-index class days. Potter (1995) conducted a regional analysis of the LASI in the United States and suggested that the Haines LASI might be modified to improve performance in the southeastern United States. Werth and Ochoa (1993) used the LASI to evaluate the growth of two forest fires in Idaho. The LASI performed well at determining the time of the most explosive fire growth. Brotak (1993) explored whether the LASI could be used in a predictive sense by using early morning data (1200 GMT). His preliminary results suggested that the LASI calculated using early morning data does not work well in predicting the fire behavior for that afternoon.

The role of atmospheric stability and vertical moisture profile is a critical aspect for fire growth. Potter (1996) examined atmospheric properties associated with large wildfires (over 400 ha) in the United States from 1971 to 1984. He compared the lower atmosphere moisture, temperature, wind, and lapse rate for the 339 large fires in the data set with climatology using the same 14-year period. The results showed that the fire-day surface air temperature and moisture variables (dewpoint depression and relative humidity) at 0000 UTC (Universal Time Coordinated) differed significantly from climatological averages of these variables. There was no difference in wind shear between fire days and climatology days. Results from wind speed and lapse rate were inconclusive for this study. Garcia Diez et al. (1993) found that the stability and the saturation deficit of the lower atmosphere were related to the number of fires occurring in the following 24 hours in Galicia (Spain). Delgado Martín et al. (1997) defined four types of days based on stability and moisture levels: dry and unstable, moist and unstable, dry and stable, and moist and stable. They found for northwestern Spain that these four types of days and their associated fire risk were related to area burned, with dry and unstable days having the highest fire risk. Additionally, they suggested that days with a high number of fires are also those that contribute to more burned area.

Future research relating all components of the weather (surface and upper air features) should be pursued using a more rigorous spatial approach. For example, using relatively high-resolution (approx. 50–100 km) gridded area burned data in the analysis along with finer temporal resolution could yield a better understanding of the role of weather in area burned and provide insight into the relationship between surface weather and upper air features from an area-burned perspective. If or when better relationships between weather and area burned are obtained, it might be possible to predict area burned for a season as seasonal climate prediction is beginning to show potential (Uppenbrink, 1997).

IV. TELECONNECTIONS

Teleconnections are significant temporal correlations between meteorological variables in widely separated points on the globe. Over the past 100 years of meteorological study, a number of these correlations have been observed between various points. Perhaps the most well-known source of teleconnections is the Southern Oscillation (SO) in the equatorial Pacific Ocean or more specifically the part of this oscillation known as El Niño (ENSO). El Niño is characterized by a strong warming of sea surface temperature in the eastern and central equatorial Pacific Ocean and is accompanied by a weakening of trade winds in this region. It occurs at irregular intervals from 2 to 7 years, though its average return period is about 3 or 4 years. During an El Niño (or warm phase of the Southern Oscillation) there is typically an area of high pressure over the tropical western Pacific and unusually low pressure in the southeastern Pacific near the coast of South America. The opposite phase of the Southern Oscillation, sometimes known as La Niña, is characterized by higher surface pressures in the eastern Pacific and cooler sea surface temperatures along the equator.

The direct effects of this persistent weather pattern are quite evident in the equatorial Pacific. For hundreds of years, the inhabitants of Peru have seen El Niño's recurrence marked with widespread fish mortality along their coast and increased rainfall and widespread flooding. In Australia, El Niño events tend to bring severe drought across large areas of the eastern part of the continent. The resulting forest fire situations can be severe as was the case during the 1982/83 El Niño and the losses in forested land and life from the "Ash Wednesday" fires (Pyne, 1991). The drought and consequent severe wildfire situation in Indonesia in 1997 were also a result of the disruption of normal monsoon rains by this phenomenon.

Effects of this widespread warming of the ocean are not limited however to the central and western Pacific. There has been a great deal of research into teleconnections of El Niño and weather patterns farther abroad (Kiladis and Diaz, 1989). Correlations have been found between El Niño occurrence and hurricane activity in the Atlantic (O'Brien et al., 1996) as well as drought occurrence in regions of Africa (Bekele, 1997). Much work has been done looking for ENSO teleconnections with weather patterns on the continent of North America as well. Horel and Wallace (1981) showed that warm waters of the equatorial Pacific were well correlated with positive 700-mb height anomalies over western Canada and 700-mb height anomalies in the eastern United States. Bunkers et al.'s (1996) study showed significant correlations between El Niño events and increased summer precipitation in the northern plains of the United States. The opposite correlation held for La Niña events. Shabbar and Khandekar (1996) and Shabbar et al. (1997) described precipitation and temperature patterns across Canada associated with the Southern Oscillation and suggested that the significant correlations they found could be used as a long-range fore-casting technique particularly for the winter season. Such forecasting of tem-perature and precipitation anomalies would be of importance to fire managers as these are two important factors in estimating fuel moisture levels and the consequent levels of forest fire potential.

Studies of fire occurrence in association with El Niño and La Niña events have been carried out as well. Simard *et al.* (1985) divided the United States into five regions and correlated fire occurrence and area burned per year in each of these regions with a relative El Niño intensity index using about 73 years of his-torical fire data. Their results imply that any relation between fire activity and El Niño occurrence in the continental United States will be regional in char-acter. They found a significant correlation between El Niño occurrence and de-creased fire activity in the southern states. However, no other statistically sig-nificant relationships could be found. Swetnam and Betancourt (1990) used fire scars, tree growth chronologies, and fire statistics from Arizona and New Mex-ico and correlated these with the Southern Oscillation Index (SOI). The SOI is defined as the normalized difference in surface pressure between Tahiti, French Polynesia, and Darwin, Australia. The SOI or its anomaly is often used as an index of the strength and duration of El Niño and La Niña events. Positive SOI values indicate a higher pressure in the eastern Pacific and, as such, a La Niña event. Low SOI values indicate a lower pressure difference and an El Niño event. Swetnam and Betancourt found that in Arizona and New Mexico large areas burn after La Niña events and smaller areas burn after El Niño events. This cor-responded to drier and wetter springs in the American southwest associated with La Niña and El Niño events, respectively. Brenner (1991) studied fire oc-currence and area burned in Florida from 1950 to 1989. He found that a higher than average area burned in La Niña years and a lower than average area burned during El Niño years. It is interesting to note that in this study the highest cor-relations he found against area burned in Florida were with sea surface tem-perature anomalies from the central Pacific. This is perhaps not surprising as one would expect continental North America to be more strongly influenced by North Pacific atmospheric circulations than those in the tropics. The potential for North Pacific teleconnections with North American climate and fire activ-ity will be discussed later in this section.

The temperature and precipitation teleconnection studies of Shabbar and Khandekar (1996) and Shabbar *et al.* (1997) suggest that a Southern Oscillation signal might be detectable in Canada. They show significant decreases in win-ter precipitation across the lower prairies and British Columbia and increases in the Northwest Territories during El Niño years. La Niña years have a corre-sponding increase in winter precipitation across the prairies. One might pre-sume that the area burned during the fire season following such an El Niño or La Niña winter precipitation anomaly might demonstrate some positive or nega-

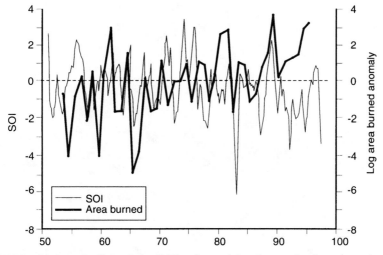

FIGURE 3 Southern Oscillation Index (SOI) and central Canada normalized area-burned anomaly for 1953 to 1995.

tive correlation, respectively. Figure 3 shows normalized area burned anomaly in central Canada plotted with smoothed monthly SOI values from 1953 to 1996. There appears, in this graph, to be no strong correlations (negative or positive) for this period. Correlation of winter levels of SOI and following summers area burned also yields no significant relationship.

Although some correlations have been found between the Southern Oscillation and fire occurrence in North America, the relative lack of signal might not be surprising since El Niño events tend to be strongest in the North American winter months and typically weaken or disappear as summer (and the fire season) approach. Indeed the correlation studies of Shabbar and Khandekar (1996) and Shabbar et al. (1997) show that links between ENSO and Canadian precipitation and temperature are strongest in the winter months. However, recent research by Bonsal and Lawford (1999) found a relationship between El Niño and La Niña events and summer extended dry spells on the Canadian prairies.

The results of Brenner (1991) suggest there may be some value in looking at North Pacific sea surface temperatures for teleconnection not just with North American climate but with fire activity. A number of studies have shown a correlation between North Pacific sea surface temperatures and El Niño events (Reynolds and Rasmusson, 1983; Trenberth, 1990; Deser and Blackmon, 1995). Studies of the North Pacific have shown Southern Oscillation-like variability in the sea surface temperature pattern, which imply that a similar North Pacific Oscillation (NPO) exists (Tanimoto et al., 1993; Trenberth and Hurrell 1993;

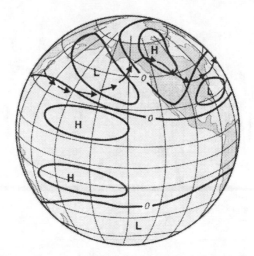

FIGURE 4   Map showing the geopotential height anomalies associated with the positive mode of the PNA teleconnection (from Horel and Wallace, 1981). The arrows depict the midtropospheric streamline as distorted by the anomaly pattern. The negative mode of the PNA has the anomalies with an opposite sign; that is, the low-high-low anomaly sequence (North Pacific-western Canada-southeast United States) of the positive mode switches to high-low-high anomaly sequence for the PNA negative mode.

Zhang *et al.*, 1997). A teleconnection has, in fact, been identified between the North Pacific and North America. Known as the Pacific North American (PNA) teleconnection (Figure 4), it is really a triple connection between an anticyclonic circulation over the North Pacific, a cyclonic circulation over western Canada, and a second anticyclonic circulation over the southeastern United States. Walsh and Richman (1981) found a signature of this triple teleconnection when they used 31 years of North Pacific sea surface temperature data and U.S. temperature data to show regional correlations. Their sea surface temperature data correlated well with air temperature fluctuations over the far western states and the southeastern states but with opposite signs in the two regions. This work also suggested that such correlations could potentially be used for seasonal predictability in some areas. Johnson and Wowchuk (1993) found a signature of the PNA triple teleconnection during a study of 35 years of area-burned data in the southern Canadian Rocky Mountains. They found that, during large-fire years, surface blocking high-pressure systems in their study area were correlated with upper level troughs in the North Pacific and in eastern North America. Skinner *et al.* (1999) found strong correlations between area burned in a number of large regions in Canada and the presence of strong 500-mb-level ridging. During extremely high years of area burned, the PNA teleconnection pattern was evident with significant correlations between the 500-mb values at the three "triplet" locations.

Teleconnections from sea surface temperatures such as the Pacific North America pattern and those from the Southern Oscillation result from complex ocean–atmosphere forcing coupled with large-scale atmospheric circulations. Climate modeling with coupled atmosphere/ocean models has shown that tropical flow patterns and consequent rainfall show little sensitivity to initial conditions of the atmosphere but are determined mainly by boundary conditions of sea surface temperature in the model (Shukla, 1998). The details of these teleconnections are not as important to us in the context of this chapter as the fact that the relatively high temporal variability of weather parameters over land masses can be shown to be correlated with the more slowly varying oceans. Weather prediction limits based on the intrinsic variability in the atmosphere are thought to be on the order of 2 weeks. However, ocean variability occurs on a much slower time scale, and, as such, teleconnections imply that seasonal climate prediction may be possible (Uppenbrink, 1997). The use of such teleconnections as long-range forecasting tools has great appeal to all those involved in the fire business. Recent research has shown that the phase of the North Pacific Oscillation has a great influence on climate anomalies in North America during the Southern Oscillation (Gershunov et al., 1999). They found correlations between the Southern Oscillation and Pacific sea surface temperatures and precipitation anomalies in North America to be much stronger during the high phase of the North Pacific Oscillation. This has strong implications for longer term predictions of North America's weather and climate as the North Pacific Oscillation varies with a 50–70-year time period (Minobe, 1997). Being able to plan a fire season's activity during the winter before, based on current sea surface temperature levels, could potentially allow fire management agencies greater efficiency in allocating resources. Indeed, coupled atmosphere/ocean modeling of the strong El Niño in 1997–1998 has shown very good results in predicting several months in advance seasonal precipitation anomalies in North America (Kerr, 1998). However, predicting fire activity is a difficult task. Forest fire occurrence and area burned tend to have a strong dependence on shorter term local weather, particularly dry spells and wind events, than they do on longer term conditions. Certainly, while an exceptionally dry or warm winter would seem to indicate a potential for much drier fuels for the upcoming fire season, small-scale synoptic events such as showers or thunderstorms tend to drive day-to-day fire potential. However, the correlations already found between area burned and the Southern Oscillation and Pacific Ocean sea surface temperatures show there is potential, with further analysis, that these teleconnections can be used to predict qualitatively the fire severity potential of the upcoming fire season. Preliminary results from Flannigan et al. (2000) show

TABLE 1 Maximum Correlation between Pacific SSTs and Seasonal Provincial
Area Burned 1953–1976 and 1977–1995[a]

Province	1953–1976	1977–1995
BC	0.53 16°S 106°E	−0.52 20°N 102°W
Yukon/NWT	0.64 20°S 178°E	0.68 20°N 122°E
Alberta	−0.72 36°N 142°W	0.55 0°N 162°E
Sask.	0.51 16°N 166°W	−0.55 12°S 162°E
Man.	0.61 24°N 174°W	0.60 12°N 130°E
W. Ont.	−0.70 4°N 150°W	0.59 4°N 118°W
E. Ont.	−0.70 20°N 134°W	0.64 4°S 146°E
Que.	−0.72 4°N 102°W	0.62 20°S 106°E

[a]The location of the maximum correlation is displayed as well.

significant correlations between the winter season sea surface temperature
(Jan.–Apr.) and provincial seasonal area burned (May–Aug.) in Canada by
segregating the analysis by North Pacific Oscillation phase (Table 1). Figures 5
and 6 show the relationship between the seasonal area burned in eastern Ontario and the lagged (Jan.–Apr.) sea surface temperature anomaly for the 1977–

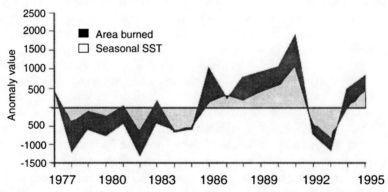

FIGURE 5 Normalized seasonal area burned in Eastern Ontario and lagged (Jan.–Apr.) SST
anomaly for 1977–1995.

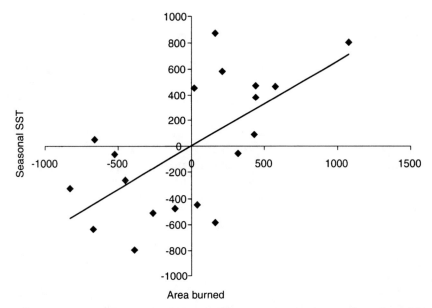

FIGURE 6 Normalized SST anomaly (Jan.–Apr.) versus normalized seasonal area burned for eastern Ontario 1977–1995.

1995 period with a time series and a scatter plot. The correlation is .64 (see Table 1) which explains about 41% of the variation in the data.

V. FUTURE WARMING AND AREA BURNED

There is consensus in the scientific community that human activities are responsible for recent changes in the climate (Intergovernmental Panel on Climate Change, IPCC, 1996). Specifically, increases in radiatively active gases such as carbon dioxide, methane, and the chlorofluorocarbons in the atmosphere are causing a significant warming of the Earth's surface. Significant increases in temperature are anticipated in the next century and are projected to occur rapidly.

There are many General Circulation Models (GCMs) that enable researchers to simulate the future climate. These models are three-dimensional representations of the atmosphere, ocean, cryosphere, and land surface. Transient simulations are available from GCMs which allow examination of the possible rates of change in the climate. Also, recent GCM simulations include sulfate aerosols. The major areas of uncertainty in GCMs include clouds and their radiative effects, the hydrological balance over land surfaces, and the heat flux at the ocean surface (IPCC, 1996). Despite these uncertainties, the GCMs provide the best means available to estimate the impact of changes in the future climate on the fire regime at larger scales. Most models agree in predicting the greatest warming at high latitudes in winter. Confidence is lower for estimates of precipitation fields, but many models suggest an increased moisture deficit particularly in the center of continents during summer. In addition to temperature, other weather variables will be altered in a changed climate such as precipitation, wind, and cloudiness. The variability of extreme events may also be altered (Mearns et al., 1989; Solomon and Leemans, 1997). Some studies suggest universal increases in fire frequency with climatic warming (Overpeck et al., 1990; IPCC, 1996). The universality of these results is questionable because an individual fire is a result of the complex set of interactions that include ignition agents, fuel conditions, topography, and weather including temperature, relative humidity, wind velocity, and the amount and frequency of precipitation. Increasing temperature alone does not necessarily translate into greater fire disturbance.

Flannigan and Van Wagner (1991) compared seasonal fire severity rating values (SSR—a component of the Canadian Forest Fire Weather Index System) from a $2xCO_2$ scenario (mid-21st century) versus the $1xCO_2$ scenario (approx. present day) across Canada. Their study used monthly anomalies from three GCMs (Geophysical Fluid Dynamics Laboratory, GFDL; Goddard Institute for Space Studies, GISS; and Oregon State University, OSU). The results suggest increases in the SSR all across Canada with an average increase of nearly 50%, which might translate roughly into an increase of area burned by 50%. Stocks et al. (1998) used monthly data from four GCMs to examine climate change and forest fire potential in Russian and Canadian boreal forests. Forecast seasonal fire weather severity was similar for the four GCMs, indicating large increases in the areal extent of extreme fire danger in both countries under a $2xCO_2$ scenario. Stocks et al. (1998) also conducted a monthly analysis using the Canadian GCM, which showed an earlier start to the fire season and significant increases in the area experiencing high to extreme fire danger in both Canada and Russia, particularly during June and July. Flannigan et al. (1998a) used daily output from the Canadian GCM to model the Fire Weather Index for both the $1xCO_2$ and $2xCO_2$ scenarios for North America and Europe. The FWI represents the intensity of fire as energy output per unit length of fire front. Figure 7 shows the ratio of the $2xCO_2$ to $1xCO_2$ values for both mean FWI and maximum FWI for the 9 years of simulation for North America and Europe. There is a great deal of regional variation between areas where FWI decreases in a $2xCO_2$ scenario (values below 1.00) to areas where the FWI increases greatly in the warmer climate. Much of eastern Canada and northwestern Canada have ratios below 1.00, indicating that the FWI decreases despite the warmer tem-

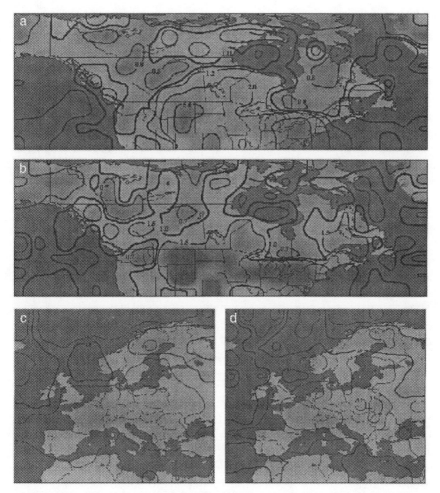

FIGURE 7   Mean and maximum FWI ratios ($2xCO_2/1xCO_2$) for North America (a, b) and Europe (c, d) (Flannigan et al., 1998a).

peratures associated with a $2xCO_2$ climate. Noteworthy is the area of decreased FWI over western and northwestern sections of Canada where historically large portions of the landscape have been burned. However, due to the coarse spatial resolution of the GCM (approx. 400 km), confidence in the results over complex terrain like mountainous areas is low. In such areas, a Regional Climate Model (RCM) should be used (Caya et al., 1995) where the finer spatial resolution (~40 km) can better resolve mountain ranges, and so on. Significant increases in the FWI are evident over parts of central North America. The ratio of extreme maximum values of the FWI for the 9-year period show a similar pattern, with higher ratios over central continental areas and lower values over portions of eastern Canada. On the other hand, there are increases in the maximum FWI over portions of western Canada. Over northern Europe, Figure 7 shows increased mean FWIs over the southern half of Sweden and extreme southeastern Finland for warmer conditions, whereas the remainder of northern Europe shows decreased mean FWIs. Interestingly, according to the simulations, fire danger will decrease in the northern sections of northern Europe where traditionally forest fires have been large and frequent, and fire danger will increase over southern Sweden where forest fires are not normally severe. Results for maximum FWIs show a similar pattern when compared with the mean FWIs, with the exception of southern Norway where FWIs increased. Consequences of climate change on fire disturbance must be viewed in a spatially dependent context.

Flannigan et al. (1998a) suggest that the reason for the decreased FWI despite the increasing temperature is due primarily to changes in the precipitation regime and, in particular, to increases in precipitation frequency. These model results (Figure 7) are in good agreement with recent fire history studies ($\sim$ the last 200 years) (Flannigan et al. 1998a), many of which show decreasing fire activity despite the warming since the end of the Little Ice Age ($\sim$1850). These modeled results are also consistent with the modeled fire weather and charcoal record anomalies for a warm period during the mid-Holocence about 6000 years BP, which was about 1°C warmer than present for Canada (Flannigan et al., 1998b).

Other factors such as ignition agents, length of the fire season, and fire management policies may greatly influence the impact of climate change on the fire regime. Ignition probabilities may increase in a warming world due to increased cloud-to-ground lightning discharges (Price and Rind, 1994). Price and Rind suggest that lightning-caused fires would increase by 44% and that the associated area burned would increase by 78% by the end of the 21st century, although they assume no changes in fuels, which may greatly influence the lightning ignitions and area burned. The longer fire season will begin earlier in the spring and extend longer into the autumn. Wotton and Flannigan (1993) estimated that the fire season length in Canada on average will increase by 22%, or 30 days in a $2xCO_2$ world. Fire management policies and effectiveness will continue to change. Also, research has suggested that the persistence of blocking ridges in the upper atmosphere will increase in a $2xCO_2$ climate (Lupo et al., 1997), which could have a significant impact on forest fires as these upper ridges are associated with dry and warm conditions at the surface and are conducive to forest fires. These are all confounding effects that may dampen or amplify the impact of a changing climate on the fire regime.

Future work might include using the improved relationships between climate/weather and area burned to estimate what the fire regime will be like in the future. This would assist land managers with long-term planning. Also, predictions of future fire regimes are critical in terms of the carbon balance where changes in the natural disturbance regime like fires could be responsible for the boreal forest becoming a source of carbon rather than a sink (Solomon and Leemans, 1997).

The important aspect of the impact of climate change on forest fires with respect to the influence on vegetation is that fire may be more important than direct effects of climate change (Weber and Flannigan, 1997). Fire can act as an agent of change to hasten the modification of the vegetation landscape into a new equilibrium with the climate if species are able to migrate fast enough.

VI. SUMMARY

Weather is a critical factor in determining the timing and size of fires. We have seen in this chapter that dry, windy, and warm surface weather conditions are conducive to area burned. However, these surface weather conditions are a reflection of the synoptic weather pattern which is a function of the dynamics of the three-dimensional atmosphere. The breakdown of the upper blocking ridge is one pattern of the upper atmospheric circulation that is a common feature of significant area-burned events. These blocking ridges typically bring 7–10 days or longer periods of warm dry weather followed by windy conditions as the ridge deteriorates. The breakdown of the upper ridge is often associated with a cold front passage at the surface which has been shown to be a common factor of large fires. Additionally, because of some large-scale circulation patterns, there are significant relationships between the weather in widely separated points on the globe, and often these circulation patterns have a cyclical nature such that seasonal forecasts may be viable. Last, preliminary indications from climate-modeling work suggest that climate change may result in increased area burned for many regions of the world.

ACKNOWLEDGMENTS

We thank two anonymous reviewers whose comments and suggestions improved this chapter. We also thank Brenda Laishley for editing this chapter.

REFERENCES

Bekele, F. (1997). Ethiopian use of ENSO information in its seasonal forecasts. *Int. J. African Stud.* 2. http://www.brad.ac.uk/research/ijas/ijasno2/ijasno2.html
Bonsal, B. R., and Lawford, R. G. (1999). Teleconnections between El Niño and La Niña events and summer extended dry spells on the Canadian Prairies. *Int. J. Climatol.* 19, 1445–1458.

Brenner, J. (1991). Southern Oscillation anomalies and their relationship to wildfire activity in Florida. *Int. J. Wildland Fire* 1, 73–78.

Brotak, E. A. (1993). Low-level weather conditions preceding major wildfires. *Fire Manage. Notes* 54, 23–26.

Brotak, E. A., and Reifsnyder, W. E. (1977a). An investigation of the synoptic situations associated with major wildland fires. *J. Appl. Meteorol.* 16, 867–870.

Brotak, E. A., and Reifsnyder, W. E. (1977b). Predicting major wildland fire occurrence. *Fire Manage. Notes* 38, 5–8.

Bunkers, M. J., Millar, Jr., J. R., and deGaetano, A. T. (1996). An examination of El Niño-La Niña-related precipitation and temperature anomalies across the northern plains. *J. Climatol.* 9, 147–160.

Caya, D., Laprise, R., Giguère, M., Bergeron, G., Blanchet, J. P., Stocks, B. J., Boer, G. J., and McFarlane, N. A. (1995). Description of the Canadian regional climate model. *Water, Air Soil Pollut.* 82, 477–482.

Daley, R. (1991). "Atmospheric Data Analysis." Cambridge University Press, Cambridge.

Delgado Martín, L., Garcia Diez, A., Rivas Soriano, L., and Garcia Diez, E. L. (1997). Meteorology and forest fires: Conditions for ignition and conditions for development. *J. Appl. Meteorol.* 36, 705–710.

Deser, C., and Blackmon, M. L. (1995). On the relationship between tropical and North Pacific sea surface temperature variations. *J. Climatol.* 8, 1677–1680.

Flannigan, M. D., and Harrington, J. B. (1987). Synoptic conditions during the Porter Lake burning experiment. *Climatol. Bull.* 21, 19–40.

Flannigan, M. D., and Harrington, J. B. (1988). A study of the relation of meteorological variables to monthly provincial area burned by wildfire in Canada 1953–80. *J. Appl. Meteorol.* 27, 441–452.

Flannigan, M. D., and Van Wagner, C. E. (1991). Climate change and wildfire in Canada. *Can. J. For. Res.* 21, 66–72.

Flannigan, M. D., Bergeron, Y., Engelmark, O., and Wotton, B. M. (1998a). Future wildfire in circumboreal forests in relation to global warming. *J. Veg. Sci.* 9, 477–482.

Flannigan, M. D., Todd, B., Wotton, B. M., Skinner, W. R., Stocks, B. J., and Martell, D. L. (2000). Pacific sea surface temperatures and their relation to area burned in Canada. *In* "Preprints of Third Symposium on Fire and Forest Meteorology," pp. 151–157. American Meteorological Society, Boston.

Flannigan, M. D., Wotton, B. M., Richard, P., Carcaillet, C., and Bergeron, Y. (1998b). Fire weather: Past, present and future. *In* "Preprints of Ninth Symposium on Global Change Studies," pp. 305–309. American Meteorological Society, Boston.

Garcia Diez, E. L., Labajo S., and de Pablo Dávila, F. (1993). Some meteorological conditions associated with forest fires in Galicia (Spain). *Int. J. Biometeorol.* 37, 194–199.

Gershunov, A., Barnett, T. P, and Cayan, D. R. (1999). North Pacific interdecadal oscillation seen as factor in ENSO-related North American climate anomalies. *EOS Trans. Am. Geophys. Union* 80, 25, 29–30.

Haines, D. A. (1982). Horizontal roll vortices and crown fires. *J. Appl. Meteorol.* 21, 751–763.

Haines, D. A. (1988). A lower atmosphere severity index for wildland fires. *Nat. Weath. Digest* 13, 23–27.

Haines, D. A., Main, W. A., Frost, J. S., and Simard, A. J. (1983). Fire-danger and wildfire occurrence in the northeastern United States. *For. Sci.* 29, 679–696.

Harrington, J. B., Flannigan, M. D., and Van Wagner, C. E. (1983). "A Study of the Relation of Components of the Fire Weather Index System to Monthly Provincial Area Burned by Wildfire in Canada 1953–80." Inf. Rep. PI-X-25. Canadian Forest Service, Petawawa National Forestry Institute, Chalk River.

Harvey, D. A., Alexander, M. E., and Janz, B. (1986). A comparison of fire-weather severity in northern Alberta during the 1980 and 1981 fire seasons. *For. Chron.* **62**, 507–513.

Hirsch, K. G., and Flannigan, M. D. (1990). Meteorological and fire behavior characteristics of the 1989 fire season in Manitoba, Canada. *In* "International Conference on Forest Fire Research," pp. B.06-1–B.06-16, Coimbra.

Horel, J. D., and Wallace, J. M. (1981). Planetary-scale atmospheric phenomena associated with the southern oscillation. *Mon. Weather Rev.* **109**, 813–829.

Intergovernmental Panel on Climate Change. (1996). "Climate Change 1995 Impacts, Adaptations and Mitigation of Climate Change: Scientific-Technical Analyses." Cambridge University Press, Cambridge.

Johnson, E. A. (1992). "Fire and Vegetation Dynamics: Studies from the North American Boreal Forest." Cambridge University Press, Cambridge.

Johnson, E. A., and Wowchuk, D. R. (1993). Wildfires in the southern Canadian Rocky Mountains and their relationship to mid-tropospheric anomalies. *Can. J. For. Res.* **23**, 1213–1222.

Kerr, R. A. (1998). Models win big in forecasting El Niño. *Science* **280**, 522–523.

Kiladis, G. N., and Diaz, H. F. (1989). Global climate anomalies associated with extremes in the southern oscillation. *J. Climatol.* **2**, 1069–1090.

Lupo, A. R., Oglesby, R. J., and Mokhov, I. I. (1997). Climatological features of blocking anticyclones: A study of Northern Hemisphere CCM1 model blocking events in present-day and double CO_2 concentration atmosphere. *Clim. Dynam.* **13**, 181–195.

Mearns, L. O., Schneider, S. H., Thompson, S. L., and McDaniel, L. R. (1989). Climate variability statistics from General Circulation Models as applied to climate change analysis. *In* "Natural Areas Facing Climate Change" (G. P. Malanson, Ed.), pp. 51–73. SPB Academic Publishing, The Hague.

Minobe, S. (1997). A 50–70 year climatic oscillation over the North Pacific and North America. *Geophys. Res. Let.* **24**, 683–686.

Nelson, R. M., Jr. (1993). Byram's derivation of the energy criterion for forest and wildland fires. *Int. J. Wildland Fire* **3**, 131–138.

Nelson, R. M. Jr., and Adkins, C. W. (1988). A dimensionless correlation for the spread of wind-driven fires. *Can. J. For. Res.* **18**, 391–397.

Newark, M. J. (1975). The relationship between forest fire occurrence and 500 mb ridging. *Atmosphere* **13**, 26–33.

Nimchuk, N. (1983). "Wildfire Behavior Associated with Upper Ridge Breakdown." ENR Rep. No. T/50. Alberta Energy and Natural Resources, Forestry Service, Edmonton.

O'Brien, J. J., Richards, T. S., and Davis, A. C. (1996). The effect of El Niño on US land falling hurricanes. *Bull. Amer. Meteorol. Soc.* **77**, 773–774.

Overpeck, J. T., Rind, D., and Goldberg, R. (1990). Climate-induced changes in forest disturbance and vegetation. *Nature* **343**, 51–53.

Potter, B. E. (1995). Regional analysis of Haines' LASI. *Fire Manage. Notes* **55**, 30.

Potter, B. E. (1996). Atmospheric properties associated with large wildfires. *Int. J. Wildland Fire* **6**, 71–76.

Price, C., and Rind, D. (1994). The impact of a $2 \times CO_2$ climate on lightning-caused fires. *J. Climatol.* **7**, 1484–1494.

Pyne, S. J. (1991). "Burning Bush: A History of Fire in Australia." Henry Holt and Company, New York.

Quintilio, D., Fahnestock, G. R., and Dube, D. E. (1977). "Fire Behaviour in Upland Jack Pine: The Darwin Lake Project." Inf. Rep. NOR-X-174. Environment Canada, Canadian Forest Service, Northern Forestry Research Centre, Edmonton.

Reynolds, R. W., and Rasmusson, E. M. (1983). The North Pacific sea surface temperature associated with El Niño events. *In* "Proceedings of the Seventh Annual Climate Diagnostics Workshop." U.S. Dept. of Commerce, National Oceanic and Atmospheric Administration, Washington, DC.

Schroeder, M. J. (1964). "Synoptic Weather Types Associated with Critical Fire Weather." USDA Forest Service, Pacific Southwest Forest and Range Experiment Station, Berkeley.

Shabbar, A., and Khandekar, M. (1996). The impact of El Niño-Southern Oscillation on the temperature field over Canada. *Atmosphere-Ocean* 34, 401–416.

Shabbar, A., Bonsal, B., and Khandekar, M. (1997). Canadian precipitation patterns associated with the southern oscillation. *J. Climatol.* 10, 3016–3027.

Shukla, J. (1998). Predictability in the midst of chaos: A scientific basis for climate forecasting. *Science* 282, 728–731.

Simard, A. J., Haines, D. A., and Main, W. H. (1985). Relations between El Niño/southern oscillation anomalies and wildland fire activity in the United States. *Agric. For. Meteorol.* 36, 93–104.

Skinner, W. R., Stocks, B. J., Martell, D. L., Bonsal, B., and Shabbar, A. (1999). The association between circulation anomalies in the mid-troposphere and area burned by wildland fire in Canada. *Theor. Appl. Climatol.* 63, 89–105.

Solomon, A. M., and Leemans, R. (1997). Boreal forest carbon stocks and wood supply: Past, present and future responses to changing climate, agriculture and species availability. *Agric. For. Meteorol.* 84, 137–151.

Stocks, B. J. (1975). "The 1974 Wildfire Situation in Northwestern Ontario." Inf. Rep. O-X-232. Canadian Forest Service, Great Lakes Forest Research Centre, Sault Ste. Marie.

Stocks, B. J., and Flannigan, M. D. (1987). Analysis of the behaviour and associated weather for a 1986 Northwestern Ontario wildfire: Red Lake 7. *In* "Proceedings of the Ninth Conference on Fire and Forest Meteorology," pp. 94–100. American Meteorological Society, Boston.

Stocks, B. J., and Walker, J. D. (1973). "Climatic Conditions Before and During Four Significant Forest Fire Situations in Ontario." Inf. Rep. O-X-187. Canadian Forest Service, Great Lakes Forest Research Centre, Sault Ste. Marie.

Stocks, B. J., Fosberg, M. A., Lynham, T. J., Mearns, L., Wotton, B. M., Yang, Q., Jin, J-Z., Lawrence, K., Hartley, G. R., Mason, J. A., and McKenney, D. W. (1998). Climate change and forest fire potential in Russian and Canadian boreal forests. *Climatic Change* 38, 1–13.

Swetnam, T. W. (1993). Fire history and climate change in giant sequoia groves. *Science* 262, 885–889.

Swetnam, T. W., and Betancourt, J. L. (1990). Fire–southern oscillation relations in the southwestern United States. *Science* 249, 1017–1020.

Tanimoto, Y., Iwasaka, N., Hanawa, K., and Toba, Y. (1993). Characteristic variations of sea surface temperature with multiple time scales in the north Pacific. *J. Climatol.* 6, 1153–1160.

Trenberth, K. E. (1990). Recent observed interdecadal climate changes in the northern hemisphere. *Bull. Amer. Metcorol. Soc.* 71, 988–993.

Trenberth, K., and Hurrell, J. W. (1993). Decadal atmosphere–ocean variations in the Pacific. *Clim. Dynam.* 9, 303–319.

Trewartha, G. T., and Horn, L. H. (1980). "An Introduction to Climate," 5th ed. McGraw-Hill, New York.

Turner, J. A. (1970). Hours of sunshine and fire season severity over the Vancouver Forest District. *For. Chron.* 46, 106–111.

Uppenbrink, J. (1997). Nota bene: Climate—seasonal climate prediction. *Science* 277, 1952.

Van Wagner, C. E. (1987). "The Development and Structure of the Canadian Forest Fire Weather Index System." Tech. Rep. 35. Canadian Forest Service, Ottawa.

Walsh, J. E., and Richman, M. B. (1981). Seasonality in the associations between surface temperature over the United States and the north Pacific Ocean. *Mon. Wea. Rev.* 10, 767–782.

Weber, M. G., and Flannigan, M. D. (1997). Canadian boreal forest ecosystem structure and function in a changing climate: Impact on fire regimes. *Environ. Rev.* 5, 145–166.

Weeks, E. R., Tian, Y., Urbach, J. S., Ide, K., Swinney, H. L., and Ghil, M. (1997). Transitions between blocked and zonal flows in a rotating annulus with topography. *Science* 278, 1598–1601.

Werth, P., and Ochoa, R. (1993). The evaluation of Idaho wildfire growth using the Haines Index. *Weather and Forecasting* 8, 223–234.

Wotton, B. M., and Flannigan, M. D. (1993). Length of the fire season in a changing climate. *For. Chron.* 69, 187–192.

Zhang, Y., Wallace, J. M., and Battisti, D. S. (1997). ENSO-like interdecadal variability: 1900–93. *J. Climatol.* 10, 1004–1020.

CHAPTER 11

Lightning and Forest Fires

DON LATHAM

USDA Forest Service, Fire Sciences Laboratory, Rocky Mountain Research Station,
Missoula, Montana

EARLE WILLIAMS

Parsons Laboratory, Massachusetts Institute of Technology, Cambridge, Massachusetts

I. INTRODUCTION

Lightning ignition of wildland fuels plays a major role in the maintenance and evolution of ecosystems. Lightning not only ignites fire but also weakens trees, facilitating insect and disease attack, causes physical damage, and kills trees and groups of trees (Taylor, 1973). Lightning ignition may also play a large part in forest response to global climate change.

Fire can also alter lightning. Cumulus clouds, called pyrocumulus, can be formed under the proper conditions by large fires. Pyrocumulus clouds have a high proportion of positive cloud-to-ground lightning, the opposite of lightning from a "normal" cumulus (Latham, 1991). One hypothesis for the cause is an interaction between the earth's electric field and the fire (Vonnegut et al.,1995). Smoke from fires can apparently cause inverted storms over a large area (Lyons et al., 1998a).

In this chapter, we will show how the predominant mechanism for lightning ignition works and how this knowledge can be used for forest health and fire management. We will discuss worldwide lightning and fire patterns. Before discussing ignition and other consequences, a brief introduction to lightning is supplied; it is a subject not usually discussed in a forestry curriculum.

II. LIGHTNING

A. THE ORIGIN AND CHARACTERISTICS OF LIGHTNING

Lightning is an energetic electrical discharge caused by the separation of positive and negative charge in clouds leading to voltage differences of order $10-100$ MV. The weight of the evidence shows that the process of charge separation requires the presence of water substance in all three phases—solid, liquid, vapor (Williams, 1989). Such a condition is called mixed phase. The mixed phase region of the atmosphere is bounded below by the $0°C$ isotherm (where ice particles melt) and above by the $-40°C$ isotherm (where supercooled liquid water is spontaneously transformed to ice). Under typical conditions, the mixed phase region is $4000-5000$ meters thick. Though a wide variety of meteorological conditions can lead to lightning, for example, isolated air mass thunderstorms, frontal storms, and snowstorms, the presence of a mixed phase region appears to be a necessary condition for lightning generation.

Separation of positive and negative charge in regions with both ice particles and water droplets (mixed phase) is thought to be caused by the collisions of ice particles prevalent in such conditions: graupel and ice crystals (or smaller graupel). Graupel are millimeter to centimeter-sized hydrometeors, which grow by the accretion of supercooled cloud droplets. Ice crystals grow by water vapor deposition in a mixed phase environment at the expense of the liquid water because the equilibrium vapor pressure with respect to ice is less than the vapor pressure with respect to liquid. In ordinary thunderstorms, for reasons that are still poorly understood, negative charge is selectively transferred to the graupel particles and positive charge, to the smaller ice crystals. The faster falling graupel particles carry negative charge downward relative to the positive charge at higher levels, creating the positive thundercloud dipole. The resulting lower negative charge region is typically in the -5 to $-20°C$ zone with the positive charge at higher (colder) levels. In summertime, the lower negative region is 5–7 km above mean sea level, with the positive region at 7–12 km. This positive dipole is responsible for the two most prevalent lightning types: intracloud (IC) lightning and cloud-to-ground (CG) lightning.

Intracloud lightning is a discharge bridging the upper positive and lower negative charge of the thundercloud dipole and is contained entirely within the cloud, therefore playing no role whatever in fire initiation at the ground. IC lightning is almost invariably the first lightning to occur in a developing storm; some storms produce only IC lightning. The IC rate can become very large during strong convective surges when deep mixed phase development is underway. Though some IC flashes can be quite energetic, the majority have smaller peak currents and smaller charge moments (the charge amount multiplied by the distance the charge moves) and involve less total energy than CG flashes. One important reason for the difference is that IC flashes can occur over small scales (hundreds of meters) whereas ordinary negative ground flashes are required to bridge the typical gap to ground of 5000–7000 m.

The lightning ground flash, as the active player in forest fire initiation, deserves more detailed discussion. CG lightning is most commonly initiated within the cloud, in the vicinity of the lower charge reservoir (usually negative). An ionized path, the stepped leader, is forged through the air toward ground in a region of high electric field. This stepped leader carries the large negative potential of the lower charge region toward Earth. As the stepped leader nears the Earth, an intense electric field develops between leader and ground. The field promotes electrical streamer propagation upward from elevated points on the ground that can connect to the approaching leader. When a connection is made, the bright, high-current (10–100 kA) return stroke is initiated and propagates upward toward the cloud at a speed approaching that of light ($1-2$ $\times$ 10^8 = m s^{-1}). For reasons not well understood, this leader-return stroke sequence is often repeated at intervals of a few tens of milliseconds, causing the characteristic flicker of lightning. The extraordinary brightness of the return stroke channel may give the impression that the lightning channel is very broad. In reality, channel diameters are in the range of millimeters to centimeters, an important consideration for the modeling discussion of Section V (e.g., Orville and Helsdon, 1974).

In the majority of ground flashes, the return stroke current peaks, in less than 1 μs, to values in the range of $5-30$ kA and then promptly decays in a few hundred microseconds. Despite the extraordinary peak power in such events (upwards of 10^{12} W), both observations and simulations have shown that the short duration of the return stroke is inadequate to raise trees and other flora in its path to kindling temperature and initiate fire (Taylor, 1969; Darveniza and Zhou, 1994).

In about 30% of return strokes, a sustained current of low amplitude, a continuing current, is observed to flow in the channel to ground immediately following the current peak for a period varying from milliseconds to hundreds of milliseconds (Uman, 1969). The conditions necessary for continuing currents are still not well resolved, but observations suggest that larger-than-usual reservoirs of electric charge in the cloud are necessary to sustain long continuing currents and that the return strokes initiating them are somewhat smaller in current amplitude than others. The early ground flashes in a developing storm, when the charge reservoir is still of modest size, seldom exhibit continuing currents. The largest and most energetic continuing currents are observed by ELF methods (Burke and Jones, 1996; Cummer and Inan, 1996; Huang *et al.*, 1999) beneath the extensive stratiform regions of precipitation common in mesoscale convective systems (Williams, 1998) and large winter snowstorms. Flashes having at least one stroke with a continuing current are called hybrid flashes. Strokes within flashes without a continuing current are called discrete strokes.

Approximately 90% of CG flashes worldwide transfer negative electric charge to ground. The 10% positive minority, whose physical origins are still poorly understood, are more prevalent in the dissipating stages of local thunderstorms, in the broad stratiform regions of mesoscale convective systems, in winter storms, and in thunderstorms ingesting smoke from fire (Latham, 1991; Lyons *et al.*, 1998a; Section VIII). The continuing current characteristics of positive and negative ground flashes are markedly different, and these differences are important in the context of fire initiation. The great majority of positive ground flashes are single stroke, and a continuing current follows almost all of these strokes. The majority of negative flashes are multistroke, and about half of these have accompanying continuing currents (Uman, 1969).

The development of statistics on parameters relevant to CG lightning has been facilitated by the availability of networks for ground flash detection, such as the National Lightning Detection Network currently operational in the United States (Global Atmospherics, Inc.) (Cummins *et al.*, 1998). This network can now locate the great majority of CG contact points within the contiguous United States and Canada to within a few kilometers spatially and within 1 μs accuracy in time. The polarity, return stroke peak current, and stroke multiplicity are now routinely archived for all ground flashes. Unfortunately, due to the limited low-frequency response of this equipment, and the limited range of the continuing current electric field change, no information on the continuing current component of ground flashes is presently available operationally. Most information on this component comes from local storm studies (Uman, 1969).

Recent studies with both the NLDN (Lyons *et al.*, 1998b) and ELF electromagnetic sensors (Boccippio *et al.*, 1995; Burke and Jones, 1996; Cummer and Inan, 1996; Huang *et al.*, 1999) have focused attention on positive ground flashes, which are likely the most energetic lightning flashes on the planet. Radar observations (Ligda, 1956) presented evidence for horizontal extension of lightning within the cloud for distances exceeding 100 km. It is now well recognized that this "spider" or "sheet" lightning ultimately connects to Earth in a positive ground flash with a long continuing current (Boccippio *et al.*, 1995; Mazur *et al.*, 1998). Such discharges are also capable of lone excitation of the electromagnetic Schumann resonances of the Earth-ionosphere cavity to levels exceeding all other lightning combined (Sentman, 1987; Boccippio *et al.*, 1995; Burke and Jones, 1996; Huang *et al.*, 1999). These spectacular discharges have also been clearly identified with sprites and elves, newly discovered luminous discharge phenomena in the mesosphere 60–90 km above the causal positive lightning. Both the energetic nature of positive flashes and their high probability to exhibit continuing currents have led to popular belief that this lightning type is the leading initiator of forest fire. This common view has been questioned, however, by Flannigan and Wotton (1991), who call attention to the large numbers of negative ground flashes with continuing current.

Lightning's potential to initiate fire has also been judged, incorrectly, on the basis of its visual appearance—"hot" lightning is bright white or bluish light and "cold" lightning has a dull red appearance. Quantitative analysis of the optical spectrum of lightning (Orville and Hendersen, 1986) has shown that the intrinsic spectra of all return stroke channels (the predominant source of lightning light) exhibit a very broad optical region, similar to sunlight. The light from very distant lightning (often referred to as "heat" lightning) is selectively filtered in the blue end of the spectrum by Rayleigh scattering, much like the light from the setting sun, leaving a predominance of red light. Lightning at closer range will suffer less selective loss of blue light and will appear whiter in

color. As discussed in Section VI.B, the most critical parameter for lightning ignition of forest fuels is the presence and duration of the continuing current and
not the temperature of the lightning plasma.

III. PREVIOUS STUDIES OF
LIGHTNING-INITIATED FIRES

In this section, observations and summaries of lightning-initiated fires are discussed. The studies can be broken arbitrarily into three categories: early, Project Skyfire, and recent. The early studies were done before the characteristics of
lightning were discovered and researched. Project Skyfire was designed to study
the possibility of reducing fire numbers by cloud seeding. Recent studies benefit from lightning location data and organized descriptors of forest fuels.

A. EARLY LIGHTNING FIRE RESEARCH

Plummer (1912) observed:

> The same flash may strike and blast a number of trees, and the results may be quite
> as curious and erratic as the lightning itself. A tree may be scorched, it may be stripped
> of its leaves, it may be cleft longitudinally, or, more rarely, severed horizontally. Pieces
> of bark or wood may be torn off in strips. One-half of a tree's crown may be withered,
> the other half remaining unharmed. Sometimes the bark is stripped from only one
> side, occasionally without a trace of burning; at other times it may be riddled, as by
> worms, with a multitude of little holes. The lightning furrow on a tree is usually
> single; but it may be double, usually in parallel lines. Furrows may be oblique or spi-
> ral, the current in such cases following the grain of the new wood. If the tree is in-
> flammable or is rendered very dry by the flash a fire may result. In other cases the dry
> duff or humus at the base of the tree is ignited by the flash.

This description is entirely adequate for today's observations.

Plummer's attempt to explain the wide range of visible effects hypothesized
distinctly different "upward-going" and "downward-going" flashes to ground,
causing different kinds of damage to trees. He does not specify what damage is
caused by which direction. As discussed in Section II, the situation is more
complex. In any case, the "direction" of flashes, and their complexity, was not
firmly demonstrated until 1934 (Schonland and Collens, 1934), so there was
no way for Plummer to know of the complexity or to formulate a test for the
hypothesis.

Another hypothesis of the time was that some species of trees were preferentially hit by lightning. Most of the research was done in Europe and on Forest
Service lands in the United States (as summarized in Plummer, 1912). Unfor-

tunately, the data were not corrected for the relative density of tree type, so the results are questionable. In fact, Plummer's conclusions state, in part, that any kind of tree is likely to be struck by lightning and that the greatest number struck in any locality will be of the dominant species, conclusions that hold to-day. As we shall see, some species do have a higher probability of ignition from lightning, due mainly to their differing foliage cast.

Another interesting note from Plummer is that "Lightning is extremely rare in Alaska, and no forest fires are known to have resulted from it" in contrast to about 200 reported lightning fires per year from modern reports (Bureau of Land Management, private communication). Lightning fires are currently important enough in Alaska that it was chosen for the first trials of lightning location systems for fire management.

Subsequent lightning ignition studies were done in the northwestern area of the United States because most lightning-caused fires occur there. The area includes western Montana, northern Idaho, and parts of eastern Washington, eastern Oregon, and northwestern Wyoming. For this area, in the years 1906–1911, about 15% (864 fires) of the fires on the National Forests were started by lightning (Plummer, 1912). So, for those years, the number of person-caused fires far exceeded lightning-caused fires.

By 1924, the percentage of lightning-caused fires had apparently increased —"Lightning causes an average of 1/3 of the fires in this region . . ." (Gisborne, 1924). He also noted that ". . . only one-fourth of the lightning storms start fires . . ." and that the storms occurred in seven "waves" (no duration timescale given) during the months of July and August. From data taken by lookouts (numbers are not given), the conclusion was drawn that ". . . one-fourth of the lightning storms ordinarily start fires . . ." and that storms could be categorized on the basis of fire starts into "fire-starting" and "nondangerous" types. Gisborne suggested that fire-starting storms had a much higher CG/incloud lightning ratio than the nondangerous ones. No numbers or statistics were given to support this viewpoint.

In a later report, Gisborne (1926) stated that for the years 1907 to 1925, the average percentage of all fires started by lightning was 39% for the years 1907 to 1925, 51% in 1924, and 80% in 1925. The reason given for this increase was an increase in the number of storms during 1924 and 1925 based on lookout reports. He noted that the Weather Service storm-days numbers, determined by thunder heard at Weather Service stations, were too small by a factor of three or so, when compared to the data from 170 lookout stations. Since there were four weather stations covering approximately the same area as the 170 lookouts, this discrepancy is no surprise.

Testing the hypothesis that storms could be separated into two categories, fire-starting and nondangerous, Gisborne (1926) found, based on 2186 reports covering 3 years that stated definitely whether or not the storm was a fire starter, "safe" storms (nondangerous) had about 30% CG flashes and the "dangerous" (fire-starting) storms had about 40% CG flashes. He did state that 1046 additional reports gave CG percentages, but the storms were not classified, so those data were not used (!). Besides rejecting data, another problem with the analysis is that more than one lookout can count a given lightning flash.

Gisborne (1931) later found, based on 8128 observations made between 1924 and 1928, that safe storms had 24% CG lightning and dangerous ones had 44%; that is, storms having high percentages of in-cloud lightning caused more fires than those having low percentages. No mention is made of how many observations were made but not included in the data used. He does note that "before the degree of lightning-fire danger can be estimated satisfactorily, fire weather forecasts must consider other factors—the characteristics of the storms and probably the seasonal and current moisture content and inflammability of the forest materials" and that ". . . thunderstorm frequency alone is not a dependable criterion of the probability of lightning-caused fires." The hypothesis that thunderstorm type alone is sufficient for fire prediction is thus not proved, and lightning counts are not directly correlated to fire starts. Rainfall accompanying safe storms was observed to last 43% longer than that from dangerous storms, according to Gisborne's data. So, are safe storms safe because ignited fires are put out by rain or because there are fewer CG flashes in them? Gisborne does not answer this question.

Morris (1934), although not proposing an ignition mechanism, came to several conclusions based on an extensive study of lightning and fire reports by fire lookouts in Oregon and Washington. Important for our exposition is this subset of conclusions:

1. ". . . No more fires per acre occur in one altitude class than in another"; that is, lightning fires, at least in Oregon and Washington, do not have a bias for high altitudes;
2. ". . . It is evident that danger zoning for lightning fires, if based solely on previous [historical] distribution of the fires is likely to be very inaccurate and unreliable";
3. "No definite lightning storm lanes or frequent 'breeding' spots were found";
4. Thunderstorm days can be classified to indicate fire-starting potential (the classes were similar to those of Gisborne (1931); and
5. Storms accompanied by high rainfall led to fewer fires than storms that had less rain.

Results 1 and 3 are contrary to anecodotal evidence that lightning strikes preferentially on ridges or peaks, rather than in valleys. Result 5 supports the "obvious"; storms often put out, through rainfall, the fires they start. We will have more to say about Results 2 and 4 later. Morris did not present IC/CG ratios or percent IC in a way usable for separating storms.

Barrows *et al.* (1977) summarized lightning-caused fire data for the years 1931–1973. For the first time, lightning fire statistics are given in terms of a stated density, rather than by arbitrary classification of land areas or by Forest Service Forest. Over this period, the average number of lightning-caused fires in the Region 1 National Forests (Montana, northern and central Idaho, and northern Wyoming) was 0.85 fires per 100 km^2/y. The standard deviation was 0.41, indicating a high variability from year to year. The density also varies over the landscape. This finding reinforces the earlier ones of this section that the lightning fire density is highly variable from year to year and place to place. No lightning statistics to compare to these numbers are available because the large lookout network studies had been discontinued, and remote lightning locations were yet to come. The percent of total fires caused by lightning is 77% averaged over the years studied. This percentage has been nearly constant since 1931.

The increase in the percentage of lightning caused fires from 15 to 77% over the years between 1912 and 1931 is curious. Plausible reasons for this increase include less and more careful logging activity, a reduction in man-caused fires, better reporting of fires in remote areas, fuel complexes becoming more fire-prone to lightning ignition due to suppression, a significant reduction in fires caused by railroad activity, and/or a general change in climate. Establishing the relative contributions of these proposed explanations is a difficult task.

All the studies in this section share one major flaw. Except for a passing remark in Gisborne's (1931) study, the tacit assumption is made of a correspondence between number of CG flashes and/or the relative amount of CG and in-cloud flashes and the number of fires that result. Rainfall is assumed to influence the number of fires through immediate wetting of the fuels either before or after ignition. The fuel type is only broadly considered, that is by predominant tree species, and fuel state is not considered. (Fuel type and fuel state are loosely defined but useful terms. Fuel type includes the biological and physical descriptions appropriate to a location. Fuel state is the moisture content of the various sizes of fuels.) Classification of thunderstorms either by IC/CG ratio or by air mass or frontal, or other schemes, was perceived to be the most useful information for prediction of lightning fire incidence.

B. PROJECT SKYFIRE

Project Skyfire began with the speculation that cloud seeding might be used to reduce lightning-caused fires—either by increasing rain or by influencing the thunderstorm to alter lightning production (Schaefer, 1949). As the observations and studies of 1912–1934 showed (see Section II.A), thunderstorms with low rainfall and low CG count were thought to be responsible for a majority of the fires in the northern Rocky Mountains, the northwest forests of Oregon and Washington, and the forests of northern California. After the cloud-seeding demonstrations of Project Cirrus (Schaefer, 1953), the Skyfire project was born in the Northern Rocky Forest and Range Experiment Station (Barrows, 1954). The aim of the project was to reduce lightning by either overseeding mature storms or by seeding early in the storm's development to initiate precipitation in the cloud. Early precipitation would cause cessation of rapid growth, reducing the amount of lightning. Enhancement of rainfall was not the aim of Skyfire, although earlier work (e.g., Gisborne, 1931) indicated that storms having high rainfall amount were accompanied by fewer fires than storms with low rainfall.

The investigation of cloud types and locations led to preliminary cloud seeding in 1956–1957 in Arizona and Montana (Barrows et al., 1958) to test the effectiveness of ground-based and airborne silver iodide cloud-seeding generators. Fuquay and Baughman (1969) found, through research carried out in Montana, that the number of CG lightning flashes might be suppressed by cloud seeding but that a larger data sample and a properly randomized experiment were necessary for a definitive conclusion.

By 1976, the final analysis of Skyfire data on randomized seeding from the 1960's was complete (Fuquay, 1974; Baughman et al., 1976). The conclusions were that cloud seeding could reduce the frequency of CG flashes by more than one-half and that the average continuing current duration might be reduced by as much as one-fourth. The possibility that reduced rainfall accompanies reduced lightning is mentioned. For lightning reduction, seeding in the early stage of cloud growth was found most effective.

A good summary of the cloud-seeding and lightning reduction experiments as well as a critique can be found in Dennis (1980), who takes issue with the analysis of Baughman et al. (1976). As with the majority of weather modification results, there is no broad consensus on the efficacy of cloud seeding for lightning suppression or rainfall enhancement. No conclusive proof was found that cloud seeding reduces either the number or severity of wildfires. As happens often in scientific research, however, the by-products of Skyfire proved at least as useful as the main product.

One of the important results was to further the conjecture (McEachron and Hagenguth, 1942; Berger, 1947; Malan, 1963; Loeb, 1966) that the continuing current might be the part of the lightning discharge responsible for ignitions in forest fuels. Relating the current behavior of an individual lightning flash to the incidence of fire at the same location is clearly a challenging task, and, for this reason, few observations are available. Fuquay et al. (1967, 1972), in two investigations, reported on 11 CG fire-starting flashes containing continuing currents varying from 40 to 280 ms in duration. They used triggered photographs of lightning as well as electric field recordings to determine the location of the flashes as well as their composition. Spotters in light aircraft searched the lightning terminus locations for fires. A ground investigation followed the aircraft observation. In addition to the fire starters, five nonstarting strikes, two hybrid and three discrete, were studied. Every fire considered was started by a flash with a continuing current longer than 40 ms. These observations led to the conclusion that earlier speculations were correct; the continuing current is the major cause of lightning-caused fires in forests.

Skyfire results also included measurements of lightning IC/CG ratios (Z values, Boccippio *et al.*, 2000) for seeded thunderstorms versus nonseeded storms. The ratios were $Z = 4.4$ for seeded storms and $Z = 3.0$ for nonseeded storms. Both of these values are higher than the Gisborne (1931) data of $Z = 2.9$ for non-fire-starting storms and $Z = 1.3$ for fire-starting storms. The Skyfire data were taken with electronic recording equipment and cameras rather than by lookouts, which might explain the difference. Boccippio *et al.* (2000) found values $Z = 3$ to $Z = 7$ for the same geographic locations. Evidently there is a high variability in the IC/CG ratio; separation of storm fire-starting effectiveness based on this ratio needs further study.

Skyfire came to a close in the early 1970s, having established that the cause of lightning fire ignitions was the continuing current in the lightning flash. Evidence was gathered that cloud seeding, if done properly, could reduce the number of CG discharges and the incidence and duration of continuing current discharges from a given storm. The research leading to the models discussed later in this chapter had its origins in Skyfire.

C. RECENT STUDIES

Recent studies of lightning-caused fires benefit from two major data sources not available to earlier researchers: organized fuel state descriptions and accurate regional CG lightning location. Fuel type and fuel state, principally moisture content in specified fuel arrays, have been combined into fire danger indices for forest managers. Generally, fuels in a given fuel type are distributed in the forest by amount of dead biomass stratified by size. Each size adsorbs water from rain and moist air. The amount of water in each size class is combined into an index indicating roughly the severity of a fire should one start. The United States uses the National Fire Danger Rating System (Deeming *et al.*, 1977), and Canada uses the Canadian Fire Weather Index System (Van Wagner, 1987). Although these systems are not identical, they are similar in that each has indices that, for various fuel types, indicate fuel moisture. The indices are routinely calculated and archived, presenting a uniform database to use with lightning locations.

Lightning location is accomplished by detecting low-frequency electromagnetic radiation emitted by lightning (Section II). Either triangulation or time-of-arrival techniques can be applied. CG (positive and negative) and IC strokes are distinguished by differences in the radiation patterns. Systems have been in use since about 1978, first in Alaska and subsequently in the United States and Canada. Early systems claimed accuracy of location in the range 1–10 km. Recent systems, using Global Positioning Systems (GPS) claim accuracy on the order of 500 m (Cummins *et al.*, 1998). Efficiency of 95% detected CG events or better is claimed for modern systems. Position accuracy and efficiency depend on the location of the CG with respect to the receiving stations in the detection network. CG locations are archived by the company that now operates the networks: Global Atmospherics, Inc., of Tucson, AZ. Several recent studies use fuel indices and lightning location together with fire reports and weather reports and are summarized later. Flannigan and Wotton (1991) studied the relationship between fires and lightning in northwestern Ontario, Canada. They found, essentially, that negative lightning ignited more fires than positive lightning and that the duff moisture code and the multiplicity of the strokes in the lightning flash, for negative strokes, were important independent variables. Although not specified in their work, it seems that stroke multiplicity has a connection because the probability of a continuing current in a flash is weakly connected to the number of strokes (Uman, 1969).

McRae (1992) found no connection between elevation, slope, aspect, or topographic unit and lightning fires for the Australian Capital Territory, supporting the much earlier findings of Morris (1934). But Van Wagtendonk (1991) did find altitude dependence for lightning-ignited fires in Yosemite National Park and justified this dependence on the grounds of vegetation type and preferred lightning storm development. Renkin and Despain (1992) found no altitude dependence for lightning fires in Yellowstone Park but did find a preponderance of lightning fires in mature stands of Douglas fir, spruce fir, and decadent lodgepole, as opposed to successional lodgepole and multiaged lodgepole stands. In addition, a threshold level of moisture in fine fuels, 13% of dry weight, was found above which fires would not continue to burn after an ignition.

Meisner (1993) studied lightning-caused fires for the period 1985 to 1991 in a small area in Southern Idaho. The lightning strikes were randomly distributed over the terrain, as were the fuel classes. The procedure used by Meisner was to break down the area by species and then calculate the efficiency on a per-species basis. Lightning-caused fire efficiencies were calculated for each 100 km^2 pixel on the map and then averaged for the fuel classes in the sampled area. The highest lightning efficiency was for logging slash, at 0.1 fires/CG flash, and the lowest was 0.003 fires/CG flash for agricultural crops. Efficiencies for mixed firs and Ponderosa pine were 0.02 fires/CG flash and 0.04 fires/CG flash, respectively. One of the findings of the study was that ". . . on days when there were 100 or more detected lightning strikes, the correlations between fire danger indices and number of fires were typically doubled." This compares nicely with Gisborne's (1931) findings summarized in Section III.A; there seem to be storm day types that correspond to high fire incidence. But according to Gisborne's observations, high CG lightning amounts result in fewer fires, the opposite conclusion. Current lightning detection systems do not presently report in-cloud flashes, so no division of storms or storm days could be made on this basis to test Gisborne's (1931) statement.

Nash and Johnson (1996) thoroughly analyzed the coupling of synoptic weather conditions with local scale weather and fuel conditions, the latter expressed in terms of the Canadian fire indices, and resulting fires in Canadian boreal and subalpine fuel types. They found, analyzing 2551 fires and 1,537,624 CG lightning flashes, that the best fuel state predictor for lightning fires was the Fine Fuel Moisture Code index. Above the moisture index of 87 (equivalent to a moisture content of 14% of dry weight; higher index value means lower moisture), very close to the 13% value found by Renkin and Despain (1992) for fires in Yellowstone National Park, the lightning-caused fire efficiency (number of fires/number of CG's in the same area) increased rapidly. The highest efficiency found was 0.06 fires/CG flash. No overall average was given, but typical values were between 0.01 and 0.04 fires/CG flash, in the same range as the Meisner (1993) values. Nash and Johnson also found that lightning efficiency of fires was higher under synoptic high pressure, when persistent rainfall was not expected. This description presents yet another "storm day type" stratification.

Rorig and Ferguson (1999) generated a discrimination rule for selecting "dry" vs. "wet" thunderstorm day occurrence in the Pacific Northwest. The rule uses the dewpoint depression at 85 kPa and the temperature difference between 85 and 50 kPa. No connection was made to fires.

For this chapter, we gathered fire and lightning locations, as well as fuel types as data for constructing map layers in a Geographic Information System (GIS). The data cover western Montana, northwestern Wyoming, and northern Idaho, as is apparent from the state outlines on Figures 1, 2, and 3 (see color insert). Summaries of lightning flash position data were obtained from the Bureau of Land Management (BLM) and Global Atmospherics, Inc., and fire data from the USDA Forest Service and Department of Interior land management agencies (primarily the BLM). Figures 1, 2, and 3 show derived spatial layers for fires, lightning density, lightning ignition efficiency, and fuel type for the years 1986–1992. In each of these figures, the fire locations are plotted as black dots. All the pixels shown on these maps are 10 km on each side (Lambert projection); the data were aggregated upward from 1-km pixels.

From Figure 1, it is evident that lightning density and fires are only loosely connected. That is, lightning is a necessary (by definition), but not a sufficient condition for ignition. Other factors, such as fuel type and moisture, must be considered. Notice that, in the very small areas of high strike density, no fires occurred over the time period covered by the maps. By inspection, roughly half

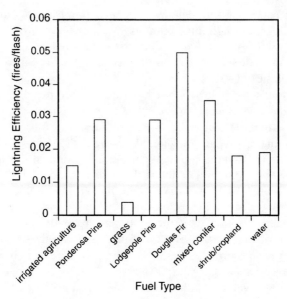

FIGURE 4   Ignition efficiency as a function of fuel type.

the fires occurred in the minimum lightning density category of 15–199 strikes/
100 km^2. The results shown in Figure 4 indicate that the average ignition efficiency over the 7-yr period is about 0.026 fires/CG flash, combining positive
and negative flashes. Because of the importance of fuel type and state, lightning
density maps cannot provide a surrogate for fire density maps, either for past
fires or for future fires, unless carefully combined with other data. This finding
echoes Gisborne's results of 1931. Unfortunately, Gisborne did not calculate an
efficiency that can be compared with the recent studies. We have not tested the
separation of storms or storm-days into safe or dangerous categories. Fuel type
and fuel state are of more importance than the incidence of lightning in determining fires (Figure 3). World fire and lightning maps (Figures 18 and 19; see
color insert) support this view. The 7-yr data span covers a wide range of fuel
states and can be considered as integrating over them. Lightning efficiency values were extracted from the map data on a 1-km pixel size. Figure 4 presents
the lightning efficiency as a function of fuel type. The highest lightning ignition
efficiencies are in the Douglas fir and spruce fir/mixed conifer fuels. Meisner's
(1993) data did not have a Douglas fir fuel type, so no direct comparison for that
fuel type can be made. His value for Ponderosa pine was 0.042 fires/CG flash
as compared to 0.029 fires/CG flash for this analysis. This discrepancy is due
at least in part to a smaller area and smaller time period in the Meisner study;
the analysis done for this chapter spans a much larger area and time period.

Nash and Johnson (1996) found ignition efficiency values ranging from 0.01 to 0.04 fires/CG flash, in the range of our analysis.

The lightning-caused fire ignition efficiency for water (Figure 4) is 0.019 fires/CG flash. This is a good example of the misleading effect of small sample size, as pointed out in Nash and Johnson (1996). The water area in our maps is 26,247 km^2 as compared to a wooded area of 690,596 km^2 and grass, shrub, and cropland area of 6,072,307 km^2. The water area is almost all in two large lakes with wooded surroundings. Reports of fire locations are often not accurate, and a small position error can place a fire on the water instead of on the shore. All these factors lead to an error that appears large because of its absurdity and comparative magnitude but that is actually very small.

More work is needed in the realm of lightning-caused fire efficiency. From the studies summarized here, the efficiency apparently depends on a multitude of factors. Among them, in order of spatial scale, are synoptic weather conditions, fuel types, thunderstorm characteristics such as rainfall and lightning rates, lightning characteristics such as positive, negative, and hybrid, and fuel state, predominately moisture content of fine fuels. Broad agreement is evident among the studies that the efficiency values for fuels and fuel conditions of interest are roughly between 0.01 and 0.05 fires/CG lightning flash. The next sections will explore the details of the interaction of lightning with fuels and the physics of the ignition process to explain the low efficiencies.

IV. INTERACTION BETWEEN LIGHTNING AND FUELS

A. LIGHTNING PATHS ON TREES

The concept of "upward" and "downward" discharge directions (Plummer, 1912; Section III.A) was generated by observations of termination of the discharge path at some specific point on the trunk of the tree. This is caused by flashover, a phenomenon that occurs on towers, electric poles and other elevated structures including trees (Darveniza, 1980). That is, at some position on the tree or other conductor, the discharge "decides" to go to ground through the air rather than continue along the conductor. The details of this phenomenon are not well understood, especially with respect to trees. But the fact that it is not universal, that sometimes the discharge does proceed all the way down the tree (or conductor), even out the root structure, could have been the source of Plummer's (1912) hypothesis of up-going and down-going strikes. The actual physics of the lightning-tree-ground interaction during a lightning discharge has not been thoroughly explained.

Plummer (1912) proposed that an increase of tree conductivity due to wetting by rain had a part in ignition. The lightning strike path usually follows the cambium layer of the tree because it has higher conductivity than either wet bark or the woody interior of the tree (Defandorf, 1955). Cambium cells transport the water necessary for the tree's operation and exhibit higher conductivity parallel to the direction of growth of the tree as opposed to perpendicular to that direction (Du Moncel, 1877). Lightning scars on trees almost invariably feature blown-off strips of bark, leaving a mark on the trunk showing that the path was under the bark. Also, the path of the strike is very often spiral, following the spiral long axis of the cambium cells. Sometimes, of course, the lightning path is straight down the tree and/or on the bark surface (Taylor, 1973).

B. ECOLOGICAL DAMAGE NOT INVOLVING FIRE

Many lightning strikes to trees do not result in fire. In fact, a strike to a tree often causes only mechanical damage, up to and including complete destruction of the tree (Plummer, 1912; Taylor, 1969). This damage could be caused by the rapid gas expansion in a return stroke, by the sudden creation of steam by a return stroke in damp rotten heartwood from a continuing current, or by both. Only the aftermath is available to show the effect, and a mechanism has not been truly identified. No statistics are available for the proportion of strikes to trees that cause damage or kill the tree without starting a fire. Because it is weakened by the strike, a lightning-struck tree can also act as a vector for insect and disease infestation, leading to insect destruction of the struck tree and infestation of trees surrounding it (McMullen and Atkins, 1962; Schmitz and Taylor, 1969). Also, there are instances, especially in the southern United States, of group-kills of trees caused by root destruction from the ground path of a discharge to one tree. Taylor (1969) provides an excellent summary of lightning effects on the forest that do not involve fire.

C. FUELS ON THE FOREST FLOOR

Anecdotal evidence implies that lightning ignition of forest fuels takes place almost exclusively in the fine fuels on the forest floor. Typically, in the forests of the northwestern United States and other forested areas of the world, lightning strikes a tree and, although the connection process for lightning to ground is complex, the result is the establishment of a path down the trunk toward the ground (Figure 5). Sometimes, the strike will flash over to ground, at a height of a meter or two. The reason for this behavior is not known. The breakdown path to mineral soil, which can be considered "ground," passes, in either case,

FIGURE 5   Lightning path on a tree (Plummer, 1912).

through the fine fuels on the forest floor. Ignitions occur along this path in the litter and duff layers on the forest floor (Fuquay *et al.*, 1967, 1972; Taylor, 1969). The composition of litter and duff layers varies considerably from place to place in forested and grassland areas, depending on the local ecosystem. For example, Ponderosa pine has, as do most pines, a needle very long in comparison to its diameter. Because of this, the litter layer in ecosystems in which this species is dominant is relatively thick compared to the duff layer. Firs, on the other hand, have needles that are much shorter, making for a compact duff layer and a very thin, almost nonexistent, litter layer.

Occasionally, the stroke will pass through the moist heartwood of a dead tree (snag) or a live tree. When that happens, the tree may be blown apart (Figure 6), with or without accompanying fire. Fires often burn from the inside of snags that are not blown apart to the forest floor fuels, causing holdover fires (fires that smolder for long times before they are discovered; Frandsen, 1987).

V. HOW IGNITION OCCURS

A. FORMULATION OF THE PROBLEM

One of the most important of the Skyfire findings was identification of the continuing current as the primary cause of lightning ignitions of forest fuels

FIGURE 6   A tree demolished by lightning (Plummer, 1912).

(Section III.B). The field studies of Taylor (1969) verified that burning debris blown off trees from discrete strokes onto litter or duff fuels does not cause ignition, and Fuquay (1974) formed the conjecture that the duration of the continuing current might be important in the ignition efficiency. This informa-tion formed the basis for a model to predict fire density (fires/100 km^2) given the presence of lightning and the type and state of the fuel. That is, real-time and predicted data from reliable, convenient, and inexpensive sources, such as weather nowcasts and forecasts, and lightning location networks, would be used with algorithms to predict the incidence of fires on the landscape.

The task, then, was to form a construct or set of questions for the model. The questions are:

- What is the thermal and radiating structure of the continuing current?
- What is the energy required to ignite fine fuels?
- Is radiation or convection the predominant mode of energy transmission from the continuing current channel to the fuel?
- What experiments are necessary to establish usable parameters?
- If a predictive model is found, what is the set of data needed to imple-ment the model for practical use?

- How should predictions be integrated with existing fire systems such as the National Fire Danger Rating System?

B. Energy Transfer from the Lightning Channel

The basic framework for ignition modeling is shown in Figure 7. A cylindrical conducting channel of hot gas, whose characteristics such as diameter and temperature are determined by a suitable physical model (Section V.C), passes into a layer of uniform fuel overlaying an electrically conductive soil layer that is the electrical ground terminal for the current in the channel. General assumptions for this framework are (1) temperature profile consistent with a hot gas arc channel model, (2) vertical penetration of the fuel, (3) infinite channel height (or at least very much longer than the depth of the fuel), and (4) no effects from horizontal current flow in the soil. Once the channel structure, particularly temperature, is established, energy transfer to the fuel depending on the mode of transfer will be calculated, with surface element A the location of maximum radiation transfer from the arc column and surface element B the location of maximum convective transfer (Figure 7). Conductive heat transfer, by diffusion within a medium, does not apply to heat transfer from the hot gas channel to the fuel (Incropera and DeWitt, 1996). Conduction does play a role in the structure of the arc channel.

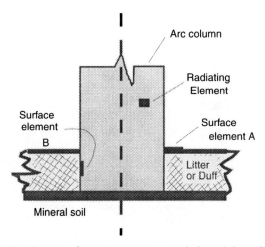

FIGURE 7   Schematic of a continuing current path through forest floor fuel.

C. Structure of the Continuing Current Channel

Details of the thermal and radiation structure of the continuing current channel were calculated using hot gas models of an electric arc channel. That is, energy balance and conservation of mass equations, together with Ohm's Law, are
solved with appropriate boundary conditions. These equations were solved numerically by Uman and Voshall (1968) for a channel with no current flow such
as a return stroke decay channel and by Latham (1986) with electric current
flow. The latter model included radiation cooling and full physical hot gas characteristics under cylindrical symmetry. Few measurements are available for long
arcs. Electrode effects dominated most measured arc characteristics (Strom,
1946; King, 1961). As a consequence, channel characteristics are poorly defined
for validating models of arcs in regions away from electrodes, the relevant situation with lightning continuing currents.

Latham's (1986) model will be used to give the details of the modeled continuing current channel. The assumptions are relatively simple and have been
justified (Uman and Voshall, 1968; Latham, 1986):

- Radial symmetry, giving a cylindrical arc;
- Unit length of channel, hence no axial dependence; and
- Optically thin gas, that is all radiation from the hot gas escapes the
  channel and is not captured and electron and ion temperatures are
  the same.

Values for hot gas physical constants as a function of temperature were available from numerous research studies (see Latham, 1986, for the extensive list).
Using the assumptions given here, the energy conservation equation for an element of the gas in cylindrical coordinates is

$$\rho c_p(\partial T/\partial t) + \rho c_p u(\partial T/\partial r) = 1/r\{[\partial/\partial r][rk(\partial T/\partial r)]\} + \sigma E^2 - S \qquad (1)$$

The conservation of mass requires

$$(\partial \rho/\partial t) + (1/r)[(\partial/\partial r)r\rho u] = 0 \qquad (2)$$

Ohm's Law stated for the gas column is

$$j = \sigma E \qquad (3)$$

The equation of state for the hot gas is

$$p = \rho RTZ \qquad (4)$$

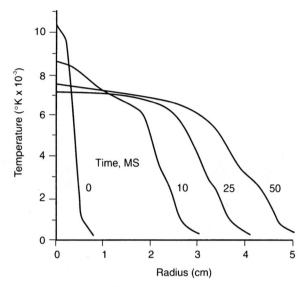

FIGURE 8   Temperature profiles of the theoretical channel as time progresses.

Boundary conditions are stated at $r = 0$ and at $r = \infty$ as

$$r = 0, \qquad \partial T / \partial r = 0; \qquad r = \infty, \qquad T = T_{\text{ambient}} \qquad (5)$$

where ρ is gas density, u is radial velocity, r is channel radius, T is gas temperature, k is thermal conductivity, Cp is specific heat at constant pressure, Z is compressibility, p is pressure, R is gas constant, σ is electrical conductivity, E is electric field, j is current density, S is radiation density, and t is time.

The initial conditions are given as a radially small (0.5 cm) high-temperature (12000 K) pulse at $t = 0$ (Figure 8), simulating the return stroke channel.

Combining Eqs. (1)–(4) describing the channel behavior results in a highly nonlinear system. Various mathematical attacks to solve the equations in closed form were not successful. As a result, finite difference computer computation was applied. As usual, when a finite difference approximation is used, some conditions on the problem cannot be actualized. In the problem at hand, we cannot integrate the equations to infinite radius. Initially, the radial boundary was placed at 100 cm for a one cm channel, leading to extensive calculation times. A boundary condition, derived from the equations, circumvented this problem (Latham, 1986), allowing more rapid calculation. With the boundary conditions properly handled, temperature profiles at successive times from 0 to 50 ms for an unbounded channel result (Figures 8 and 9). The results can be summarized qualitatively as follows: energy balance in the interior of the channel is dominated by the thermal conductivity of the hot air in the channel. On the

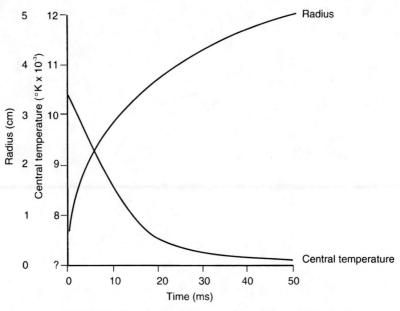

FIGURE 9   Channel central temperature as a function of time.

other hand, the radial growth rate of the channel is determined by the conductivity contrast between the ambient air and the hot channel because the edge of the channel moves outward by conductive heating at the channel edge. Basically, the radial structure at the channel edge is independent of the interior temperature (Figure 8). Also, if the current ceases, the channel temperature will decay, but the radius will continue to expand, as in the model of Uman and Voshall (1968).

Grieg et al. (1985), through observations coupled with calculations, maintained that turbulent convective mixing causes much faster (a factor of 2000) channel collapse than conduction. The expansion and collapse of that arc may resemble the collapse of the much higher energy return stroke channel. High-speed images of the arcs used in the experiments referred to in Section VI show no turbulent breakdown, but the energy densities are lower. Evidently, more exploration is necessary concerning the transition from stroke channel to continuing current channel.

Measurements have been made in other studies that support the model, at least in the prediction of the temperature of the arc. Latham (1984) presented an analysis showing that the ratio of the N_2^+ first negative radiation, an excited state of the nitrogen molecule, and the cyanogen (CN) radiation over the range 6000–9000 K can be used to estimate the channel temperature (Figures 10 and 11). Salanave (1965) and Connor (1967) reported strong radiation at these wavelengths, although the two wavelengths were not differentiated. An example

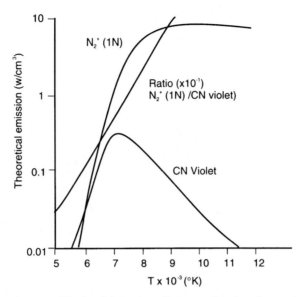

FIGURE 10 $N_2^+(1N)$ and CN violet radiation as a function of temperature.

of the ratio measurement is given in Latham (1984); the ratio gives a temperature of about 6500 K that corresponds to the model (Figure 12). The author has made no further measurements, but validation of the model should be conducted with further spectral measurements during the continuing current phase of CG flashes.

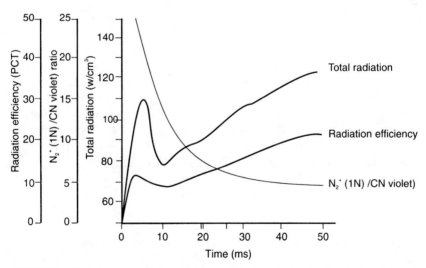

FIGURE 11   Radiation and radiation efficiency as functions of time for the calculated channel.

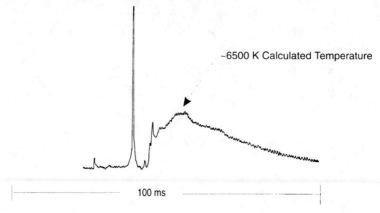

~6500 K Calculated Temperature

|—————————————— 100 ms ——————————————|

FIGURE 12 A measured $N_2^+(1N)/CN$ intensity ratio for a lightning stroke with continuing current.

D. Requirements for Fuel Ignition by Lightning

As stated earlier, forest floor fuels that are most susceptible to lightning ignition fall into three rough categories: loose, unconsolidated fuels with low bulk densities (for example Ponderosa pine or lodgepole pine litter); tightly packed small fuel particles (such as Douglas fir); or consolidated rotted fuels (such as peat). That peat can be ignited by lightning at all might be surprising at first, considering its water content and density. Not only do these ignitions occur, for example in the Florida Everglades, but the resulting fires can also smolder for long periods of time. This behavior should be explained in our model structure.

Burning (combustion) actually takes place as diffusion flames in the complex hydrocarbon gases that are "cooked" out of the fuel by heating, a process called pyrolysis. Pyrolysis is generally dependent on the temperature of the fuels (e.g., Susott, 1984). Energy transferred to the fuel by heat processes must raise the temperature to a level sufficient to initiate pyrolysis. Calculation of this temperature requires the construction of an appropriate heat (conservation of energy) equation for the heated object. These equations are usually very complex, and almost always nonlinear, because of the geometry involved and because radiation effects are proportional to T^4 (Cox, 1995; Incropera and DeWitt, 1996).

Several researchers have proposed ignition criteria for forest fuels (Martin and Broido, 1963; Simms and Law, 1967; Susott, 1984; Jones *et al.*, 1990). In general, the criteria are given in two ways: in terms of the temperature of particles at the time flaming begins and in terms of the energy flux and duration necessary to generate flaming in the material. Two ignition conditions are possible—spontaneous and piloted. In spontaneous ignition, a heat pulse is applied to the material and either ignition occurs or not. In piloted ignition, the material is subjected to either a steadily increasing or constant heat source for durations longer than a few seconds, preheating the material, and then subjected to a pulse or increase in heat. Again, ignition either occurs or not. Ignition by lightning is an example of spontaneous ignition, whereas a propagating fire is an example of piloted ignition (Cox, 1995).

In our analysis, we will rely on measured ignition criteria for short radiation pulses as a minimum of about 1.67×10^5 W m^{-2} for a 500 ms pulse (Martin and Broido, 1963), supported by measurements made by Simms and Law (1967). These ignition criteria ignore problems having to do with moisture content, physical arrangement, and the like (which act to increase the needed ignition flux) and allow us to explore the possibilities for various modes of heat transfer. The cylindrical model of Section V.C can be used to calculate theoretical heat fluxes associated with radiation and convection.

E. THEORETICAL HEAT FLUX CALCULATIONS

Referring to Figures 7 and 13, the radiation flux at patch A can be calculated. Establish a cylindrical coordinate system with its origin in the plane containing A, with the z-axis vertical and coinciding with the axis of a cylinder of hot gas with temperature T (K) and radius a (cm). Assuming a "top hat" profile for

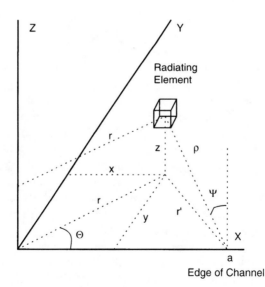

FIGURE 13   Coordinate layout for radiation and convection calculations.

the channel, if S (W m^{-3}) is the total emission per unit volume from the hot gas at constant temperature and is zero outside the cylinder, the flux at A is

$$F_A = \int_0^a \int_0^{2\pi} \int_0^\infty \frac{Sr \cos \psi \, dr d\vartheta dz}{4\pi r^2} \tag{6}$$

the variables in the problem being shown in Figure 13. With that geometry, the equation can be restated as

$$F_A = \frac{S}{4\pi} \int_0^a \int_0^{2\pi} \int_0^\infty \frac{rz \, dr d\vartheta dz}{(r^2 + a^2 + z^2 - 2ar \cos \vartheta)^{3/2}} \tag{7}$$

Integrating over z and invoking symmetry on ϑ leaves

$$F_A = \frac{S}{2\pi} \int_0^a \int_0^\pi \frac{r \, dr d\vartheta}{(r^2 + a^2 - 2ar \cos \vartheta)^{3/2}} \tag{8}$$

This integral has a singularity at $r = a$. Moving A just a little way radially from $r = a$ to avoid the singularity gives a value for the integral of $F_A = 0.3aS$ W m^{-2}.

Proceeding in the same way for a patch at B placed at $z = 0$ gives a flux of

$$F_B = \frac{S}{4\pi} \int_0^a \int_0^{2\pi} \int_0^\infty \frac{r(a - \cos \vartheta) dr d\vartheta dz}{(r^2 + a^2 + z^2 - 2ar \cos \vartheta)^{3/2}} \tag{9}$$

which reduces to $F_B = 0.5aS$ W m^{-2}.

For the present purpose, then, we take the flux from the column to be about $Sa/2$ W m^{-2} and assume some values for a growing channel. At a generously high gas temperature of 8000 K (Figure 9), the radiation heating at a radius of 1 cm is about 1×10^5 W m^{-2}, or 5×10^4 J m^{-2} for a 500-ms pulse. This value is less than the very generous required minimum energy for ignition of 8×10^4 J m^{-2} for a 500-ms pulse as established in Section V.D. Thus, although radiation heating is definitely present, it is unlikely to be a dominant part of the ignition of forest floor fuels by lightning. The relative weakness of radiation as an ignition source was also seen in the scenario developed for ignition of forest fuels by meteor fragments in Melosh et al. (1990).

Convective heating may be estimated for small-diameter fuels, such as pine needles, that are bathed in the gas flow of the arc. Suppose the gas temperature is 8000 K as in the previous example (Latham, 1986). The gas density is $0.0235 =$ kg m^{-3}, the viscosity about 1.5×10^{-3} poise (Adams, 1966). The kinematic viscosity is thus 6.4×10^{-4} m^2 s^{-1}. The diameter, d, of a pine needle from a long-needled species, such as Ponderosa pine, is about 1 mm. The velocity of the gases in an arc discharge such as a continuing current is about 2000 m s^{-1} perpendicular to the cross section of the arc (Strachan and Barrault, 1975). The Reynolds number of the gas flow is thus about 30, and the Nusselt number, Nu, is about 3 (Incropera and DeWitt, 1996). The thermal conductivity, k, of hot air at 8000 K is 2.44 W K^{-1} m^{-1} (Adams, 1966), and the heat transferred to the fuel element is about $(T_g - T_a)*$Nu$*k/d$ or 5.5×10^7 W m^{-2}, where T_g is the gas temperature and T_a is the ambient temperature. The resultant energy deposited in 500 ms would be 2.5 J m^{-2}. The convective heat transfer is roughly three orders of magnitude larger than the radiation heating calculated earlier.

If a more solid fuel is involved, such as a duff layer, consider the wall of the fuel to be the wall of a pipe with a hot gas in turbulent flow in the pipe. Calculation for heat transfer to the wall of a pipe using the same values as for small fuel heating gives a flux at the fuel surface of about 1×10^5 J m^{-2}, again larger than the heat flux from radiation as calculated above. Under the reasonable assumptions made, the importance of the duration of the current rather than its amplitude is clear.

The scenario for ignition of forest floor fuels can, then, be summarized as follows. A return stroke, or perhaps a junction process arc establishes a path through the fuel. The high-current (tens of kiloamperes) initiating arc is followed in some cases by a smaller continuing current on the order of 100 A, lasting tens to hundreds of milliseconds. The continuing current arc, which cannot shrink, expands continuously into the fuel. The center portion of the arc, under any scenario, is hot enough to ablate completely the material in its interior, and only gas is left behind. As the continuing current stops, the central temperature in the channel decays, and the channel continues to expand. The ring of unablated material at the channel edge either has received enough energy through combined radiation and convection to maintain the ignition or not.

These approximate calculations serve to show the complexity of heat transfer from lightning continuing currents to forest floor fuels. They also lead to the conclusion that theoretical solutions of the ignition problem cannot give results sufficiently accurate for routine operations. Darveniza and Zhou (1994), having done some impulse and arc heating studies using high-voltage arcs, suggested, "It was found that the power and energy transferred to the fuels from per centimeter of the arc were much greater than those of the arc itself [sic] and were current and fuel properties dependent." This statement appears to support our view that convection transfer is more important than radiation transfer. We had already turned to experiment in 1988 to lay the foundation for a useful product, as is described in the next section.

VI. IGNITION EXPERIMENTS
WITH REAL FOREST FUELS

A. APPARATUS AND EXPERIMENTAL DESIGN

As we have shown, the interaction between the hot gas channel of the lightning
continuing current and the fuels through which it passes is extremely complex.
Consequently, a series of experiments was carried out to form a model that
would allow calculation of ignition probabilities from easily obtained data.

An arc generator was constructed to simulate, as closely as possible, the discharge of a lightning continuing current. The arc generator used 42 12-V truck
batteries connected in series, controlled by a silicon-controlled rectifier switch
and initiated by either an exploding tungsten wire or by drawing the arc with a
rapidly moving carbon electrode (Latham, 1987, 1989).

In all, eight types of fuel samples (Table 1), taken from the wild and representative of most forest floor types, were subjected to a range of arcs. A series
resistance in the arc circuit controlled the current, and a computer-controlled
silicon-controlled rectifier switch controlled arc duration. The moisture content of the fuels was varied between 5 and 40% of dry weight. In all, 1600 trials
were conducted with a current range of 20–500 A and durations from 10 to

TABLE 1   Coefficients of the Probability Equation for Fuel Types
Used in the Ignition Experiments

Fuel type	A	B (1/ms)
Ponderosa pine litter	$0.97 - 0.19*Mf$	0.012
Punky wood (rotten, chunky)	$-0.59 - 0.15*Mf$	0.005
Punky wood powder (4.8 cm deep)	$1.2 - 0.12*Mf$	0.002
Punky wood powder (2.4 cm deep)	$0.13 - 0.05*Mf$	0.005
Lodgepole pine duff	$-5.6 + 0.68*d$	0.007
Douglas fir duff	$-7.1 * 1.4*d$	0.006
Englemann spruce mixed (high-altitude mixed)	$0.79 - 0.081*Mf - 8.5*\rho b$	0.011
Peat moss (commercial)	$0.42 - 0.12*Mf$	0.005

Mf is fuel moisture, % of dry weight; d is depth, cm; and ρb is bulk density
(g/cm^3). The coefficient A is dimensionless, and B (ms^{-1}) multiplies the
continuing current duration in milliseconds.

500 ms; the arc could not be maintained below 20 A, and 10 ms was a lower limit on the switch and arc initiation apparatus. A maximum current of 500 A and duration of 500 ms were chosen as high enough that extremely few continuing currents would exceed them.

Defining ignition was difficult. Pyrolysis of biomass materials unquestionably occurs at the edges of the lightning continuing current path, since pyrolysis results from temperatures near and above about 600 K (Susott, 1984), and the 7000 – 8000 K temperatures in the body of the arc (Figure 8) ablate the center material. Pyrolysis products from the material at the edge of the arc burn as linear diffusion flames, in which combustion depends on local fuel–air mixing rather than premixing as in a Bunsen burner flame (Cox, 1995).

Operationally, the difference between an ignition and a fire simply depends on whether or not the fire is seen and/or reported. Practically speaking, an ignition has taken place if the ignition source is sufficient to produce a self-sustaining combustion process once the source is gone. For the experimental results presented here, if the sample totally burned, or if smoldering continued after 5 min, an ignition was recorded.

B. EXPERIMENTAL RESULTS

The experiments were designed to develop relationships between the independent variables and the probability of ignition. Independent variables for each fuel type for the ignition tests were arc current, arc duration, fuel moisture content, fuel bulk density, and fuel depth. For each set of conditions for a given fuel type, the result was either an ignition or not, a binary value. The results were tested using logistic regression, a systematic way of turning a binary result, such as the presence or absence of an ignition, into a probability of occurrence of the phenomenon (Loftsgaarden and Andrews, 1992). Typical results are plotted in Figures 14 and 15. The former is from one series of tests on Ponderosa pine litter; the latter is from Douglas fir duff. The regression process eliminated unimportant independent variables for each fuel type. One variable, arc current, was found unimportant, and arc duration was found to be most important, as conjectured by Fuquay et al. (1972). The unimportance of the arc current arises from the temperature structure of the arc channel, as developed in Section V.E. The most significant independent variable, after duration, is plotted on the ordinate for each of the fuel types. Lines of equal ignition probability as determined from the regression are also shown.

Figure 14 presents the results of tests on Ponderosa pine. The second most significant variable for this species was moisture content. Generally, as the moisture content increased, more energy, that is a greater current duration, was required to obtain ignition, as the regression probability lines indicate. No data

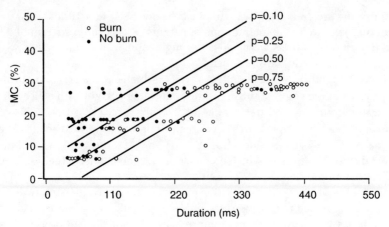

FIGURE 14   Ignition experiment data for Ponderosa pine needle fuel.

could be taken for moistures in excess of roughly 30% of dry weight because the woody material could not hold more moisture due to fiber saturation. Near fiber saturation, currents in excess of 400 ms invariably ignited the sample, and at the lower limit of the experiment, 8% moisture, currents of 50-ms duration ignited these fuels about 50% of the time.

Figure 15 shows the results of ignition tests on Douglas fir duff. For this species, the predominant secondary variable was, surprisingly, the depth of the fuel. Duff shallower than 2.8 cm did not ignite, and deeper samples ignited at small current duration, 50 ms.

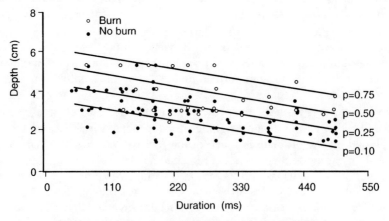

FIGURE 15   Ignition experiment data for Douglas fir duff beds.

An examination of samples that did not ignite demonstrated that the material in the high-temperature portion of the arc was ablated, as predicted from the temperatures calculated in Section V.E (Figure 8). Almost all samples with some soil showed fulgurites; glass was created from the melting of sand grains (2000 K) in the soil and dispersed through the fuels. The edges of the arc path through the fuels were charred. Clearly, the important mode of energy transfer was, as indicated previously by magnitude comparison, convective transport to fuel particles at the edge of the arc, as surmised in Section V.E. This means that the energy transferred to the material depended on the temperature structure of the edge of the arc, roughly the same for all central core temperatures (Figure 8) and durations.

The difference in the results implies a difference in heating mechanism between the more compact fuels, such as short-needled species duff (Figure 15), and the less compact litter of the long-needled species (Figure 14). In fact, the ignition probability for the former fuels did not depend on the fuel moisture or bulk density, and the latter depended on the moisture but not on bulk density or on depth.

Long-needled species litter is loosely arranged with large air spaces compared to the diameter of the needles. Short-needled species such as Douglas fir, on the other hand, have almost no litter layer, and the air spaces in the duff layer are small compared to the needle diameter. For this reason, the convection transfer to long-needled litter takes place to individual particles, and the transfer to short-needled litter takes place more in the form of the less-efficient wall transfer (Section V.E). Figure 15 shows that for Douglas fir duff less than about 2.8 cm deep, no ignitions occurred, even at fuel moisture content as low as 8% (a "very dry" duff moisture). So there may be an effect, in this dense fuel, of needing a "tube" to somehow constrain the arc in order to form a kind of muffle furnace effect (see also Chapter 13).

Ignition of spruce–fir samples having a mixed structure was dependent on both moisture and bulk density. No ignitions were observed for current durations on the order of 10 ms, and only one, in Ponderosa pine litter, at that value for fuel moisture of about 7%. Equipment limitations prevented current durations less than 10 ms. It is possible that, at least in some fuel types, currents shorter in duration than 10 ms can cause ignitions. Certainly many continuing currents are in the neighborhood of this duration (Thomson, 1980). Trends in the data for all species do show that as the current duration decreases, ignition probability decreases, and zero current should give no ignitions.

Frandsen (1987) showed that commercial peat moss, a reasonable surrogate for duff, could sustain smoldering combustion at moistures of 93–103% of dry weight. So the moisture content of the short-needled duff seems to be relatively unimportant in sustained combustion as well as in ignition. Meisner (1993) determined that the ignition efficiency in mixed fir fuels was 0.024 fires/CG flash, and the efficiency for Ponderosa pine fuel type was 0.042 fires/CG flash, about twice as large. This may be explained by the relative efficiency of heat transfer in the two fuel types. Further explanation would require further experiments. Application of logistic regression to the existing experimental results does, however, produce a useful application.

The form of the logistic regression equations obtained from the data is

$$pci(A, B, t) = [1 + \exp(-A - Bt)]^{-1} \qquad (10)$$

where $pci(A, B, t)$ is the probability of ignition. A and B are coefficients depending on the experimental material, where A is dimensionless and B has dimension (ms^{-1}), and t is the continuing current duration in milliseconds (Table 1). Unfortunately, logistic regression implies a finite probability of ignition with zero continuing current duration, inconsistent with the lightning observations discussed in Section V.A. There may be a technique or formulation that will allow rectification of this shortcoming.

The consequences of the regression artifact are slight. The number of ignitions compared to nonignitions in the experiments for current duration near 10 ms is very small (see Figures 14 and 15). Also, the observations in Fuquay *et al.* (1972) show no fires caused by continuing currents with durations less than 40 ms and an average duration of 165 ms for those continuing currents starting fires. After an ignition event, the probability of fire growth can be obtained from fire data (Latham, 1979).

VII. GENERATING MODELS FOR OPERATIONAL USE

Using experimental results for operations requires knowledge of not only the fuel parameters for the locus of a strike but also the presence and duration of a continuing current in the strike. The fuel parameters are accessible in existing operational databases, and the location of the flash is known from lightning network data (to a claimed accuracy of 500 m), at least in the United States and parts of Canada.

A. Criteria for Utility

Although ignition models can provide ignition probabilities for most of the fuels found in forested and grassy landscapes, the models still have to be accommodated to reality. According to the models as presented, we can obtain an ignition probability if we know the kind of fuel present, its state (e.g., depth, moisture content, and bulk density), and the presence and duration of a continuing current.

Fortunately, lightning location systems now cover the United States and parts of Canada. Unfortunately, the location data cannot give either the duration of a continuing current or even the presence of one. ELF methods are well suited to identifying and quantifying the continuing current (Burke and Jones, 1996; Huang *et al.*, 1999), but these methods are presently not applicable to operations. There are, however, statistics for the probability distribution of the duration of continuing currents in both positive and negative discharges, as well as the probability that a flash has a stroke with a continuing current (Latham and Schlieter, 1989). The statistics we use are valid for thunderstorms in the northern Rocky Mountains.

B. PROBABILITY DISTRIBUTIONS FOR CONTINUING CURRENTS

Application of the ignition equations requires knowledge of continuing current durations on a flash-by-flash basis. The statistics of continuing current durations have been generally thought to be lognormal (e.g., Cianos and Pierce, 1972), although Thomson (1980) found that a lognormal distribution assumption for continuing current durations failed a χ^2 test. Other distributions might work as well and are more tractable for mathematical manipulation. The Weibull is a two-parameter distribution that is the statistic of choice for failure of a whole unit contingent on the failure of a part or parts of the whole (Hahn and Shapiro, 1967). This viewpoint is attractive in that it might correspond to a continuing current channel fed by subchannels in the cloud. Weibull and lognormal distributions were each fit to continuing current duration data from the Skyfire experiments (Fuquay and Baughman, 1969), covering 141 negative hybrid flashes (duration between 20 and 520 ms) and 54 positive flashes. The two resulting distributions were tested according to a method developed by Kappenman (1988). The Weibull distribution was found to better describe the measurements. According to the Weibull fit, the distribution

$$p(t, n, s) = \left(\frac{n}{s}\right)\left(\frac{t}{s}\right)^{n-1} \exp\left[-\left(\frac{t}{s}\right)^n\right] \tag{11}$$

generates probability distribution functions, p, for continuing current duration, t in milliseconds, with the parameters $n = 1.6$, $s = 207.8$ for negative currents, and $n = 2.3$, $s = 69.3$ for positive currents. These coefficients have no physical interpretation. The distributions are shown in Figure 16 and are valid for thunderstorms in the northern Rocky Mountains. Thomson's (1980) results for Port Moresby show that there are certainly continuing current durations less than

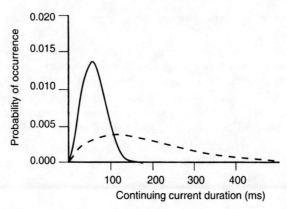

FIGURE 16 Probability distributions for negative (dashed) and positive (solid) continuing currents.

20 ms but, as the experiments summarized in Section VI show, extremely few ignitions are observed for durations less than 20 ms and none below 10 ms for laboratory experiments.

C. Conditional Probability of Ignition

Because the probability distribution function for continuing currents is known, ignorance of the characteristics of any given continuing current can be dealt with using conditional probability. The probability of occurrence of an ignition is dependent on several parameters. If a statistical description of one of the parameters, such as current duration, is known, the parameter can be effectively removed from the parameter list by integrating the probability of occurrence over the probability distribution of the parameter. In the present case, integration allows calculation of the probability of ignition per continuing current for known fuel parameters. The probabilities of ignition were given in Table 1, and the probability distributions for continuing currents in Section VII.B.

The conditional probability for lightning ignition is calculated using the integral

$$p_i(n, s, A, B) = \int_0^\infty p(t, n, s) \, pci(A, B, t) \, dt \tag{12}$$

where $p_i(n, S, A, B)$ is the probability density for ignition per continuing current event (between 0 and 1), A and B are coefficients from Table 1, and t is du-

TABLE 2   Conditional Probabilities for Ignition of Fuel Types
Used in the Ignition Experiments

Fuel	Pi neg	Pi pos
Ponderosa pine litter	$1.04\ \exp(-0.054Mf)$	$0.92*\exp(-0.087*Mf)$
Punky wood (rotten, chunky)	$0.59*\exp(-0.094*Mf)$	$0.44*\exp(-0.11*Mf)$
Punky wood powder (4.8 cm deep)	$0.9*\exp(-0.056*Mf)$	$0.86*\exp(-0.06*Mf)$
Punky wood powder (2.4 cm deep)	$0.73 - 0.011*Mf$	$0.6 - 0.11*Mf$
Lodgepole pine	$1.0/[1 + \exp(3.84 - 0.6*d)]$	$1.0/[1 + \exp(5.13 - 0.68*d)]$
Douglas fir	$1.0/[1 + \exp(5.48 - 1.28*d)]$	$1.0/[1 + \exp(6.69 - 1.39*d)]$
Englemann spruce (subalpine)	$0.8 - 0.014*Mf$	$0.62*\exp(-0.05*Mf)$
Peat moss (commercial)	$0.84*\exp(-0.058*Mf)$	$0.71*\exp(-0.72*Mf)$

Mf is fuel moisture, % of dry weight, and d is depth, cm.

ration in milliseconds. The final form of this ignition probability, in approximation, is given in Table 2.

Based on similarity of forest floor and tree characteristics, the results of the conditional probability estimates can be applied to other species (Latham and Schlieter, 1989).

D. IMPLEMENTATION OF IGNITION PROBABILITY

Figure 17 shows one way to implement the ignition probabilities into an operational system. In this scheme, a Geographic Information System map layer for ignition probability per positive and per negative lightning flash is generated using the algorithms developed in the last section. As the lightning location data come in, the location is binned into pixels that correspond to the ignition probability layer. The combination of the two becomes a projected fires layer. This scheme has not been implemented. A somewhat simpler scheme is presently employed. Lightning locations are aggregated into 1-hr files. The files are made available to dispatch users. Since the locations are points, they are used as an overlay on many different maps, including ignition efficiency, fuel type, terrain, or any other desired layer. Implementation of more sophisticated schemes is a goal for future studies.

Ignition probability information can also be used in gaming and forecasting as well as combined with forest growth programs. These uses have yet to be implemented.

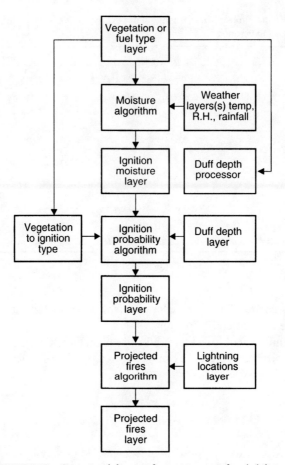

FIGURE 17   Operational diagram for ignition use of probabilities.

VIII. SMOKE, LIGHTNING, AND CLOUD MICROPHYSICS

Large-scale networks for cloud-to-ground lightning detection have disclosed an intriguing coupling between forest fires and the electrification of thunderstorms within their range of influence. As emphasized in Section II, the great majority of ordinary thunderstorms produce ground flashes with negative polarity. Within the last 20 years, several authors (Vonnegut and Orville, 1988; Latham, 1991; Vonnegut, *et al.*, 1995; Lyons *et al.*, 1998a) have documented a shift from negative to positive ground flash prevalence in association with forest fires and forest fire smoke.

At present, it is not known whether a single mechanism will afford an explanation for these several observations with different attendant conditions. When aerosol in fire smoke is ingested in a convective storm, complexity rises in areas of chemistry, cloud microphysics, and electrification. In line with these complications, multiple explanations have been put forward for the observations cited in the previous paragraph. Latham (1991) suggested a thunderstorm whose main dipole is inverted with respect to the usual positive dipole. Vonnegut *et al.* (1995) made a case for an influence-charging mechanism, with ingestion of negative space charge in the fire smoke (Latham, 1999) leading to an inverted polarity cloud. Lyons *et al.* (1998a) also considered an influence mechanism to explain their observations but discounted it because the anomalous thunderstorms they observed, though ingesting fire smoke, were more than 1000 km from the source of the smoke, casting doubt on the persistence of appreciable space charge. Lyons *et al.* (1998a) did call attention to the effect of aerosol smoke on cloud condensation and on the size of cloud droplets, with more numerous smaller droplets accompanying smoke ingestion by the cloud, an effect documented at the cloud scale in satellite observations (Kaufman and Nakajima, 1993; Rosenfeld and Lensky, 1998), and more recently by *in situ* observations (Reid *et al.*, 1999). Recent laboratory simulations (Avila *et al.*, 1998) suggest that strong positive charging of graupel particles might occur in the presence of small cloud droplets, resulting in an inverted polarity cloud, perhaps conducive to positive ground flashes. Harvey and Edwards (1991) theorized that the collection efficiency of smoke by droplets is enhanced if the smoke particles are charged.

Clearly, *in situ* measurements of clouds ingesting fire smoke are needed to shed further light on physical mechanisms. Two interesting feedback effects associated with the fire smoke/cloud coupling add incentive to such studies. First, if clouds ingesting smoke produce large numbers of positive ground flashes, and if these ground flashes exhibit the continuing currents typical of other positive flashes, then the forest fire threat is exacerbated. In fact, the Red Lake #7 fire in Canada grew a cloud that produced lightning, starting six additional fires (Stocks and Flannigan, 1987). Second, given the evidence that smoke ingestion by deep convection decreases cloud droplet size (Reid *et al.*, 1999), it is likely that droplet coalescence will be suppressed with the effect of a lower precipitation yield (Schaefer and Day, 1981) to aid in dousing the fire naturally.

IX. GLOBAL IMPLICATIONS OF LIGHTNING IGNITION CHARACTERISTICS

Global maps of lightning and forest fire incidence are shown in Figures 18 and 19, respectively (see color insert) (S.Goodman and H. Christian, personal communication, 1999; Dwyer *et al.*, 1999). As emphasized earlier in this chapter, both lightning and flammable forest fuel are required for ignition of natural fire. The areas of most prevalent lightning are the tropical continental zones— Southeast Asia, Northern Australia, Africa, Central and South America. Forests are also prevalent in these zones but as rainforests, with precipitation in abundance, flora with high moisture content, and ground beneath the rainforest canopy often completely inundated. The latter condition is likely to suppress the most common fire initiation mechanism at midlatitude (described in Sections IV and V). This claim appears to be supported by the near equatorial zones of minimum fire activity in Africa and South America, where the intertropical convergence zone is likely on an annual basis, and where lightning activity is a maximum. The majority of fires within these tropical zones are slightly displaced from the equatorial regions and are set intentionally in the respective dry seasons (associated with large-scale subsidence of the equatorial Hadley circulation) to burn back the flora of the previous wet season and prepare for crop planting.

The desert areas of the subtropics, the more permanent areas of large-scale subsidence from the intertropical convergence zone (e.g., the African Sahara, the Mexican Sonoran desert, the Chilean desert, the Namibian desert, the central area of Australia) lack both forest fuel and the moist convection necessary for lightning. As a consequence, these areas are largely devoid of naturally caused fire.

The northern boreal forests, heavily populated with flammable conifers, also experience a moderately high incidence of lightning. In fact, the observations with the Optical Transient Detector in space (Figure 18) showed a surprisingly high incidence of lightning in all mid-to-high latitude land regions [e.g., northern Canada, northern Russia (Siberia)]. Fires in a huge area of northern Russia are increasingly well documented by satellite (Kasischke *et al.*, 1999). Much of the lightning in this region had evidently been missed in earlier space-based observations (e.g., Orville and Henderson, 1986) because of the sustained daylight into the late evening at this high latitude that interferes with the optical detection of flashes (S. Goodman, personal communication 1998).

Northwestern North America, including British Columbia, is watered by ocean air mass coming in off the Pacific Ocean, dropping rain on the mostly fire-impervious high biomass density on the western slopes of the coastal mountains. The eastern slopes, on the other hand, have a generally lower biomass density, due to lower overall rainfall, but a higher lightning ignition probability, and hence more fires. The Rocky Mountains get winter snowfall that provides necessary moisture for forests and dry summers that generally cause high ignition efficiencies (Section IV.C). The forests of the American southwest, up into Utah, generally receive moisture from air from the Gulf of Mexico during the thunderstorms that arise in the summertime. The lightning fire season in this area is in the early summer (May and June), whereas the fire season in the northern Rockies is late in the summer (July and August).

Suppression of fires started by natural ignitions in this region causes fuel build-up and eventually ecosystem conversion to unhealthy forest. Natural ignition under these unhealthy conditions will eventually cause large, uncontrollable fires. Application of lightning ignition probabilities can be useful for identifying those areas that are prone to high lightning ignition probability and help decide where fires should be started, and where natural ignition can be helpful. When used in conjunction with lightning location, ignition efficiencies can indicate areas that are likely to have large numbers of holdover fires.

Moving eastward in North America, central Canada and the United States have vast cropland and very little timber. The ignition efficiency for grass is lower, and much of the area is irrigated; lightning-caused fire is of little importance even though lightning densities are high. Further east, deciduous forests have shaded, moist, dense forest floors, with concomitant low ignition efficiency. Peak fire periods are spring and fall, with few fires in the summer. The southeast is forested, but there are relatively few lightning-caused fires because of moist fuels. Lightning fires do occur in the Everglades peat (Sections V.D and VI.B).

Lightning ignition considerations were used to calculate the effect of global warming on lightning caused fires (Price and Rind, 1994). The results showed that, under a scenario of CO_2 doubling, lightning fires would increase by 30–77%, depending on the region of the United States considered. The calculations were made under the assumption that the ecosystems did not change, that is, that the fuels were constant over time. If there were a considerable increase of fires, there would be a fuel change. As with many calculations of this type, there are uncontrolled variables and no data. (See Chapter 10 in this book for more discussion of the climate issue.)

X. CONCLUSION

Lightning ignition of wildland fuels plays a major role in the maintenance and evolution of ecosystems. In this chapter, we have discussed the predominant mechanism for lightning ignition and how this knowledge can be used in fire management. There is little doubt, based on theoretical approaches, laboratory experiments, and global fire and lightning data, that the fuel type and fuel state play a much larger role than the lightning density in lightning fire ignitions. The efficiency of lightning fires seems to be in the range of 0.01–0.04 fires/CG flash in much of North America; that is, only about one to four flashes in 100 start fires. It remains to be seen whether these efficiencies are larger in other places. Certainly, in central Africa and South America, efficiencies are much lower; how much lower we have no way of knowing at present.

As we have seen, lightning fires can act upon the lightning environment through smoke, even inverting the charge structure of thunderstorms. Again, the exact mechanism remains a mystery, and there may be more than one mechanism at work. Charge separation occurs due to flaming in the presence of electric fields, and the smoke carrying that charge is carried into local pyro-cumulus. That charge may not remain on the smoke particles for long, how-ever, and inverted storms in regions far from the smoke origin in space and time may be due only to the presence of smoke particles and their interaction with thunderstorm-charging mechanisms. Finally, lightning ignition of fires may or may not have a role in ecosystems altered by climate change.

REFERENCES

Adams, P. (1966). High-temperature, equilibrium thermodynamic properties of chemically react-ing mixture of N_2, O_2, CO_2, A, and Ne. Aerospace Tech. Rept. 9, Hughes Corp., Los Angeles.

Avila, E., Caranti, G., Castellano, N., and Saunders, C. (1998). Laboratory studies of the influence of cloud droplet size on charge transfer during crystal-graupel collisions. *J. Geophys. Res.* **103**, 8985–8996.

Barrows, J. S. (1954). Lightning fire research in the Rocky Mountains. *J. For.* **52**, 845–847.

Barrows, J. S., Deterich, J. H., Odell, C. A., Kaehn, C. H., Fuquay, D. M., Schaefer, V. J., MacCready Jr., P. B., Colson, D., and Wells, H. J. (1958). Project Skyfire. *In* "Final Report of the Advisory Committee on Weather Control," Vol. II, pp. 105–125. U.S. Government Printing Office, Wash-ington, DC.

Barrows, J. S., Sandbert, D. V., and Hart, J. D. (1977). "Lightning Fires in Northern Rocky Mountain Forests." Final Report, Cooperative Agreement 16-440-CA. USDA Forest Service, Intermoun-tain Forest and Range Experiment Station, Ogden.

Baughman, R.G., Fuquay, D. M., and Mielke Jr., P. W. (1976). Statistical analysis of a randomized lightning modification experiment. *J. Appl. Meteorol.* **15**, 790–794.

Berger, K. (1947). Lightning research in Switzerland. *Weather* **2**(8), 231–238.

Boccippio, D. J., Williams, E. R., Heckman, S. J., Lyons, W. A., Baker, I. T., and Boldi, R. (1995). Sprites, ELF transients, and positive ground strokes. *Science* **269**, 1088–1091.

Boccippio, D. J., Cummins, K., Christian, H. J., and Goodman, S. J. (2001). Combined satellite and surface-based estimation of the intracloud/cloud-to-ground lightning ratio over the continen-tal United States. *Mon. Weath. Rev.*, in press.

Burke, C. P., and Jones, D. L. (1996). On the polarity and continuing current in unusually large lightning flashes deduced from ELF events. *J. Atmos. Terrestr. Physics* **58**, 531–548.

Cianos, N., and Pierce, E. T. (1972). "A Ground-lightning Environment for Engineering Usage." Tech. Rep. 1, SRI Project 1834. Stanford Research Institute, Menlo Park.

Connor, T. R. (1967). "The 1965 ARPA-AEC Joint Lightning Study at Los Alamos." Publ. LA-3754, Vol. 1. Los Alamos Scientific Laboratory, Los Alamos.

Cox, G., Ed. (1995). "Combustion Fundamentals of Fire." Academic Press, London.

Cummer, S. A., and Inan, U. S. (1996). Sprite-producing lightning using ELF radio atmospherics. *Geophys. Res. Lett.* **24**, 1731–1734.

Cummins, K. L., Murphy, M. J., Bardo, E. A., Hiscox, W. L., Pyle, R. B., and Pifer, A. E. (1998). A combined TOA/MDF technology upgrade of the U.S. national lightning detection network. *J. Geophys. Res.* **103**, 9035–9044.

Darveniza, M. (1980). "Electrical Properties of Wood and Line Design." University of Queensland Press, St. Lucia.

Darveniza, M., and Zhou, Y. (1994). Lightning-initiated fires: Energy absorbed by fibrous materials from impulse current arcs. *J. Geophys. Res.* **99**, 10663–10670.

Deeming, J. E., Burgan, R. E., and Cohen, J. D. (1977). "The National Fire Danger Rating System— 1978." Gen. Tech. Rep. INT-39. USDA Forest Service, Intermountain Forest and Range Experiment Station, Ogden.

Defandorf, F. M. (1955). A tree from the viewpoint of lightning. *J. Wash. Acad. Sci.* **45**, 333–339.

Dennis, A. S. (1980). "Weather Modification by Cloud Seeding." Academic Press, New York.

Du Moncel, T. (1877). Recherches sur la conductibilite electrique des corps mediocrament conducteurs et les phenomenes qui les accompagnet. *Annales de Chimie et de Physique* [vol.?], 471.

Dwyer, E., Pereira, J. M. C., Gregoire, J-M., and DaCamara, C. C. (1999). Characterization of the spatio-temporal patterns of global fire activity using satellite imagery for the period April 1992 to March 1993. *J. Biogeog.*, in press.

Flannigan, M. D., and Wotton, B. M. (1991). Lightning-ignited fires in northwestern Ontario. *Can. J. For. Res.* **21**, 277–287.

Frandsen, W. H. (1987). The influence of moisture and mineral soil on the combustion limits of smoldering forest duff. *Can. J. For. Res.* **17**, 1540–1544.

Fuquay, D. M. (1974). Lightning damage and lightning modification caused by cloud seeding. *In* "Weather and Climate Modification" (W. N. Hess, Ed.), pp. 604–612. Wiley, New York.

Fuquay, D. M., and Baughman, R. G. (1969). "Project Skyfire Lightning Research." Final Report to the National Science Foundation, USDA Forest Service, Intermountain Forest and Range Experiment Station, Missoula.

Fuquay, D. M., Baughman, R. G., Taylor, A. R., and Hawe, R. G. (1967). Characteristics of seven lightning discharges that caused forest fires. *J. Geophys. Res.* **72**, 6371–6373.

Fuquay, D. M., Taylor, A. R., Hawe, R. G., and Schmidt Jr., C. W. (1972). Lightning discharges that started forest fires. *J. Geophys. Res.* **77**, 2156–2158.

Gisborne, H. T. (1924). Lightning fires and storms in the northern Rocky Mountains. *In* "The Timberman," p. 2. Portland.

Gisborne, H. T. (1926). Lightning and forest fires in the northern Rocky Mountains. *Mon. Weath. Rev.* **54**, 281–286.

Gisborne, H. T. (1931). A five-year record of lightning storms and forest fires. *Mon. Weath. Rev.* **59**, 139–150.

Greig, J. R., Pechacek, R. E., and Raleigh, M. (1985). Channel cooling by turbulent convective mixing. *Physics of Fluids* **28**, 2357–2364.

Hahn, G. J., and Shapiro, S. S. (1967). "Statistical Models in Engineering." John Wiley and Sons, New York.

Harvey, T. F., and Edwards, L. L. (1991). A parametric investigation of electrical effects on aerosol scavenging by droplets over large fires. *Atmos. Environ.* **25A**, 2607–2614.

Huang, E., Williams, E., Boldi, R., Lyons, W., Nelson, T., Taylor, M., Heckman, S., and Wong, C. (1999). Criteria for sprites and elves based on Schumann resonance measurements. *J. Geophys. Res.* **104**, 1693–1694.

Incropera, F. P., and DeWitt, D. P. (1996). "Fundamentals of Heat and Mass Transfer," 4th ed. John Wiley and Sons, New York.

Jones, J. C., Ramahti, H., Fowler, D., Vorasurayakarnt, J., and Bridges, R. G. (1990). The self heating and thermal ignition propensity of forest floor litter. *J. Fire Sci.* **8**, 207–223.

Kappenman, R. F. (1988). A simple method for choosing between the lognormal and Weibull distributions. *Stat. Prob. Lett.* **7** (Sept), 123–126.

Kasischke, E. S., Bergen, K., Fennimore, R., Sotelo, F., Stephens, G., Janetos, A., and Shugart, H. H. (1999). Satellite imagery gives clear picture of Russia's boreal forest fires. *EOS* **80**, 141.

Kaufman, Y. J., and Nakajima, T. (1993). Effect of Amazon smoke on cloud microphysics and albedo analysis form satellite imagery. *J. App. Meteorol.* **32**, 729–744.

King, L. A. (1961). "The Voltage Gradient of the Free-burning Arc in Air or Nitrogen." ERA Rep. G/XT172. The British Electrical and Allied Industries Research Association, Leatherhead.

Latham, D. J. (1979). "Progress Toward Locating Lightning Fires." Res. Note INT-269. USDA Forest Service, Intermountain Forest and Range Experiment Station, Ogden.

Latham, D. J. (1984). A numerical simulation of the lightning continuing current channel. In "Proceedings, Seventh International Conference on Atmospheric Electricity," pp. 417–420. American Meteorological Society, Boston.

Latham, D. J. (1986). Anode column behavior of long vertical air arcs at atmospheric pressure. IEEE Trans. Plasma Sci. PS-14, 220–227.

Latham, D. J. (1987). "Design and Construction of an Electric Arc Generator for Fuel Ignition Studies." Res. Note INT-366. USDA Forest Service, Intermountain Research Station, Ogden.

Latham, D. J. (1989). "Characteristics of Long Vertical DC Arc Discharges." Res. Pap. INT-407. USDA Forest Service, Intermountain Research Station, Ogden.

Latham, D. J. (1991). Lightning flashes from a prescribed fire-induced cloud. J. Geophys. Res. 96, 17151–17157.

Latham, D. J. (1999). Space charge generated by wind tunnel fires. Atmos. Res. 51, 267–278.

Latham, D. J., and Schlieter, J. A. (1989). "Ignition Probabilities of Wildland Fuels Based on Simulated Lightning Discharges." Res. Pap. INT-411. USDA Forest Service, Intermountain Research Station, Ogden.

Ligda, M. G. H. (1956). The radar observation of lightning. J. Atmos. Terr. Phys. 9, 329–346.

Loeb, L. B. (1966). The mechanisms of stepped and dart leaders in cloud-to-ground lightning strokes. J. Geophys. Res. 71, 4711–4721.

Loftsgaarden, D. O., and Andrews, P. A. (1992). "Constructing and Testing Logistic Regression Models for Binary Data: Applications to the National Fire Danger Rating System." Gen. Tech. Rep. INT-286. USDA Forest Service, Intermountain Research Station, Ogden.

Lyons, W. A., Nelson, T. E., Williams, E. R., Cramer, J. A., and Turner, T. R. (1998a). Enhanced positive cloud-to-ground lightning in thunderstorms ingesting smoke from fires. Science 282, 77–80.

Lyons, W. A., Uliasz, M., and Nelson, T. E. (1998b). A climatology of large peak current cloud-to-ground lightning flashes in the contiguous United States. Mon. Weath. Rev. 126, 2217.

Malan, D. J. (1963). "Physics of Lightning." English University Press, London.

Martin, S., and Broido, A. (1963). "Thermal Radiation and Fire Effects of Nuclear Detonations." Rep. AD 422411. U.S. Naval Radiological Defense Laboratory, San Fransisco.

Mazur, V., Shao, X. M., and Krehbiel, P. (1998). "Spider" lightning in intracloud and positive cloud-to-ground flashes. J. Geophys. Res. D16, 19811–19822.

McEachron, K. B., and Hagenguth, J. H. (1942). Effect of lightning on thin metal surfaces. AIEE Trans. 61, 559–564.

McMullen, L. H., and Atkins, M. D. (1962). On the flight and host selection of the Douglas-fir beetle Dendroctomus pseudotsugae Hopk. (Coleoptera: Scolytidae). Can. Entomol. 94, 1309–1325.

McRae, R. H. D. (1992). Prediction of areas prone to lightning ignition. Int. J. Wildl. Fire 2, 123–130.

Meisner, B. N. (1993). "Correlation of National Fire Danger Rating System Indices and Weather Data with Fire Reports." Final Report, Interagency Agreement No. R500A20021. USDA Forest Service, Forest Fire Laboratory, Pacific Southwest Research Station, Riverside.

Melosh, H. J., Schneider, N. M., Zahnle, K. J., and Latham, D. J. (1990). Ignition of global wildfires at the Cretaceous/Tertiary boundary. Nature 343, 251–254.

Morris, W. G. (1934). "Lightning Storms and Fires on the National Forests of Oregon and Washington." USDA Forest Service, Pacific Northwest Forest Experiment Station, Portland.

Nash, C. H., and Johnson, E. A. (1996). Synoptic climatology of lightning-caused forest fires in subalpine and boreal forests. Can. J. For. Res. 26, 1859–1874.

Orville, R. E., and Helsdon Jr., J. H. (1974). Quantitative analysis of a lightning return stroke for diameter and luminosity changes as a function of time. *J. Geophys. Res.* **79**, 4059–4067.

Orville, R. E., and Henderson, R. W. (1986). Global distribution of midnight lightning: September 1977 to August 1978. *Mon. Wea. Rev.* **114**, 264–2653.

Plummer, F. G. (1912). "Lightning in Relation to Forest Fires." Bulletin 111. USDA Forest Service, Government Printing Office, Washington, DC.

Price, C., and Rind, D. (1994). The impact of a 2xCO$_2$ climate on lightning-caused fires. *J. Climate* **7**, 1484–1494.

Reid, J. S., Hobbs, P. V., Ragno, A. L., and Hegg, D. A. (1999). Relationships between cloud droplet effective radius, liquid water content, and droplet concentration for warm clouds in Brazil embedded in biomass smoke. *J. Geophys. Res.* **104**, 6145–6153.

Renkin, R. A., and Despain, D. G. (1992). Fuel moisture, forest type, and lightning-caused fire in Yellowstone National Park. *Can. J. For. Res.* **22**, 37–45.

Rorig, M. L., and Ferguson, S. A. (1999). Characteristics of lightning and wildland fire ignition in the Pacific Northwest. *J. Appl. Met.* **38**, 1565–1575.

Rosenfeld, D., and Lensky, I. (1998). Satellite-based insights into precipitation formation processes in continental and maritime convective clouds. *Bull. Amer. Meteorol. Soc.* **79**, 2457–2476.

Salanave, L. E. (1965). The photographic spectrum of lightning: Determination of channel temperature from slitless spectra. *In* "Problems of Atmospheric and Space Electricity" (S. C. Coriniti, Ed.), pp. 371–384. Elsevier Publishing Company, New York.

Schaefer, V. J. (1949). "The Possibility of Modifying Lightning Storms in the Northern Rockies." Pap. No. 19. USDA Forest Service, Northern Rocky Mountain Forest and Range Experiment Station, Missoula.

Schaefer, V. J. (1953). "Final Report Project Cirrus, Part 1." Report No. RL-785. General Electric Research Laboratory, Schenectady.

Schaefer, V. J., and Day, J. A. (1981). "A Field Guide to the Atmosphere." Houghton Mifflin and Co., Boston.

Schmitz, R. F., and Taylor, A. R. (1969). "An Instance of Lightning Damage and Infestation of Ponderosa Pines by the Pine Engraver Beetle in Montana." Res. Note INT-88. USDA Forest Service, Intermountain Forest and Range Experiment Station, Ogden.

Schonland, B. F. J., and Collens, H. (1934). Progressive lightning. *Proc. Roy. Soc. (Lond)* **A143**, 654–674.

Sentman, D. D. (1987). *EOS 67 (suppl.)* **67**, 1069.

Simms, D. L., and Law, M. (1967). The ignition of wet and dry wood by radiation, *Combust. Flame* **11**, 377–388.

Stocks, B. J., and Flannigan, M. D. (1987). Analysis of the behaviour and associated weather for a 1986 northwestern wildfire: Red Lake #7. *In* "Proceedings, Ninth Conference on Fire and Forest Meteorology," pp. 94–99. American Meteorological Society, Boston.

Strachan, D. C., and Barrault, M. R. (1975). "Axial Velocities in the High Current Free Burning Arc." Rep. ULAP-T34. Department of Electrical Engineering and Electronics, University of Liverpool, Liverpool.

Strom, A. P. (1946). Long 60-cycle arcs in air. *AIEE Trans.* **65**, 113–117.

Susott, R. A. (1984). "Heat of Preignition of Three Woody Fuels Used in Wildfire Modeling Research." Res. Note INT-342. USDA Forest Service, Intermountain Forest and Range Experiment Station, Ogden.

Taylor, A.R. (1969). Lightning effects on the forest complex. *In* "Proceedings, Annual Tall Timbers Fire Ecology Conference," pp. 127–150. Tall Timbers Research Station, Tallahassee.

Taylor, A. R. (1973). Ecological aspects of lightning in forests. *In* "Proceedings, Annual Tall Timbers Fire Ecology Conference," pp. 455–482. Tall Timbers Research Station, Tallahassee.

Thomson, E. M. (1980). Characteristics of Port Moresby ground flashes. *J. Geophys. Res.* **85**, 1025–1036.

Uman, M. A. (1969). "Lightning." McGraw-Hill, New York.

Uman, M. A., and Voshall, R. E. (1968). Time interval between lightning strokes and the initiation of dart leaders. *J. Geophys. Res.* **73**, 497–506.

Van Wagner, C. E. (1987). "Development and Structure of the Canadian Forest Fire Index System." Tech. Rep. 35. Canadian Forestry Service, Ottawa.

Van Wagtendonk, J. W. (1991). Spatial analysis of lightning strikes in Yosemite National Park. *In* "Proceedings, Eleventh Conference on Fire and Forest Meteorology," pp. 605–611. American Meteorological Society, Boston.

Vonnegut, B., and Orville, R. E. (1988). Evidence of lightning associated with the Yellowstone Park forest fire (abstract). *EOS Trans. AGU* **69**, 1071.

Vonnegut, B., Latham, D. J., Moore, C. B., and Hunyady, S. J. (1995). An explanation for anomalous lightning from forest fire clouds. *J. Geophys. Res.* **100**, 5037–5050.

Williams, E. R. (1989). The tripole structure of thunderstorms. *J. Geophys. Res.* **94**, 13151–13167.

Williams, E. R. (1998). The positive charge reservoir for sprite-producing lightning. *J. Atmos. Terr. Phys.* **60**, 689–692.

CHAPTER 12

Statistical Inference for Historical Fire Frequency Using the Spatial Mosaic

W. J. REED

W. J. REED

Department of Mathematics and Statistics, University of Victoria, Victoria, B.C., Canada

I. INTRODUCTION

In forested regions where the principal agent of stand regeneration is fire (e.g., in boreal forests), the spatial mosaic of stands depends on the frequency and extent of fires in the past. To some extent, knowledge of the spatial mosaic can be used to make inference about past fire frequency. This chapter deals with statistical methodology for such inference. Given complete knowledge of fire history, one could, in principle, reconstruct the spatial mosaic. In contrast, knowledge of the current spatial mosaic is not sufficient for complete knowledge of past fires. The reason for this is that evidence of fires in the past can be obliterated by subsequent fires burning over all or part of the extent of the earlier fires. The present spatial mosaic provides complete evidence of only the most recent fire or fires; it provides decreasing information about fires more distant in time; and of very early fires, little or no evidence may remain.

A *time-since-fire map* is closely related to the spatial mosaic and gives the time since the most recent fire at every point in a study area. In contrast a *time-since-fire sample* contains only the time since the most recent fire at a sample of points (drawn according to some well-defined sampling plan) from the study area. Usually because of the limitations of techniques of dating stand origins or past fires, the time-since-fire observations are grouped into classes (often of width one decade). In this chapter statistical methodologies for analyzing time-since-fire map and sample data will be discussed. The main issues addressed are:

- methods of graphical data analysis of such data;
- estimation of fire frequency at distinct epochs in the past. Likelihood methods will be used to obtain point estimates and confidence intervals.
- testing for the significance of changes in fire frequency at specified times;
- determination of when (possibly multiple) changes in fire frequency may have occurred.

The methods presented collect and summarize recent research in this area.

II. GRAPHICAL ANALYSIS

A great deal of information about historical fire frequency can be obtained by simple graphical analysis. For time-since-fire map data, one can plot cumulative proportional areas, with time since last fire $\geq t$, on a logarithmic scale, against time t (on a natural scale). The lower panels of Figures 1, 3 and 4 are examples of such plots. For sample data, the same procedure can be followed using the cumulative proportion of sample points with time since last fire $\geq t$ (Figure 2). If the hazard of burning has been both age-independent and time-homogeneous, the points on the plot should lie close to a straight line with negative slope (equal to the hazard of burning), since at any point the probability that the most recent fire occurred t or more years ago is $e^{-\lambda t}$, where λ is the hazard of burning. Abrupt changes in the plot suggest temporal changes in the hazard of burning. For example, if the hazard of burning changed at some distinct change point (as might possibly have occurred for example at the time of European intervention), then one would expect a sharp elbow in the plot at that point. Thus a plot which naturally divides into a number of straight line segments suggests a number of distinct historical epochs of roughly constant fire frequency, with the changes in fire frequency occurring at times determined by the points of intersections of adjacent segments. By comparing such plots for different subsets of the study area (e.g., broken up spatially or by aspect) one can get a good idea of whether historical fire frequency was homogeneous or not with respect to these categories.

Even though the vertices and slopes of line segments determined graphically provide rough estimates of change points and intervening fire frequencies, it is desirable to have more objective statistical methodologies for accomplishing this. Subsequent sections of this chapter describe such methodologies.

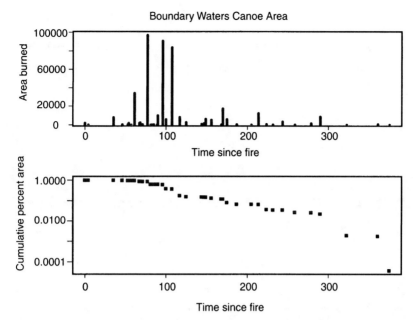

FIGURE 1 Time since last fire for Boundary Waters Canoe Area (Heinselman, 1973). The top panel shows the areas (on vertical axis in hectares) with given time since last fire (horizontal axis in years). The bottom panel plots the cumulative proportional area (on logarithmic scale) with time since fire exceeding the time (in years) on the horizontal axis.

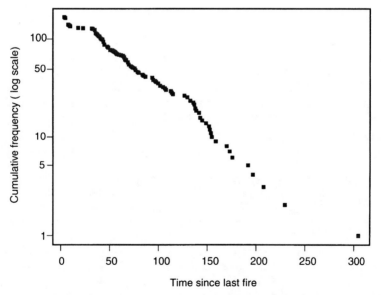

FIGURE 2 Cumulative percent of sample points (on logarithmic scale) with time since fire exceeding the time (in years) on the horizontal axis for time-since-fire sample data of Wood Buffalo National Park (Larsen, 1996).

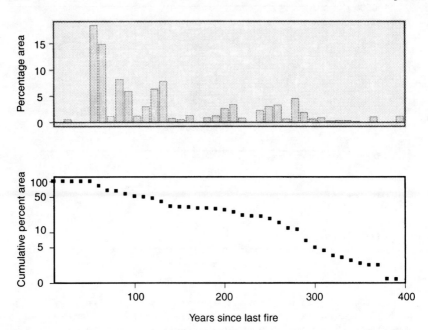

FIGURE 3   Time since last fire for Kananskis Watershed (Johnson and Larsen, 1991). The top panel shows the percentage area (on vertical axis) in given time-since-fire class (horizontal axis in years). The bottom panel plots the cumulative percentage area (on logarithmic scale) with time since fire exceeding the time (in years) on the horizontal axis.

However, the problem is approached in stages. First, the problem of obtaining maximum likelihood estimates (MLEs) of the hazards of burning prevailing between prespecified change points is addressed, along with how to obtain approximate confidence intervals and tests for the significance of the change points. This is done for both map and sample data. Following this, the question of determining change points from the data alone is addressed.

III. STATISTICAL INFERENCE WITH PRESPECIFIED CHANGE POINTS

We assume that the observations whether for map or sample data are binned into time-since-fire classes $1, 2, \ldots, m$, which are of equal width, say T years, save for the oldest which is open-ended. Specifically, assume that the classes $1, 2, \ldots, m - 1$ are defined by time since last fire in the intervals $[0, T)$, $[T, 2T), \ldots, [(m - 2)T, (m - 1)T)$, respectively and that class m is defined by time since last fire greater than or equal to $(m - 1)T$. In practice, the resolution

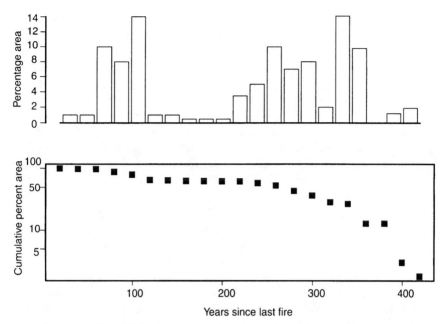

FIGURE 4 Time since last fire for Glacier National Park (Johnson *et al.*, 1990). The top panel shows the percentage area (on vertical axis) in given time-since-fire class (horizontal axis in years). The bottom panel plots the cumulative percentage area (on logarithmic scale) with time since fire exceeding the time (in years) on the horizontal axis.

of dating fires is no finer than one year, so T will be a positive integer (often $T = 10$ or 20).

Under the assumption of homogeneity, at any given location in the study area, the probability that the time since last fire belongs to class j can be determined in terms of the hazards of burning prevailing at and since that last fire date, by observing that to belong to class j, there must have been a fire between $(j - 1)T$ and jT years ago, and subsequently no fire. Specifically, it is $e^{-\sum_{i=1}^{j-1} \lambda^{(i)} T}(1 - e^{-\lambda^{(j)} T}) = \theta_j$, say, where $\lambda^{(i)}$ is the instantaneous hazard of burning between $(i - 1)T$ and iT years ago. With known change points, the parameters $\lambda^{(i)}$, for $i = 1, 2, \ldots, m$, can be expressed in terms of parameters representing the hazards of burning in the epochs between the change points. For example, with no change points, all the $\lambda^{(i)}$ are equal to the assumed constant hazard λ_0, say. If there were one change point pT years ago, with a hazard of burning λ_2, say, prevailing before that time and λ_1, since that time, then $\lambda^{(i)} \equiv \lambda_1$ for $i = 1, 2, \ldots, pT$ and $\lambda^{(i)} \equiv \lambda_2$ for $i = p + 1, p + 2, \ldots, m$, etc.

To estimate by maximum likelihood the parameters representing the hazards of burning in the epochs between the change points, one needs to construct a likelihood function [i.e., determine the probability of the observed areas (or frequencies of sample points) falling in the various classes] in terms of these parameters. If one could reasonably assume independence between points in the study area, with respect to time since last fire, the likelihood function could be obtained directly from the multinomial probabability distribution. However, such an assumption is not reasonable because of the fact that fires spread spatially. To reflect this fact, one needs a model for the distribution of areas in time-since-fire classes which exhibits a contagion effect. Two approaches have been used to date to accomplish this. One is to use a parametric model (the *Dirichlet distribution*) for the areas in time-since-fire classes. This provides a likelihood directly for map data (Reed, 1994), whereas a likelihood for sample data can be obtained by computing the (marginal) probability of observing the observed frequencies in the various classes, assuming that the proportional areas follow the Dirichlet distribution. This leads to a likelihood of the form of the *Dirichlet compound multinomial* distribution, which is the multivariate analogue of the widely used *beta-binomial* distribution (Reed, 1998).

The other approach, which in many ways is more straightforward, is to incorporate the contagion effect by using an *overdispersed multinomial* model with corresponding *quasi likelihood* (Reed *et al.*, 1998). This is based on the assumption that the contagion effect, due to spatial spread of fires, has no effect on the expected areas (or frequencies) in the time-since-fire classes, but it does have an effect on the variances–covariances, inflating them by a constant factor. This second approach is more general in the sense that the Dirichlet distribution for proportional areas, and the resulting distribution for frequencies of sample points, are both overdispersed multinomial distributions themselves (Reed, 1998). Which of the two approaches is superior in any given context is not known. This issue could probably be resolved by studying the performance of the two methods using simulated data. However, to date such a study has not been undertaken. In this chapter, we concentrate on the overdispersed multinomial approach. One advantage is that analytic expressions can be obtained for maximum likelihood estimates of epochal hazards of burning and for their standard errors. There is a simple interpretation of the MLEs, and furthermore they agree well with the graphical estimates. In contrast, using the Dirichlet model approach, numerical maximization is required to obtain MLEs of the epochal hazards of burning, and it is not possible to obtain a simple interpretation of the estimates.

A. Constructing a Quasi Likelihood for an Overdispersed Multinomial Model

For data from an overdispersed multinomial distribution in which there is a constant inflation factor (*overdispersion parameter, σ^2*) for all variances and covariances, one can construct a *quasi likelihood* function of the form

$$Q = \frac{1}{\sigma^2} \sum_{j=1}^{m} y_j \log(\theta_j) \tag{1}$$

where for map data $y_j = A_j/(\sum_{j=1}^{m} A_j)$, the proportional area in class j, and for sample data, with frequencies f_j, $y_j = f_j/(\sum_{j=1}^{m} f_j)$, the proportion of the observations in class j. The overdispersion parameter σ^2 will differ for map and sample data. In most respects regarding model parameters (other than σ^2), the quasi likelihood behaves like an ordinary log likelihood. Thus, one can find MLEs by maximizing the quasi likelihood and obtain approximate standard errors from the observed (quasi) *Fisher information matrix* (Hessian matrix of second derivatives of quasi likelihood at the maximum).

Simple analytic expressions are available for the MLEs of the hazards of burning in distinct epochs. For example if there is only one epoch (i.e., the same hazard of burning λ_0 is assumed to have prevailed at all times in the past), then from differentiation of the quasi-likelihood it is easy to show that the MLE is

$$\hat{\lambda}_0 = (-1/T) \log(\hat{q}_0)$$

where

$$\hat{q}_0 = \frac{\sum_{j=1}^{m-1} s_j}{\sum_{j=1}^{m-1} s_{j-1}}$$

and

$$s_j = \sum_{i=j+1}^{m} y_i$$

is the proportional area (or proportion of the observations, for sample data) with time since last fire at least jT, for $j = 1, 2, \ldots, m - 1$. Furthermore, the approximate variance of the estimator $\hat{q}_0$ is

$$\text{var}(\hat{q}_0) \approx \sigma^2 \frac{(\sum_{j=1}^{m-1} s_j)(\sum_{j=1}^{m-1} y_j)}{(\sum_{j=1}^{m-1} s_{j-1})^3}$$

If it is assumed that there are several epochs with distinct hazards, say λ_1 between 0 and $p_1 T$ years ago, λ_2 between $p_1 T$ and $p_2 T$ years ago, etc., then the MLEs of the separate hazards of burning can be found in a similar way. For example, if there are two epochs separated by a single change point pT years ago, then for the more recent epoch

$$\hat{\lambda}_1 = (-1/T) \log(\hat{q}_1)$$

and for the earlier epoch

$$\hat{\lambda}_2 = (-1/T) \log(\hat{q}_2)$$

where

$$\hat{q}_1 = \frac{\sum_{j=1}^{p} s_j}{\sum_{j=1}^{p} s_{j-1}} \qquad \hat{q}_2 = \frac{\sum_{j=p+1}^{m-1} s_j}{\sum_{j=p+1}^{m-1} s_{j-1}}$$

Furthermore, these estimates are independent and have approximate variances given by

$$\text{var}(\hat{q}_1) \approx \sigma^2 \frac{(\sum_{j=1}^{p} s_j)(\sum_{j=1}^{p} y_j)}{(\sum_{j=1}^{p} s_{j-1})^3}; \qquad \text{var}(\hat{q}_2) \approx \sigma^2 \frac{(\sum_{j=p+1}^{m-1} s_j)(\sum_{j=p+1}^{m-1} y_j)}{(\sum_{j=p+1}^{m-1} s_{j-1})^3}$$

To use these variance formulas to obtain standard errors and confidence intervals for the hazard of burning, etc., one needs to estimate the overdispersion parameter σ^2. This is usually done using the Pearson estimate, which for map data is

$$\hat{\sigma}^2 = \frac{1}{m - r - 1} \sum_{j=1}^{m} \frac{(y_j - \hat{\theta}_j)^2}{\hat{\theta}_j(1 - \hat{\theta}_j)}$$

where $\hat{\theta}_j$ is the MLE of θ_j and r is the number of epochs. With a single hazard of burning assumed to have prevailed at all times past, the MLEs of the cell probabilities are $\hat{\theta}_j = (1 - \hat{q}_0)\hat{q}_0^{j-1}$, $j = 1, 2, \ldots, m - 1$ and $\hat{\theta}_m = \hat{q}_0^{m-1}$. With two epochs, they are $\hat{\theta}_j = (1 - \hat{q}_1)\hat{q}_1^{j-1}$, $j = 1, 2, \ldots, p$, $\hat{\theta}_j = (1 - \hat{q}_2)\hat{q}_1^{p}\hat{q}_2^{j-p-1}$, $j = p + 1, 2, \ldots, m - 1$ and $\hat{\theta}_m = \hat{q}_1^{p}\hat{q}_2^{m-p-1}$. Similar formulas hold when there are more than two epochs.

An approximate $100(1 - \alpha)\%$ confidence interval for the hazard of burning in epoch r can be obtained from the confidence interval for q_r, which is

$$\hat{q}_r \pm z_{\alpha/2}\sqrt{(\hat{\sigma}^2/\sigma^2)\text{var}(\hat{q}_r)}$$

where $z_{\alpha/2}$ is a percentage point from the standard normal distribution. The transformation $q \to (-1/T)\log(q)$ will convert this to a confidence interval for the hazard of burning and $q \to (-T)/(\log(q))$ will convert it to a confidence interval for the fire cycle.

A test for the significance of a change point between epochs can be constructed by computing the liklelihood ratio (LR) statistic. Specifically, one computes the test statistic

$$\Lambda = 2\frac{\hat{\sigma}_1^2}{\sigma^2}(\hat{Q}_1 - \hat{Q}_0)$$

where $\hat{Q}_1$ and $\hat{Q}_0$ are the maximized quasi likelihoods assuming the change point present and not, respectively, and $\hat{\sigma}_1^2$ is the Pearson estimate of σ^2 computed with the change point present. The null distribution of this statistic from which a P-value can be obtained is approximately $F_{1,m-\nu-1}$ where ν is the number of epochs when the change point is present. An important point to note here is that, because of the problem of selection bias, this procedure is not valid if the change point is suggested from an examination of the data, e.g., from a graphical anlysis (Reed *et al.*, 1998). One can also use a LR statistic to construct confidence intervals for the epochal hazards of burning (or fire cycles) using the method described in Reed *et al.* (1997).

Example 1. Map Data. Boundary Waters Canoe Area

Figure 1 shows the time-since-fire distribution for the Boundary Waters Canoe Area, from the maps obtained by Heinselman (1973) in his pioneering fire history study. Since fires are dated to the year, the cell width here is $T = 1$. European settlement in the area first occurred in the 1860s, and there appears to be a change in the slope of the semi-log cumulative frequency plot around this time (a little more than 100 years ago). Assuming a single change point in 1865 yields the following maximum likelihood estimates of the hazards of burning and corresponding fire cycle (with approximate 95% confidence intervals):

Epoch 1 (1865–1971): $\hat{\lambda}_1 = 0.00883$ p.a., $FC_1 = 113$ (73–245) yr
Epoch 2 (before 1865): $\hat{\lambda}_2 = 0.0121$ p.a., $FC_2 = 83$ (42–3915) yr

The overdispersion parameter is estimated as 0.0593.

Example 2. Sample Data. Wood Buffalo National Park

Larsen (1996) reported a detailed study of the fire history of Wood Buffalo National Park, a 45,000 km^2 area of boreal forest straddling the border between Alberta and the Northwest Territories. Larsen determined (to the nearest year) the time since last fire at 166 randomly selected sites. These are displayed in Figure 2 with $m = 300$ classes of width $T = 1$ year. The graph suggests a change point at $p = 129$ years ago (1860). Assuming this to be the case, the maximum likelihood estimates of the hazards of burning and corresponding fire cycle (with approximate 95% confidence intervals) are:

Epoch 1 (1861–1989): $\hat{\lambda}_1 = 0.0142$ p.a., $FC_1 = 70$ (58–90) yr
Epoch 2 (1860 and earlier): $\hat{\lambda}_2 = 0.0292$ p.a., $FC_2 = 34$ (23–69) yr

The overdispersion parameter is estimated as 0.0106.

IV. THE EFFICIENCY OF SAMPLE VS. MAP DATA

Map data corresponds to a complete census of the study area. Since one can obtain estimates of past fire frequencies using observations on time since the last fire at only a random sample of points in the study area, an obvious question is how much precision is foregone by determining time since fire at a sample of points rather than everywhere in the study area. Of course, the reasons for collecting time-since-fire map data extend beyond the estimation of historical fire frequency. These issues are not addresssed here. Rather, attention is confined to assessing the relative statistical efficiency of using sample vs. map data for such estimation.

To this end, suppose that the proportional areas over the whole study area $a_j = A_j/(\sum_{j=1}^{m} A_j)$ follow an overdispersed multinomial distribution with overdispersion parameter $\sigma_0^2 < 1$. Under this assumption,

$$\text{var}(a_j) = \sigma_0^2\theta_j(1 - \theta_j); \qquad \text{cov}(a_i, a_j) = -\sigma_0^2\theta_i\theta_j$$

By first conditioning on the proportional areas a_j, one can show that for n time-since-fire observations at a random sample of points in the study area, the proportional frequencies $y_j = f_j/n$ in the m classes follow an overdispersed multinomial distribution with overdispersion parameter

$$\sigma_1^2 = \frac{1}{n}[1 + (n - 1)\sigma_0^2] > \sigma_0^2$$

The *quasi Fisher information matrix I,* which is obtained as the expected value of the Hessian matrix of second derivatives of the quasi likelihood, provides a measure of the precision of the MLEs. It is easy to show that this information matrix for map data, I_0 say, differs from that for sample data, I_1 say, only by the scalar multiplicative constant, that is,

$$I_1 = \frac{\sigma_0^2}{\sigma_1^2}I_0$$

The relative efficiency (ratio of variances of estimators) for sample data relative to map data is, thus,

$$\text{Rel. Eff.} = \frac{\sigma_0^2}{\sigma_1^2} = \frac{n\sigma_0^2}{1 + (n - 1)\sigma_0^2}$$

which is easily confirmed to be less than one and converging to one as sample size $n \to \infty$ (i.e., estimates based on sample data will have lower precision than those based on map data) but will approach the latter as sample size grows large.

One can substitute an estimate for σ_0^2 in this expression to assess the loss in precision by using sample data.

Example 1 (continued). Boundary Waters Canoe Area

The estimate of the overdispersion parameter for these data was $\hat{\sigma}_0^2 = 0.0593$. Table 1 gives estimates, using this value, of the relative efficiency of sample data

TABLE 1   Estimated Relative Efficiency of Using Sample Data (sample size n) vs. Map Data for Boundary Waters Canoe Area and Ratio of Expected Widths of Confidence Intervals

Sample size n	10	20	50	100	200	500	1000
Rel. eff.	0.39	0.56	0.76	0.86	0.93	0.97	0.98
Ratio of C.I. widths	1.61	1.34	1.15	1.08	1.04	1.02	1.01

(for various sample sizes n) vs. map data and the ratio of (expected) widths of corresponding confidence intervals. It is clear that a sample of moderate size (100 to 200) would be almost as good as a complete map, in terms of the precision of estimates. The expected width of confidence intervals would be only 8% greater with $n = 100$ and only 4% greater with a sample of size $n = 200$.

Example 2 (continued).   Wood Buffalo National Park

Larsen (1996) determined time since last fire at $n = 166$ sample points in Wood Buffalo National Park. The overdispersion parameter for a sample of this size was estimated as $\hat{\sigma}_1^2 = 0.0106$. Thus, an estimate of the overdispersion parameter for map data is $\hat{\sigma}_1^2 = (166 \times 0.0106 - 1)/(166 - 1) = 0.0046$. Note that this estimate is smaller by an order of magnitude than the corresponding estimate for the Boundary Waters Canoe Area, suggesting a much smaller contagion effect (and thus a finer patch mosaic) in Wood Buffalo than in the Boundary Waters. Using this estimate, one can assess how much precision was foregone using a sample of points rather than a complete map survey and how the precision of estimates from sample data depends on sample size. Table 2 gives information on this.

The results suggest that confidence intervals would be about one third shorter if map data had been obtained. Unlike the case of the Boundary Waters, where relatively small samples could provide precision comparable with map data, for Wood Buffalo such small samples would not be adequate. For example, for the Boundary Waters, a sample of size 50 would yield confidence intervals

TABLE 2   Estimated Relative Efficiency of Using Sample Data vs. Map Data for Wood Buffalo National Park and Ratio of Expected Widths of Corresponding Confidence Intervals (second row)

Sample size n	10	20	50	100	166	200	500	1000
Rel. eff	0.044	0.085	0.19	0.32	0.43	0.48	0.70	0.82
Ratio of C.I. widths	4.76	3.44	2.31	1.78	1.52	1.44	1.20	1.10
C.I. width rel. to actual	3.13	2.27	1.52	1.17	1.00	0.95	0.79	0.73

The last row is the ratio of width of confidence intervals using sample size n and actual sample size 166.

only about 15% wider than map data; for Wood Buffalo, such a sample size would yield confidence intervals 130% wider than map data (and 50% wider than those obtained with the actual sample size, 166). The reason for the difference lies with the order of magnitude difference in the overdispersion parameter estimates. With greater contagion, and on average larger patches in the time-since-fire mosaic in the Boundary Waters, a relatively small number of sample points can, with high probability, provide information on most patches. On the other hand, with the smaller contagion parameter and finer patch mosaic in Wood Buffalo, one needs many more sample points on average to cover most patches.

V. DETERMINING EPOCHS OF CONSTANT FIRE FREQUENCY

In the previous sections, it was assumed that the number of epochs, with constant hazard of burning, and the change points dividing them were known. Most often, this will not be the case, and one will face the problem of determining the epochs from the data. This section provides a brief description of a methodology for accomplishing this.

From the statistical point of view, determining the number and location of the change points (i.e., determining the epochs) is a problem in model identification, analogous to deciding which regressor variables should be included in a regression model. There are a number of ways to approach such problems. One is to use some sort of iterative procedure in which change points can be added or removed from the model (analogous to forward selection and stepwise procedures in regression). This approach has been used for identifying fire epochs (Reed, 1998), but there are a number of difficulties associated with it. An alternative approach which avoids these difficulties is to use the *Bayes Information Criterion* to decide on the best single (or several) models.

From one point of view, the Bayes Information Criterion (BIC) can be viewed as a log likelihood, adjusted for the number of parameters in a model (here the number of change points and intervening hazards of burning). It can also be interpreted in a Bayesian context (see later discussion).

Consider a hieararchy of models:

H_0: No change points (constant hazard of burning at all times in the past),
H_1: One change point (separating two epochs with distinct hazards of burning),
H_2: Two change points (separating three epochs with distinct hazards of burning),
   etc.

H_k: k change points (separating $k + 1$ epochs with distinct hazards of burning).

Under any model H_r there are $2r + 1$ undetermined parameters (r change points separating $r + 1$ hazards of burning λ_i or parameters $q_i = e^{-\lambda_i T}$) in addition to the overdispersion parameter. MLEs of the change points can be found by comparing the quasi likelihoods, maximized over the $(r + 1)q_i$ parameters, for all $\binom{m-1}{r}$ possible choices of r change points. A comparison of models H_0, $H_1, \ldots$, etc., can be achieved by comparing the BIC for each model. It can be shown (Reed, 2000) that an approximation for the BIC is

$$\text{BIC}_r \approx \hat{D}_r - (\text{df})_r \log\left(\frac{n}{\hat{\sigma}^2}\right) \tag{2}$$

where $\hat{D}_r$ is the minimum scaled quasi deviance for model H_r i.e.,

$$\hat{D}_r = \frac{2}{\hat{\sigma}^2}\left[\sum_{j=1}^{m} y_j \log y_j - \sum_{i=1}^{r+1}\left\{\left(\sum_{j=p_{i-1}+1}^{p_i} s_j\right)\log \hat{q}_i + \left(\sum_{j=p_{i-1}+1}^{p_i} y_j\right)\log(1 - \hat{q}_i)\right\}\right]$$

minimized over the choice of r change points $p_0, p_1, \ldots, p_r$ (with $p_0 = 0$ and $p_{r+1} = m$). The terms $\hat{q}_i$ are the MLEs of the q_i parameters for epochs $i = 1, 2, \ldots, r + 1$). Precisely,

$$\hat{q}_i = \frac{\sum_{j=p_{i-1}+1}^{p_i} s_j}{\sum_{j=p_{i-1}+1}^{p_i} s_{j-1}}$$

In Eq. (2), the term $(\text{df})_r$ is the residual degrees of freedom for model H_r [i.e., $\text{df}_r = (m - 1) - (2r + 1)$ and $n = \sum_{j=1}^{m-1} s_{j-1}$]. The estimate $\hat{\sigma}^2$ is computed under the "biggest" model contemplated, H_k. Precisely,

$$\hat{\sigma}^2 = \frac{1}{m - 2r - 2}\sum_{j=1}^{m}\frac{(y_j - \hat{\theta}_j)^2}{\hat{\theta}_j(1 - \hat{\theta}_j)}$$

where $\hat{\theta}_j$ is the MLE of θ_j under H_k.

The model with the smallest BIC can be thought of as the one with the best fit, adjusting for the number of parameters. Also one can give a Bayesian interpretation to the BIC (see Raftery, 1995). If one associates prior probabilities $\pi_1, \pi_2, \ldots, \pi_k$ to the models $H_1, H_2, \ldots, H_k$, respectively, then the posterior probabilities, after incorporating the data, are

$$P(H_r|\text{data}) = \frac{\exp(-\frac{1}{2}\text{BIC}_r)\pi_r}{\sum_{i=0}^{k}\exp(-\frac{1}{2}\text{BIC}_i)\pi_i}, \qquad r = 1, \ldots, k$$

In particular for a uniform prior (i.e., all models having same credibility a priori), the πs drop out of this expression, and then clearly the model with the smallest BIC is the one with the largest posterior probability.

In implementing this procedure, there is the problem of specifying a maximum possible number of change points, k. While, in principle, one could set $k = m - 1$, corresponding to different hazard of burning in each period, there are difficulties with doing this. First, it would lead to a considerable computational load, since under a given model H_r, the MLEs of the r change points are found by direct search. Second, there is the problem of estimating the overdispersion parameter σ^2, which is estimated under the largest model contemplated. If k is set too large, there will be few degrees of freedom for estimating σ^2. In the following examples, a maximum of $k = 6$ change points was used. In neither example was the BIC minimum with either 5 or 6 change points and the relative values of the BICs did not change much when k was reduced from 6 to 5, giving some comfort to the assumption that the models $H_0, \ldots, H_6$ cover all realistic possibilities.

Example 3. Kananaskis River Watershed

Johnson and Larsen (1991) present results of a fire history study of the 495 km^2 area of the Kananaskis watershed on the eastern side of the southern Rocky Mountains in Alberta, with a climate "transitional between plains and cordilleran types." Attempts by Johnson and Larsen to divide the map into spatial subunits with distinct fire hazard rates were unsuccessful. The time-since-fire distribution for the whole study area is displayed in Figure 3. There are $m = 40$ age classes of width $T = 10$ yr. The lower panel (a logarithmic plot of cumulative frequency against time since fire) suggests a number of possible change points (e.g., at 40, 60, 130, 230 and 280 years ago). Table 3 presents the MLEs of change points for models $H_0, \ldots, H_6$, along with the BICs and posterior probabilities assuming a uniform prior on $H_0, \ldots, H_6$. The overdispersion parameter was estimated under H_6.

The only plausible models appear to be H_2, H_3, and H_4, with H_3 being by far

TABLE 3   Maximum Likelihood Estimates of Change Points in Various Models, Associated BICs, and Posterior Probabilities of the Various Models, Assuming *a priori* That All Seven Models Are Equally Probable (for Kananaskis Watershed time-since-fire data)

Model	MLEs of change points	BIC	Posterior probability
H_0	—	−55.52	0.000
H_1	4	−149.06	0.004
H_2	4, 6	−156.35	0.171
H_3	4, 6, 24	−159.39	0.780
H_4	4, 6, 13, 23	−153.35	0.038
H_5	4, 6, 7, 13, 23	−149.42	0.005
H_6	4, 6, 7, 13, 19, 27	−136.71	0.000

TABLE 4   Maximum Likelihood Estimates and 95% LR Confidence
Intervals for the Fire Cycle in the Four Epochs between the Three
Estimated Change Points in Model H_3 (for Kananaskis Watershed
time-since-fire data)

Epoch i	Date	Fire cycle (yr)	
		MLE	95% Con. Int.
1	1940–1980	6409	969–715,000
2	1920–1940	49	34–73
3	1750–1920	136	101–189
4	pre-1750	48	30–85

the most plausible. Since the change point at 4 (1940) and 6 (1920) appear as MLEs for all these models, one can conclude with a very high degree of certainty that there were indeed changes in fire frequency at around those times. Furthermore, there is very strong support of an additional change at 23 (1750).

Of course by varying the prior probabilities of the various models, one can change the posteriors. However, a substantial skewness in the prior is required to shift the highest posterior probability from H_3. Thus, one can conclude that the data contain strong support for the three-change-points model, with the estimated changes occurring around 1940, 1920, and 1750.

The MLEs of the fire cycle (inverse of the hazard rate) and 95% likelihood ratio confidence intervals (Reed *et al.*, 1998) in the four epochs separated by the three identified change points are displayed in Table 4. The MLEs are also shown as line segments superimposed on the semilog cumulative frequency plot in Figure 5 (top panel). As one would expect, the confidence intervals for the fire cycle in adjacent epochs do not overlap. The estimated hazard for the post-1940 epoch is negligible. In fact, less than one tenth of one percent of the whole study area burned in the 40-yr period (1940–1980).

Johnson and Larsen (1991) graphically identified a change point around 1730 and estimated the pre-1730 fire cycle at about 50 years, which agrees well with the preceding results. However, they failed to identify the more recent change points identified here.

Example 4. Glacier National Park

Figure 4 presents data obtained by Johnson *et al.* (1990) from stand-origin maps for Glacier National Park (600 km² of forested land) in the Rocky Mountains of British Columbia. The vegetation is classified as being of the Interior Wet Belt Forest type. The data were presented in 20-yr age classes ($T = 20$, $m = 21$). Table 5 gives details of MLEs, BICs, and posterior probabilities under a uniform prior for models $H_0, H_1, \dots, H_6$.

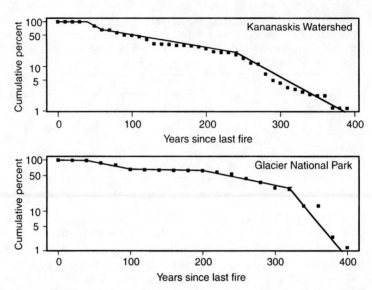

FIGURE 5 Epochs of constant hazard of burning as determined by the analysis in Section IV for Kananskis Watershed and Glacier National Park. The points are the cumulative percentage areas (on logarithmic scale) with time since fire exceeding the time (in years) on the horizontal axis. The ranges of the line segments correspond to distinct epochs of constant hazard of burning; the slopes of the segments correspond to the maximimum likelihood estimates of the hazards.

The data provide support only for models with four or more change points, with the model H_4 standing out with a very large posterior probability. To shift the posterior mode from H_4 requires a very substantial skewness in the prior distribution. Consequently the four-change-point model seems by far the most plausible, with estimated change points at 2, 5, 10, and 16 (i.e., around 1940,

TABLE 5 Maximum Likelihood Estimates of Change Points in Various Models, Associated BICs, and Posterior Probabilities of the Various Models, Assuming *a priori* That All Seven Models Are Equally Probable (for Glacier National Park time-since-fire data)

Model	MLEs of change points	BIC	Posterior probability
H_0	—	201.63	0.000
H_1	16	74.38	0.000
H_2	12, 16	42.79	0.000
H_3	5, 10, 16	28.08	0.000
H_4	2, 5, 10, 16	−9.27	0.797
H_5	2, 5, 10, 12, 16	−6.98	0.202
H_6	2, 5, 10, 12, 16,18	1.76	0.002

TABLE 6   Maximum Likelihood Estimates and 95% LR Confidence Intervals for the Fire Cycles in the Five Epochs between the Four Estimated Change Points in Model H_4 (for Glacier National Park time-since-fire data)

Epoch		Fire cycle (yr)	
i	Date	MLE	95% Con. Int.
1	1940–1980	1980	565–16700
2	1880–1940	156	40–181
3	1780–1880	1827	673–8102
4	1660–1780	151	106–224
5	pre-1660	25	17–42

1880, 1780, and 1660). There is some possibility of a fifth change point, estimated at 1740. MLEs and 95% likelihood ratio confidence intervals for the fire cycle (inverse of the hazard rate) in the five epochs separated by the four estimated change points under H_4 are displayed in Table 6. The MLEs are also shown as line segments superimposed on the semilog cumulative frequency plot in Figure 5 (bottom panel).

It is worth noting that a very similar model was identified for these data using backward elimination methods (Reed, 1998), the only difference being the change point in the 18th century was estimated at 12 (1740) rather than 10 (1780).

REFERENCES

Heinselman, M. L. (1973). Fire in the virgin forests of the Boundary Waters Canoe Area. *Quat. Res.* 3, 329–382.

Johnson, E. A., and Larsen, C. P. S. (1991). Climatically induced change in fire frequency in the southern Canadian Rockies. *Ecology* 72, 194–201.

Johnson, E. A., Fryer, G. I., and Heathcott, M. J. (1990). The influence of man on frequency of fire in the interior wet belt forest, British Columbia. *J. Ecol.* 78, 403–412.

Larsen, C. P. S. (1996). Spatial and temporal variations in boreal forest fire frequency in northern Alberta. *J. Biogeog.* 24, 663–673.

Reed, W. J. (1994). Estimating the historic probability of stand-replacement fire using the age-class distribution of old-growth forest. *For. Sci.* 40, 104–119.

Reed, W. J. (1997). Estimating historical forest fire frequency from time-since-fire sample data. *IMA J. Appl. Med. Biol.* 14, 71–83.

Reed, W. J. (1998). Determining changes in historical forest fire frequency from a time-since-fire map. *J. Agri. Biol. Environ. Stat.* 3, 430–450.

Reed, W. J. (2000). Reconstructing the history of forest fire frequency—Identifying hazard rate change points using the Bayes' Information Criterion. *Can. J. Stat.* 28, 353–365.

Reed, W. J., Larsen C. P. S., Johnson, E. A. and MacDonald, G. M. (1998). Estimation of temporal variations in fire frequency from time-since-fire map data. *For. Sci.* 44, 465–475.

CHAPTER 13

Duff Consumption

K. MIYANISHI

Department of Geography, University of Guelph, Guelph, Ontario, Canada

---

I. INTRODUCTION

Forest soils (e.g., spodosols) are often characterized by a distinct layer of partially to well-decomposed organic matter lying above the A horizon of the mineral soil. This organic layer consists of three distinguishable horizons: the top L (litter) horizon, composed of the undecomposed litter in which the materials are readily discernible as leaves, twigs, etc.; the middle F (fermentation) horizon, characterized by partially decomposed litter which is still recognizable as to its origins; and the bottom H (humus) horizon of well-decomposed organic matter in which the original structures are no longer discernible (Canada Soil Survey Committee, 1978). The boundary between the F and H horizons can sometimes be quite distinct due to a concentration of white fungal hyphae although in some cases the two horizons are almost indistinguishable from each other (Berg and Ekbohm, 1991). In the forestry and fire ecology literature, the F and H horizons together are generally referred to as duff (e.g., Van Wagner, 1972; Dyrness and Norum, 1983; Merrill and Alexander, 1987), whereas the term *forest floor* generally includes all three horizons (i.e., the litter and duff) (Merrill and Alexander, 1987), but see also Chapter 4, Section II.A.

Consumption of litter and duff by fire has been of interest for some time to foresters and ecologists (e.g., Chrosciewicz, 1959, 1967, 1974; Adams, 1966; Ahlgren, 1970), particularly those working in conifer-dominated forest in cool to cold climates (e.g., boreal and subalpine forest, Douglas fir forests of the Pacific Northwest) for a number of different reasons. Both litter and duff, especially the upper F layer, dry out more rapidly than mineral soil (Van Wagner, 1987). As a result, it has long been observed that seedlings establishing on such substrate are more subject to drought mortality than seedlings establishing on mineral soil. Charron (1998) found that, in the boreal mixedwood forest in Saskatchewan, first-year tree seedling mortality was high enough on litter and the F layer to virtually preclude successful seedling recruitment on these substrates. Successful post-fire tree regeneration in these forests is therefore determined to a large degree by the extent to which duff is removed by fire and suitable seedbed is made available (Chrosciewicz, 1974, 1976; Zasada et al., 1983; Thomas and Wein, 1985; Weber et al., 1987). The extent to which duff is consumed by fire would also influence understory plant mortality due to lethal heating (if not combustion) of roots, rhizomes, and perennating buds of shrubs and herbaceous perennials that are found within the duff (Flinn and Wein, 1977; Hungerford et al., 1995; Schimmel and Granstrom 1996; see also Chapter 14 in this book). Furthermore, as duff is consumed by fire, any dormant seeds that had accumulated within the duff would also be consumed. Therefore, the extent of duff consumption can influence the density and species composition of trees, shrubs and herbs regenerating after a fire. Duff also acts as a reservoir of nutrients for recycling (Weber, 1985; Weber et al., 1985); therefore, post-fire nutrient cycling would be influenced by the presence or absence of duff. Finally, consumption of duff by fire and subsequent mineral soil exposure can leave sites (particularly on steep slopes) vulnerable to soil erosion.

This phenomenon of organic matter consumption in forest fires is not restricted to mid- to high-latitude coniferous forests. Some humid tropical forests develop root mats which are thick layers of litter impregnated with roots (Stark and Jordan, 1978); these forests include high-altitude mountain forests and tropical lowland forests growing on highly weathered acid soils (Kingsbury and Kellman, 1997; Zech et al., 1997). Development of these root mats has been interpreted as an adaptation for nutrient retention (Jordan, 1982; Jordan and Herrera, 1981) as well as a response to aluminum toxicity of the mineral soil (Kingsbury and Kellman, 1997). Litter decomposition, mineralization, and most of the plant uptake of the released mineral nutrients are concentrated in this organic root mat layer above the mineral soil (Stark and Jordan, 1978). The generally larger seeds of tropical tree species, resulting in relatively large seedlings (Richards, 1952), and the abundance of precipitation in the humid tropics suggest that removal of this organic layer may not be a major factor in seedling survivorship as it is in the higher latitude conifer forests. Thus, Kellman and

Meave (1997) found that neither germination nor early seedling survivorship was significantly higher in burned plots than in control plots within the tropical gallery forests in Belize. However, they did find higher seedling densities in plots from which the litter layer had been manually removed and bare soil exposed. Also, given the abundance and concentration of roots in these organic layers above the mineral soil, consumption of this layer by fire would be expected to have a significant impact on plant mortality (both canopy and understory) as well as on nutrient cycling (Stark and Jordan, 1978). Although fire may have been a relatively uncommon phenomenon in undisturbed humid tropical forests (Uhl *et al.*, 1988), it is becoming a more widespread disturbance in the tropics due to changes in land use (Uhl and Kauffman, 1990; Kauffman, 1991). Kauffman *et al.* (1995) noted rootmat consumption in forest clearance fires in the Brazilian Amazon. These human-set fires as well as lightning-caused wildfires from savannas may spread to adjacent undisturbed forests, particularly during unusually dry periods (Leighton and Wirawan, 1986). Extensive rootmat consumption by wildfires has been observed recently in Venezuela (M. Kellman, personal communication) and in the cloud forests of Mexico (J. Meave, personal communication).

The physical/chemical processes involved in the consumption of organic soil horizons by fire are the same regardless of whether they are located in boreal, temperate or tropical forests. This chapter will explain these processes and present some of the models proposed in the combustion literature. However, in order to investigate combustion of any fuel, it is necessary to know something about its chemical composition and physical characteristics.

II. CHARACTERISTICS OF DUFF

Duff is derived from plant litter (leaves, twigs, etc.) which is largely composed of cellulose, hemicellulose, and lignin (Mason, 1976). Both cellulose and hemicellulose decompose much more rapidly than lignin (Alexander, 1977; Berg and Staaf, 1981; Berg *et al.*, 1982); lignin has been described as "the most recalcitrant among organic compounds in litter" (Aber and Melillo, 1991). Thus, the chemical composition of duff shows a gradient with depth of decreasing proportions of cellulose/hemicellulose and increasing proportions of lignin (Berg *et al.*, 1982). In other words, the F layer would have a greater proportion of cellulose/hemicellulose and a smaller proportion of lignin than the H layer. Unlike cellulose, lignin is a complex phenolic macromolecule whose units are not linked in an organized repeating manner; thus the precise structure of lignin is not known, and each lignin molecule may be different (Taiz and Zeiger, 1991). Furthermore, within the plant tissues, lignin is covalently bound to cellulose and other polysaccharides. Although thermal degradation (pyrolysis) and combustion chemistry of cellulose have been well studied (e.g., Lewellen *et al.*, 1976; Moussa *et al.*, 1976; Bradbury *et al.*, 1979; Shafizadeh and Sekiguchi, 1984; Ohlemiller, 1985; 1990a; 1990b), there is much less known about lignin combustion.

Plant litter also contains compounds such as cutin, suberin, and associated waxes plus a host of secondary compounds, generally grouped as terpenes/ terpenoids, phenolic compounds and alkaloids. These secondary compounds are generally specific to a plant species or a related group of species (Taiz and Zeiger, 1991). Thus, the litter and the resulting duff layer would consist of a mixture of compounds that varies spatially both within and between sites, depending on the species composition of the plant community (Berg and Ekbohm, 1991). This heterogeneity in the chemical composition of duff could present challenges in describing its combustion chemistry and attempting to develop general models of its combustion.

The F and H layers also differ in physical characteristics. The H layer is composed of smaller particles and has a higher bulk density than the F layer (Clayton *et al.*, 1977). These differences would influence their thermophysical properties such as conductivity and diffusivity and thus play a role in the rates of combustion propagation in the two layers.

Although very great depths of organic matter can accumulate in waterlogged sites, often creating large peat deposits in mid- to high-latitude wetlands, the thickness of the duff layer in better-drained upland boreal forest generally varies from 5 to 30 cm (Chrosciewicz, 1974, 1989; Dyrness and Norum, 1983; Schimmel and Granstrom, 1996). Green *et al.* (1993) similarly described F + H horizons in various types of coniferous forests on the west coast of North America ranging in depth from 4 to 25 cm and in central Siberia ranging from 8 to 10 cm. Organic root mats in tropical forests have been reported ranging from 0.2 to 10.6 cm (Kingsbury and Kellman, 1997) and from 15 to 40 cm (Jordan and Herrera, 1981). Thus, it would seem that the organic layer in non-wetland forests over a wide range of climatic conditions generally ranges from a few centimeters to 30 – 40 cm in thickness. The thickness of organic matter accumulation in any particular site depends primarily on the decomposition rate which is influenced by temperature and moisture as well as litter quality characteristics such as pH, base content, and presence of polyphenols/ tannins (Williams and Gray, 1974; Mason, 1976; Delaney *et al.*, 1996; Krause, 1998). Between-site differences in microclimate and hydrology and between species differences in litter characteristics thus result in differences in duff thickness between stand types within the same region (Potts *et al.*, 1983; Krause, 1998). For example, in boreal upland sites, *Picea mariana* stands tend to have thicker duff with distinct F and H layers, whereas *Pinus banksiana* stands have thinner duff often with an indistinct H layer (Van Wagner, 1972; Dyrness and Norum, 1983; Miyanishi *et al.* 1999). Also, duff depth depends on time since disturbance (primarily fire or logging). The high-intensity wildfires typical of the boreal forest generally result in some consumption of duff (Wein, 1983; Johnson, 1992). Krause (1998) also reported a significant decrease in duff following logging of a *Pinus banksiana* stand, with minimum levels occurring 6–9 yr after cutting. This post-logging decrease was attributed to the decreased litter input as well as increased decomposition rate due to higher ground temperatures. In the absence of disturbance, duff depths do not continue to accumulate indefinitely; eventually an equilibrium is reached between the rates of litterfall and decomposition. Krause (1998) found that duff accumulation leveled off after 12 years and that the amounts of duff in 12 to 16-year-old stands of *Pinus banksiana* in New Brunswick were not different from those found by Weber *et al.* (1985) in a 53-year-old *P. banksiana* stand in eastern Ontario. Also, Perala and Alban (1982) reported no significant differences in mean duff depths in stands ranging from 40 to 250 years old. Thus, duff depths in the boreal forest appear to reach an equilibrium within the expected interval between fires (Johnson, 1992).

III. EMPIRICAL STUDIES OF DUFF CONSUMPTION

Most studies of duff consumption have focused on measuring and attempting to predict mean values of duff consumption, either in terms of a mean decrease in depth or a mean loss of duff mass (e.g., Blackhall and Auclair, 1982; Dyrness and Norum, 1983; Little *et al.*, 1986; Weber *et al.*, 1987; Brown *et al.*, 1985, 1991; Reinhardt *et al.*, 1989). However, duff consumption within burned forest sites is extremely variable; as Dyrness and Norum (1983) observed after a fire in Alaska, "[t]he sites give the impression of many deeply burned holes scattered among broad areas of unburned or scarcely burned forest floor" (Figure 1). Similar observations have been noted by others (e.g., Zasada *et al.*, 1983). This patchiness in duff consumption appears to be more pronounced as one goes further north in the boreal forest. In terms of regeneration, the important factor is obviously not so much the overall or mean decrease in duff depth but the area of suitable seedbed exposed (Chrosciewicz, 1974, 1976). And since successful seedling recruitment is largely confined to the burned patches or holes in the duff (Charron, 1998), the total area and the spatial patterns of duff burnout would play a significant role in both the density and spatial distribution of tree seedlings in the postfire regenerating stands. These patterns of burned out duff also play a role in the spatial patterns of postfire understory species distribution and composition due to mortality of vegetatively reproducing species that are rooted in the duff as well as in the mineral soil beneath the duff (Flinn and Wein, 1977; Steward *et al.*, 1990; Schimmel and Granstrom, 1996).

To attempt to understand the areal extent and spatial patterns of duff consumption both within and between stands and to determine the factors that

FIGURE 1 Photograph of the forest floor following a fire in a boreal forest site in central Saskatchewan. The burned holes in the duff are clearly defined from the remaining unburned duff.

influence or determine these, it is necessary to understand the processes by which duff is consumed during forest fires. Fuels are consumed by either flaming combustion or smoldering combustion. Flaming combustion is the rapid oxidation of volatiles producing considerable visible radiation (see Chapter 2 in this book for a more detailed discussion). The volatiles in forest fires include terpenes, fats, oils and waxes; these are low-molecular-weight components of fuels that are readily volatilized at low temperatures. More importantly, the flames consume the gaseous pyrolysis products of the principal polymeric constituents of the woody fuels. Smoldering combustion is a nonflaming "slow combustion process in a porous medium in which heat is released by oxidation of the solid" (Schult *et al.*, 1995); it is characterized by thermal degradation and charring of the fuel (Ortiz-Molina *et al.*, 1978). Both forms of combustion involve the coupling of heat generation and heat transfer processes (Ohlemiller, 1985); however, the heat-generating process in flaming combustion is gas-phase oxidation, but it is solid-phase oxidation in smoldering combustion.

Many of the empirical studies on duff consumption make little or no mention of which combustion process is assumed to be operating. The approach in these studies is to produce strictly empirical regression models that attempt to predict duff consumption (decrease in depth or mass) from an assortment of measurable variables such as preburn duff depth and various measures and indices of duff moisture (e.g., Chrosciewicz, 1978a, 1978b; Sandberg, 1980; Blackhall and Auclair, 1982; Brown *et al.*, 1985; Little *et al.*, 1986). The major problem with this approach is that selection of the independent variables is rarely, if ever, justified on the basis of any combustion process by which duff is consumed. Even though many of these studies obtain statistically significant regressions, their empirical models are specific to the site, species, and weather conditions under which their data were collected. For example, Little *et al.* (1986) produced different predictive equations for slash burns within Pacific-Northwest Douglas fir clear-cuts depending on whether the number of days since rain was greater or less than 25 days and also on the presence or absence of a distinct dry layer of duff. Not only do the coefficients differ between these regression models but the predictive variables included in the models also differ. Thus, these studies and models provide little basis for improving our understanding of how duff is consumed or for developing more generalizable models of duff consumption. Furthermore, the lack of any process-based selection of variables in regression models can sometimes lead to trivial results and conclusions. For example, the best fit regression model obtained by Reinhardt *et al.* (1989) had only one independent variable, preburn duff depth. Since duff consumption was fairly complete in their experimental burns, they concluded that "[t]he amount of duff consumed depended mostly on how much was available for consumption."

Since duff consumption can potentially occur by both flaming and smoldering combustion, both processes will be addressed with respect to their role in duff consumption in wildfires.

IV. FLAMING COMBUSTION

The flaming front of a forest fire involves the flaming combustion of fine fuels (Johnson, 1992); flaming combustion is also discussed in Chapters 2, 3, and 5 in this book. Flaming combustion occurs when pyrolytic decomposition of the fuel supplies sufficient volatiles to fuel the flame. Lewellen *et al.* (1976) found that, if cellulose is heated rapidly and the initial products of pyrolysis do not have a chance to react and polymerize (such as if the fuel is a very thin layer of cellulose), it is completely volatilized. Thus, such rapid heating of cellulose would be expected to result in flaming combustion. However, when cellulose is heated slowly, the slow release of volatiles allows them to remain within the fuel matrix for an extended period; repolymerization of the volatiles occurs, producing a solid substance called char. While there may be some initial flaming under such circumstances, the rate of production of volatiles is insufficient to sustain flaming combustion and smoldering of the char residual occurs.

While thin layers of duff or the upper layers of deep duff, if very dry, can be consumed in the flaming front, deeper layers of more decomposed, more compacted duff are unlikely to undergo self-sustained flaming combustion in a wildfire due to their chemical and physical properties. Lignin is less readily volatilized and has a substantially higher char yield than cellulose when heated (Rothermel, 1976; Shafizadeh and DeGroot, 1976; Shafizadeh and Sekiguchi, 1984). The higher lignin:cellulose ratio of duff compared with undecomposed litter would mean a slower rate of release of volatiles. Therefore, heating of deep duff layers in a forest fire is unlikely to produce a sufficient concentration of volatiles to sustain flaming combustion. If flames from other burning objects are impinging on it, the duff can flame initially; however, it will not retain these flames once the external flames vanish.

The degree of volatilization is also influenced by the rate of heating. The rate of heating of duff in a forest fire is influenced by its moisture content. Duff dries out more slowly and therefore generally contains more moisture than litter (except shortly after a light rain that occurs following a long dry period). Thus, often when the litter has dried sufficiently to allow propagation of flaming combustion, the duff may still be quite moist. The latent heat of vaporization required to drive off the moisture in the duff would act as a large heat sink, slowing the rate of its heating by the flaming front.

Finally, according to Rothermel (1972), the rate of heat release of spreading fire per unit area decreases to a very low value as the packing ratio of the fuel approaches 10%. Packing ratio is the ratio of the fuel volume to the total volume of the fuel bed. Since duff has packing ratios greater than 10% (Frandsen, 1991a), the rate of heat generation by its oxidation is generally insufficient to sustain flaming combustion. Thus, as noted in Chapter 6 in this book, combustion of such "thermally thick" fuels has little, if any, impact on the behavior of the advancing firefront.

However, while the duff's contribution of fuel and heat to the flaming front may be relatively insignificant, heat from the flaming front can be the major factor in the smoldering combustion of duff (Van Wagner, 1972), particularly in transient (non-self-sustaining or dependent) smoldering that assumes that the duff is both ignited and consumed by heat transferred downward from the flaming front (Hawkes, 1993).

Van Wagner (1972) presented a heat budget-based model of duff consumption in upland pine stands which assumed such dependent smoldering. Before the temperature of the duff can be raised to its ignition temperature, heat from the fire must first drive off any moisture in the duff. Thus, the model assumes that the processes of heat transfer from the flaming front to the duff, evaporation of any moisture in the duff, and heating of the duff to the ignition point determine the amount of duff consumed.

Since the model assumes that the heat supply to the duff is derived only from downward radiation from the flaming front, this heat flux can be obtained using the Stefan-Boltzmann law:

$$q_{fl} = \varepsilon_{fl}\sigma T_{fl}^4 \tag{1}$$

where q_{fl} is the flux of energy radiated by the flame (W cm^{-2}), ε_{fl} is the emissivity of the flame, σ is the Stefan-Boltzmann constant (5.669 $\times$ 10^{-12} W cm^{-2} K^{-4}), and T_{fl} is the temperature (Kelvin) of the flame. Emissivity is the ratio of the radiative flux emitted by a body to that emitted by a black body at the same temperature, a black body being one that radiates at σT^4. Thus, the emissivity of virtually all heat sources would be some value less than one.

Since the duff is also radiating energy to the flame

$$q_d = \varepsilon_d \sigma T_d^4 \tag{2}$$

the net interchange of radiative energy from the flame to the duff (q) is

$$q = a_d q_{fl} - a_{fl} q_d \tag{3}$$

where a_d is the absorptivity of the duff and a_{fl} is the absorptivity of the flame. Absorptivity is the proportion of incident radiant flux that is absorbed by a body; a black body is a perfect absorber of radiation and has an absorptivity of one.

Substituting Eqs. (1) and (2) for q_{fl} and q_d, Eq. (3) becomes

$$q = a_d \varepsilon_f \sigma T_{fl}^4 - a_{fl} \varepsilon_d \sigma T_d^4 \tag{4}$$

By Kirchhoff's law, $\varepsilon_{fl} = a_{fl}$ and $\varepsilon_d = a_d$; therefore, Eq. (4) can be rewritten as

$$q = a_d \varepsilon_{fl} \sigma (T_{fl}^4 - T_d^4) \tag{5}$$

Van Wagner assumed that duff absorptivity (a_d) was 1, duff ambient temperature (T_d) was 20°C (293 K), and flame temperature (T_{fl}) was 800°C (1073 K). Substituting these values and, by convention, putting the 10^{-12} of the Stefan-Boltzmann constant into the denominator of the temperature terms, Eq. (5) becomes

$$q = 5.669\, \varepsilon_{fl} \left[\left(\frac{1073}{1000} \right)^4 - \left(\frac{293}{1000} \right)^4 \right] = 7.473\, \varepsilon_{fl} \tag{6}$$

For the duff to burn, the temperature of the duff must be raised to its ignition temperature. The quantity of heat required for this, H (measured in J g^{-1}) is given by

$$H = M c_w (1000 - 20) + lM + c_d(300 - 20) + 50.23 = 2585M + 460 \tag{7}$$

The first term represents the energy required to heat the moisture in the duff from its initial temperature (20°C) to boiling point (100°C) where M is the moisture content (% dry wt) and c_w is the specific heat of water (4.18 J g^{-1}).

The second term is the energy required to vaporize this water where l is the latent heat of vaporization $(2.25 \times 10^3 \text{ J g}^{-1})$. The third term gives the energy required to heat the duff from 20°C to its ignition temperature of 300°C, where c_d is the specific heat of duff $(1.47 \text{ J g}^{-1} \text{ °C}^{-1})$. The final term is the heat of desorption of bound water in the duff (50.23 J g^{-1}).

If the net energy from the flame given by Eq. (6) is equal to the energy required to burn the duff given by Eq. (7), the resulting energy balance can be written as

$$q = \frac{DVH}{w} \tag{8}$$

where D is the mass of duff consumed (g cm^{-2}), V is the rate of advance (cm s^{-1}) of the flaming front, and w is the horizontal thickness or width of the flaming front (cm). Solving for D gives

$$D = \frac{qw}{VH} \tag{9}$$

Since $w/V = \tau$, the residence time (s) of the flaming front, Eq. (9) can be rewritten as

$$D = \frac{\tau q}{H} \tag{10}$$

Although residence time (τ) varies with the fire behavior, for his experimental burns of pine stands Van Wagner assumed a value of 60 s. Substituting this value and Eqs. (6) and (7) into Eq. (10) gives

$$D = \frac{7.473\varepsilon_{fl} \cdot 60}{2585M + 460} \tag{11}$$

In order to be able to predict duff consumption (D) from duff moisture content (M), it is necessary to obtain an expression for ε_{fl}. In lieu of determining ε_{fl} from actual measurements of downward heat flux, Van Wagner used measured values of D and M to obtain estimates of ε_{fl} and then plotted these estimates of ε_{fl} against M, obtaining a straight line relationship:

$$\varepsilon_{fl} = 0.770 - 0.00543M \tag{12}$$

Finally, substituting Eq. (12) for ε_{fl} in Eq. (11) gives

$$D = \frac{7.473(0.770 - 0.00543M) \cdot 60}{2585M + 460} = \frac{345.253 - 2.435M}{2585M + 460} \tag{13}$$

The model showed a reasonable fit to data from 12 experimental fires, 6 in *Pinus resinosa* and *P. strobus* stands and 6 in *P. banksiana* stands (Figure 2). How-

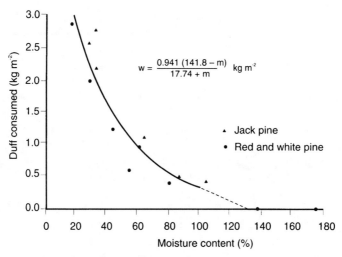

FIGURE 2   Empirical results and model prediction of duff consumption ($kg\ m^{-2}$) as a function of percent moisture content. From Van Wagner (1972).

ever, as Van Wagner noted, this fit of the empirical data to the model is not a test of model validity since the estimate of ε_{fl} was obtained empirically from measured values of D and M and then this estimate was used to obtain the final equation for determining D from M.

One limitation of the model is its prediction of 7.55 kg m^{-2} as the maximum amount of duff that could be consumed (even at 0% duff moisture which would almost never occur in most forests even under the driest weather conditions). This value converts to a maximum depth of duff that could be consumed of 10.6 cm, assuming a duff bulk density of 0.071 g m^{-3} (the standard Duff Moisture Code layer in the Canadian Fire Weather Behaviour System is 7 cm deep and 5 kg m^{-2} in weight). This maximum depth exceeds the duff depths found in the *Pinus* stands of the Great Lakes–St. Lawrence forest for which the model was developed and, therefore, may not be unreasonable. However, without some modification, the model would obviously not be generalizable to boreal forest stand types with deeper duff depths, such as *Picea mariana* stands. This limit also suggests that heat transfer from the flaming front may not be the main process driving duff consumption observed in stands with deep duff layers.

Despite some of the questionable assumptions and limitations, Van Wagner's (1972) model is notable for its process approach to modeling duff consumption, in contrast to almost all of the previous and subsequent studies in the forestry literature that have taken a strictly empirical approach.

Hawkes (1993) noted that this model did not take into account any heat generated by the combustion of the duff itself. Therefore, he modified Van Wagner's model by adding heat transfer from the combustion zone in the duff. Based on six temperature histories of smoldering in peat, he assumed an average temperature of 400°C for the duff combustion zone and used the Stefan-Boltzmann law to estimate this second heat source. As in the Van Wagner model, Hawkes' model assumes that all heat transfer is by radiation and does not consider conduction or convection. In testing this revised model in the lab using a propane burner as the heat source and peat as the fuel, Hawkes obtained a reasonably good fit between the predicted and measured depths of burn (converted from the weight of duff consumed). However, because he was interested only in dependent smoldering (such as might occur in slash burns), he limited the moisture levels of his samples to those that would not allow self-sustaining smoldering combustion (i.e., a minimum duff moisture content of 220%). Van Wagner's (1972) model predicted no duff consumption when the moisture content is greater than 134% because beyond that moisture level, the surface litter fuels would not carry a burn. Thus, the usefulness of Hawkes' model to more realistic conditions in fires (whether natural or prescription set) may be questionable. The problem was that at lower fuel moistures, self-sustained smoldering occurred in the deep fuel beds. This suggests that in most forest fires duff would not be consumed only during passage of the flaming front. In fact, studies have found that duff consumption more typically occurs by self-sustained smoldering combustion *after* passage of the flaming front (e.g., Wein, 1981; 1983; Hungerford *et al.*, 1995). This may also partially explain why studies of experimental burns have not been able to find a strong relationship between fire intensity and duff consumption (e.g., Alexander, 1982).

Since the physical and chemical characteristics of duff, in fact, make smoldering combustion more likely than flaming combustion (as discussed earlier), and since smoldering is now generally recognized as the major process responsible for duff consumption in forest fires (Frandsen, 1991b; Johnson, 1992; Hungerford *et al.*, 1995), the next section will explain the process of smoldering and discuss some of the models that have been proposed for this process.

V. SMOLDERING COMBUSTION AND PYROLYSIS

Smoldering has been defined as "a self-sustaining, low temperature combustion process involving pyrolysis of the substrate ahead of a solid-phase combustion front" (Shafizadeh *et al.*, 1982). It is characterized by thermal degradation and charring of the solid material with evolution of smoke (Moussa *et al.*, 1976). Because it sometimes involves emission of visible glow, smoldering has also been referred to by some as glowing combustion (Williams, 1977; Johnson, 1992).

However, Drysdale (1985) makes a distinction between glowing combustion and smoldering and, although glowing combustion "is associated with the surface oxidation of carbonaceous materials," it differs from smoldering in that "thermal degradation of the parent fuel does not occur, nor is it required." On the other hand, if glowing combustion specifically refers only to the process of surface oxidation of a solid, it may be viewed as the final stage of the smoldering process (Simmons, 1995). Only porous materials which form a solid carbonaceous char upon thermal degradation are capable of self-sustained smoldering combustion (Drysdale, 1985). Duff as well as organic peat soils are capable of such sustained smoldering (Frandsen, 1987, 1991a,b; Hawkes, 1993; Hungerford et al., 1995).

A large number of studies have been conducted on smoldering in various materials such as dust/sawdust (Cohen and Luft, 1955; Palmer, 1957), cigarettes (Egerton et al., 1963; Muramatsu et al., 1979; Yi and Kim, 1996), rolled cardboard/paper and incense sticks (Kinbara et al., 1967; Sato and Sega, 1985), polyurethane foam (Ohlemiller et al., 1979) and cellulose (Lewellen et al., 1976; Moussa et al., 1976; Ohlemiller, 1985, 1990a,b). Few, if any, studies appear to have been conducted on smoldering in duff, although one study was conducted on packed beds of plant litter (Jones et al., 1994). Therefore, any understanding of smoldering combustion of duff must come from empirical studies and theoretical models of smoldering in other materials that have been better studied. A number of lab studies on smoldering have used peat as a substitute for forest duff with the justification that peat has similar particle sizes and bulk densities as duff (Frandsen, 1987, 1991a,c; Hartford, 1989; Hawkes, 1993). Dried commercial peat generally consists of 95% organic matter (Hawkes, 1993). Using Curie-point pyrolysis-mass spectrometry, Bracewell and Robertson (1987) showed that the chemical composition of peat was very similar to that of the F duff layer. Unfortunately, most of the literature on duff consumption (cited in an earlier section of this chapter) as well as the studies of smoldering in peat mentioned here indicate little awareness of the numerous studies of the smoldering process published in the combustion literature. What follows is an introduction to this literature and is not meant to be a comprehensive review.

Smoldering differs from flaming combustion in its less complete oxidation of the fuel, its lower temperature, and its slower rate of propagation (Ohlemiller, 1985). Ohlemiller noted that, despite wide variation in fuel type and fuel configuration, smoldering velocities do not vary greatly, suggesting that the process is limited by the oxygen supply rate rather than by the fuel oxidation kinetics. Moussa et al. (1976) and Williams (1977) had also concluded that smoldering velocities were generally governed by the rate of oxygen diffusion rather than by reaction kinetics. If the rate of oxygen supply is increased sufficiently, smoldering can undergo a transition to flaming combustion (e.g., the use of bellows to revive a dying fire).

As with flaming combustion, smoldering occurs through the coupling of heat release and heat transfer mechanisms. Smoldering is initiated by heating of a porous solid such as cellulosic polymers, resulting in its decomposition and yielding char and volatiles. Char is "the black solid residue of a variable aromatic nature remaining after the end of the initial rapid weight loss upon heating of a polymer" (Ohlemiller, 1985). Pyrolytic degradation of polymers such as cellulose is exothermic if oxygen is available or endothermic in the absence of oxygen; both processes produce char which then reacts with oxygen in an exothermic reaction. Sustained propagation of smoldering depends on sufficient heat from the exothermic reactions being transferred ahead of the smoldering zone to result in pyrolysis of the virgin fuel and exposure of fresh char which can then oxidize (Drysdale, 1985). Thus, pyrolysis and char production are essential steps in smoldering combustion.

Pyrolysis temperatures for the three main constituents of duff differ: hemicellulose 200–260°C, cellulose 240–350°C, and lignin 280–500°C (Roberts, 1970). Lignin requires higher temperatures and yields less volatiles and more char than cellulose (Shafizadeh and DeGroot, 1976; Drysdale, 1985). The pyrolysis of cellulosic materials has been extensively studied (e.g., Broido, 1976; Lewellen et al., 1976; Shafizadeh and DeGroot, 1976; Bradbury et al., 1979; Shafizadeh et al., 1982; Shafizadeh and Sekiguchi, 1984); therefore, most of the following section on pyrolysis will come from studies of cellulose pyrolysis. Pyrolysis is also discussed in Chapters 2 and 3 in this book.

Low temperature pyrolysis of cellulose produces a large variety of end products including CO_2, CO, CH_4, C_2H_4, H_2, H_2O, low-molecular-weight organic acids, and levoglucosan (Shafizadeh and DeGroot, 1976). Lewellen et al. (1976) proposed that, with decreased heating rate, these products of pyrolysis have a longer residence time within the pyrolyzing matrix. This extended period of contact between the primary products allows their repolymerization, cracking or cross-linking to produce char. Thus, Lewellen et al. (1976) and Bradbury et al. (1979) found that char yields decreased with increased heating rate (e.g., from 20% of original cellulose weight at $\sim$300°C s^{-1} to 2% at $\sim$1500°C s^{-1}). With an appropriate heating rate, thin samples of cellulose can be completely volatilized with no production of residual char. Besides rate of heating, char formation is also a function of particle size and the presence of moisture and inorganics; larger particle size and the presence of moisture and inorganics increase the yield of char (Shafizadeh and DeGroot, 1976). Thus, the relatively large particle sizes in duff, the presence of duff moisture, and the relatively slow heating rate in a forest fire would be expected to result in high char yields from duff.

Although studies of pyrolysis of cellulosic materials indicate the extreme complexity of the process, Bradbury et al. (1979) found that the pyrolytic gasification of cellulose could be described by either a two- or three-step reaction scheme, depending on the temperature range (Figure 3). At higher tempera-

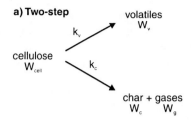

a) Two-step

cellulose
W_{cell}

k_v → volatiles W_v

k_c → char + gases W_c W_g

b) Three-step

cellulose W_{cell} $\xrightarrow{k_i}$ active cellulose W_A

k_v → volatiles W_v

k_c → char + gases W_c W_g

FIGURE 3 (a) Two-step and (b) three-step reaction schemes for pyrolytic gasification of cellulose where W refers to the normalized weights of the respective materials—cellulose (cell), activated cellulose (A), volatiles (v), char (c), and gases (g)—and k refers to the rate constants. From Bradbury *et al.* (1979). A kinetic model for pyrolysis of cellulose. *J. Appl. Polm. Sci.* **23**, 3271–3280. Copyright © 1979 John Wiley & Sons, Inc. Reprinted by permission of John Wiley & Sons, Inc.

tures, pyrolysis of cellulose can be modeled as two competitive first-order reactions (Figure 3a) with the following rate equations:

$$\frac{-dW_{cell}}{dt} = (k_v + k_c)[W_{cell}] \tag{14}$$

$$\frac{dW_c}{dt} = 0.35k_c[W_{cell}] \tag{15}$$

where W_{cell}, W_v, W_c, and W_g are the normalized weights of cellulose, volatiles, char, and gases respectively (i.e., $W_{cell} + W_v + W_c + W_g = 1$). The rate constants, k_v and k_c, can be expressed by the following Arrhenius relationships:

$$k_v = 1.9 \times 10^{16} \exp\left(-\frac{47,300}{RT}\right) \text{min}^{-1} \tag{16}$$

$$k_c = 7.9 \times 10^{11} \exp\left(-\frac{36,600}{RT}\right) \text{min}^{-1} \tag{17}$$

where R is the universal gas constant and T is temperature. The numerical values are the preexponential factor and activation energy for each reaction and would be specific to the reaction. Temperature (T) at time t is given by

$$T = T_f - (T_f - T_0) \exp(-2.92t) \tag{18}$$

where T_f is the final pyrolysis temperature and T_0 is the initial temperature.

For lower temperatures where an initial period of accelerating weight loss is observed, Bradbury et al. (1979) proposed a model involving three reactions (Figure 3b); first, an active cellulose is formed by an initiation reaction; then the active cellulose decomposes by two competitive first-order reactions, the first yielding volatiles and the second yielding char and a gaseous fraction. The rate equations for this model are

$$\frac{-d(W_{cell})}{dt} = k_i[W_{cell}] \tag{19}$$

$$\frac{d(W_A)}{dt} = k_i[W_{cell}] - (k_v + k_c)[W_A] \tag{20}$$

$$\frac{d(W_c)}{dt} = 0.35k_c[W_A] \tag{21}$$

where W_A is the normalized weight of active cellulose.

The rate constant k_i can be expressed by the following Arrhenius relationship:

$$k_i = 1.7 \times 10^{21} \exp\left(-\frac{58,000}{RT}\right) \text{ min}^{-1} \tag{22}$$

Shafizadeh and Sekiguchi (1984) also found that the chemical composition and combustion behavior of cellulosic chars varied with the temperature at which the fuel was heated. Chars formed at lower temperatures had higher concentrations of aliphatic carbons, whereas chars formed at higher temperatures had higher concentrations of aromatic carbons. Aliphatic carbons are more reactive and burn at lower temperatures than the more stable aromatic carbons.

There is some empirical evidence for distinct endothermic pyrolysis and exothermic oxidation zones in smoldering (Muramatsu et al., 1979) which have been incorporated into some smoldering models (e.g., Moussa et al., 1976; Leisch, 1983; Peter, 1992). Separation of pyrolysis and oxidation zones depends on the relative movement of the oxygen supply gas and solid reaction zones. However, Ohlemiller (1985) also cited evidence for both endothermic nonoxidative pyrolysis and exothermic oxidative pyrolysis and proposed that the two processes compete for the virgin fuel (Figure 4). Ohlemiller concluded that "the available evidence points to the likelihood of a significant role for oxidative pyrolysis as an appreciable heat source in many [but not all] smoldering processes." Thus, Kashiwagi and Nambu (1992) and Di Blasi (1995) separated the degradation of cellulosic fuel to char into an endothermic nonoxidative reaction and an exothermic oxidation reaction.

Following pyrolysis of the virgin fuel and the production of char, the next major process involved in smoldering is the exothermic oxidation of the char, producing ash and gases. This process is the primary heat source for the propa-

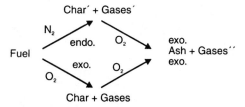

FIGURE 4 Scheme of thermal behavior in smoldering showing the two competing processes by which fuel is pyrolyzed—endothermic pyrolysis in the absence of oxygen and exothermic pyrolysis in the presence of oxygen. Subsequent oxidation of char is exothermic. Reprinted from *Prog. Energy Combust. Sci.*, Ohlemiller, T. J., Modeling of smoldering combustion propagation, pp. 277–310, 1985, with permission from Elsevier Science.

gation of smoldering; therefore, an understanding of smoldering propagation requires an understanding of the factors determining the rate of char oxidation. To identify the mechanisms controlling the propagation of smoldering, Ohlemiller (1990a) looked at the structure (spatial distribution of temperature, oxygen mole fraction, and residual organic mass fraction) of the smoldering wave as it moved laterally through a thick (18-cm) permeable horizontal layer of cellulose insulation; this is an inherently two-dimensional propagation process. Smoldering was initiated on one end face of the fuel layer, and the movement of the smoldering zone was followed by three sets of thermocouples embedded at two depths and at three positions along the fuel as well as by a movable oxygen probe. The distribution of the residual organic mass fraction was obtained by rapidly extinguishing smoldering by smothering with nitrogen and then sectioning the fuel layer. Figure 5 shows the profiles of temperature, oxygen mole fraction and residual organic mass fraction. The length of the smoldering wave was found to be about two to three times greater than the fuel depth (which was 18 cm). Since Palmer (1957) found a comparable ratio of wavelength to fuel thickness for sawdust layers only a few centimeters deep, Ohlemiller (1990a) concluded that similar factors, mainly the rate of oxygen supply, control the rate of smoldering over this range of fuel thicknesses. Figure 6 shows the oxygen inflow paths and heat release zones in a smoldering wave that Ohlemiller inferred from his analysis of the wave structure. He presented a further argument that the shape of the smoldering wave is determined by the rate of oxygen diffusion from above. Leisch (1983) also showed that combustion wave velocity generally decreased with depth in smoldering beds of grain dust and wood dust and concluded that the velocity of smoldering was determined by the net effect of heat generation (determined by oxygen diffusion rate from the surface) and heat loss from the surface.

A distinction is made in the smoldering literature between forward and reverse smoldering, the two limiting cases of one-dimensional smoldering propagation. The reactants in forward (or co-flow) smoldering enter the reaction zone

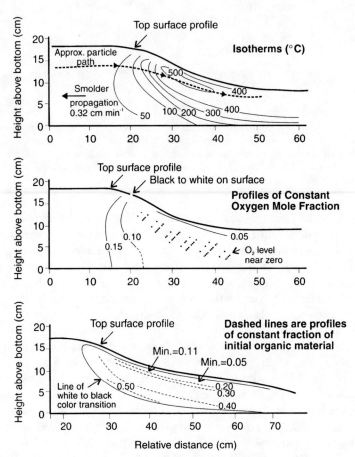

FIGURE 5 Steady state profiles of temperature, oxygen mole fraction, and fraction of remaining organic material superimposed on a cross-sectional view of the smolder wave. Reprinted by permission of Elsevier Science from Smoldering combustion propagation through a permeable horizontal fuel layer, Ohlemiller, T. J., *Combustion & Flame* 81, 341–353, Copyright 1990 by The Combustion Institute.

from opposite sides, somewhat analogous to diffusion flames (Figure 7). The reactants in reverse (or opposed flow) smoldering enter the reaction zone from the same side, somewhat analogous to a premixed flame (Buckmaster and Lozinski, 1996). The horizontal propagation process (Figure 6) has a mixed character. Smoldering studies also distinguish between situations of free flow or forced flow of air/oxygen to the fuel. Smoldering of duff following a forest fire would thus be typically categorized as a free flow forward smoldering process, except near the outward spreading front where it approaches a reverse smoldering character.

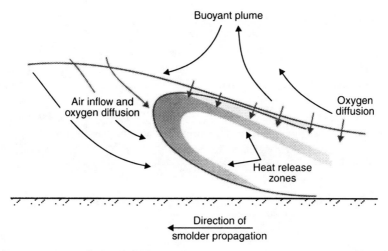

FIGURE 6   Oxygen inflow paths and heat release zones in a spreading smolder wave. Reprinted by permission of Elsevier Science from Smoldering combustion propagation through a permeable horizontal fuel layer, Ohlemiller, T. J., *Combustion & Flame* 81, 341–353, Copyright 1990 by The Combustion Institute.

In summary, the major heat sources involved in smoldering combustion are the processes of oxidative pyrolysis (polymer degradation) and char oxidation, whereas the major heat sinks are the processes of nonoxidative pyrolysis and water movement (Ohlemiller, 1985). Any modeling of smoldering combustion would thus require rate expressions for these reactions as well as for the transfer of heat between the sources and sinks.

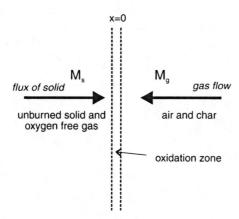

FIGURE 7   Illustration of forward smoldering in which the reactants enter the reaction zone from opposite sides, analogous to a diffusion flame. Reprinted by permission of Elsevier Science from An elementary discussion of forward smolder, Buckmaster, J. and Lozinski, D., *Combustion & Flame*, 104, 300–310, Copyright 1996 by The Combustion Institute.

VI. MODELS OF SMOLDERING COMBUSTION

Various heat transfer models have been proposed for the propagation of smoldering; Drysdale (1985) gives a highly readable description of a couple of the models (Moussa et al., 1976; Williams, 1977), and Ohlemiller (1985) provides an extensive review and evaluation of a large number of models. The simplest models are ones that are based on a greatly simplified heat balance and that derive a simple algebraic equation showing the dependence of smoldering velocity on various parameters (e.g., Cohen and Luft, 1955; Kinbara et al., 1967; Egerton et al., 1963).

As a model of smoldering, Williams (1977) used the fundamental equation for horizontal fire spread through any porous fuel layer, based on application of the energy-conservation principle:

$$q = \rho V \Delta h \tag{23}$$

where q is the net energy flux (W cm^{-2}) across the surface of fire inception, ρ is the bulk density of the fuel (g cm^{-3}), V is the rate of spread of smoldering (cm s^{-1}), and Δh is the thermal enthalpy change (J g^{-1}) in raising unit mass of fuel from ambient temperature to the ignition temperature. To determine the rate of smoldering, this equation can be rewritten as

$$V = \frac{q}{\rho \Delta h} \tag{24}$$

Obviously, in this model, there is no separate consideration of the endothermic/exothermic production of char and the exothermic oxidation of char, nor any consideration of the relative flow direction of air and the reaction zone. The model only considers the net energy flux from these processes. As the net energy flux (q) decreases, the rate of propagation of smoldering decreases.

Assuming that the ignition temperature is not too different from the maximum temperature in the exothermic oxidation zone (T_{max}), then

$$\Delta h = C_{fu}(T_{max} - T_0) \tag{25}$$

where C_{fu} is the heat capacity of the fuel (J g^{-1} °C^{-1}) and T_0 is the ambient temperature (°C). Also, assuming that heat transfer from the exothermic oxidation zone to the endothermic pyrolysis zone is by conduction and that a quasi steady state exists, Eq. (24) becomes

$$V \approx \frac{\lambda(T_{max} - T_0)}{x} \cdot \frac{1}{\rho C_{fu}(T_{max} - T_0)} \tag{26}$$

The term ($T_{max} - T_0$) cancels out, leaving:

$$V \approx \frac{\lambda}{\rho C_{fu}} \cdot \frac{1}{x} \tag{27}$$

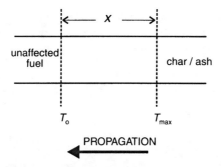

FIGURE 8   Simple heat transfer model for the propagation of smoldering combustion. From Drysdale (1985). Copyright John Wiley & Sons. Reproduced by permission of John Wiley & Sons Limited.

where λ is thermal conductivity of the fuel (W cm^{-1} °C^{-1}) and x is the distance (cm) over which heat is being transferred (Figure 8), which has been found by Palmer (1957) to be on the order of 0.01 m or 1 cm. Thus, according to this model, the rate of propagation is independent of the maximum temperature in the oxidation zone. Furthermore, since thermal diffusivity (α) is defined as

$$\alpha = \frac{\lambda}{\rho C} \tag{28}$$

Eq. (27) can be rewritten simply as

$$V \approx \frac{\alpha}{x} \tag{29}$$

Both Ohlemiller (1985) and Drysdale (1985) note that this model gives very crude estimates and can only be used to give an order of magnitude estimate of the rate of propagation of a smoldering wave; for example, for dry duff, $\alpha \approx$ 10^{-3} cm^2 s^{-2} (Marshall and Holmes, 1979). Thus, the estimated rate of propagation of smoldering within dry duff would be of the order of 10^{-3} cm s^{-1} (10^{-5} m s^{-1}). Despite this criticism of the model, Jones et al. (1994) applied it to smoldering in packed beds of *Casuarina* needles and found that the rate of propagation estimated by the model (4×10^{-4} m s^{-1}) was remarkably close to the mean propagation rate obtained from six experimental burning trials (4.7×10^{-4} m s^{-1}). Even though this value is an order of magnitude greater than those reported for smoldering of other lignocellulosic materials, Jones et al. attributed this difference to the configuration of the material and the effect of this on the width of the burning zone which they found to be approximately two needle diameters, about 0.001 m or 1 mm, in contrast to the 0.01 m given by Palmer (1957).

A second smoldering model is that presented by Kinbara et al. (1967); this model considered downward propagation of smoldering on vertical cylindrical samples (e.g., circular rods of rolled paper or incense sticks) that are exposed to air on all sides. Their model has three assumptions:

1. The heat produced by combustion per unit time is proportional to the rate of air supplied by diffusion which is proportional to $(T - T_0)$ where T is the temperature of the sample and T_0 is the ambient temperature (°C);
2. This heat is inversely proportional to $(T_{ig} - T_0)$, where T_{ig} is the ignition temperature of the sample;
3. A stagnant layer of combustion gases covers the surface of the sample, and both the fresh-air supply to the sample and the heat dissipation from the sample to the air take place at the surface of this layer.

Given these assumptions, Kinbara *et al.* proposed an equation for the propagation velocity (V) of smoldering:

$$V^2 = \left(\frac{4\lambda}{c^2\rho^2}\right)\left(\frac{p}{S}\right)\left(\left[\frac{q}{T_{ig} - T_0}\right] - h\right) \tag{30}$$

where λ is thermal conductivity, c is specific heat, ρ is density of the fuel, S is cross-sectional area of the smoldering solid, p is the periphery of the stagnant layer, h is the heat transmission coefficient, and q is the heat produced by combustion of the solid. As with Williams' model, this model uses only the net heat flux from the endothermic and exothermic reactions involved in smoldering. Kinbara *et al.* found that their model gave a good fit to data obtained from downward smoldering of small (less than 20 mm diameter) fuels (rolled paper and incense sticks). This model was specifically developed to describe smoldering in these very small cylindrical fuels and was not intended to be generalized to other very different fuel configurations such as that found on the forest floor. Thus, the area of the smoldering zone in this model is constrained by the diameter of the fuel, and duff on the forest floor would represent a fuel of virtually infinite diameter. Also, due to convection currents bringing air upward around the periphery of the cylindrical fuel, a cone-shaped smoldering zone develops; this would not be the case for downward smoldering of a semi-infinite fuelbed. However, Kinbara *et al.*'s model represents another relatively simple model of smoldering based on energy conservation principles.

A third model reviewed by Drysdale (1985) and Ohlemiller (1985) is an example of a model that does not use a net heat flux but deals separately with the endothermic pyrolysis of cellulose and exothermic char oxidation. This steady state model of smoldering (again in cellulose cylinders) developed by Moussa *et al.* (1976) assumes that these processes occur in two zones separated by an interface with suitable matching requirements (Figure 9). Also, unlike the pre-

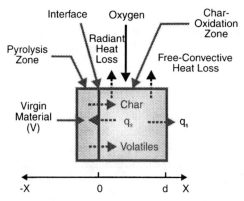

FIGURE 9   Idealized model of smoldering combustion showing heat and mass transports to and from the pyrolysis and char-oxidation zone. From Moussa *et al.* (1976).

vious two models which do not deal with extinguishment, this model can determine the extinguishment limit.

In the first step of this model, the temperature and speed of a steadily propagating pyrolysis zone are determined for a given heat flux at the interface. The major process is conductive heat transfer within the fuel which is undergoing thermal degradation (assumed to be an endothermic process). The governing equations for the cellulose pyrolysis model can be found in the appendix of Moussa *et al.* (1976). The following equation is an approximate closed-form solution that they present for their pyrolysis model:

$$V = \left\{ \frac{\lambda}{\rho_\infty c} \frac{F_c}{\ln \Omega} \left[Ei\left(-\frac{E_c}{RT_s}\right) + \frac{F_g}{F_c} Ei\left(-\frac{E_g}{RT_s}\right) - Ei\left(-\frac{E_c}{RT_x}\right) - \frac{F_g}{F_c} Ei\left(-\frac{E_g}{RT_x}\right) \right] \right\} \tag{31}$$

where V is propagation speed, λ is thermal conductivity, ρ is density, c is specific heat of the fuel, F_c and F_g are the frequency factors for the char and volatile forming reactions (assumed to occur in parallel), Ω is the ratio of virgin fuel density at the interface over that at $x \to -\infty$, E_c and E_g are the activation energies of the char and volatile formation reactions, R is the universal gas constant, and the interface temperature $T_s = T_\infty + q_s/(\rho_\infty c V)$. $Ei(x)$ is the exponential integral function whose values can be obtained from books of standard mathematical tables (e.g., Beyer, 1991). The first term (outside the square brackets) represents heat transfer which is assumed to be solid-phase conduction within a semi-infinite slab and is dependent on the thermophysical properties of thermal conductivity (λ), density (ρ), and specific heat (c) of the fuel (see further discussion of conduction in Chapter 14 in this book). The terms within the square brackets are a result of the pyrolysis kinetics.

In the next step of the model, the heat flux generated by char oxidation is determined for specified ambient oxygen mole fractions and partial pressures. The reaction rate (k) for char oxidation is represented by

$$k = FP_{O_{2,s}}^{1/2} \exp\left(-\frac{E_c}{RT_s}\right) \qquad (32)$$

where F is the frequency factor for char oxidation, $P_{O_{2,s}}$ is the partial pressure of oxygen at the char surface, E_c is the activation energy for char oxidation, R is the universal gas constant, and T_s is the temperature at the char surface.

The oxygen flux at the surface (J_{O_2}) is governed by diffusion through the free-convective boundary layer surrounding the char and is given by

$$J_{O_2} \simeq 2.2 \times 10^{-5} \sqrt{\frac{T_m}{\delta}} \frac{(P_{O_{2,\infty}} - P_{O_{2,s}})}{P} \qquad (33)$$

where T_m is the mean temperature at the boundary layer, δ is the average thickness of the boundary layer, $P_{O_{2,\infty}}$ and $P_{O_{2,s}}$ are the partial pressures of oxygen at $x \to -\infty$ and $x = 0$, respectively, and P is pressure.

The steady partial pressure of oxygen at the char surface can be obtained by equating its rate of arrival [Eq. (33)] with its rate of consumption, which is given by Eq. (32) times a stoichiometric ratio, ϑ, according to the reaction $C + O_2 \to CO_2$. To obtain the heat flux generated by char oxidation, the rate of char reaction [Eq. (32)] is multiplied by its heat of combustion (32 kJ g^{-1}).

Finally, the heat flux (q_s) available for pyrolysis is the heat flux generated by char oxidation (q) minus heat loss by radiation and free convection to the surroundings through the circumferential area and by conduction to the residual char through the cross-sectional area:

$$q_s \pi \frac{d^2}{4} = \frac{1}{2}(\pi d^2 \cdot (q - \varepsilon \sigma T_s^4 - h(T_s - T_\infty))) \qquad (34)$$

where d is the diameter of the fuel sample, T_s is the temperature at the oxidizing surface ($x = 0$), T_∞ is the temperature at $x \to -\infty$, and h is the free-convection heat transfer coefficient (McAdams, 1942; see also Chapter 14 in this book). The heat loss by conduction to the residual char is assumed to equal the feedback heat flux available for pyrolysis. By matching the available heat flux with that required for pyrolysis, the steady smoldering speed and temperature and the extinguishment limit can be determined as shown in Figure 10. When the feedback heat flux does not intersect the curve for the heat flux required for propagation of pyrolysis, extinguishment occurs (as shown with the inverted U-shaped dashed curve). When the two curves are tangential, the extinguishment limit is predicted. The intersections of the two curves give two solutions; only the upper solution is stable. Thus, the usefulness of this model is

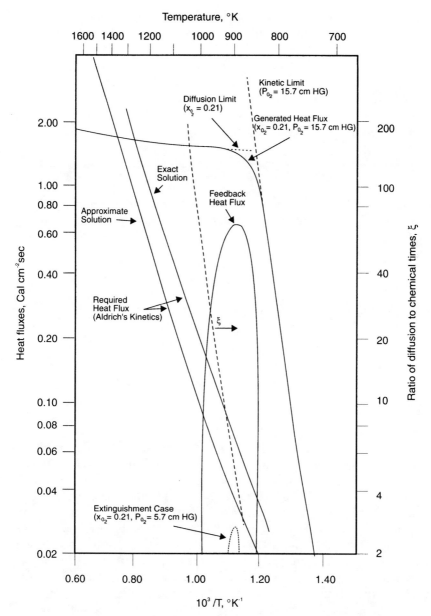

FIGURE 10 Heat fluxes versus inverse temperature where X_{O_2} is oxygen mole fraction and P_{O_2} is oxygen partial pressure (cm Hg). The feedback heat fluxes are calculated for the case of standard atmospheric conditions ($X_{O_2} = 0.21$, $P_{O_2} = 15.7$) and for the extinguishment case ($X_{O_2} = 0.21$, $P_{O_2} = 5.7$). Also shown is the approximate solution for the required heat flux for Aldrich's kinetics. From Moussa *et al.* (1976).

that, unlike the previous two models that only determine velocity of the smoldering wave, this one can be used to determine the extinguishment limit.

However, as with the Kinbara et al. (1967) model, the direct applicability of the model by Moussa et al. (1976) for studies of duff consumption is constrained by the fact that these models were developed for small vertical cylindrical fuels. Smoldering combustion in horizontal dust layers, as modeled by Leisch (1983), would be closer to the situation of smoldering in duff. In Leisch's model, velocity of smoldering is given by

$$V = \frac{\lambda}{\rho C} \frac{T_{max} - T_c}{T_c - T_0} \gamma_{ox} \tag{35}$$

where λ, ρ, and C are the thermal conductivity, density, and heat capacity of the fuel; T_{max} is the maximum temperature in the oxidation zone; T_c is charring temperature, and T_0 is ambient temperature. The value of γ_{ox} is given by

$$\gamma_{ox} = \frac{(\text{Surface Area})}{\text{Mass}} P_{O_2}^{1/2} k_{ox} \tag{36}$$

The rate of char oxidation, k_{ox}, is determined by

$$k_{ox} = F_{ox} \exp\left(-\frac{E_{ox}}{RT_{avg}}\right) \tag{37}$$

where $F_{ox} = F_{oxa} + F_{oxg}$, F_{oxa} is the preexponential factor for the oxidation of char to ashes, F_{oxg} is the preexponential factor for the oxidation of char to gases, $E_{ox} = E_{oxa} + E_{oxg}$ (the activation energies for the oxidation of char to ashes and gas), and $T_{avg} = (T_c + T_{max})/2$.

A limitation of Leisch's (1983) model in attempting to understand the development of burned holes in duff is that it is a one-dimensional model. Di Blasi (1995) presented a two-dimensional model of smoldering propagation through horizontal layers of cellulosic particles. In this model, the chemical processes involved in smoldering are modeled as one-step reactions for each main chemical pathway: pyrolytic degradation of cellulose to char and gases, oxidative degradation of cellulose to char and gases, and oxidation of char to ash and gases. The model does not include any gas phase reactions since they are generally not considered of importance in smoldering (Ohlemiller, 1985). The physical processes are then described by energy, momentum, and mass balance equations. The model essentially consists of ten governing equations that are solved numerically (a method of solution is given by Di Blasi, 1994). Simulation results of the model (Di Blasi, 1995) gave profiles of the structure of the smoldering wave in terms of temperature and char density, oxygen mass fraction and solid density, and gas overpressure (Figure 11). These profiles indicated that both

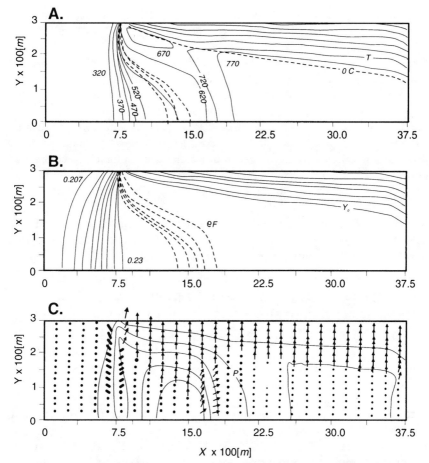

FIGURE 11 Simulated structure of the smoldering wave after 1280 s from ignition: (a) solid lines are contours of temperature (K) in steps of 50 K and dashed lines are contours of char density (kg m^{-3}) in steps of 0.75 kg m^{-3}; (b) solid lines are contours of oxygen mass fraction in steps of 0.023 and dashed lines are contours of solid density (kg m^{-3}) in steps of 6.8 kg m^{-3}; (c) contours of gas overpressure (atm) in steps of 1.5 × 10^{-3}. From Di Blasi (1995). Copyright Overseas Publishers Association N.V. with permission of Gordon and Breach Publishers.

pyrolytic and oxidative degradation of the fuel occur down to a depth of about 1 cm. Below that depth, only endothermic pyrolysis occurs due to depletion of oxygen. Thus, the solid degradation front advances more rapidly near the surface. However, the maximum reaction rates (and temperatures) occur just beneath the surface where significant amounts of oxygen are still available for both oxidation reactions but where heat loss from the surface is less than at the surface.

Numerous other one- and two-dimensional models of smoldering combustion can be found in the literature (e.g., Egerton *et al.*, 1963; Ohlemiller *et al.*, 1979; Moallemi *et al.*, 1993; Buckmaster and Lozinski, 1996; Yi *et al.*, 1998). Ohlemiller (1985) presented a general model of the thermophysics of propagation which he used to compare and evaluate various models of smoldering combustion propagation. He concluded that all of the models he reviewed, including that of Moussa *et al.* (1976), represented great simplifications of the general model as is evident by the lists of assumptions accompanying each of the models. The usefulness of these models, according to Ohlemiller, is not so much in being good predictive models but "in rationalizing certain experimentally-observed trends." Furthermore, they provide a basis for determining and testing the important factors that influence the rate of smoldering and the extinguishment limits. As noted earlier, without such a physical process basis, empirical studies of duff consumption can provide little justification for selection of variables to measure and can lead to spurious or trivial relationships that do not advance our understanding of duff consumption.

VII. CONTRIBUTION OF SMOLDERING COMBUSTION MODELS TO UNDERSTANDING OF DUFF CONSUMPTION

The purpose of this chapter has been to introduce the reader to some of the combustion literature that would be useful for developing an understanding (and potentially models) of duff consumption by fire. It was not intended to be a comprehensive review or evaluation of the smoldering literature (see Ohlemiller, 1985). There has also been no discussion here of ignition of duff (see Chapters 2 and 11 in this book for more on this topic), and the focus has been on the propagation/extinction of smoldering. It bears repeating that, with few exceptions, empirical studies of duff consumption reported in the forestry and ecological literature indicate no awareness of this extensive body of knowledge on smoldering, the major process by which duff is consumed by fire. Since the typical purely empirical approach results in very site-specific predictive models of duff consumption, the justification for some of these studies appears to be simply that a particular forest type in a particular region has not yet been studied. As mentioned in an earlier section, rarely, if ever, can the empirically derived predictive equations developed for a particular site, forest type, and set of fire weather conditions be used elsewhere and under different conditions. Therefore, rather than more empirical studies of duff consumption using a more or less arbitrary assortment of variables, what is required now is to make use of the available models of the smoldering process as a guide to select more appropriate variables for study and to work on further developments/adaptations of existing models of smoldering to make them useful for the study of duff consumption in forest fires.

Heat transfer and energy-balance-based models all indicate that smoldering can only be propagated and sustained if sufficient heat for pyrolysis is transferred to the virgin fuel from the oxidation zone. Thus, factors that increase heat loss between the oxidation zone and the pyrolysis zone would be important in determining the extinguishment limit. One such factor is the thickness of the fuel layer (Bakhman, 1993). The layer of duff on mineral soil may be seen as a semiopen system consisting of a flat layer of fuel on an inert bed with the other surface (the top) bordering gas (air). For continued propagation of smoldering in any system, heat losses from the reaction zone (char oxidation) must not be too high since sufficient heat for the endothermic pyrolysis of the next layer of duff must be supplied from this reaction zone. The fuel itself acts as insulation, and the thinner the fuel layer, the greater the heat loss from the system. Thus, Bakhman (1993) found that the temperature at which a smoldering wave was initiated decreased with increasing thickness of the fuel. Palmer (1957) found that, at depths greater than the critical minimum depth necessary for sustained smoldering, the rate of smoldering varied with the depth of the fuel layer. He also found that the minimum depth for sustained smoldering increased with increasing mean particle diameter. For cork dust with a mean particle diameter of 2.0 mm, the minimum depth for sustained smoldering was 47 mm. The mean particle size of duff collected from a *Pinus contorta–Picea engelmannii* stand in Alberta was found to be 2.15 mm diameter (Hawkes, 1993). Thus, smoldering may not be sustainable in such duff with depths less than about 50 mm. Jones *et al.* (1994) also reported minimum depths of packed needle fuels that were required for sustained smoldering. This heat loss explanation might also partially account for the finding in some empirical studies that preburn duff depth is a significant factor in duff consumption (Blackhall and Auclair, 1982; Brown *et al.*, 1985; Chrosciewicz, 1978a,b); however, see the earlier discussion of this with regard to the Reinhardt *et al.* (1989) study. Furthermore, it may help to explain the results of Miyanishi *et al.* (1999) who found within the same fire significantly larger burned holes in the duff of *Picea mariana* stands with thicker preburn duff layers than in adjacent *Pinus banksiana* stands with thinner duff. If thickness of the fuel and the subsequent effect on heat loss is a critical factor as proposed by Bakhman (1993), we might expect propagation of smoldering to be more likely in stands with deeper duff, resulting in larger patches of burned out duff. The question then arises: to what extent can within-stand variation in duff depth account for extinguishment of smoldering and thus contribute to the limits of the burned patches?

All the smoldering models require as inputs parameter values for the particular fuel (e.g., density, thermal conductivity, heat capacity, heats of reaction). While values of these properties are available for a similar fuel, peat, and for some of the constituents of duff (particularly cellulose but also lignin to some extent), little information is available on values for duff from various forest types. How much do these properties of duff vary spatially and at what scales is there significant variation? Is within-stand variation as great as between-stand variation? Furthermore, how does spatial variation in these fuel properties compare with the spatial and temporal variation in other significant factors such as fuel moisture? It may turn out that, despite the variation in these fuel properties, the variation in fuel moisture may be so much greater that, for all intents and purposes, some constant values (e.g., mean values for particular forest types) can be used in the models (as is the case with values for fuel chemistry in fire behavior models).

Most of the models also indicate that the rate of oxygen supply (i.e., diffusion of oxygen through the fuelbed from the surface) is the rate-limiting factor for smoldering (Ohlemiller, 1985, 1990a). Also, Moussa *et al.* (1976) concluded that the extinguishment limit would be set by the rate of oxygen diffusion to the char zone. In the case of duff smoldering in a forest fire situation, ignition of the duff would normally occur on the surface (unlike in many of the lab situations described in this chapter where the whole surface of one end of the fuelbed is ignited). Thus, in a forest fire, the smoldering zone would progress both downward and laterally in the duff, creating a hole that extends in size as smoldering continues (Figure 12). The distance between the oxygen source at the surface and the char oxidation zone might thus remain relatively constant, and we would not generally expect smoldering to be extinguished due to a lack of oxygen. One situation in which oxygen might become limiting would be as the smoldering zone approached a region of saturated (water-logged) duff. However, in this situation, it is also unlikely that the heat generated by the char

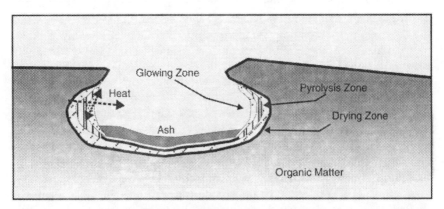

FIGURE 12 Cross-sectional diagram of duff showing the development of a burned hole from smoldering combustion. From Hungerford *et al.* (1995).

oxidation zone would be sufficient to evaporate all the water and raise the temperature of the duff high enough to produce pyrolysis and char formation.

Thus, as has long been recognized, most likely the key factor that determines the extinguishment limit of smoldering in the forest fire situation is duff moisture. Moisture in the duff would result in latent heat flux for evaporation of the water. As noted by Frandsen (1987), the large heat of vaporization of water (2.25×10^3 J g^{-1}) provides a very effective heat sink. Sufficient latent heat flux could result in insufficient heat available for endothermic pyrolysis (charring) of duff, thus resulting in extinguishment of smoldering. According to Yi and Kim (1996), pyrolysis in the region where water evaporation is occurring can be considered negligible. In most experimental situations, the moisture content of the fuel is typically uniform throughout the fuel. However, duff moisture content varies spatially, both within stands and between stands. Thus, further study and modeling of the spatial variation in duff moisture at various scales would be useful.

Bridge and Johnson (2000) showed that the organization of plant communities along moisture and nutrient gradients could be used to relate the spatial distribution of tree species (stand types) to hillslope position and surficial geology. Thus, on glaciofluvial substrate in the mixedwood boreal forest of central Saskatchewan, the ridge tops (which are drier and more nutrient poor) are dominated by *Pinus banksiana,* and the mid to bottom slopes (which are wetter and more nutrient rich) are dominated by *Picea mariana.* Furthermore, duff depths in stands dominated by *Pinus banksiana* are typically significantly less than duff depths in stands dominated by *Picea mariana* (Johnson, 1981; Dyrness and Norum, 1983). Thus, we might expect some correlation between duff depth and hillslope position (i.e., distance from the ridge line). Furthermore, studies have shown that a significant proportion of duff moisture is due to upward transfer of water from the mineral soil to the duff (e.g., Samran *et al.,* 1995). Since soil moisture is influenced by substrate (e.g., glacial till has a higher field capacity than glaciofluvial substrate) and since, within substrate type, soil moisture shows a gradient along hillslopes, it may be possible to use a hydrologic model to predict between-stand variation in duff moisture. With information on between-stand variation in both duff depth and duff moisture, it may be possible to begin to develop a model of variation in duff consumption at the hillslope scale.

At the stand level, Bajtala (1999) found significantly drier duff beneath compared to beyond tree crowns within stands of *Pinus banksiana* and *Picea mariana.* Also, Miyanishi *et al.* (1999) reported a significant spatial correlation between burned holes in the duff and standing boles of trees killed by fire. If duff moisture is the major factor determining extinction of smoldering in forest fires, the patchiness of duff consumption within stands often observed in forest fires would indicate significant variation in duff moisture at this scale. Such

variation could potentially be explained by differences in precipitation input
due to interception by tree crowns, differences in evaporation as well as dew
formation due to crown interception of radiation, and differences in bulk density and moisture-holding capacity of the ground cover (herbaceous/moss cover
versus needle litter). Thus, a hydrologically based model of within-stand spatial
variation in duff moisture could also be developed. Since such a model would
involve the role of the tree canopy in duff moisture dynamics, it could indicate
a significant relationship between the areal extent of duff consumption within
burned forests and density of trees, particularly those that have a higher interception efficiency (e.g., *Picea* spp. compared with *Pinus* spp.). Finally, the proposed models for duff moisture at the two scales could be combined.

Another aspect that bears further investigation is the interaction between the
effects of duff depth and moisture. Recall from the earlier discussion that there
exist critical minimum depths of fuelbed required for smoldering propagation
resulting from the balance between heat generation from the char oxidation
zone and convective heat loss from the surface. Since the process of evaporation of water from the duff is also a heat sink, we might expect the critical depth
of fuelbed to increase with increasing moisture content. Thus, McMahon *et al.*
(1980) reported sustained smoldering in organic soil blocks with moisture contents as high as 135%. Some preliminary lab studies using varying depths of a
peat fuelbed with varying moisture content have, in fact, shown a significant
positive relationship as predicted (Miyanishi and Johnson, unpublished data).

Although not exhaustive, the preceding discussions illustrate how an appreciation and understanding of the processes involved in duff consumption
and use of smoldering models available in the combustion literature can lead to
a somewhat different approach to the study of duff consumption than has been
used in the past. This physical process-based approach provides a means of going beyond empirical case studies, leads to more appropriate research questions, and ultimately should result in a better understanding of the phenomenon of duff consumption.

NOTATION

ROMAN LETTERS

C	heat capacity	$J\ kg^{-1}\ K^{-1}$
D	duff consumption	$kg\ m^{-2}$
E	activation energy	$J\ mol^{-1}$
F	frequency factor	s^{-1}
H	heat of reaction	$J\ kg^{-1}$

J	mass flux	kg m^{-2} s^{-1}
M	moisture content	% dry weight
P	pressure	J m^{-3}
R	universal gas constant	8.314 J mol^{-1} K^{-1}
S	cross-sectional area	m^2
T	temperature	°C or K
V	rate of flaming front advance	m s^{-1}
V	wave speed	m s^{-1}
W	normalized weight	
X	mole fraction	
a	absorptivity	dimensionless
c	specific heat	J kg^{-1} K^{-1}
d	sample diameter	m
h	free-convection heat transfer coefficient	W m^{-2} K^{-1}
k	specific rate of reaction	s^{-1} or min^{-1}
l	latent heat of vaporization	2.25 $\times$ 10^6 J kg^{-1}
p	periphery	m
q	heat flux	W m^{-2}
t	time	s
w	width of combustion zone	m
x	spatial coordinate	m

Greek Letters

α	thermal diffusivity	m^2 s^{-1}
δ	boundary layer thickness	m
ε	emissivity	dimensionless
λ	thermal conductivity	W m^{-1} K^{-1}
ρ	density	g cm^{-3} or kg m^{-3}
σ	Stefan-Boltzmann constant	5.668 $\times$ 10^{-2} W m^{-1} K^{-1}
τ	residence time	s
Ω	ratio of the virgin cellulose density at the interface over that at x $\rightarrow$ $-\infty$, taken to be 10^{-5}	

Subscripts

c	char
cell	cellulose
A	active cellulose
d	duff
f	final
fu	fuel
fl	flame
g	gas
i	initial
ig	ignition
max	maximum
v	condensable volatiles
w	water
s	conditions at $x = 0$ (surface)
0	ambient
∞	conditions at $x \to -\infty$

REFERENCES

Aber, J. D., and Melillo, J. M. (1991). "Terrestrial Ecosystems." Saunders College Publishing, Philadelphia.

Adams, J. L. (1966). Prescribed burning techniques for site preparation in cutover jack pine in southeastern Manitoba. *Pulp Pap. Mag. Can.* 67 **WR**: 574–584.

Ahlgren, C. E. (1970). "Some Effects of Prescribed Burning on Jack Pine Reproduction in Northeastern Minnesota." Miscellaneous Report No. 94. Agricultural Experiment Station, University of Minnesota, St. Paul.

Alexander, M. (1977). "Introduction to Soil Microbiology." John Wiley and Sons, New York.

Alexander, M. E. (1982). Calculating and interpreting forest fire intensities. *Can. J. For. Res.* 60, 349–357.

Bajtala, M. J. (1999). "Spatial Patterns of Duff Consumption in Black Spruce and Jack Pine Stands in the Boreal Mixedwood Forest." M.Sc. thesis, University of Guelph, Guelph.

Bakhman, N. N. (1993). Smoldering wave propagation mechanism. 1. Critical conditions. *Combustion, Explosion and Shock Waves* 29, 14–17.

Berg, B., and Ekbohm, G. 1991. Litter mass-loss rates and decomposition patterns in some needle and leaf litter types. Long-term decomposition in a Scots pine forest. VII. *Can. J. Bot.* 69, 1449–1456.

Berg, B., and Staaf, H. (1981). Leaching, accumulation and release of nitrogen in decomposing forest litter. *Ecol. Bull. (Stockholm)* 33, 163–178.

Berg, B., Hannus, K., Popoff, T., and Theander, O. (1982). Changes in organic chemical components of needle litter during decomposition. Long-term decomposition in a Scots pine forest. I. *Can. J. Bot.* **60**, 1310–1319.

Beyer, W. H. (1991). "CRC Standard Mathematical Tables and Formulae," 29th ed. CRC Press, Boca Raton.

Blackhall, J. W., and Auclair, A. D. (1982). Best solution models of prescribed fire impact in subalpine *Picea glauca-Abies lasiocarpa* forests of British Columbia. *In* "Problem Analysis of Prescribed Burning: A Synthesis Text" (A. D. Auclair, Ed.), p. 32. Department of Fisheries and Forestry, Canadian Forest Service, Pacific Forest Research Centre, Victoria.

Bracewell, J. M., and Robertson, G. W. (1987). Characteristics of soil organic matter in temperate soils by Curie-point pyrolysis-mass spectrometry, III. Transformations occurring in surface organic horizons. *Geoderma* **40**, 333–344.

Bradbury, A., Sakai, Y., and Shafizadeh, F. (1979). A kinetic model for pyrolysis of cellulose. *J. Appl. Polym. Sci.* **23**, 3271–3280.

Bridge, S. R. J., and Johnson, E. A. (2000). Geomorphic principles of terrain organization and vegetation gradients. *J. Veg. Sci.* **11**, 57–70.

Broido, A. (1976). Kinetics of solid-phase cellulose pyrolysis. *In* "Thermal Uses and Properties of Carbohydrates and Lignins" (F. Shafizadeh, K. V. Sarkanen and D. A. Tillman, Eds.), pp. 19–35. Academic Press, New York.

Brown, J. K., Marsden, M. A., Ryan, K. C., and Reinhardt, E. D. (1985). "Predicting Duff and Woody Fuel Consumed by Prescribed Fire in the Northern Rocky Mountains." Research Paper INT-337. USDA Forest Service, Intermountain Forest and Range Experiment Station, Ogden.

Brown, J. K., Reinhardt, E. D., and Fischer, W. C. (1991). Predicting duff and woody fuel consumption in northern Idaho prescribed fires. *For. Sci.* **37**, 1550–1566.

Buckmaster, J., and Lozinski, D. (1996). An elementary discussion of forward smolder. *Combust. Flame* **104**, 300–310.

Canada Soil Survey Committee. (1978). "The Canadian System of Soil Classification." Publication 1646. Research Branch, Canadian Department of Agriculture, Ottawa.

Charron, I. (1998). "Sexual recruitment of trees following fire in the southern mixedwood boreal forest of Canada." M.Sc. thesis, Concordia University, Montreal.

Chrosciewicz, Z. (1959). "Controlled Burning Experiments on Jack Pine Sites." Technical Note No. 72. Canada Department of Northern Affairs and National Resources, Forestry Branch, Forest Research Division, Ottawa.

Chrosciewicz, Z. (1967). "Experimental Burning for Humus Disposal on Clear-Cut Jack Pine Sites in Central Ontario." Publication No. 1181. Canada Department of Forestry and Rural Development, Forestry Branch, Ottawa.

Chrosciewicz, Z. (1974). Evaluation of fire-produced seedbeds for jack pine regeneration in central Ontario. *Can. J. For. Res.* **4**, 455–457.

Chrosciewicz, Z. (1976). Burning for black spruce regeneration on a lowland cutover site in southeastern Manitoba. *Can. J. For. Res.* **6**, 179–186.

Chrosciewicz, Z. (1978a). "Slash and Duff Reduction by Burning on Clear-Cut Jack Pine Sites in Southeastern Manitoba." Information Report NOR-X-199. Northern Forest Research Centre, Canadian Forest Service, Edmonton.

Chrosciewicz, Z. (1978b). "Slash and Duff Reduction by Burning on Clear-Cut Jack Pine Sites in Central Saskatchewan." Information Report NOR-X-200. Northern Forest Research Centre, Canadian Forest Service, Edmonton.

Chrosciewicz, Z. (1989). Prediction of forest-floor moisture content under diverse jack pine canopy conditions. *Can. J. For. Res.* **19**, 1483–1487.

Clayton, J. S., Ehrlich, W. A., Cann, D. B., Day, H. H., and Marshall, I. B. (1977). "Soils of Canada. Volume 1—Soil Report." Canada Department of Agriculture Research Branch, Ottawa.

Cohen, L., and Luft, N. W. (1955). Combustion of dust layers in still air. *Fuel* **34**, 154–163.

Delaney, M. T., Fernandez, I. J., Simmons, J. A., and Briggs, R. D. (1996). "Red Maple and White

Pine Litter Quality: Initial Changes with Decomposition." Technical Bulletin 162. Maine Agricultural and Forest Experiment Station, Orono.

Di Blasi, C. (1994). Smolder spread through thin horizontal fuel layers. *In* "Advanced computational methods in heat transfer III. Proceedings of the Third International Conference" (L. C. Wrobel, C. A. Brebbia and A. J. Nowak, Eds.), pp. 323–330. Computational Mechanics Publications, Southampton.

Di Blasi, C. (1995). Mechanisms of two-dimensional smoldering propagation through packed fuel beds. *Combust. Sci. Tech.* 106, 103–124.

Drysdale, D. (1985). "An Introduction to Fire Dynamics." John Wiley and Sons, Chichester.

Dyrness, C. T., and Norum, R. A. (1983). The effects of experimental fires on black spruce forest floors in interior Alaska. *Can. J. For. Res.* 13, 879–893.

Egerton, A., Gugan, K., and Weinberg, F. (1963). The mechanism of smoldering in cigarettes. *Combust. Flame* 7, 63–78.

Flinn, M. A., and Wein, R. W. (1977). Depth of underground plant organs and theoretical survival during fire. *Can. J. Bot.* 55, 2550–2554.

Frandsen, W. H. (1987). The influence of moisture and mineral soil on the combustion limits of smoldering forest duff. *Can. J. For. Res.* 17, 1540–1544.

Frandsen, W. H. (1991a). Burning rate of smoldering peat. *Northwest Sci.*, 65, 166–172.

Frandsen, W. H. (1991b). Smoldering spread rate: A preliminary estimate. *In* "Proceedings of the Eleventh Conference on Fire and Forest Meteorology" (P. L. Andrews and D. F. Potts, Eds.), pp. 168–172. Society of American Foresters, Bethesda.

Frandsen, W. H. (1991c). Heat evolved from smoldering peat. *Int. J. Wildland Fire* 1, 197–204.

Green, R. N., Trowbridge, R. L., and Klinka, K. (1993). "Towards a Taxonomic Classification of Humus Forms." Forest Science Monograph 29. Society of American Foresters, Bethesda.

Hartford, R. A. (1989). Smoldering combustion limits in peat as influenced by moisture, mineral content, and organic bulk density. *In* "Proceedings of the Tenth Conference on Fire and Forest Meteorology" (D. C. MacIver, H. Auld, and R. Whitewood, Eds.), pp. 282–286. Forestry Canada, Environment Canada, Ottawa.

Hawkes, B. C. (1993). Factors that influence peat consumption under dependent burning conditions: A laboratory study. Ph.D. dissertation, University of Montana, Missoula.

Hungerford, R. D., Frandsen, W. H., and Ryan, K. C. (1995). Ignition and burning characteristics of organic soils. *In* "Proceedings of the Tall Timbers Fire Ecology Conference in Wetlands, a Management Perspective" (S. I. Cerulean and R. T. Engstrom, Eds.), pp. 78–91. Tall Timbers Research Station, Tallahassee.

Johnson, E. A. (1981). Vegetation organization and dynamics of lichen woodland communities in the Northwest Territories, Canada. *Ecology* 62, 200–215.

Johnson, E. A. (1992). "Fire and Vegetation Dynamics: Studies from the North American Boreal Forest." Cambridge University Press, Cambridge.

Jones, J. C., Goh, T. P. T., and Dijanosic, M. J. (1994). Smoldering and flaming combustion in packed beds of *Casuarina* needles. *J. Fire Sci.* 12, 442–451.

Jordan, C. F. (1982). Amazon rain forests. *Am. Scientist* 70, 394–401.

Jordan, C. F., and Herrera, R. (1981). Tropical rain forests: Are nutrients really critical? *Am. Nat.* 117, 167–180.

Kashiwagi, T., and Nambu, H. (1992). Global kinetic constant for thermal oxidative degradation of a cellulosic paper. *Combustion and Flame* 88, 345–368.

Kauffman, J. B. (1991). Survival by sprouting following fire in tropical forests of the eastern Amazon. *Biotropica* 23, 219–224.

Kauffman, J. B., Cummings, D. L., Ward, D. E., and Babbitt, R. (1995). Fire in the Brazilian Amazon: 1. Biomass, nutrient pools, and losses in slashed primary forests. *Oecologia* 104, 397–408.

Kellman, M., and Meave, J. (1997). Fire in the tropical gallery forests of Belize. *J. Biogeog.* **24,** 23–34.

Kinbara, T., Endo, H., and Sega, S. (1967). Downward propagation of smoldering combustion through solid materials. *In* "Eleventh Symposium (International) on Combustion," pp. 525–531. The Combustion Institute, Pittsburgh.

Kingsbury, N., and Kellman, M. (1997). Root mat depths and surface soil chemistry in southeastern Venezuela. *J. Tropical Ecol.* **13,** 475–479.

Krause, H. (1998). Forest floor mass and nutrients in two chronosequences of plantations: Jack pine vs. black spruce. *Can. J. Soil Sci.* **78,** 77–83.

Leighton, M., and Wirawan, N. (1986). Catastrophic drought and fire in Borneo tropical rain forest associated with the 1982–1983 El Niño Southern Oscillation Event. *In* "Tropical Rain Forests and the World Atmosphere" (G. T. Prance, Ed.), pp. 75–102. AAAS Selected Symposium 101, Westview, Boulder.

Leisch, S. O. (1983). "Smoldering Combustion in Horizontal Dust Layers." PhD dissertation, University of Michigan, Ann Arbor.

Lewellen, P. C., Peters, W. A., and Howard, J. B. (1976). *In* "Sixteenth Symposium (International) on Combustion," pp. 1471–1480. The Combustion Institute, Pittsburgh.

Little, S. N., Ottmar, R. D., and Ohmann, J. L. (1986). "Predicting Duff Consumption from Prescribed Burns on Conifer Clearcuts in Western Oregon and Western Washington." Research Paper PNW-362. USDA Forest Service, Pacific Northwest Forest and Range Experiment Station, Portland.

Marshall, T. J., and Holmes, J. W. (1979). "Soil Physics." Cambridge University Press, New York.

Mason, C. F. (1976). "Decomposition." Edward Arnold (Publishers) Ltd., London.

McAdams, W. H. (1942). "Heat Transmission," 2nd ed. McGraw-Hill, New York.

McMahon, C.K., Wade, D. D., and Tsoukalas, S. N. (1980). Combustion characteristics and emissions from burning organic soils. *In* "Proceedings of the 73rd Annual Meeting of the Air Pollution Control Association," Paper No. 80-15.5.

Merrill, D. F., and Alexander, M. E. (1987). "Glossary of Forest Fire Management Terms." Canadian Committee on Forest Fire Management, National Research Council of Canada, Ottawa.

Miyanishi, K., Bajtala, M. J., and Johnson, E. A. (1999). Patterns of duff consumption in *Pinus banksiana* and *Picea mariana* stands. *In* "Proceedings of Sustainable Forest Management Network Conference, Science and Practice: Sustaining the Boreal Forest," pp. 112–115. Sustainable Forest Management Network, Edmonton.

Moallemi, M. K., Zhang, H., and Kumar, S. (1993). Numerical modeling of two-dimensional smoldering process. *Combust. Flame* **95,** 170–182.

Moussa, N. A., Toong, T. Y., and Garris, C. A. (1976). Mechanism of smoldering of cellulosic materials. *In* "Sixteenth Symposium (International) on Combustion," pp. 1447–1457. The Combustion Institute, Pittsburgh.

Muramatsu, M., Umemura, S., and Okada, T. (1979). A mathematical model of evaporation-pyrolysis processes inside a naturally smoldering cigarette. *Combust. Flame* **36,** 245–262.

Ohlemiller, T. J. (1985). Modeling of smoldering combustion propagation. *Prog. Energy Combust. Sci.* **11,** 277–310.

Ohlemiller, T. J. (1990a). Smoldering combustion propagation through a permeable horizontal fuel layer. *Combust. Flame* **81,** 341–353.

Ohlemiller, T. J. (1990b). Forced smolder propagation and the transition to flaming in cellulosic insulation. *Combust. Flame* **81,** 354–365.

Ohlemiller, T. J., Bellan, J., and Rogers, F. (1979). A model of smoldering combustion applied to flexible polyurethane foams. *Combust. Flame* **36,** 197–215.

Ortiz-Molina, M. G., Toong, T-Y., Moussa, N. A., and Tesoro, G. C. (1978). Smoldering combustion of flexible polyurethane foams and its transition to flaming or extinguishment. *In* "Seventeenth Symposium (International) on Combustion," pp. 1191–1200. The Combustion Institute, Pittsburgh.

Palmer, K. N. (1957). Smoldering combustion in dusts and fibrous materials. *Combust. Flame* 1, 129–154.

Perala, D. A., and Alban, D. H. (1982). "Rates of Forest Floor Decomposition and Nutrient Turn-over in Aspen, Pine, and Spruce Stands on Two Different Soils." Research Paper NC-227. USDA Forest Service, North Central Forest Experiment Station, St. Paul.

Peter, S. (1992). "Heat Transfer in Soils Beneath a Spreading Fire." Ph.D. dissertation, University of New Brunswick, Fredericton.

Potts, D. F., Zuuring, H., and Hillhouse, M. (1983). Spatial analysis of duff moisture and structure variability. *In* "Proceedings of the Seventh Conference on Fire and Forest Meteorology," pp. 18–21. American Meteorology Society, Boston.

Reinhardt, E. D., Brown, J. K., and Fischer, W. C. (1989). Fuel consumption from prescribed fire in northern Idaho logging slash. *In* "Proceedings of the Tenth Conference on Fire and Forest Meteorology" (D. C. MacIver, H. Auld, and R. Whitewood, Eds.), pp. 155–160. Forestry Canada, Environment Canada, Ottawa.

Richards, P. W. (1952). "The Tropical Rain Forest: An Ecological Study." Cambridge University Press, Cambridge.

Roberts, A. F. (1970). A review of kinetics data for the pyrolysis of wood and related substances. *Combust. Flame* 14, 261–272.

Rothermel, R. C. (1972). "A Mathematical Model for Predicting Fire Spread in Wildland Fuels." Research Paper INT-115. USDA Forest Service, Intermountain Forest and Range Experiment Station, Ogden.

Rothermel, R. C. (1976). Forest fires and the chemistry of forest fuels. *In* "Thermal Uses and Properties of Carbohydrates and Lignins." (F. Shafizadeh, K. V. Sarkanen, and D. A. Tillman, Eds.), pp. 245–259. Academic Press, New York.

Samran, S., Woodard, P. M., and Rothwell, R. L. (1995). The effect of soil water on ground fuel availability. *Forest Sci.* 41, 255–267.

Sandberg, D. V. (1980). Duff reduction by prescribed underburning in Douglas-fir. Research Paper PNW-272. USDA Forest Service, Pacific Northwest Forest and Range Experiment Station, Portland.

Sato, K., and Sega, S. (1985). The mode of burning zone spread along cylindrical cellulosic material. *J. Fire Sci.* 3, 26–34.

Schimmel, J., and Granstrom, A. (1996). Fire severity and vegetation response in the boreal Swedish forest. *Ecology* 77, 1436–1450.

Schult, D. A., Matkowsky, B. J., Volpert, V. A., and Fernandez-Pello, A. C. (1995). Propagation and extinction of forced opposed flow smolder waves. *Combust. Flame* 101, 471–490.

Shafizadeh, F., and DeGroot, W. F. (1976). Combustion characteristics of cellulosic fuels. *In* "Thermal Uses and Properties of Carbohydrates and Lignins." (F. Shafizadeh, K. V. Sarkanen, and D. A. Tillman, Eds.), pp. 1–17. Academic Press, New York.

Shafizadeh, F., and Sekiguchi, Y. (1984). Oxidation of chars during smoldering combustion of cellulosic materials. *Combust. Flame* 55, 171–179.

Shafizadeh, F., Bradbury, A. G. W., DeGroot, W. F., and Aanerud, T. W. (1982). Role of inorganic additives in the smoldering combustion of cotton cellulose. *Ind. Eng. Chem. Prod. Res. Dev.* 21, 97–101.

Simmons, R. F. (1995). Fire chemistry. *In* "Combustion Fundamentals of Fire" (G. Cox, Ed.), pp. 405–473. Academic Press, London.

Stark, N. M., and Jordan, C. F. (1978). Nutrient retention by the root mat of an Amazonian rain forest. *Ecology* 59, 434–437.

Steward, F. R., Peter, S., and Richon, J. B. (1990). A method for predicting the depth of lethal heat penetration into mineral soils exposed to fires of various intensities. *Can. J. For. Res.* **20**, 919–926.

Taiz, L., and Zeiger, E. (1991). "Plant Physiology." Benjamin/Cummings Publishing Company, Redwood City.

Thomas, P. A., and Wein, R. W. (1985). The influence of shelter and the hypothetical effect of fire severity on the postfire establishment of conifers from seed. *Can. J. For. Res.* **15**, 148–155.

Uhl, C., and Kauffman, J. B. (1990). Deforestation, fire susceptibility, and potential tree responses to fire in the eastern Amazon. *Ecology* **71**, 437–449.

Uhl, C., Kauffman, J. B., and Cummings, D. L. (1988). Fire in the Venezuelan Amazon 2. Environmental conditions necessary for forest fires in the evergreen rainforest of Venezuela. *Oikos* **53**, 176–184.

Van Wagner, C. E. (1972). Duff consumption by fire in eastern pine stands. *Can. J. For. Res.* **2**, 34–39.

Van Wagner, C. E. (1987). "Development and Structure of the Canadian Forest Fire Weather Index System." Forestry Technical Report 35. Canadian Forestry Service, Ottawa.

Weber, M. G. (1985). Forest soil respiration in eastern Ontario jack pine ecosystems. *Can. J. For. Res.* **15**, 1069–1073.

Weber, M. G., Hummel, M., and Van Wagner, C. E. (1987). Selected parameters of fire behavior and *Pinus banksiana* Lamb. regeneration in eastern Ontario. *For. Chron.* **63**, 340–346.

Weber, M. G., Methven, I. R., and Van Wagner, C. E. (1985). The effect of forest floor manipulation on nitrogen status and tree growth in an eastern Ontario jack pine ecosystem. *Can. J. For. Res.* **15**, 313–318.

Wein, R. W. (1981). Characteristics and suppression of fires in organic terrain in Australia. *Aust. For.* **44**, 162–169.

Wein, R. W. (1983). Fire behavior and ecological effects in organic terrain. *In* "The Role of Fire in Northern Circumpolar Ecosystems" (R. W. Wein and D. A. MacLean, Eds.), pp. 81–95. John Wiley, New York.

Williams, F. (1977). Mechanisms of fire spread. *In* "Sixteenth Symposium (International) on Combustion," pp. 1281–1294. The Combustion Institute, Pittsburgh.

Williams, S. T., and Gray, T. R. G. (1974). Decomposition of litter on the soil surface. *In* "Biology of Plant Litter Decomposition" (C. H. Dickinson and G. J. F. Pugh, Eds.), pp. 611–632. Academic Press, London.

Yi, S-C., and Kim, Y. C. (1996). Modeling of smoldering cellulosic materials—Evaporation and pyrolysis processes. *Preprints of the Division of Fuel Chemistry of the American Chemical Society* **41**, 1183–1186.

Yi, S-C., Bae, S. Y., and Kim, S. H. (1998). Heat transfer of a smoldering flammable substrate. Part 2. A theoretical model and its application. *J. Fire Sci.* **16**, 32–45.

Zasada, J. C., Norum, R. A., Van Veldhuizen, R. M., and Teutsch, C. E. (1983). Artificial regeneration of trees and tall shrubs in experimentally burned upland black spruce/feather moss stands in Alaska. *Can. J. For. Res.* **13**, 903–913.

Zech, W., Senesi, N., Guggenberger, G., Kaiser, K., Lehmann, J., Miano, T. M., Miltner, A., and Schroth, G. (1997). Factors controlling humification and mineralization of soil organic matter in the tropics. *Geoderma* **79**, 117–161.

CHAPTER 14

Fire Effects on Trees

M. B. DICKINSON AND E. A. JOHNSON

*Kananaskis Field Stations and Department of Biological Sciences,
University of Calgary, Calgary, Alberta, Canada*

I. INTRODUCTION

There has been considerable interest among ecologists and foresters in the effects fires have on populations, communities, and ecosystems. However, relatively little attention has been paid to understanding *how* wildfires cause these effects. Needed is a more thorough understanding of the physical processes of heat transfer between the fire and the organism and in turn how the resulting patterns of individual damage affect population and ecosystem processes. To achieve such insight, ecologists and foresters must borrow liberally from the work of physicists and engineers. The preceding chapters in this book will be of considerable help.

In this chapter, we will explain some of the ways to model heat transfer between fires and trees. While our focus is primarily on trees, the discussion applies with suitable modification to herbs and shrubs. We also have used *relatively simple models* so that both their underlying mechanisms and predictive

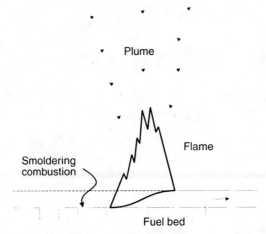

FIGURE 1 Cross-sectional view of a fire line burning in surface fuels (e.g., leaf litter, small-diameter downed-woody material, herbs, and shrubs). The flame gives rise to the buoyant plume and both are characterized by a turbulent flow regime. Smoldering combustion may proceed at low rates in organic soil layers and decomposed downed-woody material after the flame has passed.

powers will be intuitively clear. The heat transfer models introduced in this chapter are most applicable to trees damaged in surface fires with relatively low fire-line intensities (kW m^{-1}). The heat sources in the models are the flame, the buoyant plume arising from the flame, and the smoldering combustion of organic material (Figure 1).

Plants can be divided into components that have similar heat transfer characteristics (Figure 2). Since the components are primarily constructed of cellulose, their heating is determined mainly by their physical dimensions and spatial position relative to the heat sources. Plants must fulfill four principal functions: mechanical support, photosynthesis, reproduction, and hydraulics (Niklas, 1992). As a consequence, trees have a rather stereotyped spatial arrangement of parts. The stem and branches position thin photosynthetic structures high enough to intercept light. The sizes of the branches are scaled by the weight of the foliage they support (e.g., Corner, 1949; Ackerly and Donoghue, 1998). The trunk, which elevates the canopy, will be larger in diameter and scaled roughly as a power of height, 1.5 for dicots and 1.1 for conifers (Niklas, 1994). The whole structure is attached to the ground by roots scaled by their functions of attachment (Ennos, 1993) and soil-resource acquisition (Russell, 1977).

Thus, crown components will generally be small in diameter, meaning that they will have little internal resistance to heating by the flame or buoyant plume and can be killed and combusted easily. Trunks as a whole will have considerable resistance to heating by the flame but can be damaged because their vascular cambium is near the outside of their diameter. Trunks, except when small

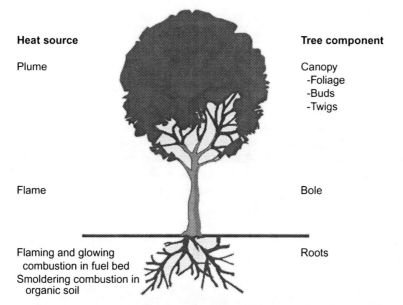

Heat source   **Tree component**

Plume

Canopy
  -Foliage
  -Buds
  -Twigs

Flame   Bole

Flaming and glowing   Roots
  combustion in fuel bed
Smoldering combustion in
  organic soil

FIGURE 2   The spatial arrangement of tree components and the sources of heat which affect them in surface fires. The foliage, buds, and twigs are heated by the plume and the live bark and vascular cambium in the bole are heated by the flame. The base of the bole and roots may be heated by the downward heat flux from flaming and glowing combustion in the fuelbed or by smoldering combustion in organic soil layers after the fire line has passed.

in diameter, are not consumed by the fire. The roots are insulated from the heat of the flame but are heated by smoldering combustion after the fire line has passed. Roots are vulnerable to smoldering combustion because heating is often of long duration and the vascular cambium is near the surface.

In this chapter, we will begin by discussing tissue necrosis in the crown, bole, and roots. Then, we will introduce an approach by which patterns of tissue necrosis in these components might be used to understand the death of individual trees. As background for this chapter, Chapter 2 describes flames, Chapters 5 and 6 describe spreading fires, and Chapters 7 and 8 describe plumes above fires. Chapter 13 addresses in more detail smoldering combustion processes and their effects on tree recruitment.

A. HEAT TRANSFER MECHANISMS

Heat transfer occurs by conduction, convection, and radiation. Heat transfer texts include an introduction to these heat transfer mechanisms, providing models that can be applied to the heating of trees (e.g., Sucec, 1985; Holman,

1986). The introductory material presented in this section will serve as a basis for understanding the more complex models used in later sections to describe necrosis of individual plant components.

1. Conduction

Conduction heat transfer occurs along a gradient from high to low temperatures (as required by the second law of thermodynamics). The rate of heat transfer is determined by the driving force (the temperature difference) and the heating properties of the material. Conduction in solids occurs by collisions between molecules and by the migration of free electrons. In a fluid (such as air), conduction is due to collisions of molecules that are in constant, random motion.

The simplest case of conduction is when heat flow occurs along one space coordinate (i.e., is one dimensional). For instance, consider the central region of a semi-infinite slab (i.e., a slab that is thin relative to its height and width) that is homogeneous in terms of its temperature and thermal properties (Figure 3A). If the temperature of one of the slab's surfaces is raised, heat will be transferred inward, perpendicular to the surface. If the surface temperature is held constant for a long enough time, heat transfer is steady state and is described by Fourier's law of conduction

$$q = -kA\frac{dT}{dx} \tag{1}$$

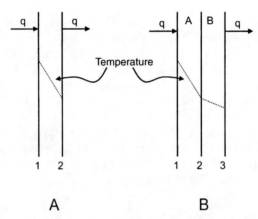

FIGURE 3 One-dimensional steady state conduction through a semi-infinite slab. (A) A homogeneous slab, the left surface of which (1) is maintained at a higher temperature than the right (2). (B) A slab composed of two layers, the left layer, A, having a higher thermal conductivity than the right [B, see Eq. (4)]. In both cases, the temperature difference has been maintained for a sufficiently long period that the rate of heat flow q (W) into the left surface is equal to the rate of heat flow out of the right surface.

which upon integration gives

$$q = -kA\frac{(T_2 - T_1)}{\Delta x} \tag{2}$$

where q is the heat transfer rate (W), k is the thermal conductivity (W m$^{-1} \cdot$ °C), A is area (m^2), Δx is the thickness of the slab (m), and T_1 and T_2 (°C) are the temperatures of the front (heated) and back side of the slab, respectively. The negative sign is inserted in Eq. (1) because, by definition, the heat flow q is positive when the temperature derivative dT/dx is negative (i.e., $T_2 - T_1$ is negative). The heat flux (i.e., q/A, the heat flow per unit area) is thus a result of the gradient in temperature through the slab (dT/dx or $[T_2 - T_1]/\Delta x$) and the thermal conductivity k. Equation 1 defines thermal conductivity and, given the heat flux and the temperature gradient, can be used to derive estimates of thermal conductivity for a material.

Thermal conductivity of wood, bark, and foliage varies with their density, moisture content, and temperature (e.g., Byram *et al.*, 1952a; Martin, 1963; Reifsnyder *et al.*, 1967). If a material does not have homogeneous conductivity values (e.g., bark and wood) but can be divided into parts connected in series, the heat transfer equations can be solved simultaneously. For instance, for steady state heat transfer through a material composed of two layers of different conductivity (Figure 3), the rate of heat flow is equal across each layer:

$$q = -k_A A\frac{T_2 - T_1}{\Delta x_A} = -k_B A\frac{T_3 - T_2}{\Delta x_B} \tag{3}$$

where Δx_A and Δx_B are the thicknesses of layers A and B, respectively. These equations can be solved simultaneously resulting in

$$q = \frac{T_1 - T_3}{\dfrac{\Delta x_A}{k_A A} + \dfrac{\Delta x_B}{k_B A}} \tag{4}$$

wherein the resistances to conduction heat transfer for each layer are added together in the denominator. Notice that the temperature change is constant within each layer, confirming that the rate of heat transfer q is also constant.

2. Convection

Convection heat transfer between solid objects (e.g., foliage, buds, and twigs) and fluids (e.g., air) occurs as a result of relative motion. In the heat transfer literature, free convection occurs when the motion of the fluid is driven by gradients in temperature (and hence density) within the fluid near the surface of a solid caused by heat transfer between the solid and fluid. For example, heat waves rising off deserts during windless and sunny days are a manifestation of free convection of heat from the warm ground to the overlying and cooler air. When the relative motion driving convection heat transfer is not caused by localized temperature gradients at the surface of an object, the process is termed forced convection. For instance, forced convection heat transfer occurs when wind flowing through a canopy carries heat away from the surface of radiatively heated foliage (e.g., Gates, 1980). In this chapter, we are primarily interested in forced convection, wherein the causes of relative motion are buoyant flames and plumes.

Convection heat transfer is inherently more complicated than conduction because it depends on complex processes occurring within the boundary layer at the surface of a structure. Convection heat transfer is proportional to (1) the temperature difference between the free-moving fluid and the surface of a structure and (2) the characteristics of the boundary layer as described by the convection heat transfer coefficient. Steady state convection heat transfer is defined by Newton's law of cooling

$$q = -hA(T_s - T_\infty) \tag{5}$$

where q is the rate of heat transfer (W), h is the convection heat transfer coefficient (W m^{-2} · °C), A is the surface area over which convection is occurring (m^2), T_s is the temperature at the surface of the structure, and T_∞ is the temperature of the free-moving fluid. The negative sign indicates that heat flow (q) into the structure is positive when the temperature of the free-moving fluid is greater than that of the structure's surface (i.e., $T_s - T_\infty$ is negative). Equation (5) is the defining equation for the convection heat transfer coefficient.

3. Thermal Radiation

All bodies above absolute zero emit thermal radiation according to the Stefan-Boltzmann law:

$$E = \varepsilon \sigma T^4 \tag{6}$$

where E is the emitted radiant energy flux (W m^{-2}) integrated over all wavelengths, ε is the emissivity (dimensionless), σ is the Stefan-Boltzman constant (5.669 × 10^{-8} W m^{-2} · °K^4), and T is the temperature (°K). Emissivity is the ratio of the emissive power of a body to that of an ideal emitter (blackbody) and is assumed to be a constant over all temperatures and wavelengths.

Because the emitted radiant flux is directly proportional to temperature [Eq. (6)], a structure that is heated above a fire line reradiates a portion of the heat it gains from the flame or plume causing a proportional reduction in its temperature. Under the assumption that a structure is being heated only by radiation, net heat transfer into a structure is proportional to the temperature difference between the structure and its surroundings:

$$q_n \propto (T^4 - T_s^4) \tag{7}$$

where q_n is the net heat transfer rate (W), T is the temperature of the structure, and T_s is the temperature of the surroundings.

II. EFFECTS OF FIRE ON THE TREE BOLE

In low-intensity surface fires, the fire line passes the bole of a tree, bathing it in flame (Figure 2). Heat is conducted through the bark into the underlying vascular cambium and, if the temperatures are high enough and flame residence time is long enough, live tissues within the bole are killed. Low-intensity fires will often have little effect on the canopy because the temperatures in the plume at the height of the canopy are too low. However, as fire-line intensity increases, vascular cambium damage will be accompanied by necrosis of canopy components. Later in this chapter, we discuss flame characteristics and model the process of heat transfer from the flame into the bark at the base of a tree. Then, we discuss how one would predict the extent of vascular cambium necrosis around the bole of a tree. To make such a prediction, one must consider the increase in flame residence times and temperatures that occur on the leeward sides of trees in the presence of wind, a process that often causes fire scars in trees that survive fires.

A. FLAME CHARACTERISTICS

Raising the temperature of the bole involves convection and radiation heat transfer between the flame and the surface of the bark and conduction through the bark and underlying wood (e.g., Fahnestock and Hare, 1964). The relatively simple heat transfer model we present is based on the assumption that the main source of heat will be flame gases, not radiation, and that the convection heat transfer coefficient between the flame and the surface of the bark will be very large because of the steep gradient in temperatures at the bark's surface (see discussion that follows). With these simplifying assumptions, flame gas temperatures become the primary focus. Unfortunately, predicting and measuring flame gas temperatures are not straightforward tasks.

Flames propagate through a natural fuelbed primarily by convective and radiative heating of the fuel particles ahead of the fire (e.g., Chapters 5 and 6 in this book; Thomas, 1967; Drysdale, 1990). As the flames approach, fuel particle temperatures increase, moisture is driven off, and solid fuel is converted to gas (see review in Albini, 1980, and Chapter 3 in this book). Piloted ignition occurs when the rate of gas liberation is sufficiently high. The rate of mixing of fuel with ambient air limits the rate of combustion in the fuelbed. The remaining gaseous fuel enters the zone above the fuelbed, is mixed with air by turbulence and diffusion, and combusts.

Flame temperatures are dependent on several processes. The amount of heat produced by combustion is dependent on the chemical composition of the gaseous fuel and the completeness of combustion. Part of the heat from combustion goes to raising the temperatures of the gases in the mix (including air and water vapor), and another part is lost through radiation and mixing with ambient air. Because of the number of variables involved (e.g., Albini, 1980), it is not yet practical to predict the temperatures of flames burning in natural fuels. The incorporation of chemical kinetics in fire behavior models is expected to lead to advances in predicting flame temperatures and other aspects of fire behavior (Weber, 1991).

The alternative to predicting flame temperatures from models is to measure them. However, flame temperatures reported in the literature should be treated with caution (Martin *et al.*, 1969; Vines, 1981; Gill and Knight, 1991). The temperature measured is that of the measuring device and not of the flame gases. Correctly estimating gas temperatures requires balancing heat gained by the measuring device (e.g., a thermocouple) with heat lost:

$$q_{cv} + q_r = q_{rr} + q_{cd} \tag{8}$$

where q_{cv} is heat gained by convection, q_r is heat gained by radiation, q_{rr} is heat lost by re-radiation, and q_{cd} is heat lost by conduction along the leads of the thermocouple sensor away from the sensing junction (Martin *et al.*, 1969). Because it generally has a small effect on the heat budget of the thermocouple, conduction along the leads can be ignored. Shielded-aspirated thermocouples are used to estimate flame gas temperatures because they maximize heat gain by convection, minimize heat gain by radiation from the flame, and minimize the effect of radiative heat loss from the thermocouple. Convection heat transfer scales with the velocity of air flow and is increased as flame gases are drawn past the thermocouple. Shielding the thermocouple minimizes heat gain from radiation from the flame. To minimize net radiative heat transfer [Eq. (7)] between the thermocouple and the shield, the flame gases are drawn through concentric layers of material surrounding the thermocouple, heating the material and thereby increasing q_r to a value comparable to q_{rr}.

Martin *et al.* (1969) report temperatures of flames above a burning crib of white fir sticks measured either by a shielded-aspirated thermocouple or a thermocouple that was neither shielded nor aspirated. Maximum temperatures measured were approximately 1100 and 800°C, respectively, the lower value being one often reported for natural fires.

Because temperatures vary vertically through a flame, it is not immediately clear at which height temperatures should be estimated. Vertical temperature distributions through the centerline of the flame and plume (i.e., maximum temperatures) have been described by Weber *et al.* (1995); see also Chapter 5. Maximum temperatures are highest in the fuelbed. Above the fuelbed, temperatures decline exponentially with height through the flame and plume because of

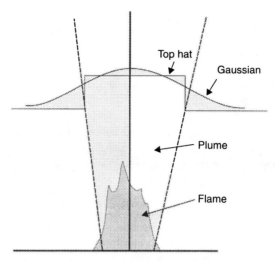

FIGURE 4   Top-hat and Gaussian temperature profiles in a plume above a fire line. Similar profiles could be fit to the fire line. The top-hat profile can be thought of as an average temperature.

turbulent mixing. Weber *et al.* (1995) closely described the vertical temperature
profile with semiempirical equations that required minimal field measurements.

To predict heat transfer into the bole, not only must one consider maximum
temperatures but also the time course of temperatures as the flame passes the
bole. Turbulent mixing at the margin of flames leads to a Gaussian rise and fall
in temperatures as a fire passes a given point (Figure 4; see Thomas, 1963; Mc-
Caffrey, 1979). The limit of the visible flame occurs where the temperature rise
above ambient falls below about 500°C (e.g., Thomas, 1963). Flame residence
time τ_f (s) is the duration of flaming combustion. Flame residence time can be
calculated from flame width x_f (m) and rate of spread R (m s^{-1}) by the following formula:

$$\tau_f = \frac{x_f}{R} \tag{9}$$

One can measure residence time, flame width, and rate of spread directly (e.g.,
Rothermel and Deeming, 1980; Simard *et al.*, 1984; McMahon *et al.*; 1986, Gill
and Knight, 1991) or estimate one or more of them from fire-behavior models
(see review in Weber, 1991).

Most interest in fire modeling has been on rate of spread and heat output of
the fire line (i.e., fire-line intensity); as a consequence, models of the residence
time of flaming combustion are not as well developed. Peterson and Ryan (1986)
give an empirical model showing that the residence time of flaming combustion increases with fuel moisture content and decreases with the average fuel

particle surface area to volume ratio. For a given fuel moisture content, fuelbeds in which high surface area to volume fuels predominate (e.g., grass) will exhibit the shortest residence times.

The highest rates of heat transfer into the bark of a tree will occur where flame temperatures are highest and last the longest. Flames are roughly triangular in shape, decreasing in width toward the tip (e.g., Thomas, 1963; Nelson, 1980). Because of the greater width and higher maximum temperatures, one will generally be interested in the lower portion of the flame. As discussed later, however, bark thickens toward the base of the bole leading to increases in residence times required to cause vascular cambium necrosis.

B. HEAT TRANSFER INTO THE BOLE

The vascular cambium is killed when subjected to a heat pulse of sufficiently high temperatures and duration. The vascular cambium is the meristematic tissue that gives rise to the secondary vascular tissues (xylem and phloem). The bark is usually defined as all the tissue, live and dead, outside the vascular cambium (i.e., including phloem). As a fire line passes a tree, it will rapidly heat the exterior of the bark. Since wood has low thermal conductivity and the vascular cambium is relatively near the surface, we assume that the important heating of the vascular cambium occurs through the overlying bark and not through the bole itself. The heating will not occur over a long enough time period for a constant temperature gradient to be established in a tree bole, and the bark will cool after the fire has passed. The lack of a constant temperature gradient characterizes unsteady state heating.

For this general situation, the one-dimensional heat transfer equation (e.g., Holman, 1986) is

$$\frac{\partial^2 \theta}{\partial x^2} = \frac{1}{\alpha} \frac{\partial \theta}{\partial \tau} \tag{10}$$

where θ is the temperature rise from ambient (°C), x is depth of the vascular cambium (m), α is the thermal diffusivity of the bark and underlying wood $(m^2 \, s^{-1})$, and τ_c is the duration of heating (s). As opposed to the steady state case where only position x within the material is required [Eq. (1)], both time τ and position must be considered in unsteady state conduction [Eq. (10)].

Equation (10) is one-dimensional in the sense that the temperature gradient driving heat transfer is directed into the bole from the bark's surface and at right angles to the bark's surface. The model is applicable to boles to the extent that bark surface temperatures are even over the surface of the bole and that bole curvature has minimal effects. Trees with fissured bark (see discussion that follows) or small diameters may prove to require the consideration of more than one dimension of heat flow.

Equation (10) can be solved by the Laplace transform technique for the following boundary conditions (e.g., Holman, 1986): $-h\theta = -k\partial\theta/\partial x$ at $x = 0$ and $\tau > 0$, θ remains finite as $x \to \infty$ and $\tau > 0$, and $\theta = \theta_0 = T_f - T_0$ at $\tau = 0$ and $0 \le x < \infty$ where h is the convection heat transfer coefficient (W m$^{-2} \cdot$ °C) and k is the thermal conductivity (W m$^{-1} \cdot$ °C). Because the temperature rise is minimal at the center of the bole, boles heated by flames can be considered to have infinite depth x (i.e., $0 \le x < \infty$ is reasonable). The result is

$$\frac{\theta}{\theta_0} = \mathrm{erf}\left(\frac{x}{2\sqrt{\alpha\tau}}\right) + \exp\left(\frac{hx}{k} + \frac{h^2\alpha\tau}{k^2}\right)$$
$$\cdot \left[1 - \mathrm{erf}\left(\frac{x}{2\sqrt{\alpha\tau}} + \frac{h\sqrt{\alpha\tau}}{k}\right)\right] \tag{11}$$

where erf is the Gauss error function whose argument can be found in mathematical tables (e.g., Abramowitz and Stegun, 1964) for a given value of the excess temperature ratio, defined as

$$\frac{\theta}{\theta_0} = \frac{T - T_f}{T_0 - T_f} \tag{12}$$

where T is the temperature at depth x (i.e., at the vascular cambium), T_f is the average flame temperature, and T_0 is the ambient temperature. Equation (11) is called the semi-infinite solid model of transient conduction.

Equation (11) can be simplified by assuming that the convection heat transfer coefficient, h, is infinitely large (i.e., the bark surface temperature is equal to the flame temperature); it then reduces to

$$\frac{\theta}{\theta_0} = \mathrm{erf}\left(\frac{x}{2\sqrt{\alpha\tau}}\right) \tag{13}$$

The vertical velocity of flames and their turbulence ensure that convective heat transfer rates are high (see discussion that follows), making this assumption reasonable as a first approximation. Future work on both convection heat transfer between the flame and bole and bark combustion (e.g., Gill and Ashton, 1968; Vines, 1968; Uhl and Kaufmann, 1990; Pinard and Huffman, 1997) will clarify how bark surface temperatures should be treated.

The semi-infinite solid model of transient conduction is a special case of unsteady state conduction wherein an object's surface temperature increases instantaneously and then remains constant for the duration of heating (this is the transient temperature trace). In the case of bole heating in fires, the transient temperature trace has been approximated by an estimate of the average flame temperature. The temperature profile below the surface of an object at different times after exposure to a transient temperature trace is shown in Figure 5. After the flame has passed the bole of a tree, the temperature of the vascular cambium will continue to rise until the gradient in temperatures at the vascular cambium

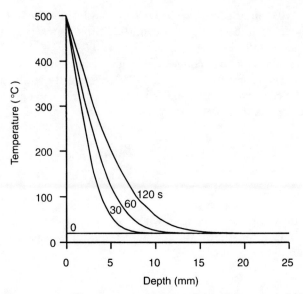

FIGURE 5   Temperature profile within a semi-infinite solid (such as a tree bole) 0, 30, 60, and 120 s after being exposed to a transient temperature pulse (i.e., a sudden increase in surface temperature that is sustained indefinitely). Equation (14) was used to calculate temperature profiles for a bark surface temperature of 500°C, an ambient temperature of 25°C, and a thermal diffusivity of $1.5 \times 10^{-7} \, \mathrm{m^2 \, s^{-1}}$.

is reversed [i.e., dT/dx in Eq. (1) becomes positive]. This cooling process can be included in more complicated models (e.g., Carslaw and Jaeger, 1959; Costa *et al.*, 1991).

Along with a transient temperature trace, the semi-infinite solid model of transient conduction [Eq. (13)] assumes that the thermal diffusivities of the outer and inner bark and underlying wood are equal. Whole bark (inner and outer) thermal diffusivities appear to be very similar to wood for the species that have been tested (Reifsnyder *et al.*, 1967). Unsteady state heat transfer models for layered composites can be used to determine the effects of variability in thermal diffusivity (e.g., see Carslaw and Jaeger, 1959, for analytical models and heat transfer texts, Holman, 1986, for numerical approaches).

Equation (13) has been used to explain heating of the vascular cambium in fires (e.g., see Spalt and Reifsnyder, 1962; Martin, 1963; Vines, 1968; Peterson and Ryan, 1986; Brown and DeByle, 1987; Gutsell and Johnson, 1996). The model can be used to predict the time required to cause necrosis of the vascular cambium in a tree of a given bark thickness. To illustrate, we will set T to the temperature at which vascular cambium necrosis would occur over a short period (e.g., 65°C), the ambient temperature to 20°C, and the average flame temperature to 500°C. The excess temperature ratio becomes

$$\frac{\theta}{\theta_0} = \frac{65 - 500}{20 - 500} = 0.8958 \tag{14}$$

and, from the tables for this value of the error function,

$$\left(\frac{x}{2\sqrt{\alpha\tau_c}}\right) = 1.15 \tag{15}$$

where τ_c is the flame residence time that would cause vascular cambium necrosis (i.e., the critical residence time). After inserting a value for the thermal diffusivity, (e.g., 1.35×10^{-7} m^2 s^{-1}; Spalt and Reifsnyder, 1962) and rearranging, Eq. (15) reduces to

$$\tau_c = 1.4 \times 10^6 x^2 \tag{16}$$

From Eq. (16), one would expect that the critical residence time τ_c would scale as the square of bark thickness. Hare (1965), Vines (1968), and Gill and Ashton (1968) present vascular cambium temperature data from heating trials that can be roughly compared with the model's prediction (see Figure 6 for details). Vines (1968) used a radiant heater and attempted to prevent bark combustion; the averaged data from five species yielded a scaling exponent of 1.2. Gill and Ashton (1968) used a radiant heater and allowed bark combustion to occur for three species; the average scaling exponent was 1.1 (range 1.0–1.2). Hare (1965) used a propane torch to heat the bark of a number of species; the average scaling exponent was 2.3 (range 2.1–2.5; Figure 6). Hare's (1965) study best fits the transient temperature trace assumption of the semi-infinite solid model; thus, it is not entirely surprising that his scaling exponents are closest to the model's prediction. Costa et al. (1991) related Pinus pinaster vascular cambium temperatures during surface fires to predictions from a numerical model that used bark surface temperature traces and included the effect of bole curvature. Their model suggests that the time to reach a given temperature scales as the 3.2 power of bark thickness.

1. Thermal Tolerance of Cells and Tissues

Tissue and cell necrosis occur at exponentially increasing rates as temperatures rise. These rates have been described empirically using the Arrhenius equation of physical chemistry (e.g., Johnson et al., 1974; Castellan, 1983):

$$\frac{d\Omega}{d\tau} = A \exp -\left[\frac{E}{\Re T}\right] \tag{17}$$

where Ω is a damage index (dimensionless), τ is time (s), A (s^{-1}) is called the frequency factor, E (J mol^{-1}) is termed the activation energy, $\Re$ is the gas constant (8.314 J mol$^{-1} \cdot$ K), and T is temperature (K). The activation energy E and

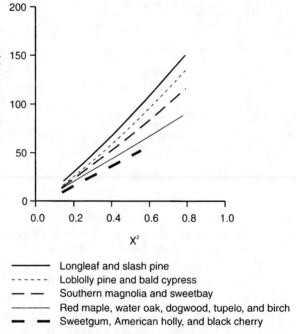

FIGURE 6   Heating times required to raise vascular cambium temperatures to 40°C scaled by the square of bark thickness as suggested by the semi-infinite solid model of transient conduction [Eq. (17); data from Hare (1965), reprinted from the *Journal of Forestry* 63 (4): 248–251, published by the Society of American Foresters, 5400 Grosvenor Lane, Bethesda, MD 20814-2198. Not for further reproduction]. Heating was done with a propane torch and species whose bark heated at similar rates were grouped. The near linearity of the relationships suggests that, for this situation, the model captures the most important aspects of the heat transfer problem.

the frequency factor A are typically estimated from regression analyses and have been found to be nearly constant within limited temperature ranges (e.g., from ambient to ~70°C). The group $E/\Re T$ is dimensionless and is called the Arrhenius number. Damage Ω to plants from exposure to elevated temperatures has been expressed as cell death (e.g., Lorenz, 1939; Dewey *et al.*, 1977; Levitt, 1980), reduction in cellular respiration (Caldwell, 1993), cessation of cytoplasmic streaming (Alexandrov, 1964), and protein coagulation resulting from protein denaturation (Levitt, 1980). When applied to complex systems such as cells and tissues, the meaning of the activation energy E and the frequency factor A are not entirely clear (e.g., Levitt, 1980), though, for animal cell death, these parameters, when suitably scaled, correspond to those expected for protein denaturation (Rosenberg *et al.*, 1961).

The Arrhenius equation [Eq. (17)] can be integrated over a given temperature profile to provide predictions of damage Ω (see Martin *et al.*, 1969, and

Chapter 7 in this book for a related approach). Predicting the effects of unsteady state temperature traces in this way have been of particular interest in the study of human skin burns (e.g., Henriques, 1947; Takata, 1974; Diller and Hayes, 1983; Diller *et al.*, 1991). When suitably parameterized, Eq. (17) could serve as the basis for predicting such things as the proportion of phloem cells surviving in the live bark or the extent to which cellular respiration rates are reduced by a heat pulse. Equation (17) provides a basis for the finding that the rates of damage rise approximately exponentially with temperature to the point at which tissue death is effectively instantaneous (e.g., Nelson, 1952; Dewey *et al.*, 1977; Levitt, 1980). It is this critical temperature that justifies the use of the relatively simple semi-infinite solid model of transient conduction [Eq. (13)] to predict vascular cambium necrosis.

2. Bark Thickness

Bark thickness varies among species, with tree age, with height along the bole, and with bole diameter. Bark thickness measured at breast height is often nearly linearly related to stem diameter and, consequently, can be expressed as a percentage of stem diameter (Table 1; see also Spalt and Reifsnyder, 1962; Hengst and Dawson, 1994; Pinard and Huffman, 1997). Age has been found to affect bark thickness with older stems having thicker bark than younger stems for a given diameter (Hale, 1955; Spalt and Reifsnyder, 1962).

Bark near the base of trees may often be thicker than one would predict from a linear relationship between diameter and bark thickness (e.g., Hale, 1955; Glasby *et al.*, 1988). Because of the power-law scaling between the critical time for vascular cambium necrosis and bark thickness (see earlier discussion), basal thickening of the bark leads to substantially longer critical residence times (Figure 7; see also Brown and DeByle, 1987).

The one-dimensional, semi-infinite solid model of transient conduction assumes that bark fissuring has minimal effects. During fires, lower bark surface temperatures have been measured in fissures than on bark plates (e.g., Fahnestock and Hare, 1964). It follows that fissuring must lead to deeper boundary layers and, thus, lower convection heat transfer rates [Eq. (5)]. Consequently, lower bark surface temperatures in fissures counteract the effects of the thinner bark. Two- or three-dimensional numerical methods could be used to explore the effects of bark fissuring (e.g., Holman, 1986), allowing one to determine when it must be included in models of the bole-heating process.

3. Thermal Diffusivity

Thermal diffusivity quantifies the rate at which a temperature wave penetrates a material during transient conduction or, in other words, the ease with which a material absorbs heat from its surroundings. When an object is subjected to

TABLE 1   Bark Thickness as a Percentage of Diameter at Breast Height
for Selected Species from Two Regions

Region	Species	Bark (%)
Northern Rocky Mountains,	Larix occidentalis	7.4
United States[a]	Pseudotsuga menziesii	6.7
	Populus balsamifera	5.9
	Abies grandis	4.3
	Pinus ponderosa	4.1
	Thuja plicata	2.5
	Pinus contorta	1.6
	Abies lasiocarpa	1.5
	Picea engelmannii	0.7
Eastern Amazonia, Brazil[b]	Manilkara huberi	9.1
	Lecythis lurida	8.5
	Lecythis idatimon	7.7
	Cordia sericalyx	7.6
	Cecropia sciacophylla	5.8
	Xylopia aromatica	5.6
	Inga alba	4.8
	Dipopyros duckei	4.7
	Macrolobium angustifolium	4.4
	Metrodorea flavida	4.4
	Tetragastris altissima	4.3
	Jacaranda copaia	4.1
	Pourouma guianensis	3.9
	Inga sp.	3.5
	Ecclinusa sp.	2.3

[a]Finch (1948) in Spalt and Reifsnyder (1962).
[b]Uhl and Kauffman (1990).

a rapid increase in temperature (transient conduction), it is necessary to quantify not only the temperature gradient and thermal conductivity [as in the steady
state, Eq. (1)] but also the heat required to raise the temperature of the material. This is determined by the material's heat capacity c ($J kg^{-1} \cdot °C$) and density ($kg m^{-3}$), where heat capacity c is the amount of heat required to raise the
temperature of a unit mass of material by one degree. Thus, thermal diffusivity
($m^2 s^{-1}$) is

$$\alpha = \frac{k}{\rho c} \tag{18}$$

where the denominator is referred to as the volumetric heat capacity ($J m^{-3} \cdot °C$).

For tree boles, thermal diffusivity in the semi-infinite solid model of transient
conduction [Eq. (13)] is that of the bark and underlying wood. Wood and bark

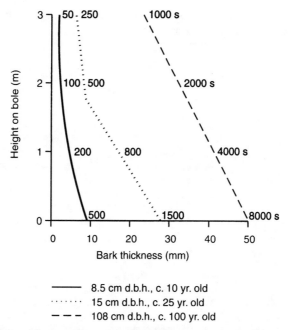

FIGURE 7   Predicted flame residence times required to cause necrosis of the vascular cambium at increasing heights along the boles of three *Eucalyptus oreades* trees of different sizes and ages. Bark thickness data from Glasby *et al.* (1988). Critical residence times were calculated from Eq. (14) for a bark surface temperature of 500°C, a critical temperature for tissue necrosis of 65°C, an ambient temperature of 25°C, and a thermal diffusivity of 1.5×10^{-7} m^2 s^{-1}.

have been found to be similar in their thermal properties (Martin 1963; Reifsnyder *et al.*, 1967); as such, we will focus on bark. Bark is a mixture of materials including outer bark, inner bark, air, and water. The different effects of these components on bark thermal conductivity, density, and heat capacity must be accounted for in calculations of thermal diffusivity (see Martin, 1963; Reifsnyder *et al.*, 1967). Table 2 contains values of the properties of barks and other materials. Air has a relatively low thermal conductivity but a high thermal diffusivity because of its low density. On the other hand, water and bark have similarly low thermal diffusivities, water because of its high volumetric heat capacity and bark because of its low thermal conductivity.

Standard methods for measuring thermal diffusivity and its components are given in the American Society for Testing and Materials handbooks (e.g., ASTM, 1998). Thermal diffusivity can be estimated by the flash method in which one side of a thin sample is subjected to a short pulse of radiant energy (e.g., Taylor, 1979). The ensuing temperature rise on the back side of the sample is measured

TABLE 2   Thermal Conductivity k, Heat Capacity c, Density ρ, and Thermal Diffusivity α of Various Materials at Ambient Temperatures ($\sim$20°C)[a]

Material	k (W m$^{-1}\cdot$°C)	c (J kg$^{-1}\cdot$°C)	ρ (kg m^{-3})	α (m^2 s^{-1} × 10^7)
Water	0.614	4179	998	1.47
Bark				
Shortleaf pine	0.100	1348	497	1.5
Longleaf pine	0.105	1268	520	1.6
Martin (1963)	—	—	—	1.2–1.6
Reifsnyder et al. (1967)	—	—	—	0.6–1.6
Wood				
Longleaf pine	0.133	1411	540	1.7
Shortleaf pine	0.116	1369	458	1.8
Moist sand	0.52	—	—	4.81
Moist silty clay loam	0.82	1720	1250	3.81
Iron	73	452	7897	204
Air	0.026	1006	1.18	219

[a]Single values for shortleaf and longleaf pine bark and wood are averages for ovendried conditions (Reifsnyder et al., 1967). The ranges in thermal diffusivity of bark are for moisture contents from <1 to about 30% moisture. Martin's (1963) range in thermal diffusivities is for samples from a group of temperate zone species, and Reifsnyder et al's (1967) is again for shortleaf and longleaf pine samples. Sand properties are from Van Wijk and Bruijn (1964), and silty loam properties are from Asrar and Kanemasu (1983). Values for other materials are from Holman (1986).

and, with these data, an unsteady state heat transfer equation is solved for thermal diffusivity. A benefit of the flash method is that the size of the sample can be small. As far as we are aware, this convenient method has not been used for bark.

Alternatively, thermal conductivity and heat capacity can be estimated separately, and thermal diffusivity can be calculated from Eq. (18). Thermal conductivity can be measured by steady state or transient methods. Martin (1963) and Reifsnyder et al. (1967) used the transient "hot wire" method in which the temperature rise in an electrically heated wire imbedded within the sample is greatest when the thermal conductivity is least. The value of thermal conductivity is calculated by solving a transient heat transfer model of radial heat flow containing the thermal properties of the probe and sample. Steady state methods include the use of a guarded-hot-plate apparatus in which a given heat flux produces a measured temperature difference across a sample (ASTM, 1998). A steady state conduction equation based on Fourier's Law [e.g., Eq. (1)] is then solved to determine the thermal conductivity. Heat capacity is generally measured by means of a differential scanning calorimeter (ASTM, 1998) wherein a sample and a reference material are heated at a constant rate by using separately controlled resistance heaters. The differential heat flow into or out of a sample is compared with the reference material to determine the heat capacity or the heat capacity plus any applicable latent heats.

In studies of a group of temperate conifer and deciduous species, Martin (1963) and Reifsnyder et al. (1967) found that bark density and moisture content are the primary correlates of bark thermal diffusivity. Martin's (1963) equation for thermal conductivity is based on data from the inner and outer bark of ten species (both deciduous and conifer) over a range in moisture contents from 0 to 114% of dry weight. Tests were done at 25°C. The equation (converted to $W\ m^{-1}\cdot {}^{\circ}C$) is

$$k = -0.00846 + 0.000210\rho + 0.000554\rho_m \qquad (19)$$

where ρ is bulk density ($kg\ m^{-3}$) calculated as the oven dry weight divided by the volume at the moisture content at which the test was done (so as to account for shrinkage upon drying). Martin (1963) assumed a 13.6% expansion in volume between dry and wet bark. The moisture density ρ_m ($kg\ m^{-3}$) is the portion of bark density accounted for by moisture. Equation (19) accounted for 98% of the variation in thermal conductivity. Reifsnyder et al.'s (1967) equation for the whole bark of three *Pinus* species (longleaf, shortleaf, and red pine) gives very similar results. Thermal conductivity within bark is nearly equal along the radial and transverse planes (Martin, 1963).

Heat capacity of bark is affected by moisture content because of the high heat capacity of water (Table 2) and because, when a sample is heated, energy is absorbed as the chemical bonds between water and cellulose are broken (this is the heat of desorption, Byram et al., 1952b; Martin, 1963; Reifsnyder et al., 1967). Thus, bark heat capacity ($J\ kg^{-1}\cdot {}^{\circ}C$) is composed of the following components (Martin, 1963):

$$c = c_{db} + Mc_w + \Delta c \qquad (20)$$

where c_{db} is the heat capacity of dry bark, M is the moisture content as a fraction of dry weight, c_w is the heat capacity of water (from tables), and Δc is the elevation of heat capacity from the desorption of bound water. The heat capacity of dry bark was similar to that of wood and was estimated by Martin (1963) as

$$c_{db} = 1105 + 4.856T \qquad (21)$$

where T is temperature (°C). The elevation in heat capacity was given as follows:

$$\Delta c = 1277M, \qquad 0 \le M \le 0.27 \qquad (22)$$

$$\Delta c = 348, \qquad M > 0.27 \qquad (23)$$

Martin (1963) found no significant differences among species in heat capacity of dry bark.

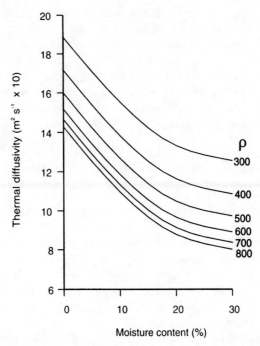

FIGURE 8   The effect of moisture and density ρ (kg m^{-3}) on the thermal diffusivity of shortleaf and longleaf pine bark. From Reifsnyder *et al.* (1967).

Bark thermal diffusivity decreases as moisture and density increase (Figure 8). Bark moisture varies among species depending on the relative amount of inner and outer bark and the drying characteristics of the outer bark (e.g., Reifsnyder *et al.*, 1967). Bark densities can vary considerably, for example, Martin and Crist (1967) give a range of 280–1290 kg m^{-3} for individual oven dry samples from 19 species. Variation in moisture content has less of an effect on thermal diffusivity than might be imagined. Bark thermal conductivity and volumetric heat capacity both increase with moisture content, but volumetric heat capacity increases at a faster rate resulting in a small net decrease in thermal diffusivity (Figure 8).

In our discussion of thermal diffusivity, we have implicitly assumed that it will be constant as temperatures rise over the course of heating [this is also an assumption of the semi-infinite solid model of transient conduction, Eq. (13)]. This is clearly not true; for instance, a substantial heat sink is inherent in both the desorption of bound water and the evaporation of free water (e.g., Byram *et al.*, 1952b). In addition, thermal conductivities and heat capacities of bark, water, and air increase while their densities decrease with temperature. A first approximation of the heat sink associated with the evaporation of free water is to cap the surface temperature to which an object is exposed at 100°C (Mercer *et al.*, 1994). In fact, bound water undergoes desorption, and free water is evaporated continuously as temperatures rise (e.g., Byram *et al.*, 1952b). Modeling continuous variation in thermal diffusivity requires more complicated heat transfer models than Eq. (13) and would appear to yield relatively little predictive benefit at present.

4. Predicting the Extent of Vascular Cambium Necrosis around the Bole

Given that the assumptions of the semi-infinite solid model of transient conduction [Eq. (13)] are approximately met, it is possible to estimate the residence time of the flame that would cause necrosis of the vascular cambium around the entire circumference of the bole (i.e., girdling). Girdling is predicted if the residence time of flaming combustion is greater than the critical residence time for vascular cambium necrosis. In Figure 9, residence times [Eq. (9)] were calculated from flame width and rate of spread for 50 experimental fires in Australian

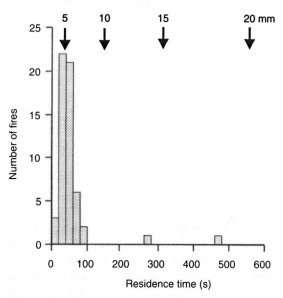

FIGURE 9 Comparison of fire residence times and critical residence times for bole vascular cambium necrosis. The frequency distribution gives the residence times of 50 experimental fires in *Eucalyptus sieberi* forest. Data from Cheney *et al.* (1992). The arrows indicate the critical residence times that would girdle trees with bark of the indicated thickness. Critical residence times were calculated from Eq. (14) for a bark surface temperature of 500°C, a critical temperature for cambial necrosis of 65°C, an ambient temperature of 20°C, and a thermal diffusivity of $1.35 \times 10^{-7} \, \text{m}^2 \, \text{s}^{-1}$.

Eucalyptus sieberi forests (data from Cheney *et al.*, 1992). Given our model of vascular cambium necrosis [e.g., Eq. (13)] and our chosen values of flame temperatures and thermal diffusivities (Figure 9), only two of these fires had residence times sufficient to have girdled trees with bark thickness >1 cm.

If the residence time of a given fire is insufficient to girdle a tree, partial vascular cambium necrosis may still occur. When a fire is burning in wind, heating on the downwind (leeward) side of a tree bole is either equal to or greater than heating on the upwind (windward) side. If the unequal heating is sufficient to cause vascular cambium necrosis on the leeward side, but not on the windward side, and if the tree survives the fire, a fire scar results. Gutsell and Johnson (1996) provide a mechanistic hypothesis of fire-scar formation based on fluid dynamics and heat transfer processes that fits various observations about standing leeward flames and fire scars.

A model of fire-scar formation must account for three observations about fire scars: (1) fire scars are found only on the leeward sides of trees; (2) small trees rarely have fire scars; and (3) fire scars are usually triangular in shape, becoming narrower with height. Fire scars are created when fires burn in an ambient wind because, it is hypothesized, the airflow around the bole of a tree creates a pair of vortices. In the presence of the vortices, a standing leeward flame develops as the free-moving flame passes a tree. The standing leeward flame has a longer residence time and higher temperatures than the free-moving flame, accounting for the greater heat transfer into the leeward bark. Below, leeward vortices and the standing leeward flame are discussed in sequence.

As air flows around the bole of a tree, a reverse flow occurs producing a pair of vortices (Figure 10a–10c). The formation of the vortices depends on the magnitude of the bole Reynolds number

$$\text{Re} = \frac{uD\rho}{\mu} \tag{24}$$

where D is the characteristic dimension (tree diameter, m), u is the horizontal wind speed (m s^{-1}), ρ is the density of the air (kg m^{-3}), and μ is the dynamic viscosity of the air (kg m$^{-1} \cdot$s). The dynamic viscosity of air increases with temperature as the fire approaches the tree, but, because they vary over a much larger range, the wind speed and tree diameter have much larger effects on the Reynolds number. Note that, with a constant wind speed, trees with smaller diameters will have lower Reynolds numbers than larger trees.

When the Reynolds number is less than about 5, the flow around the tree is laminar and unseparated from the bole (Figure 10a; see, e.g., Swanson, 1970). At higher Reynolds numbers, the flow separates from the bole and a pair of adjacent vortices forms against the leeward side of the tree (Figure 10b). The points of separation define the maximum size of the fire scar. At a Reynolds

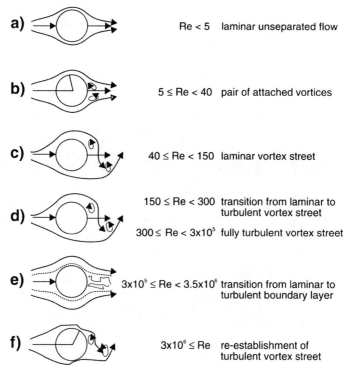

a)   Re < 5   laminar unseparated flow

b)   5 ≤ Re < 40   pair of attached vortices

c)   40 ≤ Re < 150 laminar vortex street

d)   150 ≤ Re < 300 transition from laminar to turbulent vortex street

$300 \leq$ Re $< 3 \times 10^5$ fully turbulent vortex street

e) $3 \times 10^5 \leq$ Re $< 3.5 \times 10^6$ transition from laminar to turbulent boundary layer

f) $3 \times 10^6 \leq$ Re   re-establishment of turbulent vortex street

FIGURE 10   Flow regime around a cylinder as a function of Reynolds number. From Gutsell and Johnson (1996).

number of about 40, a waviness in the wake behind the bole increases in amplitude until it becomes unstable and breaks. The result is a "vortex street" wherein each vortex of the pair is alternately shed into the wake behind the tree to be reformed anew against the bole (Figure 10c). The frequency of the shedding is dependent on the Reynolds number. As the Reynolds number increases further, there are other transitions, for example to turbulent flow (Figures 10d and 10e) and to a more downwind point at which the flow separates from the bole (Figure 10f).

The patterns of flow in each leeward vortex are fundamentally the same as those in fire whirls, tornados, and dust devils (Emmons and Ying, 1967). Ideally, each vortex consists of two zones of flow (Figure 11): an outer zone of irrotational flow and an inner zone, or core, of rotational flow (see, e.g., Eskinazi, 1962). In the outer zone, the flow is translated around the core with no change in the orientation of the air particles. For this orientation to be maintained, the

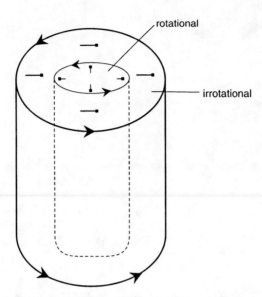

FIGURE 11 Irrotational and rotational zones of flow in a vortex. In the flow, particles change orientation in the rotational zone, while their orientation does not change in the irrotational zone. From Gutsell and Johnson (1996).

tangential velocity of the air particles farthest away from the core must be lower than that of the air particles closest to the core (Figure 12). For irrotational flow outside the vortex core, the radius r and tangential velocity u_t yield a constant (i.e., ru_t).

In the rotating core, the tangential velocity and radius have the opposite relationship. Velocity increases linearly with radius and u_t/r is a constant. Consequently, particles in the rotational zone change their orientation as they rotate (Figure 11). Rotation in the core is termed solid body rotation because, as in a rotating solid, the relative distances among particles do not change with time. The pattern of rotational and irrotational flow in a vortex play an important role in explaining the development of the standing leeward flame as a free-moving flame passes a tree.

The residence time of the standing leeward flame is approximately double that of the free-moving flame. Figures 13a–e shows the development of the standing leeward flame as the free-moving flame passes by a tree. As soon as the front of the flame reaches the center of the periphery of the tree (Figure 13b), the flame is drawn horizontally into the cores of the leeward vortices, increasing the residence time by approximately 0.5 tree diameters. The standing leeward flame persists until the trailing edge of the fire leaves the region of the vortices

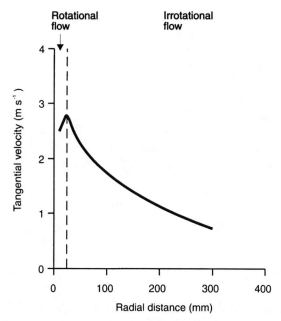

FIGURE 12   The tangential velocity (u_t) distribution in a vortex. Tangential velocity increases from the center of the vortex to the outer boundary of the rotational zone and then decreases through the zone of irrotational flow. Reprinted from Chigier *et al.*, *Combustion and Flame* **14**, 171–180, Copyright 1970, with permission from Elsevier Science.

on the leeward side of the tree (Figure 13e), increasing flame residence time by a further 1.5 tree diameters. Consequently, the residence time of the standing leeward flame is given by

$$\tau_f = \frac{x_f}{R} + \frac{2D}{R} \tag{25}$$

where $2D/R$ is the increase in residence time accounted for by the effects of the leeward vortices [cf. Eq. (9)].

The horizontal draw that increases the residence time of the standing leeward flame results from a "centrifugal pump effect" caused by the relatively low pressures at the axes of each vortex core (see Chapter 5). The low pressure within a vortex core is produced by the increase in angular momentum from the axis to the edge of the rotating core. Friction within the boundary layer slows the rotation at the base of the vortex allowing boundary-layer air to be drawn into the relatively buoyant vortex cores.

The centrifugal pump effect is augmented by two additional sources of buoyancy. The first source is the warm gases drawn into the core through the bound-

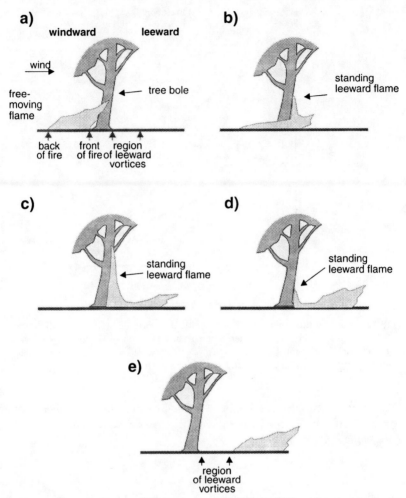

FIGURE 13 The effect of leeward vortices on a free-moving flame (from Gill, 1974, reprinted from *Forest Science* 20 (3): 198–205, published by the Society of American Foresters, 5400 Grosvenor Lane, Bethesda, MD 20814-2198. Not for further reproduction). When the leading edge of a free-moving flame reaches the center of the periphery of a tree (a), the flame is drawn into the leeward vortices, producing a standing leeward flame (b). The standing leeward flame continues to increase in height (c) and then recedes as the free-moving flame passes the tree (d). Once the trailing edge of the free-moving flame is beyond the leeward vortices, the standing leeward flame has completely receded (e).

ary layer as the free-moving flame approaches the tree. Second, after the flame is drawn into the vortices, combustion adds substantially to the buoyancy within the vortex cores; this mechanism has been called the chimney effect (Fahnestock and Hare, 1964).

The chimney effect is made possible by a reduction in turbulent mixing between the core and outer irrotational zone. Angular momentum increases from the core's axis to its outer edge, dampening turbulence within the core. Concomitantly, horizontal movement of air into the core is opposed. As a consequence, the buoyancy force within the core is not rapidly decreased by mixing, resulting in a considerable increase in flame height. In addition, flame height increases because the flow of gaseous fuel in the vortex cores is greater than the rate of mixing with the surrounding air, and, hence, there is a greater height along which combustion can take place.

Differential heating around the base of a tree results not only from the increased residence time of the standing leeward flame relative to the free-moving flame but also from its elevated temperatures (Figure 14). The cause of the elevated temperatures appears to be that the rotation of the fuel in the vortex cores increases the combustion rate of fuel and air.

The typical triangular shape of fire scars can be explained by considering the three-dimensional temperature profile of the standing leeward flame. Based on

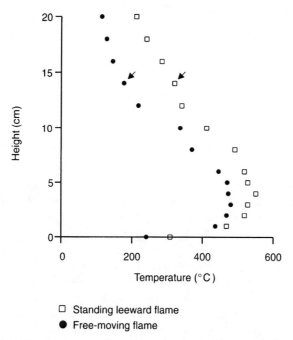

FIGURE 14   Vertical temperature profiles of a standing leeward flame and a free-moving flame. Both flames are of the same average height (indicated by the arrows). Average temperatures in the standing leeward flame are higher than those of the free-moving flame at any given height. From Gutsell and Johnson (1996).

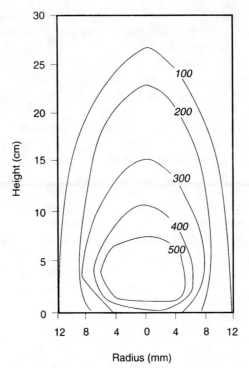

FIGURE 15   The two-dimensional temperature distribution (°C) in a cross section of a standing leeward flame. The outer margin of the flame extends approximately to the outermost isotherm. The temperature is highest at the center of the flame and decreases toward the sides and top. From Gutsell and Johnson (1996).

experimental results in a wind tunnel (Gutsell and Johnson, 1996), the temperatures at the surface of a cylinder were found to be highest at the middle of the standing leeward flame (Figure 15). The temperatures decreased with height after reaching a peak at around 40% of flame height. This resulted in triangular isotherms, wide near the base of the cylinder and decreasing in width towards the flame tip.

Fire scars will not occur on small trees and at low wind speeds because the Reynolds numbers are too low for the formation of leeward vortices [Eq. (24)]. Unequal heating between the windward and leeward sides of trees will also not occur if the rate of spread of the free-moving flame is high. This can be appreciated by considering Eq. (25). As the rate of spread increases, the difference between the residence time of the free-moving flame and that of the standing leeward flame disappears because $2D/R$ (the increase in residence time on the leeward side of the tree because of vortex formation) becomes small.

Ultimately, it may prove that relatively simple fluid dynamics and heat transfer models will be adequate to predict the extent of vascular cambium necrosis around tree boles during wildfire. Thermal tolerance models [e.g., Eq. (17)] will also be required when estimates of such things as cell survivorship and physiologial impairment are desired.

III. EFFECTS OF FIRE ON CANOPY COMPONENTS

To understand how fires cause tissue necrosis in tree canopies, one needs to consider not only the process of heat transfer into canopy components (foliage, buds, twigs, fruits, and seeds) but also the characteristics of the plume (Figure 2). Modeling necrosis of canopy components is more involved than modeling bole cambial necrosis because it is not reasonable to assume, as we did for tree boles being heated by flames, that convection heat transfer rates are high. Convection heat transfer coefficients are expected to vary considerably with the characteristics of the air flow at different heights within a plume and with the size and shape of the canopy component. Predicting convection heat transfer into canopy components requires estimates of plume temperatures and vertical velocities at different heights and the residence times of those temperatures.

A. PLUME CHARACTERISTICS

Buoyant plumes are wedge-shaped, turbulent columns of hot, low-density air that rise above heat sources. As plumes rise, they mix with the cooler and denser ambient air (e.g., Morton *et al.*, 1956; Taylor, 1961). Because of this turbulent mixing, plume width increases with height, and plume temperatures and average gas velocities decrease with height. Ambient winds bend plumes, reducing the temperatures reached at a given height above ground while, at the same time, increasing the residence times of those temperatures. Plumes above fire lines burning in surface fuels are assumed to approximate line sources of heat at sufficient heights within the canopy (e.g., Van Wagner, 1973). Two-dimensional plume models form the basis of the following discussion and are more thoroughly described in Chapter 7. Three dimensions are required to describe more complex plume behaviors such as vorticity that are key to understanding the dynamics of blowup fires (see Chapters 7 and 8).

Steady state, two-dimensional plumes can be approximately described by a top-hat profile wherein plume characteristics are averaged over the width of the plume (Figure 4; see also Morton *et al.*, 1956). More realistically, average temperatures and vertical velocities rise gradually from the margin of the plume to the centerline (the Gaussian profile). The Gaussian profile results from the lateral entrainment of ambient air into the rising plume.

In the relatively simple heat transfer model we introduce to describe heating of canopy components, a steady state, top-hat temperature profile drives heating, and the associated vertical velocity profile is used to estimate convection heat transfer coefficients. The top-hat plume structure has appeal for heat transfer modeling because it is mathematically simpler to treat top-hat profiles than Gaussian ones. For plumes in an ambient wind, Gaussian and top-hat models have been shown to give very similar predictions of plume characteristics (Davidson, 1986).

Mercer and Weber (1994; see Chapter 5 in this book) provide models of a line-source, top-hat plume in both still air and a cross wind. Their steady state, two-dimensional model employs coupled partial differential equations that balance the fluxes of mass, momentum, and heat between the plume and the ambient air. The equations are called conservation equations because the dynamic and thermodynamic properties of a rising, entraining plume are governed by conservation of mass, momentum, and energy. These equations can be solved to derive estimates of plume temperatures, velocities, and widths at different heights which are then used to evaluate the heating of canopy components.

B. Heat Transfer into Canopy Components

The height at which foliage necrosis (i.e., crown scorch) occurs can be predicted from an equation that estimates the height within the canopy at which a lethal temperature for foliage necrosis is reached (see Van Wagner, 1973). The lethal temperature was defined as that at which foliage necrosis would occur at short exposure times (e.g., 1 minute). No heat transfer model is employed in this approach. The assumption, of course, is that foliage has little resistance to convection and conduction and, thus, that the plume temperature is a good approximation of the temperature of the foliage.

This approach has been surprisingly useful despite differences in foliage shape, size, display, and, consequently, convection heat transfer coefficients (e.g, Peterson and Ryan, 1986; Reinhardt and Ryan, 1988; Finney and Martin, 1993; Gould et al., 1997). For example, Van Wagner (1973) found that the following equation (see Taylor, 1961; Thomas, 1964) closely predicted scorch height z (m) for a broadleaf and several pine species (see Figure 16):

$$z = \frac{jI^{2/3}}{T_p - T_0} \tag{26}$$

where I is fire-line intensity $(kW\ m^{-1})$, j is a proportionality constant that must be estimated from experimental data, and T_p and T_0 are the plume and ambient

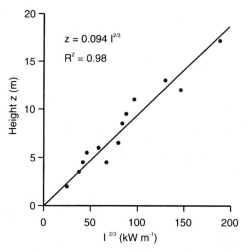

FIGURE 16 The height of crown scorch z as a function of the two-thirds power of fire-line intensity [see Eq. (27)]. The plume temperature was set to 60°C, the temperature assumed to cause foliage necrosis after approximately one minute of heating. Data from Van Wagner (1973).

temperatures, respectively. Often, the temperature difference is implicitly included in the proportionality constant j (e.g., Figure 16).

Because no heat transfer model is employed, using Van Wagner's (1973) approach has several limitations. First, plume temperature alone does not provide any understanding of how the plume is causing necrosis of the canopy component. All the biology is in the heat transfer coefficients of conduction and convection and in the response of the tissue to the temperature pulse! Second, validation is difficult because the proportionality constant incorporates both heat transfer processes and plume behavior. Third, several kinds of canopy components are more resistant to heating than foliage (e.g., buds, branches, and fruits).

Let us now formulate a simple heat balance for a canopy component so that we can use it to determine the time required for necrosis in a wildfire plume. We have specifically chosen a simple model which should be a good first approximation for small canopy components (e.g., twigs, foliage, buds, flowers, fruits, and seeds). In the heat balance, heat is convected onto the component's surface at a rate equal to the increase in internal energy of the component. The convection (i.e., Newton's law of cooling) and volumetric heat capacity terms in the heat budget are, respectively,

$$-hA(T - T_\infty) = c\rho V \frac{dT}{d\tau} \tag{27}$$

where c is the heat capacity (kJ kg^{-1} · °C), ρ is the density (kg m^{-3}), V is the volume to be heated (m^3), τ is time (s), h is the convection heat transfer coefficient $(W\ m^{-2} \cdot {}^\circ C)$, A is the surface area over which convection occurs (m^2), T is the temperature of the structure, and T_∞ is the plume temperature $({}^\circ C)$.

Defining $\theta = T - T_\infty$ and $dT/d\tau = d\theta/d\tau$ and then rearranging gives

$$\frac{d\theta}{d\tau} = -\frac{hA}{\rho cV}\theta \tag{28}$$

a first order, ordinary differential equation that can be solved for θ as a function of time for the initial condition $T = T_0$ at $\tau = 0$. Separating variables and integrating gives

$$\theta = B \exp\left[-\frac{hA}{\rho cV}\tau\right] \tag{29}$$

where B is a constant of integration. At $\tau = 0$, B must be equal to the difference between the ambient temperature T_0 and the plume temperature T_∞. Thus, the equation can be expressed as

$$\frac{T - T_\infty}{T_0 - T_\infty} = \frac{\theta}{\theta_0} = \exp\left[-\frac{hA}{\rho cV}\tau\right] \tag{30}$$

The equation can be solved for the time required for the canopy component to reach the critical temperature for tissue necrosis

$$\tau = \frac{\rho cV}{hA}(\ln \theta_0 - \ln \theta) \tag{31}$$

when estimates of the dimensions and heating properties of the component are available, the plume and ambient temperatures are known, and T is set to a critical temperature for tissue necrosis (e.g., $65^\circ C$).

The coefficient $hA/\rho cV$ in Eq. (30) is the thermal time constant (s^{-1}) because it gives the time required for θ to reach 37% of θ_0. The thermal time constant will vary depending on the biological and physical properties of the canopy component. For example, the heat capacity and density depend on the kind of material and its water content, the surface area and volume depend on the size and shape of the component, and the convection heat transfer coefficient depends on the size, shape, and orientation of the component along with its arrangement relative to other components. Ecologists thus will be interested in how these values change relative to each other because they describe the means by which individuals and species differ in how they are affected by fire.

How good is such a simple model? As far as we are aware, there are no tests of the model for canopy components in fires. However, using results derived from laboratory heating trials, Johnson and Gutsell (1993) used the model to explain patterns of cone opening in two serotinous-coned *Pinus* species. In cones, resin bonds between scales hold the cones closed until they are heated, generally by fire. When the bonds are broken, the scales reflex allowing seeds to be

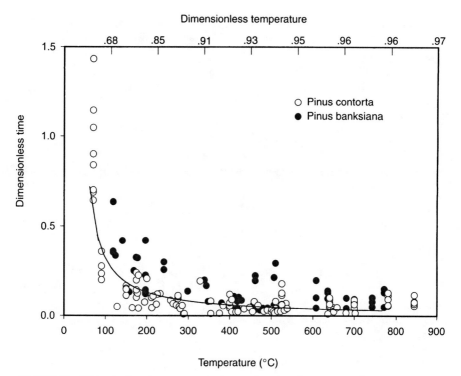

FIGURE 17   Dimensionless time to cone opening as a function of temperature for two serotinous-coned conifer species. From Johnson and Gutsell (1993).

released. The resins have a melting point (slightly higher than 100°C) that is substituted for T in Eq. (30). Other parameter values in Eq. (30) were also estimated from the literature. The dimensionless time to cone opening ($hA/\rho cV \cdot \tau$) at different temperatures in a furnace followed an exponential relationship as suggested by the model (Figure 17).

The simplicity of Eq. (30) derives from the fact that only the time dependence of temperature is considered and the dependence of temperature on position within the object is ignored. Because temperature is treated as uniform, the thermal conductivity is not included in Eq. (30), and the model is called the lumped-heat-capacity model. Foliage and small branches, twigs, buds, flowers and seeds are thermally thin (i.e., their surface temperatures are close to their interior temperatures under heating). Accordingly, the lumped-parameter method may often be a useful approximation considering errors in measurement and in estimation of some of the parameters, particularly the convection heat transfer coefficient (h).

As an object becomes larger, it becomes more resistant to being heated by conduction. We can follow a formal procedure for deciding if the temperature gradient within an object is small enough to ignore and, consequently, whether Eq. (30) is applicable. For there to be a minimal internal temperature gradient, the resistance to conduction must be small relative to the resistance to convection (e.g., Sucec, 1985); that is,

$$\frac{R_i}{R_o} \ll 1 \tag{32}$$

where R_i is the internal resistance to heating by conduction and R_o is the resistance to convection heat transfer from the outside. These resistances can be calculated by the following equations:

$$R_i = \frac{L}{kA}, \qquad R_o = \frac{1}{hA_s} \tag{33}$$

where L is an average path length over which conduction occurs from the center of the object to the surface, and A is the surface area over which convection occurs (not necessarily the entire surface area A_s). The term hA_s determines the convection heat transfer rate and is familiar from Eq. (5). Substituting Eq. (33) into Eq. (32) gives

$$\frac{h}{k}\left(\frac{A_s L}{A}\right) = \frac{hD}{k} \ll 1 \tag{34}$$

where D has the dimension of length and is called the characteristic dimension of the canopy component. For our purposes, we define D as the maximum linear distance along which heat is conducted. For example, for a twig, D would be bark thickness (i.e., depth of the vascular cambium), and, for a bud, D would be the depth of the meristem.

The expression on the left of the inequality in Eq. (34) is called the Biot number (Bi) and is dimensionless. How much less than one should the Biot number be? When Bi ≤ 0.1, the difference in temperature between a given depth within the object of interest and its surface is always $\leq 5\%$ percent (e.g., Sucec, 1985). When the temperature difference is so small, there is little accuracy lost in using Eq. (30).

1. Convection Heat Transfer Coefficients

Modeling convection heat transfer [e.g., Eqs. (5) and (31)] requires estimates of convection heat transfer coefficients h for the canopy components of interest. For simple geometries and flow characteristics, analytical and semianalytical calculations of h can be made. For the somewhat irregular canopy components in the turbulent flow of the plume, the use of correlations derived from extensive experimentation would be expected to give more reliable results. We show one approximate analytical solution and then introduce the experimental approach.

As air flows past an object, its velocity approaches zero at the object's surface. Accordingly, the convection heat transfer rate [Newton's law of cooling, Eq. (5)] must equal the rate of conduction [Fourier's law, Eq. (1)] through this still layer:

$$-hA(T_s - T_\infty) = -kA\frac{dT}{dx}\bigg|_{x=0} \tag{35}$$

where A is the surface area over which heat transfer is occurring, T_s is the surface temperature, T_∞ is the free-stream temperature, k (W m$^{-1}\cdot$C) is the thermal conductivity of the air, and dT/dx is the temperature gradient at the surface of the object (i.e., the depth x within the object is zero). Rearranging and simplifying gives

$$h = \frac{-k\dfrac{dT}{dx}\bigg|_{x=0}}{T_\infty - T_s} \tag{36}$$

When the gradient in temperature dT/dx is steep at the surface of the object, h is large. The steepness of the temperature gradient dT/dx depends on conduction heat transfer and fluid dynamics within the boundary layer.

Equation (36) can be solved after a treatment of the boundary-layer conditions (e.g., see Holman, 1986). For instance, the solution for the average convection heat transfer coefficient over the length of a flat plate when flow within the boundary layer is laminar is given in dimensionless form as

$$\text{Nu} = 0.664\,\text{Re}^{1/2}\,\text{Pr}^{1/3} \tag{37}$$

where Nu is the average Nusselt number, Re is the Reynolds number, and Pr is the Prandtl number (all three variables are dimensionless). The Nusselt number contains the average convection heat transfer coefficient h (W m$^{-2}\cdot$°C):

$$\text{Nu} = \frac{hD}{k} \tag{38}$$

where D (m) is the characteristic dimension relative to the flow of air (e.g, length of the plate in this case) and k is the thermal conductivity of air (W m$^{-1}\cdot$°C).

The Reynolds number [Eq. (24)] gives the ratio of inertial forces to viscous forces. At high Reynolds numbers, inertial forces predominate and fluid flow in the boundary layer is turbulent, resulting in steep temperature gradients and high convection heat transfer rates.

The Prandtl number relates the thickness of the fluid dynamic and thermal boundary layers:

$$\text{Pr} = \frac{\nu}{\alpha} = \frac{\dfrac{\mu}{\rho}}{\dfrac{k}{\rho c}} = \frac{c\mu}{k} \tag{39}$$

where ν is the kinematic viscosity (μ/ρ, m^2 s^{-1} and α is the thermal diffusivity of the fluid ($k/\rho c$, m^2 s^{-1}). High kinematic viscosities (ν) mean that viscous forces will be felt farther away from the surface of an object. At the same time, high thermal diffusivities (α) mean that temperature influences from the molecular transport of heat will be felt farther out in the flow field. The Prandtl number is thus the connecting link between the velocity field and the temperature field above the surface of an object.

Equation (37) can be solved for the convection heat transfer coefficient by

$$h = 0.664 \left(\frac{k}{D} \right) \text{Re}^{1/2} \, \text{Pr}^{1/3} \tag{40}$$

Because Eq. (40) includes the dimensionless variables that are most important in determining convection heat transfer rates, it serves as the point of departure for experimental determinations of convection heat transfer coefficients. For example, Gates (1980) uses Eq. (40) and experimental determinations of the proportionality constants to estimate convection heat transfer coefficients for small leaves

$$\text{Nu} = 1.86 \, \text{Re}^{1/2} \, \text{Pr}^{1/3} \tag{41}$$

and large leaves

$$\text{Nu} = 1.18 \, \text{Re}^{1/2} \, \text{Pr}^{1/3} \tag{42}$$

Although the equations fit data from a variety of sources (including field data), there is some disagreement on the extent to which the equations are valid over the range of wind turbulence seen within canopies (Schuepp, 1993). Wind turbulence can increase boundary-layer turbulence and, thus, momentum and convection heat transfer rates.

Few studies have been conducted on convection heat transfer for canopy components other than broad leaves. However, branches and conifer needles approach a cylindrical shape, and buds and fruits approach a spherical shape. Experimental correlations for smooth cylinders and spheres are available in the literature (e.g., see references in Sucec, 1985). The equations are similar in form to Eq. (37).

Mutual sheltering by aggregated canopy components would tend to reduce convection heat transfer coefficients for individual leaves, needles, twigs, and buds. For example, long leaf pines, *Pinus palustris* (e.g., Andrews, 1917; Wahlenberg, 1946; Byram, 1948) and Australian grass trees, a branching, upright monocot, *Xanthorrhoea australis* (Gill and Ingwersen, 1976), are famous for the protection of their meristems by dense foliage. An empirical solution to this problem has been suggested for convection heat transfer between wind flowing through the canopy and conifer needles (Monteith and Unsworth, 1990; Nikolov *et al.*, 1995):

$$Nu_s = \frac{Nu_i}{2.1} \qquad (43)$$

where the subscript s refers to sheltered and i to isolated needles. The denominator is termed the shelter factor and is an index of the reduction in flow velocity experienced by foliage for forced convection in laminar flow.

2. Predicting Height of Foliage, Bud, and Twig Necrosis

To predict the height of foliage, bud, and twig necrosis, one must take into account the fact that the fire moves through the fuelbed resulting in a certain residence time of plume temperatures. The residence time of a top-hat plume in still air is proportional to the width of the plume (x_p) and the rate of spread of the fire line [cf. Eq. (9)]:

$$\tau_p(z) = \frac{x_p}{R} \qquad (44)$$

where τ_p is plume residence time (s) at height z (m) and R is the rate of spread of the fire line (m s^{-1}). See Chapter 7 for a discussion of residence times for plumes in a cross wind.

Necrosis of a given canopy component will depend on plume temperature, the velocity of the plume (through its effect on convection heat transfer coefficients), and plume residence time at a given height above a fire line. Plume models (see Chapter 7) can be used in conjunction with heat transfer and thermal-tolerance models to calculate the height within the canopy below which necrosis of a given canopy component would occur. The dimensions of canopy components will follow some distribution (e.g., Corner, 1949; Ackerly and Donoghue, 1998), and, as a first approximation, the average dimension of the canopy component may suffice in calculations of the height of necrosis.

Vascular cambium necrosis in the bole is expected to co-vary with damage to canopy components. Within species, the thickness of the bark often scales positively with the height of the canopy. This, along with the scaling of flame and plume characteristics, would be expected to result in a positive relationship between the extent of damage to the vascular cambium and the extent of damage to canopy components. The relationship between vascular cambium and canopy damage would be expected to differ among species because species vary considerably in (1) the heating properties of their bark and canopy components and (2) the allometric relationships among tree size, bark thickness, and the distribution of canopy components.

IV. ROOT NECROSIS

Because soils have low thermal diffusivities (Table 2) relative to the duration of flaming, glowing, and smoldering combustion, soil heating, like bark heating, has been modeled as an unsteady state process (e.g., Aston and Gill, 1976; Steward et al., 1990; Peter, 1992; Oliveira et al., 1997). To predict soil heating, one must first describe heat flux at the soil surface. Where smoldering does not occur, this flux is driven by flaming and glowing combustion (Peter, 1992; Oliveira et al., 1997). From an analysis of convection and radiation heat transfer, Oliveira et al. (1997) conclude that radiation will dominate downward heat flux. Van Wagner (1972) estimated downward radiation heat flux from the Stefan-Boltzman law and assumptions about flame and ambient surface temperatures [e.g., Eqs. (6) and (7)]. After the flame has passed, the soil surface cools by convection and radiation (Peter, 1992; Oliveira et al., 1997).

Because of their porosity, heat is transferred through dry soils by conduction, convection, and radiation. In saturated soils, heat is transferred by conduction alone (De Vries, 1963; Peter, 1992). The thermal conductivity of dry and saturated soils can be approximated by averaging the effects of minerals, air, and water (De Vries, 1963). In moist soils (i.e., soils that are not saturated), downward heat transfer from the soil surface occurs not only by conduction but also by transport of latent and sensible heat through mass transfer of water in both its liquid and vapor phases (e.g., De Vries, 1958; Westcot and Wierenga, 1974; Aston and Gill, 1976). Water vapor diffusion through porous soils can account for a considerable fraction of the downward heat flux (e.g., Westcot and Wierenga, 1974). Near the surface in moist soils, the latent heat of vaporization constrains soil temperature rise to 80–100°C until liquid water is evaporated (Aston and Gill, 1976; Peter, 1992).

Where an organic layer is absent or, where present, does not smolder, empirical measurements and mechanistic models indicate that maximum soil temperatures typically do not rise above 60°C below 5 cm depth (Aston and Gill, 1976; Raison et al., 1986; Hungerford, 1989; Steward et al., 1990; Bradstock and

Auld, 1995; Oliveira *et al.*, 1997). An organic layer that does not combust insulates the underlying mineral soil, but smoldering increases the downward heat flux (e.g., Swezy and Agee, 1991; Hartford and Frandsen, 1992; Peter, 1992). Where organic soil layers are present, predictive models of root and basal cambium necrosis will depend on an understanding of the processes that determine the rates of spread and patchiness of organic layer combustion (see Chapter 13).

There are few studies on necrosis of roots and basal vascular cambium from soil heating and organic layer combustion. Using soil heat transfer models, Steward *et al.* (1990) and Peter (1992) simulated the depth of lethal heat penetration by assuming that roots would die instantaneously if the temperatures in the surrounding soil reached 60°C. Swezy and Agee (1991) found extensive fine-root necrosis in surface soils after smoldering combustion, although it is not clear how much of the necrosis was from smoldering combustion itself and how much was an indirect response to bud necrosis in the canopy or necrosis of live bark and sapwood tissues. Long durations of elevated temperatures at the bases of Ponderosa pines accompanied smoldering combustion of duff, but the heating did not lead to extensive vascular cambium necrosis (Ryan and Frandsen, 1991). Ryan and Frandsen hypothesized that temperatures at the vascular cambium were constrained by the removal of heat in the mass flow of water in the sapwood (see also Vines, 1968).

V. TREE MORTALITY

Even though mechanistic models are being developed to link fire behavior and tissue necrosis (e.g., Peterson and Ryan, 1986; Gutsell and Johnson, 1996), little advancement has been made on drawing mechanistic links between patterns of tissue necrosis and tree death. Current models are typified by logistic regressions in which a continuous probability function is derived from binary tree death data (i.e., trees are either dead or alive at some time after fire). The equations have the following general form:

$$P_d = \frac{1}{1 + \exp[-(b_0 + b_1 x_1 + b_2 x_2 + \cdots + b_i x_i)]} \tag{45}$$

where P_d is the probability of tree death, the x_i's are independent variables, and the b_i's are empirically derived coefficients. Generally, independent variables are descriptors or surrogates of fire behavior (e.g., fire-line intensity, fuel consumption) and fire effects (e.g., crown scorch, bark char). Equation (45) is not a mechanistic model because the independent variables and the relationships among independent variables are chosen with little or no regard to the actual processes by which either fire causes tissue necrosis or tissue necrosis causes tree death. Because they do not include processes, current models lack generality and cannot be applied beyond the specific conditions on which they were based.

To understand the process of tree death, we expect that a useful starting point will be to consider the effects of fires on tree carbon budgets (not a new suggestion, e.g., Ryan et al., 1988; Harrington, 1993; Glitzenstein et al., 1995). Trees die when they are unable to acquire or mobilize sufficient resources to maintain normal function in the face of injury and stress (see Waring, 1987). It is hypothesized that trees allocate available carbohydrates in a hierarchical manner: new foliage, buds, and roots have the highest priority while remaining resources are allocated to stored reserves, growth, and defensive compounds (Waring and Pitman, 1985; Christiansen et al., 1987; Waring, 1987). Trees that are likely to die often show departures in carbon allocation patterns relative to healthy trees.

Imbalances in carbon budgets have been found either to cause death directly or to precipitate cascades of events that ultimately result in death. In a direct manner, Douglas fir death following tussock moth defoliation was shown to depend on the extent and duration of the defoliation and the amount of stored carbohydrate reserves that could be mobilized to renew lost foliage (e.g., Webb, 1981). Imbalances in a tree's carbon budget caused by such stresses as prolonged drought, defoliation, and suppression may often predispose a tree to being killed (e.g., Pedersen, 1998; Waring and Pitman, 1985; Christiansen et al., 1987; Dewar, 1993; Gottschalk et al., 1998; Davidson et al., 1999). For example, through effects on carbon budgets, defoliation by Gypsy moths makes trees more susceptible to stem boring insects and root pathogens, albeit by mechanisms that are not entirely straightforward (see review in Davidson et al., 1999).

As with tree death resulting from causes other than fire, tree death from fire-caused necrosis in the canopy, roots, and bole can be considered from a carbon-budget perspective. Foliage, bud, and branch necrosis, in analogy to insect defoliation, would be expected to reduce carbon fixation rates and thereby increase the likelihood that a tree will die. Indeed, the degree of crown scorch emerges from statistical models [e.g., Eq. (45)] as one of the more consistent predictors of tree death (e.g., Van Wagner, 1973; Ryan et al., 1988; Ryan and Reinhardt, 1988; Finney and Martin, 1993; Harrington, 1993; Regelbrugge and Conard, 1993). Root necrosis from fire would be expected to affect rates of soil-resource acquisition and postfire carbon allocation patterns. Phloem and vascular cambium necrosis in the bole of a tree would be expected to impede transport processes. For example, fire scars were found to be related to crown die back in mature giant Sequoia trees (*Sequoiadendron giganteum*), presumably because fire scars reduced the cross-sectional area of sapwood (Rundel, 1972). Various features of fire-caused tree death, including season-of-burn effects (e.g., Ryan et al., 1988; Harrington, 1993; Glitzenstein et al., 1995) and differences among taxa in susceptibility to fire, may prove to be quantitatively predictable by linking heat transfer, tissue necrosis, and carbon-budget models.

VI. DISCUSSION

In this chapter, we have illustrated with relatively simple models the kinds of mechanistic linkages that we believe are required to develop an understanding of tissue necrosis and tree death from fires. In the following, we will look at a series of papers from the mainstream ecological literature that we believe would have profited from a more mechanistic approach. If anything, we hope to reiterate that simple mechanistic models help to clarify the study of fire effects.

Mutch (1970) suggested that ". . . fire-dependent plant communities burn more readily than non-fire-dependent communities because natural selection has favored development of characteristics that make them more flammable." Testing the Mutch hypothesis requires an understanding of flammability. Flammability has a precise definition for premixed flames (i.e., flames in which the fuel and air are intimately mixed before ignition; see Chapter 2 in this book). For premixed flames, flammability refers to successful ignition of the fuel–air mixture by a pilot flame, ignition being dependent on pressure, temperature, and the ratio of fuel to air. Flammability is not defined for diffusion flames, such as those of wildfires, where the fuel and air are initially separated. In the context of forest fires, flammability is a vague term that encompasses most aspects of fire dynamics (see discussion in Anderson, 1970). Being so broad, it has little practical use and appears only to generate confusion.

Williamson and Black (1981) and Rebertus and Williamson (1989) found that, after surface fires, oak stem and whole-plant mortality were greater adjacent to canopy pines compared with areas away from pines where the canopy was dominated by oaks. They attributed this result to the higher maximum temperatures measured near pines during fires. Their conclusions are not strongly supported because their maximum temperatures are ambiguous, and maximum temperatures constitute a weak link between fire behavior and tree-stem and whole-plant mortality.

It is not clear which of several possible maximum temperatures Williamson and Black (1981) and Rebertus and Williamson (1989) intended to measure. Accurately estimating temperatures requires that one balance a heat budget for the measuring device [see Eq. (8) and Chapter 2 in this book]. Maximum temperatures were estimated from waxes of a range of melting points applied to the surface of ceramic tiles strung at different heights above the ground. There was no attempt to account for the heat budget of the tiles, consequently; the temperatures have an unknown relationship to flame and plume temperatures. If one were interested in the maximum temperatures of canopy components (e.g., foliage and buds), one would have to match the heating properties of the measuring device with those of the canopy component. However, the heating properties of the tiles do not appear to mimic any tree component.

Maximum temperatures, even if accurately measured, have limited meaning in terms of heat transfer into trees. Given the tacit assumption by Williamson and Black (1981) and Rebertus and Williamson (1989) that bole heating was not important, we must determine heat transfer from the plume into the foliage, buds, and branches. The lumped-heat-capacity model [Eq. (30)] suggests that average temperatures, vertical velocities, and residence times determine heat transfer from plumes into canopy components. Maximum temperatures constitute an incomplete description of plumes relative to heat transfer and, consequently, tell us little about the temperatures experienced by canopy components. With data of unknown relevance to tissue necrosis, the papers provide a weak link between fire behavior and tree-stem and whole-plant mortality.

Jackson *et al.* (1999) explore the relationship between bark thickness and tree survival after fires. In ecosystems characterized by short-interval, low-intensity fires, species tended to develop thicker bark early in their ontogeny relative to species occupying ecosystems in which fires occurred at long intervals. As suggested by a simulation model, the explanation for this result was that higher investment in bark where fires were of low intensity would increase tree survival but would not affect tree survival where fires were intense because, in intense fires, crown necrosis would render bole tissue necrosis irrelevant. A key component of the simulation model was the relationship between bark thickness and the probability that a tree would survive a fire. Heat transfer models would appear to have allowed Jackson *et al.* (1999) to constrain the number of possible relationships. Power-law scaling between bark thickness and critical residence time (e.g., Figure 6) suggests that minimal investment in bark results in large increases in protection.

Using mechanistic models requires that ecologists and foresters acquire skills outside the traditional realm of biology. However, by looking more closely at, for example, heat transfer into the vascular cambium around the boles of trees during fires, we find that the biology is in the thickness of the bark, the shape of the bark surface, the combustion properties of the bark, and the thermal properties of the bark and underlying wood. When mechanistic models guide our research, we can hope to pose clear and testable hypotheses regarding fire effects.

NOTATION

Roman Letters

A	area	m^2
A	frequency factor	s^{-1}
b	regression coefficient	undefined
c	heat capacity	$J\ kg^{-1} \cdot {}^{\circ}C$
D	characteristic dimension	m

E	radiant energy flux	$W\ m^{-2}$
E	activation energy	$J\ mol^{-1}$
h	heat transfer coefficient	$W\ m^{-2} \cdot {}^{\circ}C$
I	fire-line intensity	$kW\ m^{-1}$
k	thermal conductivity	$W\ m^{-1} \cdot {}^{\circ}C$
L	length	m
M	moisture fraction	dimensionless
P_d	probability of death	dimensionless
q	heat transfer rate	W
R	fire-line rate of spread	$m\ s^{-1}$
R_i	resistance to conduction	${}^{\circ}C\ W^{-1}$
R_o	resistance to convection	${}^{\circ}C\ W^{-1}$
T	temperature	${}^{\circ}C, {}^{\circ}K$
u	velocity	$m\ s^{-1}$
V	volume	m^3
x	depth or width	m
z	height	m

GREEK LETTERS

α	thermal diffusivity	$m^2\ s^{-1}$
ε	emmisivity	dimensionless
θ	temperature difference	${}^{\circ}C, {}^{\circ}K$
μ	dynamic viscosity	$kg\ m^{-1} \cdot s$
ρ	density	$kg\ m^{-3}$
ρ_m	moisture density	$kg\ m^{-3}$
τ	time	s
v	kinematic viscosity	$m^2\ s^{-1}$
Ω	damage index	dimensionless

DIMENSIONLESS GROUPS

Bi	Biot number
Nu	Nusselt number

Pr   Prandtl number
Re   Reynolds number

CONSTANTS

B   integration constant
j   proportionality constant
$\mathfrak{R}$ gas constant $8.314 \, J \, mol^{-1} \cdot K$
σ Stefan-Boltzman constant $5.668 \times 10^{-8} \, W \, m^{-2} \cdot K^4$

SUBSCRIPTS

c   critical value
cd   conduction
cv   convection
db   dry bark
f   flame condition
i   isolated condition
n   net
p   plume condition
r   radiation
rr   re-radiation
s   surface or sheltered condition
w   water
0   ambient condition
∞ free stream condition

ADDITIONAL READINGS

Monteith, J. L., and Unsworth, M. H. (1990). "Principles of Environmental Physics." Edward Arnold Publishers, London.
Sucec, J. (1985). "Heat Transfer." Wm. C. Brown Publishers, Dubuque.
Swanson, W. M. (1970). "Fluid Mechanics." Holt, Rinehart, and Winston, New York.

REFERENCES

Abramowitz, M., and Stegun, I. A. (1964). "Handbook of Mathematical Functions." Applied Mathematics Series No. 55. U.S. Department of Commerce, National Bureau of Standards, Washington, DC.

Ackerly, D. D., and Donoghue, M. J. (1998). Leaf size, sapling allometry, and Corner's Rules: phylogeny and correlated evolution in maples (*Acer*). *Am. Naturalist* **152**, 767–791.

Albini, F. A. (1980). "Thermochemical Properties of Flame Gases from Fine Wildland Fuels." Research Paper INT-243. USDA Forest Service, Intermountain Forest and Range Experiment Station, Ogden.

Alexandrov, V. Y. (1964). Cytophysiological and cytoecological investigations of heat resistance of plant cells toward the action of high and low temperature. *Quart. Rev. Biol.* **39**, 35–77.

Anderson, H. E. (1970). Forest fuel ignitibility. *Fire Technol.* **6**, 312–319.

Andrews, E. F. (1917). Agency of fire in propagation of longleaf pines. *Bot. Gaz.* **64**, 497–508.

Asrar, G., and Kanemasu, E. T. (1983). Estimating thermal diffusivity near the soil surface using Laplace transform: Uniform initial conditions. *Soil Sci. Soc. Amer. J.* **47**, 397–401.

ASTM. (1998). "Annual Book of ASTM Standards." American Society for Testing and Materials, West Conshohocken, PA.

Aston, A. R., and Gill, A. M. (1976). Coupled soil moisture, heat and water vapour transfers under simulated fire conditions. *Aus. J. Soil Res.* **14**, 55–66.

Bradstock, R. A., and Auld, T. D. (1995). Soil temperatures during experimental bushfires in relation to fire intensity: Consequences for legume germination and fire management in south-eastern Australia. *J. Appl. Ecol.* **32**, 76–84.

Brown, J. K., and DeByle, N. V. (1987). Fire damage, mortality, and suckering in aspen. *Can. J. For. Res.* **17**, 1100–1109.

Byram, G. M. (1948). Vegetation temperature and fire damage in the southern pines. *Fire Control Notes* **9**, 34–36.

Byram, G. M., Fons, W. L., Sauer, F. M., and Arnold, R. K. (1952a). "Thermal Conductivity of Some Common Forest Fuels." USDA Forest Service, Division of Fire Research, California Forest and Range Experiment Station, Berkeley.

Byram, G. M., Fons, W. L., Sauer, F. M., and Arnold, R. K. (1952b). "Thermal Properties of Forest Fuels." USDA Forest Service, Division of Fire Research, California Forest and Range Experiment Station, Berkeley.

Caldwell, C. R. (1993). Estimation and analysis of cucumber (*Cucumis sativus* L.) leaf cellular heat sensitivity. *Plant Physiol.* **101**, 939–945.

Carslaw, H. S., and Jaeger, J. C. (1959). "Conduction of Heat in Solids." Clarendon Press, Oxford.

Castellan, G. W. (1983). "Physical Chemistry," 3rd ed. Addison-Wesley, London.

Cheney, N. P., Gould, J. S., and Knight, I. (1992). "A Prescribed Burning Guide for Young Regrowth Forests of Silvertop Ash." Research Paper No. 16. Forestry Commission of New South Wales, Research Branch, Sydney.

Chigier, N. A., Beér, J. M., Grecov, D., and Bassindale, K. (1970). Jet flames in rotating flow fields. *Combustion and Flame* **14**, 171–180.

Christiansen, E., Waring, R. H., and Berryman, A. A. (1987). Resistance of conifers to bark beetle attack: Searching for general relationships. *For. Ecol. Manage.* **22**, 89–106.

Corner, E. J. H. (1949). The Durian theory of the origin of the modern tree. *Ann. Bot.* **13**, 367–414.

Costa, J. J., Oliveira, L. A., Viegas, D. X., and Neto, L. P. (1991). On the temperature distribution inside a tree under fire conditions. *Int. J. Wildland Fire* **1**, 87–96.

Davidson, G. A. (1986). Gaussian versus top-hat profile assumptions in integral plume models. *Atmos. Environ.* **20**, 471–478.

Davidson, C. B., Gottschalk, K. W., and Johnson, J. E. (1999). Tree mortality following defoliation by the European gypsy moth (*Lymantria dispar* L.) in the United States: A review. *For. Sci.* **45**, 74–84.

De Vries, D. A. (1958). Simultaneous transfer of heat and moisture in porous media. *Trans. Amer. Geophys. Union* **39**, 909–916.

De Vries, D. A. (1963). Thermal properties of soils. *In* "Physics of Plant Environment" (W. R. Van Wijk, Ed.), pp. 210–235. North-Holland Publishing Company, Amsterdam.

Dewar, R. C. (1993). A mechanistic analysis of self-thinning in terms of the carbon balance of trees. *Ann. Bot.* **71**, 147–159.

Dewey, W. C., Hopwood, L. E., Sapareto, S. A., Gerweck, M. S., and Gerweck, L. E. (1977). Cellular responses to combinations of hyperthermia and radiation. *Radiology* **123**, 463–474.

Diller, K. R., and Hayes, L. J. (1983). A finite element model of burn injury in blood-perfused skin. *J. Biomech. Eng.* **105**, 300–307.

Diller, K. R., Hayes, L. J., and Blake, G. K. (1991). Analysis of alternate models for simulating thermal burns. *J. Burn Care Rehab.* **12**, 177–189.

Drysdale, D. (1990). "An Introduction to Fire Dynamics." John Wiley and Sons, New York.

Emmons, H. W., and Ying, S. J. (1967). The fire whirl. *In* "Eleventh Symposium (International) on Combustion," pp. 475–488. The Combustion Institute, Pittsburgh.

Ennos, A. R. (1993). The scaling of root anchorage. *J. Theoret. Biol.* **161**, 61–75.

Eskinazi, S. (1962). "Principles of Fluid Mechanics." Allyn and Bacon, Boston.

Fahnestock, G. R., and Hare, R. C. (1964). Heating of tree trunks in surface fires. *J. For.* **62**, 799–805.

Finch, T. L. (1948). Effect of bark growth in measurement of periodic growth of individual trees. Research Note 60. USDA Forest Service, Northern Rocky Mountain Forest and Range Experiment Center, Missoula.

Finney, M. A., and Martin, R. E. (1993). Modeling effects of prescribed fire on young-growth coast redwood trees. *Can. J. For. Res.* **23**, 1125–1135.

Gates, D. M. (1980). "Biophysical Ecology." Springer Verlag, New York.

Gill, A. M. (1974). Toward an understanding of fire-scar formation: Field observation and laboratory simulation. *For. Sci.* **20**, 198–205.

Gill, A. M., and Ashton, D. H. (1968). The role of bark type in relative tolerance to fire of three central Victoria eucalypts. *Aus. J. Bot.* **16**, 491–498.

Gill, A. M., and Ingwersen, F. (1976). Growth of *Xanthorrhoea australis* R. Br. in relation to fire. *J. Appl. Ecol.* **13**, 195–203.

Gill, A. M., and Knight, I. (1991). Fire measurement. *In* "Conference on Bushfire Modeling and Fire Danger Rating Systems" (N. P. Cheney and A. M. Gill, Eds), pp. 137–146. CSIRO Division of Forestry, Yarralumla, ACT.

Glasby, P., Selkirk, P. M., Adamson, D., Downing, A. J., and Selkirk, D. R. (1988). Blue Mountains ash (*Eucalyptus oreades* R. T. Baker) in the western Blue Mountains. *Proc. Linnean Soc. New South Wales* **110**, 141–158.

Glitzenstein, J. S., Platt, W. J., and Streng, D. R. (1995). Effects of fire regime and habitat on tree dynamics in north Florida longleaf pine savannas. *Ecol. Monogr.* **65**, 441–476.

Gottschalk, K. W., Colbert, J. J., and Feicht, D. L. (1998). Tree mortality risk of oak due to gypsy moth. *Eur. J. For. Pathol.* **28**, 121–132.

Gould, J. S., Knight, I., and Sullivan, A. L. (1997). Physical modeling of leaf scorch height from prescribed fires in young *Eucalyptus sieberi* regrowth forests in south-eastern Australia. *Int. J. Wildland Fire* **7**, 7–20.

Gutsell, S. L., and Johnson, E. A. (1996). How fire scars are formed: Coupling a disturbance process to its ecological effect. *Can. J. For. Res.* **26**, 166–174.

Hale, J. D. (1955). Thickness and density of bark: Trends of variation for six pulpwood species. *Pulp Paper Mag. Can.* **56**, 113–117.

Hare, R. C. (1965). Contribution of bark to fire resistance of southern trees. *J. For.* 63, 248–251.

Harrington, M. G. (1993). Predicting *Pinus ponderosa* mortality from dormant season and growing season fire injury. *Int. J. Wildland Fire* 3, 65–72.

Hartford, R. A., and Frandsen, W. H. (1992). When it's hot, it's hot . . . or maybe it's not! (Surface flaming may not portend extensive soil heating). *Int. J. Wildland Fire* 2, 139–144.

Hengst, G. E., and Dawson, J. O. (1994). Bark properties and fire resistance of selected tree species from the central hardwood region of North America. *Can. J. For. Res.* 24, 688–696.

Henriques, F. C., Jr. (1947). Studies of thermal injury V: The predictability and the significance of thermally induced rate processes leading to irreversible epidermal injury. *Arch. Pathol.* 43, 489–502.

Holman, J. P. (1986). "Heat Transfer." McGraw-Hill Book Company, New York.

Hungerford, R. D. (1989). Modeling the downward heat pulse from fire in soils and in plant tissue. *In* "Proceedings of the 10th Conference on Fire and Forest Meteorology," pp. 148–154. Forestry Canada, Ottawa.

Jackson, J. F., Adams, D. C., and Jackson, U. B. (1999). Allometry of constitutive defense: A model and a comparative test with tree bark and fire regime. *The American Naturalist* 153, 614–632.

Johnson, E. A., and Gutsell, S. L. (1993). Heat budget and fire behavior associated with the opening of serotinous cones in two *Pinus* species. *J. Veg. Sci.* 4, 745–750.

Johnson, F. H., Eyring, H., and Stover, B. J. (1974). "The Theory of Rate Processes in Biology and Medicine." John Wiley and Sons, New York.

Levitt, J. (1980). "Responses of Plants to Environmental Stresses, Volume I, Chilling, Freezing, and High Temperature Stresses," 2nd ed. Academic Press, New York.

Lorenz, R. W. (1939). "High Temperature Tolerance of Forest Trees." Technical Bulletin 141. Minnesota Agricultural Experiment Station, St. Paul.

Martin, R. E. (1963). Thermal properties of bark. *For. Prod. J.* 13, 419–426.

Martin, R. E., and Crist, J. B. (1967). Selected physical-mechanical properties of eastern tree barks. *For. Prod. J.* 18, 54–60.

Martin, R. E., Cushwa, C. T., and Miller, R. L. (1969). Fire as a physical factor in wildland management. *In* "Proceedings of the Ninth Annual Tall Timbers Fires Ecology Conference," pp. 271–288. Tall Timbers Research Station, Tallahassee.

McCaffrey, B. J. (1979). "Purely Buoyant Diffusion Flames: Some Experimental Results." NBSIR 79-1910. United States Department of Commerce, National Bureau of Standards, Washington, DC.

McMahon, C. K., Adkins, C. W., and Rodgers, S. L. (1986). A video mage analysis system for measuring fire behavior. *Fire Manage. Notes* 47, 10–15.

Mercer, G. N., and Weber, R. O. (1994). Plumes above line fires in a cross wind. *Int. J. Wildland Fire* 4, 201–207.

Mercer, G. N., Gill, A. M., and Weber, R. O. (1994). A time-dependent model of fire impact on seed survival in woody fruits. *Aus. J. Bot.* 42, 71–81.

Monteith, J. L., and Unsworth, M. H. (1990). "Principles of Environmental Physics." Edward Arnold, London.

Morton, B. R., Taylor, G. I., and Turner, J. S. (1956). Turbulent gravitational convection from maintained and instantaneous sources. *Proc. Roy. Soc. Lond. A* 234, 1–23.

Mutch, R. W. (1970). Wildland fires and ecosystems—A hypothesis. *Ecology* 51, 1046–1051.

Nelson, R. M. (1952). "Observations on Heat Tolerance of Southern Pine Needles." Research Paper 14. USDA Forest Service, Southeast Forest Experiment Station, Asheville.

Nelson, R. M., Jr. (1980). "Flame Characteristics for Fires in Southern Fuels." Research Paper SE-205. USDA Forest Service, Southeast Forest Experiment Station, Asheville.

Niklas, K. J. (1992). "Plant Biomechanics: An Engineering Approach to Plant Form and Function." The University of Chicago Press, Chicago.

Niklas, K. J. (1994). The scaling of plant and animal body mass, length, and diameter. *Evolution* **48**, 44–54.

Nikolov, N. T., Massman, W. J., and Schoettle, A. W. (1995). Coupling biochemical and biophysical processes at the leaf level: An equilibrium photosynthesis model for leaves of C$_3$ plants. *Ecological Modeling* **80**, 205–235.

Oliveira, L. A., Viegas, D. X., and Raimundo, A. M. (1997). Numerical predictions on the soil thermal effect under surface fire conditions. *Int. J. Wildland Fire* **7**, 51–63.

Pedersen, B. S. (1998). The role of stress in the mortality of midwestern oaks as indicated by growth prior to death. *Ecology* **79**, 79–93.

Peter, S. J. (1992). "Heat Transfer in Soils Beneath a Spreading Fire." PhD dissertation, University of New Brunswick, Fredericton.

Peterson, D. L., and Ryan, K. C. (1986). Modeling postfire conifer mortality for long-range planning. *Environ. Manage.* **10**, 797–808.

Pinard, M. A., and Huffman, J. (1997). Fire resistance and bark properties of trees in a seasonally dry forest in eastern Bolivia. *J. Tropical Ecol.* **13**, 727–740.

Raison, R. J., Woods, P. V., Jakobsen, B. F., and Bary, G. A. V. (1986). Soil temperatures during and following low-intensity prescribed burning in a *Eucalyptus pauciflora* forest. *Aus. J. Soil Res.* **24**, 33–47.

Rebertus, A. J., and Williamson, G. B. (1989). Longleaf pine pyrogenicity and turkey oak mortality in Florida xeric sandhills. *Ecology* **70**, 60–70.

Regelbrugge, J. C., and Conard, S. G. (1993). Modeling tree mortality following wildfire in *Pinus ponderosa* forests in the central Sierra Nevada. *Int. J. Wildland Fire* **3**, 139–148.

Reifsnyder, W. E., Herrington, L. P., and Spalt, K. W. (1967). "Thermophysical Properties of Bark of Shortleaf, Longleaf, and Red Pine." Bulletin No. 70. Yale University School of Forestry, New Haven.

Reinhardt, E. D., and Ryan, K. C. (1988). How to estimate tree mortality resulting from underburning. *Fire Management Notes* **49**, 30–36.

Rosenberg, B., Kemeny, G., Switzer, R. C., and Hamilton, T. C. (1961). Quantitative evidence for protein denaturation as the cause of thermal death. *Nature* **232**, 471–473.

Rothermel, R. C., and Deeming, J. E. (1980). "Measuring and Interpreting Fire Behavior for Correlation with Fire Effects." General Technical Report INT-93. USDA Forest Service, Intermountain Forest and Range Experiment Station, Ogden.

Rundel, P. W. (1972). The relationship between basal fire scars and crown damage in giant sequoia. *Ecology* **54**, 210–213.

Russell, R. S. (1977). "Plant Root Systems: Their Function and Interaction with the Soil." McGraw-Hill, London.

Ryan, K. C., and Frandsen, W. H. (1991). Basal injury from smoldering fires in mature *Pinus ponderosa* Laws. *Int. J. Wildland Fire* **1**, 107–118.

Ryan, K. C., and Reinhardt, E. D. (1988). Predicting postfire mortality of seven western conifers. *Can. J. For. Res.* **18**, 1291–1297.

Ryan, K. C., Peterson, D. L., and Reinhardt, E. D. (1988). Modeling long-term fire-caused mortality of Douglas-fir. *For. Sci.* **34**, 190–199.

Schuepp, P. H. (1993). Tansley review no. 59: Leaf boundary layers. *New Phytologist* **125**, 477–507.

Simard, A. J., Eenigenburg, J. E., Adams, K. B., Nissen, R. L., Jr., and Deacon, A. G. (1984). A general procedure for sampling and analyzing wildland fire spread. *For. Sci.* **30**, 51–64.

Spalt, K. W., and Reifsnyder, W. E. (1962). "Bark Characteristics and Fire Resistance: A Literature Survey." Occasional Paper No. 193. USDA Forest Service, Southeastern Forest Experiment Station, Asheville.

Steward, F. R., Peter, S., and Richon, J. B. (1990). A method for predicting the depth of lethal heat penetration into mineral soils exposed to fires of various intensities. *Can. J. For. Res.* **20**, 919–926.

Sucec, J. (1985). "Heat Transfer." Wm. C. Brown Publishers, Dubuque.

Swanson, W. M. (1970). "Fluid Mechanics." Holt, Rinehart, and Winston, New York.

Swezy, D. M., and Agee, J. K. (1991). Prescribed-fire effects on fine-root and tree mortality in old-growth ponderosa pine. *Can. J. For. Res.* 21, 626–634.

Takata, A. (1974). Development of criterion for skin burns. *Aerospace Med.* 45, 634–637.

Taylor, G. I. (1961). Fire under the influence of natural convection. *In* "The Use of Models in Fire Research" (W. G. Berl, Ed.), pp. 10–31. Publication 786. National Academy of Sciences, National Research Council, Washington, DC.

Taylor, R. E. (1979). Heat-pulse thermal diffusivity measurements. *High Temperatures–High Pressures* 11, 43–58.

Thomas, P. H. (1963). The size of flames from natural fires. In "Ninth Symposium (International) on Combustion," pp. 844–859. Academic Press, New York.

Thomas, P. H. (1964). "The Effect of Wind on Plumes from a Line Heat Source." Fire Research Note 572. Joint Fire Research Organization, Fire Research Station, Boreham Wood.

Thomas, P. H. (1967). Some aspects of the growth and spread of fire in the open. *Bibliog. Agric. Sec. E For.* 40, 139–164.

Uhl, C., and Kauffman, J. B. (1990). Deforestation, fire susceptibility, and potental tree responses to fire in the eastern Amazon. *Ecology* 71, 437–449.

Van Wagner, C. E. (1972). Duff consumption by fire in eastern pine stands. *Can. J. For. Res.* 2, 34–39.

Van Wagner, C. E. (1973). Height of crown scorch in forest fires. *Can. J. For. Res.* 3, 373–378.

Van Wijk, W. R., and Bruijn, P. J. (1964). Determination of thermal conductivity and volumetric heat capacity of soils near the surface. *Soil Sci. Soc. Amer. Proc.* 28, 461–464.

Vines, R. G. (1968). Heat transfer through bark, and the resistance of trees to fires. *Aus. J. Bot.* 16, 499–514.

Vines, R. G. (1981). Physics and chemistry of rural fires. *In* "Fire and the Australian Biota," (A. M. Gill, R. H. Groves, and I. R. Noble, Eds), pp. 129–149. Australian Academy of Science, Canberra.

Wahlenberg, W. G. (1946). "Longleaf Pine: Its Use, Ecology, Regeneration, Protection, Growth and Management." Charles Lathrop Pack Forest Foundation, Washington, DC.

Waring, R. H. (1987). Characteristics of trees predisposed to die. *BioScience* 37, 569–574.

Waring, R. H., and Pitman, G. B. (1985). Modifying lodgepole pine stands to change susceptibility to mountain pine beetle attack. *Ecology* 66, 889–897.

Webb, W. L. (1981). Relation of starch content to conifer mortality and growth loss after defoliation by the Douglas-fir tussock moth. *For. Sci.* 27, 224–232.

Weber, R. O. (1991). Modeling fire spread through fuel beds. *Prog. Energy Combustion Sci.* 17, 67–82.

Weber, R. O., Gill, A. M., Lyons, P. R. A., Moore, P. H. R., Bradstock, R. A., and Mercer, G. N. (1995). Modeling wildland fire temperatures. *CALMScience Suppl.* 4, 23–26.

Westcot, D. W., and Wierenga, P. J. (1974). Transfer of heat by conduction and vapor movement in a closed soil system. *Soil Sci. Soc. Amer. Proc.* 38, 9–14.

Williamson, G. B., and Black, E. M. (1981). High temperature of forest fires under pines as a selective advantage over oaks. *Nature* 293, 643–644.

CHAPTER 15

Forest Fire Management

DAVID L. MARTELL

Faculty of Forestry, University of Toronto, Toronto, Ontario, Canada

I. Introduction
II. The Relationship between Fire and Forest Land
   Management Objectives
III. Assessing Fire Impacts
   A. Least Cost Plus Damage Model of
      Fire Economics
   B. Assessing the Impact of Fire on Timber Supply
IV. Forest Fire Management Organizations
   A. A Simple Fire Suppression Model
   B. Fire Load Management
   C. Fire Suppression
V. Level of Fire Protection Planning
VI. Some Challenges
   A. Relating Outcomes to Strategies
   B. Dealing with Uncertainty
   C. Moving beyond Fire Exclusion
   Further Reading
   References

I. INTRODUCTION

Societies that are confronted with potentially destructive forest or wildland fires develop fire management organizations to modify fire's impact on people, the things they value, and the ecosystems about which they are concerned. A fire management organization's objectives are determined by the social, economic, and political institutions that control it. The extent to which it achieves its objectives depend upon (1) how well the fire and ecosystem processes and the impact of fire management are understood; (2) the degree to which the social and economic impacts of fire are understood; (3) the technology and resources society puts at the organization's disposal; (4) the knowledge, skills, and experience of the people in the organization; and (5) the challenges posed by nature.

Most of the earlier chapters of this book have dealt with the physics, chemistry, and biology of wildland fire and its ecological impact on forest stands and landscapes. The focus has been largely what wildland fire specialists describe as fire science. Fire scientists carry out basic research (e.g., fire spread), applied research (fire danger rating system development and implementation), and a great deal of research that cannot readily be classified on the basic/applied research spectrum. Previous chapters illustrated the importance of moving beyond the development of relatively simple empirical curve-fitting models and furthering our basic understanding of fire processes and their impact on ecosystems. In this chapter, we turn our attention to the ways in which societies attempt to shape wildland fire and influence its impact on them and surrounding ecosystems. We study wildland fire from a management perspective and explore what fire managers do, why they do what they do, how they decide to do what they do, and the short, intermediate, and long-term social, economic, and ecological consequences of their actions. We focus on what is commonly referred to as *fire management*.

> *Fire Management:* The Canadian Glossary of Forest Fire Management Terms (Merrill and Alexander, 1987) defines *forest fire management* as the "activities concerned with the protection of people, property, and forest areas from wildfire and the use of prescribed burning for the attainment of forest management and other land use objectives, all conducted in a manner that considers environmental, social, and economic criteria."

Most North American wildland fire agencies were developed to combat fire in the early 1900s and for many decades they had names that included terms like *fire control* and *fire protection*. They were unabashedly fire suppression organizations, and their names reflected their view of wildland fire and its impact on the world. The current widespread use of the term *fire management* reflects a more enlightened view that wildland fire is a natural phenomenon that has a range of social, economic, and ecological impacts that should be considered when it is managed.

Fire management programs typically include prevention measures to reduce the number of people-caused fires that occur, detection systems to find fires while they are small, initial attack systems to contain fires before they burn over large areas, and large fire management systems that are designed to minimize the damage that results from large fires that are not controlled by the initial attack system. They also include fuel modification measures to mitigate the impact of fires that do occur and the use of prescribed fire to fulfill silviculture, wildlife habitat management, and other land management objectives. Fire management also calls for (albeit very infrequently in most jurisdictions) conscious decisions to allow some wildfires to burn freely or to be subjected to limited suppression action if and when the net benefit of doing so is thought to be positive.

Fire management researchers carry out basic research to further our understanding of fire management processes (e.g., the influence of suppression action on fire growth) and applied research to develop decision support systems (e.g., initial attack system models) that fire managers can use to help predict the consequences of their actions and to evaluate alternative policies, strategies, and tactics. Many fire management systems researchers use Operations Research/Management Science (OR/MS) methodologies which entail the use of mathematical models and information technology. Pollock and Maltz (1994, p. 4) characterize Operations Research "as a *process* of developing mathematical models in order to understand (and possibly influence) complex operational systems." Martell (1982) and Martell *et al.* (1998) review many of the fire management operations research publications that have appeared over the last four decades.

Physical and biological scientists can and often do study systems that are relatively isolated from human intervention, but fire management researchers must deal with large complex systems composed of natural processes, people, and machines. Fire managers must resolve decision-making problems (e.g., initial attack dispatching) as circumstances dictate and, even though they base their decisions in part on the understanding of basic physical and biological processes developed by scientists, they frequently have to deal with processes that are, at best, poorly understood. Although they can postpone some decisions (e.g., the purchase of airtankers) while they gather more information and reflect upon the possible consequences of their actions, they seldom have the luxury of postponing their decisions until researchers have developed a thorough understanding of the pertinent processes. Fire management researchers that develop decision support systems to meet their needs must therefore incorporate in their systems models what is currently known about physical, biological, social, and economic processes, empirical curve-fitting models, and the subjective assessments of designated experts. In this chapter, we will describe how they do so.

A forest fire can burn over a large area during a short period of time. When it does so, it can leave tremendous impacts on public safety, property, forest resources, and forest ecosystems in its wake. Although many fire management activities are directed at specific fires, fire management programs are designed to modify the impact of fire across vast landscapes and administrative regions over very long planning horizons that are embedded in larger ecoclimatic and administrative hierarchies that span many forest management units, districts, regions, provinces, nations, and even continents. There is no definitive spatial and temporal scale at which fire can or should be managed. One can focus on a portion of an active fire's current perimeter or one can deal with fire regimes across very large spatial and temporal scales. We will approach fire and its management from a range of spatial and temporal scales, and we will study the impact of fire and fire management on fire regimes at landscape levels embedded within political and administrative structures that span many such landscapes. We begin by clarifying our use of the term *fire regime*.

> *Fire Regime:* Whalen (1995) described a *fire regime* in terms of the following five basic elements: (1) fire intensity, (2) the season during which burning takes place, (3) fire size or extent, (4) fire type, and (5) fire frequency.

Fire management programs are designed to modify one or more of those five basic elements of a fire regime, either directly or indirectly. Prevention measures, for example, reduce fire frequency, whereas fire detection and suppression activities reduce final fire size. Prescribed burning can either increase or decrease fire intensity, size, and frequency.

In Section II we describe the relationship between fire management objectives and higher level forest land management objectives, and in Section III we deal with fire economics and the assessment of fire impacts. Section IV deals with fire management organizations, and in Section V we describe how fire managers carry out level of fire protection planning to balance the benefits of mitigating the detrimental impacts of fire and reaping its benefits with the cost of doing so. We conclude with Section VI in which we describe some of the more significant challenges that will need to be addressed as fire management agencies increasingly move away from fire exclusion to the adoption of policies that call for modified fire suppression and the use of prescribed fire.

Since we are primarily interested in fire management and its impact on fire regimes, we describe the activities associated with each subsystem, their potential impact on the basic elements of a fire regime, and their potential social, economic, and ecological impacts. We discuss the decision-making problems associated with each fire management subsystem and the management information systems and decision support systems that are designed to enhance fire management decision-making.

Figure 1 is a hierarchical view of fire management decision making which will be used to structure our investigation of fire management systems. The list of decisions included in Figure 1 is not exhaustive but is characteristic of most large North American wildland fire management agencies. Most of those agencies are still very heavily committed to fire control, but the use of prescribed fire to achieve ecological objectives is a major concern of many agencies, particularly those responsible for fire management in national parks and wilderness areas in Canada and the United States.

Sound fire management is based, in part, on the extent to which the decisions associated with any particular level in the hierarchy are compatible with the decisions taken above and below. Consider daily airtanker deployment, for example. The number of airtankers to be deployed on a particular day will be

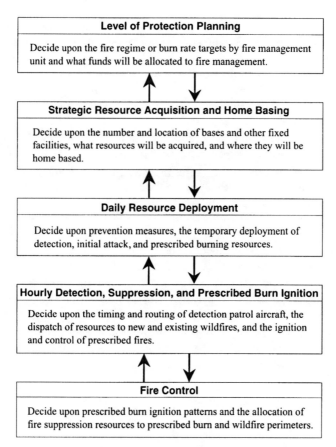

FIGURE 1   Fire management decision hierarchy.

influenced by the number of airtankers the agency owns or leases for the fire season (a consequence of higher level strategic decision making) and the cost of temporarily borrowing more costly airtankers from other agencies. High borrowing costs could deter fire managers from borrowing airtankers from others, which could produce daily airtanker shortages, longer initial attack response times, and more escaped fires in the lower initial attack level of the hierarchy. Fire management poses many such challenges.

We draw heavily on our own experience with the Ontario Ministry of Natural Resources (OMNR) in the province of Ontario in Canada, and what we know about other North American wildland fire management agencies to illustrate our discussion. Most wildland fire management agencies are, for the most part, emergency response systems; however, they vary from jurisdiction to jurisdiction around the globe, and we do not attempt to capture the rich diversity of approaches, philosophies, and practices that are influenced by local ecosystems, climate, culture, economics, and political systems. For comprehensive historical accounts of the development of forest fire management in some North American jurisdictions, see, for example, Holbrook (1943), Lambert and Pross (1967), and Pyne (1982). Pyne (1997) reviews fire and fire management from a more global perspective.

It is important to note that, although private enterprise plays a significant role in wildland fire management in North America (e.g., many fire management agencies contract flying services from private companies and public corporations), fire management is ultimately a public sector endeavor and management and decision making in the public sector differs in many ways from management in the private sector. Gass (1994) describes some of the many ways in which public sector OR/MS differs from its practice in the private sector including the relative importance of political concerns, the difficulties associated with evaluating costs and benefits, and the "freedom of information" environment that makes public sector policy analysis and decision-making processes very open. He points out that, unlike their private sector counterparts that tend to focus on the "bottom line," public sector decision makers must address efficiency (the proper use of resources), effectiveness (the attaining of specified goals and objectives), and equity (the extent to which all citizens are treated alike).

II. THE RELATIONSHIP BETWEEN FIRE AND FOREST LAND MANAGEMENT OBJECTIVES

A fire management program is usually but one of many components of a broader forest land management program. Societies develop and implement land management plans to increase the likelihood that the human activities which take place on or near designated parcels of land will be compatible with the needs and desires of the people and institutions that have jurisdiction over that land. Forest land management plans might, for example, call for some areas to be treated as wilderness in which the impact of humans is minimized to protect natural ecosystem processes, others to be designated as recreational parks in which heavy human use is tempered to reduce its impact on natural processes, and yet others that are managed primarily for industrial fiber production. Land management plans should reflect the diverse needs and interests of the many groups of people who have interests in the forest. Consequently, even though forest land management is frequently a source of bitter conflict in modern industrial societies, we will assume that societies can and ultimately do develop land management plans that reflect the way they want their forest lands to be

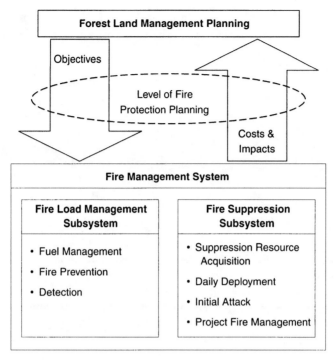

FIGURE 2   The relationship between fire and forest management planning, a refinement of the fire management systems framework presented in Martell *et al.* (1999).

managed. The fire manager's role is to develop and implement fire management plans that are compatible with and contribute to the achievement of the objectives of the higher level forest land management plans.

Figure 2 is a refinement of the fire management systems framework presented in Martell *et al.* (1999) and illustrates the hierarchical relationship between fire management and forest land management. Forest land management planners should begin by identifying fire management objectives such as average annual burned area targets and transmitting them to the fire organization. Fire managers should then determine if they can achieve those targets and develop strategies that will minimize the cost of doing so. They should then transmit their predictions concerning the cost of achieving the specified fire management targets and their assessment of the social, economic, and ecological impacts of the fire behavior that might ensue, back up to the forest land management planners. If the forest land management planners and their clients are not satisfied with those predicted costs and impacts, they should revise their fire management objectives and send the revised targets back down to the fire organization. Area burned is not simply the fortunate or unfortunate consequence of a chance process but rather the result of explicitly planned human intervention in complex natural processes. The iterative process of balancing forest land management fire needs with fire management costs and impacts to achieve a workable solution is what is known as *level of fire protection planning*, a topic we will address in Section V.

Just as forest management plans must vary across the landscape, so too must fire management objectives vary from time to time and place to place. Clough (1963) said that "management has been defined as the art of 'guiding the activities of a group of people toward the achievement of a common goal.'" One can think of forest fire management as *the art and science of guiding people in efforts to modify fire regimes that vary across spatial and temporal landscapes in accordance with forest land management objectives, in a cost-effective manner.* This might, for example, call for some fires to be allowed to burn reasonably freely in remote wilderness areas, the use of prescribed fire to fulfill ecological and other land management needs, the aggressive suppression of fire in areas where they pose significant threats to people and their property, and the attempts to limit fire losses in timber production areas. In Section IV, we will study the structure and functions of fire management organizations that are designed to achieve such objectives, but, before we do so, we will investigate how the social, economic, and ecological impacts of fire can be assessed.

III. ASSESSING FIRE IMPACTS

Our discussion of forest fire management has, to this point, been predicated on an assumption that forest fire managers and their land management clients can assess the impact of fire and fire management programs on people, property, forest resources, and ecosystems, and earlier chapters addressed the impact of fire on ecosystems. Assessing the social and economic impact of fire poses many complex, largely unresolved problems and is the subject of what forest fire management specialists refer to as *fire economics*. Gorte and Gorte (1979) and Baumgartner and Simard (1982) review much of the work that has been carried out in this area. In this section we describe how such impacts are currently assessed, and we identify some of the many significant gaps that remain in this area.

Consider a forest fire management agency that strives to minimize the net destructive impact of fire subject to constraints on the resources that society puts at its disposal and the ways in which they are to be used. Funds are allocated to fire management on the assumption that the subsequent benefits will exceed the value of the money spent. Fire management costs can be assessed and expressed in monetary terms by using standard accounting procedures. Salaries and equipment costs, for example, are easily assessed as is the cost of supporting fire management personnel working in the field. Standard accounting practices will produce reasonably accurate estimates of the annual cost of fixed assets such as aircraft, fire suppression equipment, buildings, communications, and information systems. The benefits of fire management activities include the reduced losses that result from limiting the number and size of destructive wildfires and the increased forest resource productivity and enhanced environment that result from the use of prescribed fire and successfully monitoring and modifying the suppression of beneficial wildfires.

Each year forest fire managers and the governments and other organizations that fund them must decide how much money to allocate to fire management. Many attempts have been made to develop formal procedures for determining optimal, or appropriate, levels of fire protection. Fire management specialists use what is referred to as the least cost plus damage or LCD fire economics model, first developed by Sparhawk (1925), to explore such issues. We will use the simplest variant of the basic LCD model to explain the basic principles of fire economics.

A. Least Cost Plus Damage Model of Fire Economics

Forest fire managers use the term presuppression to describe the activities that take place before fires actually occur; for example, the hiring of fire fighters and the commitment to pay their regular salaries for the entire fire season regardless of the number of fires that actually occur. The extra costs, for example, the overtime wages earned once a fire has been reported and is being fought, are referred to as suppression expenditures.

Let us assume that a forest fire management agency must decide how much money to allocate to presuppression efforts prior to the start of the fire season. Those funds could be used to maintain fire lookout towers and other infrastructure and fire suppression equipment such as portable power pumps, hose, and hand tools. They could also be used to pay the regular salaries of fire fighters and other staff; to charter detection aircraft, airtankers, and transport helicopters; and to lease or purchase trucks and other resources that can be used to combat fires once the season begins. The more fire fighters the agency hires, the less likely shortages will occur. That will make it easier to deploy initial attack crews at bases close to areas where fires are expected to occur, in advance, before the fires are reported. The reduced response times that would result should decrease the likelihood that fires will escape initial attack and the net result should be a reduction in fire damage and suppression costs. It is reasonable to assume that suppression costs will decrease as presuppression expenditures

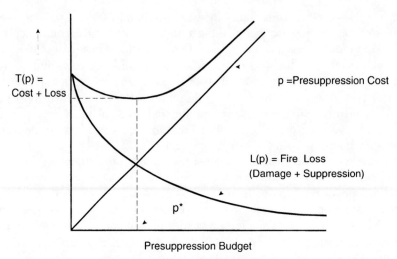

FIGURE 3   Sparhawk's (1925) least cost plus damage fire economics model.

increase but, like most human endeavors, at a decreasing rate due to decreasing marginal returns.

Figure 3 depicts the decision-making problem faced by fire managers. Let p denote the presuppression budget and $L(p)$ the fire loss which includes the suppression cost (e.g., airtanker flying costs and the overtime wages of fire fighters) and the fire damage. The total cost plus loss (T) is $p + L(p)$ which is the sum of the $L(p)$ curve and the presuppression cost p. The optimum presuppression budget (p^*) is the value of p that minimizes the total cost plus loss. Since fire is natural and sometimes beneficial, some writers allow $L(p)$ to be positive (a loss) or negative (a benefit) and refer to the LCD model as the least cost plus net value change model, but the result will be the same if benefits are treated as negative costs. The LCD principle, in its simplest form, is that a forest fire management system should be operated at such a level that the total cost of operating the system plus the fire loss (or more precisely, the net loss) is minimized (i.e., minimize cost plus loss).

In the simple case depicted in Figure 3, the total cost plus loss (T) is a convex function of the presuppression cost so one can easily identify its minimum by taking its derivative with respect to p, equating the result to zero, and solving for p^*, the optimum presuppression budget. Real fire management systems are far more complicated.

1. LCD Limitations

The LCD model illustrates the basic principles of fire economics but has many drawbacks that limit its practical use for fire management planning. It is difficult to specify a production function that relates the annual area burned to the presuppression budget. Analysis of the impact of fire management on area burned is confounded by the fact that the historical data may no longer be relevant due to changes in land use, the forest fuel complex, or even climate and fire suppression technology. Furthermore, the production function is not a univariate function of the presuppression budget as area burned depends not only on the amount of money spent on presuppression but also on how those funds are allocated to the many presuppression activities that can take place.

Consider, for example, a very simple fire management organization that has no prevention program and no airtankers. It has a fixed presuppression budget and must decide how much money to use to hire fire fighters and how much money to spend on helicopters and trucks to transport fire fighters to fires. If it chooses to charter a large number of helicopters and hire a small number of fire fighters, then fire fighters will be able to fly to fires as soon as they are reported. However, on days when many fires are reported, the supply of fire fighters will be quickly exhausted, and the fires that occur later in the day will burn freely for long periods of time while they wait for fire fighters to be released from the earlier fires or borrowed and flown in from other agencies at increased cost. If the agency hires many fire fighters and charters few helicopters, then the first few fires reported each day will be helitacked quickly, but the later fires will wait and burn freely for long periods of time as crews travel to them by truck or wait for helicopters to be released from earlier fires.

What is needed, of course, is a multidimensional fire management production function that relates the annual area burned to, for example, prevention efforts, the number and location of fire lookout towers, the number and type of detection aircraft and detection observers hired, the number and type of airtankers chartered, the number and type of transport helicopters chartered, and the number of fire fighters hired. It is not possible to develop such functions to be used in the LCD optimization framework at the present time, and it is doubtful that it will be possible to do so in the foreseeable future.

Another critical limitation of the basic LCD model is that its deterministic structure does not capture the stochastic elements that are so characteristic of forest fire management systems. Fire management success varies significantly from year to year, not only due to changes implemented by managers but more often due to variability in weather, fire ignition, and fire behavior processes.

The most significant problem that limits the practical significance of the LCD model is one that extends far beyond fire management and that is the difficulty of assessing fire impacts and expressing them in monetary terms so that the net impact of alternative fire management strategies can be assessed in monetary terms. Most forest fire management agencies have a hierarchy of objectives that guide their activities. For example, the Ontario Ministry of Natural Resources' fire management objectives are, in order of priority: (1) to enhance public safety, (2) to reduce the detrimental impact of fire on property and forest resources, and (3) to utilize the beneficial impacts of fire which may include the use of pre-scribed fire for silviculture purposes and modified suppression of fires in some remote areas to benefit from the impact of fire on natural ecosystems.

Historical data clearly indicate that fire management saves lives, but it is dif-ficult to assess how it contributes to public safety. Even if it were possible to do so, it would be neither possible nor socially acceptable to assign a monetary value to the lives saved. Property loss reductions are also difficult to assess as that would entail identifying physical structures such as houses and transpor-tation facilities that might have burned had fires not been suppressed, and us-ing their monetary value to assess the damage averted. Even if that could be done, the human cost incurred when, for example, homes are burned exceeds the assessed cost of the structures destroyed, and it would not be possible to as-sess such costs.

Growing recognition that fire is a natural component of many ecosystems poses even greater challenges. Previous chapters make it eminently clear that it is difficult to assess the impact of fire on ecosystems and that such problems will persist for the foreseeable future. Add to that the need to express such im-pacts in monetary terms, and it is clear that such problems will not be resolved for a very long time, if ever.

Given such limitations, the LCD model provides valuable insight into the ba-sic principles of fire economics but is of little or no practical significance. Fire managers and the political institutions that support them, therefore, rely on heuristic approaches to fire management budgeting. In the past, they typically implemented slight changes from year to year, particularly after "bad" fire sea-sons when people reacted to severe losses and funding levels were increased or when governments grew complacent after a few "good" years and attributed small fire losses to an "overabundance" of fire suppression resources (rather than some fortuitous combination of adequate resources, sound fire manage-ment, good weather, and good luck) and diverted funds from fire programs to other pressing needs. They relied on their intuition and allocated what they thought might be the best amount of money to spend on presuppression based on their experience and understanding. In Section V, we describe how level of fire protection planning is currently practiced, but first we will describe one fire impact that can be assessed in both physical and monetary terms—the impact of fire on industrial fiber production or timber supply.

B. ASSESSING THE IMPACT OF FIRE ON TIMBER SUPPLY

North American forests have been, and continue to be, used extensively for in-dustrial timber production, and most North American forest fire management agencies were developed to protect people and timber from fire damage. It is therefore not surprising that a great deal of effort has been devoted to assessing the impact of fire on timber supply and that is the impact that is perhaps best understood. Forests are now viewed as far more than industrial fiber sources and that will become even more evident in the future, but we do not know how to measure or assess the impact of fire on those other "nontimber" values. Since the methodologies developed to assess the impact of fire on timber supply might ultimately be extended to some nontimber values, this subsection is devoted to a detailed analysis of the impact of fire on timber supply.

We will use a simple hypothetical 100,000-ha jack pine forest that is used solely for industrial timber production to illustrate how one can assess the impact of fire on timber supply. Assume our hypothetical forest has the stand age class structure shown in columns 1 and 2 of Table 1 and in Figure 4.

We will assume that the merchantable volumes produced when stands are harvested vary with age as shown in column 3 of Table 1, which contains the gross merchantable volumes for Site Class II jack pine stands in the province of Ontario taken from Plonski (1974).

Suppose a fire burned 4000 ha of age class 7, 3000 ha of age class 8, 2000 ha of age class 9, and 1000 ha of age class 10. Using the gross merchantable volume

TABLE 1   Age Class Distribution and Yield of Merchantable Timber Volume for the Hypothetical Forest

Age class (years)	Area (ha)	Gross merchantable volume (m³/ha)
0–10	18,127	0
10–20	14,841	0
20–30	12,151	23
30–40	9,948	68
40–50	8,145	110
50–60	6,669	147
60–70	5,460	174
70–80	4,470	190
80–90	3,660	199
90–100	2,996	203
100–110	2,453	204
110–120	2,009	204
120–130	1,644	204
130–140	1,346	204
140–150	1,102	204
> 150	4,979	204
Total	100,000	

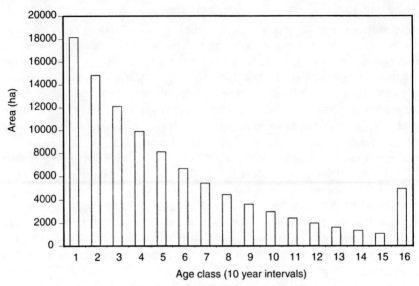

FIGURE 4   Age class distribution of the hypothetical forest.

figures in column 3 of Table 1, we find that the volume burned would be 1,867,000 m³. Our first task is to assess the impact of such a fire.

Suppose the company that harvests wood from the hypothetical forest sells it at a rate of $5.00/m³. A superficial *burn site* assessment of the fire could be obtained by multiplying the merchantable volume of the wood consumed by the fire by the selling price of wood and that would produce an estimated cost of $9,335,000 for the 10,000-ha fire. The true economic cost of the timber destroyed by the fire can differ significantly from this figure as we explain later.

The harvesting activities that take place in a forest that is managed for timber production will be governed by a timber harvest scheduling plan that stipulates when each stand is to be cut. The economic value of the timber is the present net worth of all the economic returns associated with the harvesting activities that are scheduled to occur over the planning horizon. Suppose future costs and revenues are discounted at an annual rate of 3%. The present value of a cubic meter of wood that is to be sold 5 years from now is not $5 but rather $5/(1.03)⁵ or $4.31. The economic value of the timber in a forest is therefore determined by the harvest schedule, but when fire burns stands that are scheduled to be harvested, that schedule will have to be revised. The cost of a fire is therefore the economic value of the timber as determined by the preburn harvest schedule minus the economic value of the timber determined by the revised postburn harvest schedule.

Van Wagner (1979) was one of the first to focus attention on forest-level fire and timber supply issues, and he used computer simulation models to do so. Forest managers have developed and used a myriad of techniques to develop timber harvest schedules, and linear programming optimization models are often used for such purposes. Johnson and Scheurman (1977) describe what they refer to as the standard Model I and Model II forms of the timber harvest scheduling model. Timber harvest scheduling linear programming (LP) models are deterministic models, but the need to account for the variability of fire leads to stochastic forest level problems that are complex, largely mathematically intractable problems. Fortunately, Reed and Errico (1986) developed a mean value formulation of the forest-level problem in which one assumes some constant average fire loss occurs each year. Although this approach ignores the variability that complicates forest management under uncertainty, their mean value formulation of the problem produces timber harvest scheduling solutions that are close to those solutions which more detailed stochastic models would produce for sufficiently large forest management units. The Reed and Errico (1986) model is similar in structure to Model II, but it is different enough to merit recognition as a new form, Model III. One of the most appealing aspects of Model III is that it is well suited for dealing with flammable forests.

The form of Model III we use here is identical to the model presented in Martell (1994) which is a variant of the basic Reed and Errico (1986) model. A 300-yr planning horizon is partitioned into 30 10-yr periods, and the forest stands are aggregated into 10-yr age classes. Let $x_{i,t}$ denote the area of the forest in age class i at the start of period t, and let $h_{i,t}$ denote the area of age class i harvested during period t. Assume that some constant fraction of the forest (0.2%) burns each year, that harvesting takes place before burning in each period, and that the burned area is uniformly distributed over the forest. Stands are assumed to regenerate naturally to jack pine after harvesting or burning. The harvest flow is constrained to be constant and a terminal constraint that the merchantable volume growing in the forest at the end of the 300-yr planning horizon average $40.2 \ m^3 \ ha^{-1}$ is imposed to ensure the entire forest is not clear cut at the end of the planning horizon.

The Model III LP model can be used to assess the impact of a fire (and fire management) on our hypothetical forest. The economic impact of a fire on the timber value of a forest is the timber value of the forest given the prefire timber harvest schedule less the postfire value under a revised plan. Neither plan has to be optimal, but it makes little sense to evaluate fire impacts using suboptimal timber harvest schedules. The optimal preburn Model III timber harvest schedule for our 100,000-ha hypothetical forest produces an annual allowable cut of $2.41 \ m^3 \ ha^{-1} \ yr^{-1}$ and a present net worth of \$407/ha. The optimal postburn timber harvest schedule produced by running Model III with a revised age class structure (all the burned area is moved into age class 1) produced an annual allowable cut of 2.37 m^3 ha^{-1} yr^{-1} and a present net worth of \$399/ha, a loss of 2%. Thus a fire that burns 10% of the area and 21% of the volume reduces the value of the forest by only 2%. It is important to note, however, that this numerical result is specific to this particular hypothetical forest and the relative sizes of the burn fraction and that loss cannot be generalized to other fires and forests with different age class structures, growth and yield functions, and harvest flow constraints.

Timber harvest scheduling models can also be used to help evaluate and set initial and extended attack priorities. Suppose that two fires were burning out of control and threatening to grow even larger in two distinctly different parts of a forest. One could assess the pre- and postburn values of the forest for both fires and determine which of the two fires would have a greater impact on the economic value of the timber in the forest. Such information could be used to inform decisions concerning the allocation of scarce suppression resources to escaped fires.

Timber supply models can also be used to evaluate fire management programs. Figure 5 is a graph that relates the annual allowable cut to the average annual burn fraction in our hypothetical forest. Suppose that the average annual burn rate of the natural fire regime in our hypothetical forest was 2% and that fire management reduced that to 0.2%/yr. That would increase the an-

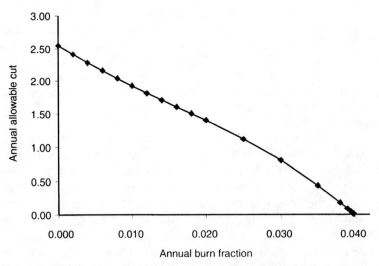

FIGURE 5   The impact of the annual burn fraction on the annual allowable cut in the hypothetical forest.

nual allowable cut from 1.40 m^3 ha^{-1} yr^{-1} (a present net worth of $236/ha) to 2.41 m^3 ha^{-1} yr^{-1} (with a present net worth of $407), an increase of 72% in the timber production value of the forest. Since that present net saving of $171/ha would be realized over a 300-yr planning horizon, it can be transformed (using a 3% discount rate) into an equivalent annual saving of $5 ha^{-1} yr^{-1}.

Timber is but one of many forest-based values that benefit people, but some of the lessons we have learned about evaluating the impact of fire on timber supply may be applicable to other values. Boxall *et al.* (1996), for example, have studied the impact of fire on the attractiveness of recreational canoe routes. When a fire burns through an area through which a canoe route passes, that fire will alter the vegetation of some portion of the larger network of canoe routes in ways that are not evident by looking at the burn site itself. To assess the impact of such a fire, one would have to evaluate the canoe route network before and after the fire. Clearly, scale is important when assessing fire impacts and stand- or burn-site specific assessments are not appropriate for assessing the impact of fire on timber supply. The management unit is an appropriate scale at which to assess the impact of fire, but it is important to recognize that it is based on an assumption that management units are independent entities. That may well be the case when they are owned or managed by competing enterprises, but companies or governments can pool management units for risk management purposes, and that possibility means that there is nothing inherently sacrosanct about management boundaries.

If one wishes to assess the ecological impact of fire, it is clear that one must look beyond the burn itself, but it is not obvious how far one must look. One can, as is the case with timber production and management units, identify the habitat utilized by some species of interest and attempt to identify a scale which captures the true impact of fire on those populations. Clearly, the appropriate scale will vary by species and the communities of interest.

It is also important to note that, when the impact of a specific fire is assessed, assessment should be carried out within a context of the underlying on-going fire regime. Using a Model III LP model pays explicit recognition to the fact that, were you to save some forest stand from destruction due to fire today, your "saving" will be measured with respect to the fact that the stand might well burn in the near future despite your most recent intervention. The fire regime embedded in Model III accounts for that factor. The larger the average annual burn fraction, the smaller the economic impact of a specific fire, since increased annual burn fractions decrease the values of both the pre- and postburn forests. The ecological impact of fire must also be assessed in such contexts.

Finally, it is important to note that fire impacts need not be expressed in monetary terms. The impact of fire on timber supply can be evaluated in terms of its impact on projected harvest flow or dollars, and both are legitimate measures that may be of interest to fire management clients. Since it would be difficult, if not impossible, to assess the ecological impact of fire or its impact on environmental services in monetary terms, fire managers and their clients need to identify what measure or measures to use and then develop multi-attribute evaluation procedures for evaluating fire management programs, a challenge that lies well beyond the scope of this chapter.

IV. FOREST FIRE MANAGEMENT ORGANIZATIONS

Forest fire management organizations are emergency response systems but, unlike urban fire, police, and ambulance services, they deal with natural processes that can have both beneficial and detrimental impacts on people, property, forest resources, and ecosystems. Since it is virtually impossible to exclude fire from most forests and attempts to do so can be very costly, fire management agencies use the resources allocated to them to minimize what they and their land management clients collectively judge to be the net detrimental impact of fire.

A fire regime will produce a spatial and temporal distribution of burn patches embedded in a larger forest landscape composed of many patches that have been shaped and influenced by both human and natural processes. Consider the simplest case in which we focus on one very simple result of a fire regime, the average annual fraction of the fire management area or fire region burned each year. That burn fraction should be compatible with the forest management objectives, and the role of the fire manager is to develop and implement a fire management plan that will achieve that average annual burn fraction at a reasonable cost. Fire managers often partition the area under their jurisdiction (the *fire region*) into zones or compartments that are reasonably homogeneous with respect to ecosystem processes, land use activities, and values at risk and then specify their fire management objectives in terms of fire regimes or the level of fire protection they will attempt to achieve in each zone.

The forest fire management planning process can then be viewed as comprising four basic steps:

1. Partition the fire region into zones or compartments that are reasonably homogeneous with respect to forest ecosystems, land use patterns, and values at risk;
2. Assess the potential beneficial and detrimental impacts of fire in each zone;
3. Select an appropriate level of protection or fire regime for each zone and develop a plan to minimize the cost of achieving that objective;
4. Implement, monitor, and revise the fire management plan over time.

Forest fire management objectives vary from agency to agency, a reflection of the variability of both societies and the ways in which they interact with the ecosystems that envelop them. The Aviation, Flood, and Fire Management Branch of the Ontario Ministry of Natural Resources, the agency responsible for forest fire management in the province of Ontario in Canada, is representative of many large North American forest fire management agencies. The OMNR is responsible for fire management on more than 85 million hectares and spends roughly $85 million a year on its activities. A permanent staff of 220 is augmented by 640 fire fighters that are hired each fire season. The agency also operates an aircraft fleet that includes 9 large CL-415 airtankers and 5 smaller Twin Otters that can serve as airtankers. Approximately 14 helicopters, 15 detection aircraft, and 7 bird-dog aircraft (that carry air attack officers that direct airtanker operations over the fire) are hired each season, but those figures vary from year to year. The OMNR participates in national and international mutual aid agreements that make it possible for it to share its fire suppression resources with other North American fire management agencies and to borrow resources from other agencies when the need arises.

During the years 1976–1999, the number of forest fires in Ontario ranged from 735 to 3970 per year with an average of 1713 fires per year and a median of 1588 fires per year. The area burned each year during that period ranged from 9444 to 611,939 ha with an average of 241,734 ha/yr and a median of 176,462 ha/yr. The year-to-year variation in the number of fires and area burned are illustrated in Figures 6 and 7.

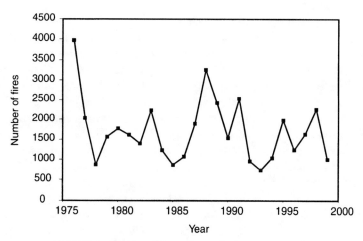

FIGURE 6   Number of forest fires in Ontario, 1976–1999.

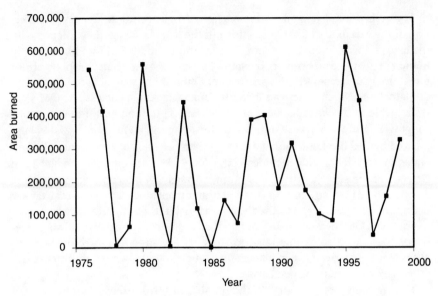

FIGURE 7   Area burned by forest fires in Ontario, 1976–1999.

A. A SIMPLE FIRE SUPPRESSION MODEL

Societies allocate funds to fire management on the assumption that the allocation will reduce the damage that results from wildfire and enable them to reap
some of fire's benefits. In this section, we present a simple conceptual model of
the fire suppression process which will illustrate how such activities can influence final fire size that, in turn, influences the economic impact of fires.

Forest fires are ignited by people or some natural agent such as lightning, and
they smolder or burn with an active open flame and emit smoke until they are
either extinguished due to a lack of combustible fuel or detected and reported
to the duty officer or initial attack dispatcher who decides what suppression resources will be dispatched to the fire. The initial attack crew travels to the fire
by truck or by air, sets up its suppression equipment at or near the fire, and begins suppression action as soon as possible. The ground crew may or may not
be assisted by airtankers that drop water or fire retardant on, or just ahead of,
the active fire front.

The effectiveness of the initial attack force will depend upon its composition, how soon it arrives at the fire, the fuel and terrain in which the fire is burning, and the fire's behavior. Most North American forest fires are contained when
they are relatively small, but some escape initial attack and can grow to very
large sizes. Such fires are usually classed as escaped fires and, when they become
large, they can seldom be controlled until favorable weather conditions arrive

and persist long enough for the fire fighters to establish control lines that are strong enough to contain the fire if and when weather conditions subsequently deteriorate.

Forest fire management agencies use fire control status classification systems to identify a fire's control status and important points and intervals of time in its life cycle. We combined the fire control status classification system used by the Ontario Ministry of Natural Resources with the formal mathematical model and graphical representation of the life cycle of a fire developed by Parks (1964) and modified the combined system slightly to produce the formal description of the life cycle of a wildland fire presented in Table 2 and Figure 8.

The sooner a fire is detected and reported, the larger the initial attack force; and the quicker it arrives at the fire, the more likely the fire will be contained at a small size. Some fires, however, are so large and/or intense that they cannot be contained regardless of the size of the initial attack force, and they can burn freely until a significant change in fuel, weather, topography, or some combination of these factors occurs and the fire intensity decreases to levels where fire fighters can work safely to contain their spread.

TABLE 2   Fire Control Status Classification System

Event	Description	Time	Time interval
Ignition	Fire is ignited.	T_{IGN}	—
Detection	Fire is detected by the public or the fire management agency.	T_{DET}	Detection interval $= T_{DET} - T_{IGN}$
Report	Fire is reported to the duty officer or the initial attack dispatcher.	T_{REP}	Report interval $= T_{REP} - T_{DET}$
Dispatch	Initial attack resources dispatched to the fire.	T_{DISP}	Dispatch interval $= T_{DISP} - T_{REP}$
Arrive	Initial attack forces arrive at or near the fire perimeter.	T_{ARR}	Travel time $= T_{ARR} - T_{DISP}$
Attack	Initial attack suppression action begins.	T_{ATT}	Response time $= T_{ATT} - T_{REP}$
Being Held	Fire classed as being held (BHE). The fire is no longer spreading but has the potential to resume spreading in the near future.	T_{BHE}	Time to containment $= T_{BHE} - T_{ATT}$
Under Control	Fire classed as under control (UCO). The fire is no longer spreading and is not expected to resume spreading in the future.	T_{UCO}	Control interval $= T_{UCO} - T_{ATT}$
Out	Fire is declared out (OUT).	T_{OUT}	Mop up $= T_{OUT} - T_{UCO}$

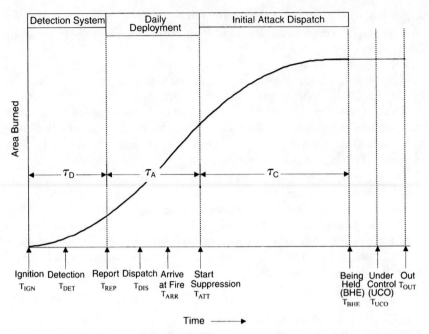

FIGURE 8   The life cycle of a forest fire.

Parks (1964) developed a simple single fire suppression model that can be used to illustrate how the components of a fire management system interact to determine the final area burned. He assumed that fires were reported as soon as they were detected, and he defined the detection interval $\tau_D = T_{DET} - T_{IGN}$ as the time between ignition and detection. He assumed that the initial attack force begins suppression action as soon as it arrives at the fire, and he defined the attack interval, $\tau_A = T_{ATT} - T_{REP}$, as the elapsed time from detection (report) until fire suppression action begins. He defined the control interval, $\tau_C = T_{BHE} - T_{ATT}$, as the time to contain the spread of the fire.

Parks began by ignoring fire geometry and developed a very simple aspatial fire spread model. He assumed that, in the absence of suppression activity, the growth of a free-burning fire (area/unit time) at the time it is detected (T_{DET}) can be represented by the following differential equation:

$$\frac{dy}{dt} = G_D + Ht$$

$$y(0) = y_D$$

(1)

where y is the area burned, dy/dt is the rate of fire growth, t is the elapsed time measured from the time the fire is detected, $y(0)$ is the size of the fire when it is detected, G_D is the fire's growth rate when it is detected, and H is the acceleration component of fire growth.

Let y_A denote the size of the fire at time T_{ATT} when suppression action begins. Then

$$y_A = y(\tau_A) = y_D + G_D\tau_A + \tfrac{1}{2}H\tau_A^2 \tag{2}$$

and at that time the fire will continue to grow at a rate

$$G_A = y'(\tau_A) = G_D + H\tau_A \tag{3}$$

Consider, for example, the simple case in which a fire burns freely on a flat area in the absence of wind in the shape of a circle with a radius that increases at a constant rate which depends on the current condition of the forest fuel complex and the weather. Let $R = r\tau_D + rt$ denote the radius of the fire at time t where r is its constant radial rate of spread and t denotes the elapsed time from the time the fire was detected (and reported). If $A(t)$ denotes the area of the fire at time t, then

$$A(t) = \pi r^2\tau_D^2 + 2\pi r^2\tau_D t + \pi r^2 t^2 \tag{4}$$

and

$$\frac{dA(t)}{dt} = 2\pi r^2\tau_D + 2\pi r^2 t \tag{5}$$

so

$$y_D = \pi r^2\tau_D^2 \tag{6}$$

$$G_D = 2\pi r^2\tau_D \tag{7}$$

and

$$H = 2\pi r^2 \tag{8}$$

Parks (1964) modeled the growth of a fire that is being fought as follows. Let x denote the number of units of fire suppression resource allocated to the fire, and assume that those resources decelerate the growth of the fire. Let E denote the efficiency factor for the fire suppression resources. Parks assumed that the effect of the suppression force is to decelerate the rate of growth of the fire at some rate which is a linear function of the size of the force and the time it works. The growth of the fire at time t after T_{ATT}, the time suppression begins, is, therefore,

$$\frac{dy}{dt} = G_D + Ht - E(t - \tau_A)x \qquad t \geq \tau_A$$

$$= G_A - (Ex - H)(t - \tau_A) \tag{9}$$

Let $\tau_C = T_{BHE} - T_{ATT}$ denote the elapsed time interval from the start of initial attack until the fire is declared as being held or contained (BHE). It can be shown that the area burned from the time suppression action begins until the fire ceases to spread is

$$y_C - y_A = \frac{G_A^2}{2(Ex - H)} \tag{10}$$

The growth of the fire is composed of two parabolic portions, one when the fire is burning freely and accelerating until suppression action begins, and a second after suppression action begins when it begins to decelerate. Note that the number of suppression units allocated to the fire must be greater than H/E to bring the fire under control. Parks' model is illustrated in Figure 8. In his case, there is no delay between detection and report so $T_{DET} = T_{REP}$ and no delay between when the crew arrives at the fire and starts suppression so $T_{ARR} = T_{ATT}$.

Others have studied the fire suppression process and developed mathematical models to predict the final size and shape of a fire depending upon its behavior, the productivity of the suppression force, and their interaction. Fried and Fried (1996) reviewed some of the many models that have been developed and developed a containment model that overcomes many of the limitations inherent in earlier models. They showed how their model can be applied and solved numerically to predict the final size and shape of an elliptical fire that is subjected to direct attack by ground crews.

One of the most appealing features of the Parks (1964) model is that the final size of the fire can be expressed as an explicit function of fire behavior and fire suppression productivity parameters. It indicates that the more resources devoted to detection, the faster the fire is detected, the smaller τ_D is, and the smaller the final size of the fire is. The more money allocated to initial attack transport (the use of helicopters rather than trucks, for example), the smaller τ_A is and the smaller the final fire size is, assuming that there are always fire fighters available to be dispatched whenever a fire is reported (an issue we address later in this chapter). The more suppression resources dispatched to the fire, the shorter τ_C is and the smaller the final fire size is. The more resources devoted to mop up, the less likely the fire will escape once it has been declared BHE. Such predictions are compatible with what fire managers believe about real fire suppression processes.

The next step is to explore the economics of fire suppression, and Parks (1964) did just that. He defined the following components of the cost of a fire:

C_F = the fixed cost of maintaining the fire suppression organization ($ per fire)

C_S = cost of dispatching a unit of fire suppression resource to a fire (e.g., the cost of transporting the resources to the fire) ($ per unit of suppression resource)

C_T = cost incurred by the need to increase the fire organization's readiness while the fire is burning out of control ($ per unit time)

C_B = cost of fire damage incurred per unit area burned ($ per unit area)

C_X = suppression cost incurred while the fire is burning out of control ($ per unit resource per unit time)

Let $x_0 = H/E$ denote the minimum suppression force required to contain the fire, x be the size of the suppression force, and $z = x - x_0$ denote the number of suppression units above x_0 sent to the fire.

The fire will be contained and declared BHE or reach its final size when it stops growing or $dy/dt = 0$. Given Eq. (9), it can be shown that the fire stops growing when $\tau_C = t - \tau_A$ is given by the following expression:

$$\tau_C = \frac{G_A}{(Ex - H)} \tag{11}$$

The total cost of the fire is, therefore,

$$C = C_F + C_S x + C_T \tau_C + C_B y_C + C_X \tau_C x \tag{12}$$

or

$$C = C_F + C_S x + C_X \frac{G_A}{(Ex - H)} x + C_B \left[y_A + \frac{G_A^2}{2(Ex - H)} \right]$$

$$+ C_T \left[\frac{G_A}{(Ex - H)} \right] \tag{13}$$

If we let

$$C_R = C_B + \left[2\frac{C_T}{G_A} \right]$$

then

$$C(z) = C_0 + C_S z + \frac{(C_X G_A H/E^2) + (C_R G_A^2/2E)}{z} \tag{14}$$

where

$$C_0 = C_F + \frac{C_S H}{E} + \frac{C_X G_A}{E} + C_B y_A$$

Parks (1964) showed that the optimal number of suppression resources to allocate to a fire is given by the following expression:

$$z^* = \sqrt{\frac{C_R G_A^2 E + 2C_X G_A H}{2E^2 C_S}} \tag{15}$$

1. A Numerical Example

Consider a fire for which $G_D = 2.0$ ha hr^{-1}, $H = 1.5$ ha hr^{-2}, $E = 0.5$ ha hr^{-2} per suppression unit, $y_D = 0.25$ ha, and $\tau_A = 0.57$ hr. Suppose that $C_F = 0$, $C_S = 100$, $C_X = 3.0$, $C_B = 500$, and $C_T = 0$. Using the expression for the optimum suppression effort given above, $x_0 = 3.0$, $z^* = 7.0$, and the optimum suppression effort for this fire is 10 units. Figure 9 illustrates how the total cost plus loss for this fire varies as the fire suppression effort varies.

Parks (1964) applied his model to some specific fire shapes (e.g., direct and indirect attack on rectangular fires that burn freely on one side like a fire burning up a canyon). He also extended his model to deal with the optimization of two simultaneous fires under certainty, and he solved a simple stochastic fire suppression problem which he modeled as a Markov decision process.

Fire managers must weigh the costs and potential benefits of deciding how many resources to allocate to each activity and each fire, and, when they do so, they must consider other fires that are burning or might be reported in the near future. Given the number of factors that must be considered and the uncertainty involved, this clearly is a daunting task. Although the Parks (1964) single and multiple fire suppression process models provide valuable insight into the behavior of fire suppression systems, they are not robust enough to serve the prac-

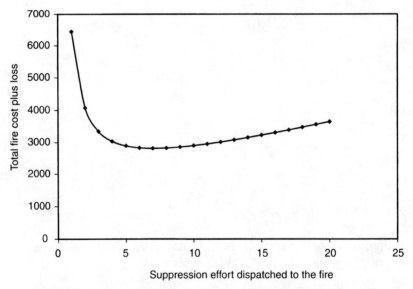

FIGURE 9 Numerical example of Parks' (1964) single fire suppression model.

tical needs of fire managers. We will now describe the basic components of a modern forest fire management system and study how fire managers deal with such problems. We begin with the fire load management subsystem depicted in Figure 2.

B. FIRE LOAD MANAGEMENT

The term *fire load* refers to the magnitude of the suppression task associated with fires that occur in a designated area during some specified time interval, typically a day. The more fires that occur and the larger and more intense they are when detected, the greater the fire load is. The objective of the fire load management subsystem is to reduce the fire load to be managed. This can be accomplished by preventing fires from occurring, modifying the forest fuel complex to temper the behavior of fires that do occur, and detecting and reporting wildfires soon after they are ignited so that they are small when the initial attack fire suppression force arrives.

1. Fuel Management

The objective of the fuel management subsystem is to modify forest vegetation or fuel complexes to reduce the likelihood that fires will occur and to stop or slow the spread of fires and thereby reduce the social, economic, and biological impacts of fires that do occur. It includes the construction and maintenance of fuel breaks, extensive understory thinning or fuel modification, and the use of prescribed fire to modify fuels to reduce their flammability, for silvicultural purposes, or to enhance wildlife habitat. Fire is a natural component of many forest ecosystems, and agencies that are responsible for fire management in some parks and wilderness areas may use prescribed fire in an attempt to sustain natural ecosystem processes. Fuel management influences the *type* and *intensity* aspects of a fire regime directly, but it also indirectly influences its *fire size* or *extent*. Martell (1982) described some of the decision support systems that have been developed to enhance fuel management decision making, but our understanding of fuel management is ultimately based on the extent to which we understand how fire spreads through heterogeneous fuel complexes that include patches and strips that have been modified to influence fire behavior.

2. Fire Prevention

Since the objective of the fire prevention program is to reduce the number of people-caused fires that occur in designated areas, it can reduce the *fire frequency* characteristics of a fire regime. Fire prevention specialists must decide how to encourage people to refrain from starting fires intentionally or accidentally. They use media advertising campaigns to transmit prevention messages to the public, invoke site-specific land use restrictions such as forest closures that prohibit travel in designated areas, or ban the use of campfires by recreationists to reduce the likelihood that accidental fires will occur. They carry out fire investigations to determine how specific fires were started and sometimes invoke law enforcement measures to sanction people who start fires. Martell (1982) described some of the prevention studies that have been carried out in the past. However, human behavior is influenced by media advertising that reaches them throughout their lives, beginning with fire prevention programs directed at young school children as well as localized special advertising and land use restrictions invoked during hazardous fire weather conditions. Therefore, it is very difficult to assess the effectiveness of prevention measures. As a consequence, fire prevention processes are not well understood, and there are few decision support systems to enhance the cost effectiveness of prevention programs.

3. Predicting Daily Fire Occurrence

Prevention planners need to know when and where people-caused fires might occur so that they can focus their efforts where they will have their greatest impact. Detection planners need to know when and where both people and lightning-caused fires are likely to occur so that they can direct their detection patrols to fly near those fires soon after they ignite. Later in this chapter, we will illustrate the importance of daily fire occurrence predictions to fire managers who must deploy their initial attack resources close to fires before they are reported to minimize initial attack response times. In this section, we will discuss daily fire occurrence prediction, one of the key components of a fire management information system. We will deal primarily with people-caused fires to explore the basic principles of daily fire occurrence prediction, but we will also summarize very briefly some of the lightning-caused fire occurrence and fire arrival prediction systems that have been developed.

a. Predicting Uncertain Events

The daily occurrence of forest fires can be characterized as a random or chance process. Just as we are uncertain whether heads or tails will result when a coin is tossed, we are almost always uncertain how many fires will occur each day. On some days fire occurrence predictions can be made with greater certainty than on others; for example, that no fires will occur on a rainy day with no lightning. Nevertheless, there is almost always some degree of uncertainty involved.

b. Forest Fire Ignitions and Arrivals

It is important to distinguish between the occurrence and arrival of forest fires. The occurrence time is the time a fire is ignited. Another important time is the time the fire is first reported to the fire management agency or it "arrives" and demands the organization's attention. Given the hold-over behavior of some fires that may smolder undetected for several days before they are detected, particularly those caused by lightning, it is important to distinguish between predictions concerning fire occurrences and fire arrivals. Not all fires that occur will eventually arrive since some of them will be extinguished naturally by weather or a lack of fuel.

c. Probabilistic Models of People-Caused Fire Occurrence

The binomial probability distribution can be used to model the daily occurrence of people-caused fires in a specific area. The binomial distribution is applicable to processes with repeated independent trials (e.g., Blake, 1979) such as people subjecting forest fuels to firebrands. Suppose that there are n people in a forested area during a particular day and that each of those people subjects the forest fuel complex to a single firebrand. Let each firebrand represent an independent trial which results in success (a fire occurs) or failure (a fire does not occur). Let θ denote the probability that a trial results in a fire, and assume that it is the same for all the people in the forest. If the probability of success is constant and the trials are independent (people do not influence each other), the probability distribution of the number of fires that will occur is binomial with parameters n and θ:

$$b(x; n, \theta) = P\{x \text{ successes in } n \text{ trials}\}$$
$$= \frac{n!}{x!(n - x)!}\theta^x(1 - \theta)^{n-x} \qquad x = 0, 1, 2, \ldots n \tag{16}$$

In most jurisdictions, it would be very difficult to determine how many people are present in a designated area each day. It is, therefore, difficult to use historical data to estimate θ and to determine n to predict fire occurrence during any particular day. Fortunately, it can be shown (e.g., Blake, 1979), that the Poisson probability distribution is the limiting form of the binomial distribution as θ becomes very small, n becomes very large, and $n\theta$ remains constant. That is, lim as $\theta \to 0$, $n \to \infty$, and $n\theta$ remains constant,

$$b(x; n, \theta) \cong \frac{\lambda^x e^{-\lambda}}{x!} \tag{17}$$

where $\lambda = n\theta$ is the expected number of fires per day.

Since there are usually many people in the forest and the probability that any one of them will start a fire is very small, it is reasonable to use the Poisson distribution to model daily people-caused forest fire occurrence. Cunningham and Martell (1973) studied daily people-caused forest fire occurrence in the Sioux Lookout area of northwestern Ontario and found that it was reasonable to use the Poisson distribution to model fire occurrence there.

The Poisson distribution is a single parameter distribution and

$$P(x) = \frac{\lambda^x e^{-\lambda}}{x!} \qquad \text{for} \quad x = 0, 1, 2 \ldots \tag{18}$$

where $P(x)$ = probability that x fires will occur and λ = expected number of fires per day. If one accepts the validity of the Poisson model, the task is to develop operational procedures for estimating λ for each fire management compartment each day.

d. A Simple People-Caused Forest Fire Occurrence Model

Daily people-caused forest fire occurrence is influenced by many factors including the number of people present in the forest, their behavior which is influenced by the land use activities in which they are engaged and the prevention measures they have encountered, and the condition of the forest fuel complex.

The major factors which influence the occurrence of people-caused forest fires are the ease of ignition of the fine fuel and the potential number of sources of ignition to which that fuel is exposed. The Fine Fuel Moisture Code (FFMC) is one of the component indices of the Canadian Forest Fire Weather Index (Van Wagner, 1987). Since the FFMC was designed as a measure of the moisture content of the fine fuels, one would expect it to have an important effect upon people-caused fire occurrence. Previous research results (e.g., Cunningham and Martell, 1973) indicate that this is the case.

One of the simplest procedures for predicting daily people-caused fire occurrence is to use daily records of the observed FFMC and the corresponding number of fires that occurred to estimate λ, the expected number of fires per day. This can be accomplished by compiling historical fire occurrence and fire weather data into the format shown in the following numerical example.

e. A Numerical Fire Occurrence Prediction Example

Suppose that one analyzed a set of historical data and obtained the results shown in Table 3. If the current day's forecast FFMC is 83, then using the for-

TABLE 3 Relationship between Average Daily People-Caused Fire Occurrence and the FFMC for a Hypothetical Area

Category	FFMC	Days	Fires	Average number of fires per day
1	0–74	200	2	0.01
2	75–79	150	9	0.06
3	80–84	120	18	0.15
4	85–89	100	24	0.24
5	90–100	90	32	0.36

mula for the Poisson distribution with an average of 0.15 fires per day, one would obtain the probabilistic prediction shown in Table 4.

Some forest fire management agencies use more sophisticated daily fire occurrence predictions such as the logistic regression analysis procedure developed by Martell et al. (1987) which can be used to predict daily people-caused fire occurrence by subseason and the procedure developed by Martell et al. (1989) which models seasonality explicitly. Poulin-Costello (1993) used Poisson regression techniques to relate the expected number of fires per day to fire danger rating indices. The development of lightning stroke counters and lightning strike location systems has resulted in a substantial improvement in the quantity and quality of information available for predicting lightning-caused forest fire occurrence, and analogous lightning-caused fire occurrence prediction models have been developed. For example, Kourtz and Todd (1992) developed a daily lightning fire occurrence prediction model that uses fuel moisture and lightning stroke data to predict fire ignitions. The holdover smoldering process is modeled to predict how many "detectable" fires are burning undetected in an area each day.

TABLE 4 Probabilistic People-Caused Fire Occurrence Prediction for a Category 3 FFMC Day with an Expected Value of 0.15 Fires per Day

Number of fires	Probability
0	.8607
1	.1291
2	.0097
3 or more	.0005

4. Fire Detection

The objectives of the forest fire detection system are to find and report fires while they are small and to provide the initial attack dispatcher with information that will enable him or her to prioritize fires as they are reported and to tailor the initial attack response to the potential impact of each fire. The sooner the fire is reported and the better the information concerning its precise location, accessibility, its current size, its spread rate, its intensity, the fuel complexes ahead of the fire, and the values at risk, the more likely that threatening fires will be contained while they are still small.

The fire detection system's primary effect is on the *size* or *extent* component of the fire regime. However, since fire detection measures increase the likelihood that a fire will be contained soon after it is ignited and while it remains small, they reduce the likelihood that fire will continue to burn during later periods when more severe fire weather prevails or through more hazardous fuel complexes located elsewhere on the landscape. Fire detection activities, therefore, indirectly decrease the area burned by intense fire and thereby influence the *intensity* aspects of the fire regime as well.

During the years 1976–1999 the detection size of forest fires in Ontario ranged from less than 0.1 ha to 20,000 ha, but 0.1 ha or less is the smallest detection size that is assigned to a fire on an official fire report form in Ontario. The average detection size was 6.4 ha, but the distribution was highly skewed with a median of 0.1 ha and a mode of 0.1 ha. Of the fires, 95% were less than or equal to 3.4 ha when first detected. The skewness is largely a result of a few large fires in the extensive protection zone in the far north.

Forest fire management agencies use fixed lookout towers and patrol aircraft to detect fires, but they also depend upon the public to find and report fires in and near populated areas. Each agency's strategy is driven by the relative cost effectiveness of the different modes of detection and the values at risk. Towers and patrol aircraft comprise what is commonly referred to as the *organized* detection system, and the public constitutes the *unorganized* detection system. Those terms reflect the belief that, although fire managers can influence the behavior and effectiveness of the public, they have much more direct control over the performance of the organized detection system.

Satellites equipped with infrared scanners and other remote sensing devices can detect forest fires, but the resolution of currently publicly available satellite technology is such that satellites cannot find fires until they grow to sizes that are considered too large for effective initial attack. Satellites are, therefore, used primarily to monitor large, on-going fires in some remote low-priority areas where they can provide fire size information with some delay at less cost than conventional fixed-wing aircraft.

a. The Unorganized Detection System

In most jurisdictions, the public views fire as a potentially destructive force and assumes it has a civic duty to report fires that are burning out of control in forested areas. Given such attitudes, fire management agencies need not devote scarce resources to searching for fires that will be detected and reported by the public while they are still small. It makes little sense, for example, to build a fire lookout tower at the edge of a forested community or to fly a detection patrol aircraft along a highway as the public will find and report most of the fires that occur in such areas before they are detected by trained detection observers, but at little or no cost. Fire management agencies are therefore free to let the public detect fires in and near populated areas and to devote their efforts to more remote areas where fires might burn undetected for long periods of time.

Fire managers can and should devote some resources to the unorganized detection system and enhance its cost-effectiveness by influencing public attitudes and enhancing the communications infrastructure to facilitate the reporting of fires, thereby reducing the time interval between detection and formal reporting. Fire management agencies, therefore, publicly stress the importance of reporting fires and use both print and broadcast media to publicize how to report forest fires. The unorganized detection system has benefitted from the increasing availability of cellular telephone technology in many rural areas in recent years. Since the cost-effectiveness of the unorganized detection system has not been well studied, fire managers must rely upon their intuition and experience when they decide how to influence the behavior of the people and thus the performance of the unorganized detection system in their area.

b. The Organized Detection System

Most forest fire management agencies use fire towers or lookouts located atop high hills or mountains, patrol aircraft, or mixed systems with both towers and patrol aircraft, for fire detection purposes. Towers are fixed and expensive to construct and operate, but they provide constant surveillance of an area when the observer is in the tower. Tower systems, therefore, provide excellent but costly coverage.

Aerial detection systems are very flexible because patrol routes can vary from day to day or even hour to hour as fire weather conditions vary, and a patrol aircraft can be diverted from its planned route to enable the onboard observer to check suspicious smokes reported by the public. Although aircraft are expensive to charter or own and operate, most North American forest fire management agencies that rely on aircraft are satisfied with aerial coverage levels that cost less than the former towers would cost had they remained in service. However, detection aircraft provide intermittent surveillance, and once they pass over a specific area, any fires they miss or any new fires that occur before they pass near that area again can burn freely undetected and have the potential to remain undetected so long that they cannot be controlled by the initial attack force. While this problem can, to a large extent, be addressed by using many aircraft to fly almost continuous patrols over areas where potentially destructive fires are likely to occur, detection patrols never really provide the continuous coverage possible with extensive tower networks.

Kourtz and O'Regan (1968) studied the economics of fire detection systems and concluded that tower systems are best for high-value areas under intensive forest management and that aircraft are best for less-valuable extensive forest management areas with relatively small detection budgets.

Strategic detection planners must decide how many lookout towers are required and where they will be constructed. Computer-based spatial analysis techniques (see Mees, 1976, for example) can be used to assess the "seen area" surrounding each potential tower site and the number and type of fires that have occurred within the seen areas in the past.

Detection Patrol Route Planning   Many forest fire management agencies use patrol aircraft to search for fires, and each day detection planners must decide the times at which airborne patrols will leave specified airports and the routes they will follow. Since the time and location of fires is uncertain, this constitutes a very difficult stochastic combinatorial planning problem.

Forest fire managers use the term *visibility* to describe the ability of an observer to see a *smoke,* the smoke plume that rises above a small fire burning under a forest canopy. In Ontario's boreal forest region, the visibility is usually assumed to be roughly 24 km either side of a patrol aircraft in good weather. Visibility varies from day to day and depends upon many factors including vegetation, atmospheric haze conditions, the location of the sun with respect to the observer and the fire, the altitude of the observer, the characteristics of the fire (fuel, weather, size, diurnal variation), and the observer.

Aerial detection observers do not detect every smoke they pass, and the fire detection process can be modeled as a stochastic process for planning purposes. Detection planners can divide the fire region into many small cells and define the *detection probability* as the conditional probability that the observer finds a fire when he or she looks in a cell, if there is one fire burning undetected in that cell. The detection probability depends upon the visibility factors described earlier. The following problem is a very simplified example of a daily detection patrol route planning problem which illustrates some of the basic principles of daily detection patrol aircraft management.

A Simple Detection Patrol Routing Problem   Suppose that a fire is known to be burning in one of a number of cells some distance from the initial attack

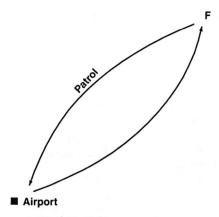

FIGURE 10 Detection patrol routing.

base but the precise cell in which the fire is located is unknown [as depicted in Figure 10]. The duty officer must know the precise location and behavior of the fire before dispatching an initial attack force. Should he/she use a patrol aircraft or rely on the public to provide that information?

The detection planner must decide whether to send a detection aircraft to look for the fire and, if so, at what time that patrol should be dispatched. To simplify the problem, we will assume that (1) the fire started at 08:00, (2) the fire spreads in the shape of a circle and the radius increases at a constant rate of 36 m hr^{-1}, and (3) the fire damage is \$200/ha based on the size of the fire when it is detected. Given these parameters, the fire loss depending upon the time the fire is found is shown in Table 5.

Assume that the detection probability varies throughout the day as shown in Table 6 and Figure 11. The expected cost if a patrol is dispatched early enough to look in the cell at 10:00 A.M. is

$$(1000 + 320)(0.2) + (1000 + 11{,}720)(1 - 0.2) = 10\,440$$

TABLE 5 Fire Loss Assuming Fire Is Circular

Time found (hr)	Hours burned	Area (ha)	Fire cost ($)
10:00	2	1.6	320
12:00	4	6.5	1,300
14:00	6	14.7	2,940
16:00	8	26.1	5,220
18:00	10	40.7	8,140
20:00	12	58.6	11,720

TABLE 6   Detection Probability Function

Look time	Aircraft detection probability	Public detection probability
10:00	0.2	—
12:00	0.4	—
14:00	0.6	—
16:00	0.8	—
18:00	0.6	—
20:00	—	1

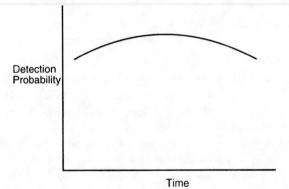

FIGURE 11   Detection probability function.

The expected costs associated with detection flights dispatched at later times are shown in Table 7. The expected cost plus loss is convex, as shown in Figure 12. The optimum solution is to dispatch a detection patrol to look in the cell at 14:00 hours.

The tactical detection patrol routing problems faced by detection planners are much more complex than this simple patrol timing problem. One can ap-

TABLE 7   Expected Cost Plus Loss

Time the detection observer looks in the cell	Flying cost ($)	Expected cost plus loss ($)	
10:00	1000	10,440	
12:00	1000	8,552	
14:00	1000	7,452	Optimum
16:00	1000	7,520	
18:00	1000	10,572	
Public finds fire at 20:00	0	11,720	Do not fly

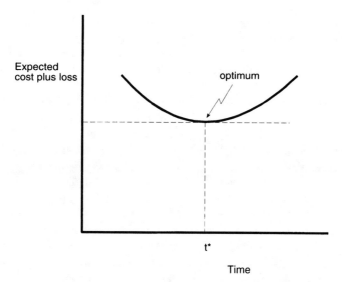

FIGURE 12   Expected detection cost plus loss function.

proach such problems by partitioning the fire region into a large number of small rectangular cells and predicting the expected number of fires or the parameters of the probability distribution of the number of new fires that occur in each cell each hour of the day. Forest vegetation or fuel type maps, weather and fire danger rating indices, and "values at risk" maps can be used to help identify potentially critical cells that "must" be visited. The task is then to design "good" sets of patrol routes that will route aircraft over or close to the critical cells.

Kourtz developed and field tested many novel approaches to daily detection routing problems, many of which were reviewed by Martell (1982). His most recent approach (Kourtz and Mroske, 1991) was to formulate the problem as what operational researchers refer to as a multiple salesperson-traveling salesperson problem, and he used a simulated annealing algorithm to solve such problems on a daily basis.

C. FIRE SUPPRESSION

Free-burning fires grow in size, become more difficult to control, and, if they are destructive, cause more damage over time. In the simplest case, a fire on flat terrain covered with a homogeneous fuel complex will burn in the shape of a circle in the absence of wind. The perimeter and therefore the difficulty of containing the fire will increase as a linear function of time. One can view fire suppression as a race between a fire that produces burning perimeter and fire fighters that produce fire line. The fire is contained when the amount of fire line constructed by the initial attack crew is greater than or equal to the perimeter of the fire. Since the perimeter of a free-burning fire increases as a linear function of time, the greater the head start of the fire and the faster it is spreading, the longer it takes to contain the fire.

If one assumes fire damage is a linear function of the area burned, then fire damage increases as a quadratic or nonlinear function of time, as illustrated in Figure 8. Fire management systems are, therefore, designed to predict when and where those fires are likely to occur, deploy fire fighters, helicopters, airtankers, and other suppression resources close to areas where fires are expected to occur, find and report fires soon after they are ignited, dispatch initial attack resources to fires soon after they are reported, carry out suppression action to contain fires as quickly as possible, and minimize the damage of large potentially destructive escaped fires that result when fires are not controlled by the initial attack force. In the following sections, we describe each of these fire suppression subsystems, the decision making associated with their management, and some of the planning models and decision support systems that have been developed to enhance their performance.

Fire suppression systems act on individual fires, but we are primarily interested in their impact on landscape fire regimes. Presuppression planning measures to support decision making concerning fire suppression resource acquisition, home basing, and daily deployment help fire managers design and operate their initial attack system to minimize response times and decrease the number of fires that escape initial attack and grow to large sizes. They influence the *fire size* or *extent* component of a fire regime directly but, like detection, they also have an indirect impact on *fire intensity*.

1. Fire Suppression Resource Acquisition and Strategic Deployment

Strategic fire suppression resource management entails deciding how many permanent initial attack bases and other facilities will be established and where they will be located, the number and type of airtankers and transport aircraft that will be purchased and kept for long periods of time, the number of fire fighters that will be hired, and the number and type of aircraft that will be chartered for a particular fire season. The more resources acquired, the less likely the agency will have to pay premium rates for aircraft chartered for short periods of time and the less likely they will have to incur the transportation costs and waiting times associated with short-term aircraft charters and borrowing fire fighters and equipment from other agencies. Once a base has been established, it is difficult to relocate it. Airtankers are very effective but costly to own and operate, and, since they do not fly long hours each year, they can last for many years. Well-trained fire fighters are essential skilled personnel that cannot readily be replaced. The way such strategic decision-making problems are resolved, there-fore, has significant impacts that can ripple throughout a fire organization for many years.

It is difficult to determine how much of each type of fire suppression resource to acquire and where to base those resources so that they are readily available to be dispatched to fires as they are reported. Fire management agencies often use large computer-based decision support systems to help resolve such decision-making problems. Martell (1982) described some of the many strategic plan-ning models, most of which focused on airtanker operations, that have been de-veloped. We discuss the use of such models in Section V where we study level of fire protection planning.

Fire management planning is a hierarchical process, and decisions concern-ing the mix of resources required depends upon where those resources will be permanently based, how they will be repositioned or deployed each day as fire occurrence ebbs and flows across the protected area, how they will be dis-patched to fires, and how they will be used on each fire. Fire management sys-tems are so large and complex that managers must decompose them into hier-archical systems of decisions such that decisions made at any level within a hierarchy are compatible with decisions being made above and below that level in the hierarchy, as described in Section I.

Consider, for example, the decision of where to home-base airtankers. Fire occurrence processes vary significantly over both time and space, and the de-mand for fire suppression resources shifts from day to day and place to place. Fire suppression resources are very mobile, and fire managers move them from their home bases to areas where they are most needed on a daily basis. In the province of Ontario, for example, it is not uncommon to require a large num-ber of fire fighters and airtankers in different areas, and, when such needs arise, resources are quickly moved or deployed from their home bases or their cur-rent locations to the areas where they are most needed. Once the agency has de-cided how many and what type of airtankers to purchase or lease, it must de-cide where to home-base them. Given their mobility and the fact that they can be shared with other agencies, fire managers attempt to develop home-basing strategies that will minimize the time and flying costs incurred while meeting shifting daily deployment needs.

Greulich (1976) developed an integrated airtanker home-basing/daily de-ployment model for two bases in California, but his model would not have been mathematically tractable for the number of airtankers and bases used in Ontario. MacLellan and Martell (1996) decomposed the problem into separate home-basing and daily deployment decision-making systems and developed a mathe-matical programming model which the Ontario Ministry of Natural Resources used to help decide how it should home base its airtankers. They dealt with daily airtanker deployment needs by consulting fire managers and asking them to express their daily airtanker deployment needs in terms of the number of air-tankers required at each base as a function of the forecasted fire weather and the number of fires expected to occur in the area surrounding each base. Their mathematical programming model accounted for historical fire weather and fire occurrence patterns and identified home-basing strategies (how many air-tankers should be home-based at each airport) to minimize the cost of satisfying the daily airtanker deployment rules identified by the fire managers. That home-basing system is only as good as the daily deployment rules embedded in the model and, since they and the Ontario fire managers that developed the model opted to use subjective deployment rules, it begs the question: How should air-tankers be deployed at each base each day? Daily airtanker deployment poses many complex challenges due to the need to deal with stochastic fire occur-rence and behavior processes that vary over both time and space, so it is rea-sonable to approximate the deployment practices fire managers will use as an interim measure. In the next section, we describe how daily suppression de-ployment needs can be formally modeled and assessed.

2. Initial Attack Resource Deployment

Fire occurrence rates vary over both time and space, and each day fire managers must decide where to deploy their initial attack resources to minimize initial attack response times. They must, for example, decide how many fire fighters to place on initial attack standby at each base each day. The more crews they place on alert, the more costs they will incur, but the less likely they will expe-rience crew shortages. They can reduce the number of crews on initial attack standby but, if they do so, they run the risk of experiencing shortages which will mean delays in acquiring crews from other bases or agencies, extra costs in transporting them in to the area experiencing the shortage, and an increased likelihood that fires will escape initial attack. They face similar decision-making problems associated with the daily deployment of airtankers and transport air-craft. Martell (1982) described some of the models that have been developed to enhance daily deployment decision-making.

Initial attack resource deployment analysis is complicated by uncertainty concerning the timing and location of fires and interactions *between* fires. For ex-ample, if the response time to one fire is excessive, that fire will grow as it waits, and its size at attack will increase. Initial attack resources will be required on that fire for a longer period of time so that the *next* fire that arrives may have to wait even longer. A delayed response to one fire can therefore produce im-pacts that can ripple over both time and space and affect the ability of the ini-tial attack system to respond to far-distant fires several days later. That calls for systems-level approaches to deployment analysis.

Many service systems are designed such that, when the number of customers that require service is greater than the number of servers, the surplus customers join a queue where they wait for service. Queues are prevalent in our society, and we are accustomed, for example, to queueing for supermarket cashiers, automatic banking machines, and airline ticket agents. Service system managers must balance the cost of service (more servers cost more money) with the cost (or inconvenience) to customers associated with waiting for service. For an introduction to the basic principles of queueing theory, see Ross (1985).

a. Initial Attack Queues

There are many queueing systems in forest fire management systems, one example of which is an airtanker initial attack system in which fires are customers and airtankers are servers. The initial attack queue is a list of fires that require initial attack action. Airtanker systems are complex queueing systems since fires arrive at rates that vary over the course of the day, fires grow and their service times can increase as they wait in the queue, and the airtanker servers must fly out to serve their customers where they are burning.

To model the behavior of a queueing system, we need to model the customer arrival process, the queue, the service discipline, and the service process. The arrival of customers that need service (fires) can be modeled as a Poisson process.

One must also describe the queue where customers wait for service. The initial attack queue is a list of fires with an infinite capacity and the queueing discipline describes how waiting customers are selected for service. It may be a first-come-first-served (FCFS) discipline, or it may be a priority queue. If it is a priority queue, the high-priority customers may preempt lower priority customers that are in service when they arrive.

The description of the service process includes the number of service channels and the service time distribution which is typically exponential or Erlang. Interarrival times and service time distributions are often modeled as exponential distributions because their Markovian or memoryless property simplifies the mathematical analysis of queueing systems (see Ross, 1985, p. 190).

Let T be a random variable with an exponential distribution.

$$f_T(t) = \mu e^{-\mu t} \qquad t \geq 0 \qquad \text{Probability density function}$$
$$= 0 \qquad\qquad t < 0 \tag{19}$$

$$F_T(t) = 1 - e^{-\mu t} \qquad t \geq 0 \qquad \text{Cumulative distribution function}$$
$$E(T) = 1/\mu \qquad\qquad \text{Var}(T) = 1/\mu^2 \tag{20}$$

Queueing models can provide a rich variety of performance measures including the expected time a customer waits in the queue, the probability that a customer has to wait for service, and the expected number of customers in the queue.

The simple *M/M/s* queueing system model is a model of a system in which customers arrive according to a Poisson process which has parameters that remain constant over time, the service time is exponential, and there is one queue in front of s servers. More complex models can be developed to deal with situations in which the fire arrival rate varies over the day due to diurnal variation in fire occurrence and behavior processes and long travel times make it inappropriate to use exponential service time distributions.

A Simple Daily Airtanker Deployment Problem   Consider a simple case in which the fire region depicted in Figure 13 is partitioned into two sectors, and each sector has one airtanker base that houses all the airtankers that respond to fires within the sector. Assume that airtankers cannot be dispatched to fight fires outside the sector in which they are based.

We ignore diurnal variation in the fire arrival rates and the fact that finite travel time is incompatible with the use of an exponential service time distribution and model each sector as an independent *M/M/s* queueing system. It can be shown that W_q, the expected waiter time in the queue for an *M/M/s* queue, is given by the following formula (see Blake, 1979, p. 346):

λ = average fire arrival rate (fires/hr)
$1/\mu$ = average time to fight (service) a fire (hr)
$\rho = \lambda/\mu$
P_0 = probability the system is empty

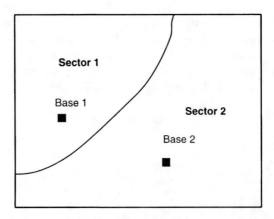

FIGURE 13   A hypothetical fire region with two airtanker bases.

$$P_0 = \left[\sum_{k=0}^{s-1} \frac{\rho^k}{k!} + \frac{\rho^s}{(s-1)!\,(s-\rho)} \right]^{-1} \tag{21}$$

W_q, the expected waiting time in the queue, is

$$w_q = \frac{P_0}{\mu} \left(\frac{\rho^s}{(s-1)!\,(s-\rho)^2} \right) \tag{22}$$

Suppose that the fire management policy for the area calls for a fire waiting time of 0.10 hr or less. The expression for W_q can be used to determine how many airtankers (servers) are required in sectors 1 and 2 to achieve that goal. Response time targets by sector are $W_{q1} \leq 0.10$ and $W_{q2} \leq 0.10$. The results presented in Table 8 indicate we would need 5 + 9 or 14 airtankers.

To demonstrate the importance of resource sharing, we could aggregate the sectors 1 and 2 into 1 region with $\lambda_R = \lambda_1 + \lambda_2$. The results for the aggregate region are shown in Table 9.

We would need only 12 airtankers to achieve the same level of service for two cooperating bases. It should be noted, however, that we have ignored the "extra" travel time required to move between sectors, so 12 is actually a lower bound estimate of the number of airtankers required and the true number may be greater than 12.

The queueing models we have described are for simple steady state $M/M/s$ queueing systems with single bases. It is possible to develop queueing models of more realistic fire management systems. Martell and Tithecott (1991) applied an $M(t)/M/s$ for airtankers at a single base when the fire arrival rate varies over the course of the day. They used numerical methods to solve the differential equations that described the system and produced time-dependent performance measures. They developed software that enabled them to field-test the

TABLE 8   Airtanker Response Time by Sector

Sector	Fire arrival rate (λ fires/hr)	Fire service rate (μ fires/hr)	Number of airtankers, s	Expected waiting time in the queue (W_q hr)
1	2	0.75	1–2	Undefined
			3	3.19
			4	0.38
			5*	0.09
2	4	0.75	1–5	Undefined
			6	1.43
			7	0.33
			8	0.11
			9*	0.04

TABLE 9   Airtanker Response by Region

Fire arrival rate (λ fires/hr)	Fire service rate (μ fires/hr)	Number of airtankers, s	Expected waiting time in the queue (W_q hr)
6	0.75	1–8	Undefined
		9	0.87
		10	0.27
		11	0.11
		12	0.05

use of the model in the northwestern region of Ontario. Managers indicated that, since it did not model base interaction, it would not satisfy their needs. Islam (1998) subsequently studied initial attack airtanker systems and developed several models for spatial airtanker queueing systems with time-dependent arrivals and interaction between bases, one of which is described in Islam and Martell (1998).

3. Initial Attack Dispatching

As soon as a fire has been reported, the initial attack dispatcher or duty officer must decide what suppression resources will be dispatched to that fire and when they will depart for the fire. Fires are placed in the initial attack queue as they are reported, and some have higher priorities than others depending upon their location and the threats they pose to public safety, property, and other values. Initial attack dispatching is complicated by uncertainty concerning what might happen later in the day. While airtankers are quite mobile and can readily be diverted from one fire to another as the need arises, initial attack crews are not as flexible. If an initial attack crew is allocated to a fire, it will not be available to fight higher priority fires that might be reported later that same day and, since crews are usually tied up on fires for several days, today's initial attack dispatching effects can ripple throughout the organization for the next several days.

4. Initial Attack

Once the initial attack crew arrives at the fire, the initial attack fire boss must assess the situation and devise a suppression strategy. The current fire behavior, surrounding vegetation, values at risk, number and type of resources currently available, and the forecast weather will influence his or her decision concerning the attack strategy. He or she may, for example, opt to have the initial attack crew set up a power pump, lay hose up to the fire, and gradually work around the fire perimeter while extinguishing the flame front. If there is insufficient water available, the crew may use hand tools such as axes, shovels, and other specialized line building equipment. Airtankers may drop water or special fire-spread-retarding chemicals on the fire while the initial attack crew is traveling to the fire, and helicopters slinging buckets that can hover and drop water precisely might work with them as they construct fire line and attack hot spots near the fire perimeter. The decision making associated with initial attack operations is the domain of skilled and experienced fire fighters, and there have been few attempts to develop decision-making aids to enhance their effectiveness.

During the years 1976–1999 the fire size at the start of initial attack in Ontario ranged from less than 0.1 ha to 27,774 ha, but 95% of the fires were attacked at less than 5.5 ha. The average fire size at attack was 24.8 ha, but the distribution is highly skewed with a median of 0.1 ha and a mode of 0.1 ha. Again, the skewness is largely a result of a few large fires in the extensive protection zone in the far north.

5. Large Fire Management

In many jurisdictions, a very small proportion of the fires burn a large portion of the area that is burned each year. In the province of Ontario, for example, 95% of the fires burned to a final size of 33 ha or less during the years 1976–1999. The average final fire size was 141.1 ha, but the distribution was highly skewed with a median of 0.2 ha and a mode of 0.1 ha. Fires that are not controlled by the initial attack force are classed as escaped fires, and they have the potential to grow to very large sizes and cause considerable damage. However, the small number of large fires can have very significant impacts, some of which are beneficial, on natural forest ecosystems.

Large fires are difficult and very costly to contain, and observers sometimes question what may appear to be futile suppression activities. One reason, of course, is that most agencies operate under what are essentially fire exclusion policies in much of their area, and they and the governments that fund them might be found legally liable if they simply admitted defeat and decided not to attempt to control large fires. Nevertheless, many fire managers believe that they can limit the damage caused by large fires in a cost-effective manner. Their rationale is as follows.

Large fires are driven across the landscape by weather and in many regions (e.g., the boreal forest region of Canada) suppression action may not only be ineffective but also very dangerous to fire fighters while high intensity fast-spreading fires are moving on wide fronts. Large fires can move as fast as 7 km hr^{-1} on fronts as wide as several kilometers with flame fronts in excess of

50 m in the boreal forest. It makes little or no sense to attempt to contain fires while they are making such runs. Under such circumstances, fire fighters sometimes carry out what they refer to as value protection measures. For example, they may use portable power pumps, hose, and irrigation sprinklers to establish sprinkler lines around isolated cottages and other valuable structures in the paths of such fires.

Fire fighters also combat running wildfires by burning out from natural or human-made barriers. For example, they may use helicopters equipped with helitorches to fly close to a lake or river in the fire's path, ignite fire at the edge of the lake, and then continue to light fire progressively closer to the advancing fire. Their objective is to augment an existing barrier, the lake or river, and thereby create a much wider fuel break in the hope that they can stop the progress of the fire when it reaches that barrier.

Fire suppression organizations often simply monitor and project the growth of running fires but, while they do so, they identify strategic areas where they hope to establish control lines. They move suppression resources up close to the fire and establish base camps from which they can launch their attack when weather conditions turn in their favor and the fire intensity subsides. Fires that are allowed to burn freely in the largely continuous fuel complexes of the boreal forest have the potential to continue burning until the advent of winter snowfall and, in some cases, they have even been known to smolder in the deep duff under the snow during particularly bad drought years. But such fires usually move in short fast-paced bursts. As fire weather conditions worsen, the fire becomes more active. If the drying continues for a sufficiently long time, the fire will, usually under the influence of a strong wind, move rapidly for part of a day or more until the weather changes and it settles down to smolder until burning conditions worsen yet again. The fire manager's objective is to mobilize his or her resources so that the fire can be contained during one of those lulls. If and when such efforts are successful, the damage that might have been incurred during subsequent high-burning-hazard periods will be averted.

Large fire suppression teams may have to determine what values to sacrifice during burning out operations to enhance the likelihood that they can control the fire at a later date. They carry out what they refer to as escaped fire situation analyses (EFSA) to evaluate alternative strategies for dealing with large escaped fires.

V. LEVEL OF FIRE PROTECTION PLANNING

The least cost plus damage model provides valuable insight into the potential long-term economic impact of fire management, and supplementary models like the fire and timber supply model augment that understanding with detailed information concerning the timber supply implications of fire management from landscape and regional perspectives. But fire management objectives should be based on the potential social, economic, and ecological impacts of fire, and we have, at best, only begun to scratch the surface in terms of our understanding and ability to quantify such impacts. Fire management policy analysts and others will ultimately have to provide land managers and the public with comprehensive planning models that can be used to assess the impact of fire and fire management on social systems and forest ecosystems. It will take considerable time and effort to develop our understanding sufficiently to assess the social, economic, and ecological impacts of fire management. But the public safety, property, and timber supply concerns that lead to the development of variants of current policies remain, and fire managers must continue to suppress fire, albeit with more discretion than was the case in the past, until the policies that govern their agencies direct them to do otherwise. Fire managers and their many diverse clients, therefore, need pragmatic interim solutions that can be refined as time passes and our knowledge and understanding grow.

In 1935, the U.S. Forest Service developed an effective fire exclusion policy by using what it referred to as the 10:00 A.M. policy—that all fires should be contained by 10:00 A.M. the day following the day the fire is detected (Pyne, 1982, p. 116). That surrogate initial attack objective guided their prevention, detection, and suppression activities to minimize area burned for many years. Most North American forest fire management agencies developed and implemented variants of the 10:00 A.M. rule. Some agencies recognized explicitly that they need not and cannot achieve such objectives on all fires, particularly those that occur in remote areas and do not threaten people, property, and timber supplies. The Ontario Ministry of Natural Resources, for example (see Martell, 1994), partitioned their fire region into intensive, measured, and extensive protection zones. All wildfires that occur in the intensive protection zone are attacked aggressively until they are controlled. Fires in the extensive protection zone are not attacked unless they pose significant threats to people or property. Initial attack action is taken on all fires that occur in the measured protection zone. Measured protection zone fires that are not controlled by the initial attack force are subjected to an escaped fire situation analysis that may call for continued aggressive suppression, modified suppression, or continued monitoring without suppression. Even though such zoning schemes suited the needs of the OMNR and other fire management organizations well in the past, the small number of very large zones are no longer adequate to meet the complex needs of fire management agencies that will have to assess social, economic, and ecological impacts of their activities in far more detail than was ever the case in the past.

Martell and Boychuk (1994) described conceptually how zoning schemes can be refined to suit the needs of the many diverse clients that want fire management on tracts of land they deem as being important. One can begin by partitioning the fire region into a large number of zones that are reasonably homogeneous with respect to forest ecosystems, land use, and values at risk and by specifying an average annual burn fraction for each zone. Such fractions could then be used to help quantify the potential impacts of fire on public safety and property, and they could readily be incorporated into timber supply models to assess the timber supply implications of the proposed burn fractions by zone. Ecologists who study fire regimes recognize the average fire return interval and burn fraction as important components of fire regimes and will no doubt develop models that relate ecological impacts to burn fractions.

The interested stake holders in each zone could then consider the potential impact of the proposed burn fraction and the ecosystems in those zones and arrive at a decision, preferably a consensus, as to what average annual burn fraction is acceptable in each zone. The next step would be for the fire management agency to use a strategic level of protection planning model to determine how they could meet that objective and the cost of doing so.

Many fire management agencies use modern information technology and OR/MS methodologies to guide their planning. One approach is to decompose the fire management system into many small relatively simple subsystems and then develop large comprehensive computer simulation models that predict the consequences of specified fire management program alternatives and are used for fire management budget planning purposes. Examples include LEOPARDS (McAlpine and Hirsch, 1999), National Fire Management Analysis System (NFMAS), and California Fire Economics Simulator (CFES) (Fried and Gilless, 1988). Figure 14 is a schematic representation of the basic structure of a fire suppression computer simulation model. Such models are designed to enable managers to specify what fire suppression resources they would hire and then simulate how the proposed system would perform were those resources to be used to fight historical fires or representative sets of hypothetical fires. The key issue is model validity or the extent to which such models can truly represent the real world, and increased effort is expected to be devoted to the development, testing, and use of such models in the near future.

Martell et al. (1984) developed an initial attack simulation model (IAM) which the Ontario Ministry of Natural Resources used to help determine the number and type of airtankers required in Ontario. That model predicts several measures of system performance including the fraction of fires that escape initial attack depending upon the number and type of airtankers, fire fighters, and transport aircraft allocated to fire suppression. Martell et al. (1995) later modified IAM to produce LANIK, a modern desktop computer implementation of IAM, to facilitate its use for strategic level of protection planning in Ontario.

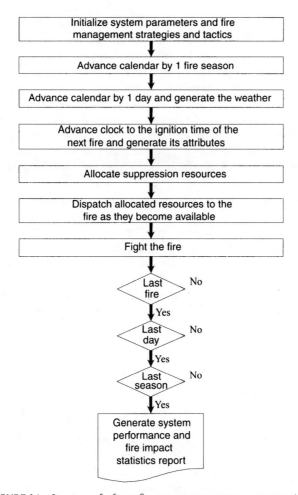

FIGURE 14 Structure of a forest fire management system computer simulation model.

McAlpine and Hirsch (1999) later extended LANIK and embedded it in a GIS to produce a Windows'95 model called LEOPARDS.

LEOPARDS has since been used for a variety of planning purposes including an assessment of the potential implications of climate change, changes in initial attack fire crew staffing levels, and the upgrading of the OMNR's airtanker fleet with modern CL-415s. Many other forest fire management agencies have developed similar systems (e.g., the U.S. Forest Service's NFMAS, the California Division of Forestry's CFES, and Chile's KITRAL [Pedernera and Julio, 1999]).

There is no reason to believe that all the stake holders will agree with the fire management costs that will emerge when they engage in detailed zonal level of protection planning exercises, so iterative procedures that hopefully converge to an acceptable agency wide strategy will have to be used.

After the burn fractions have been specified by zone, they collectively can be used as a strategic objective to be met by the fire organization. Its task then becomes one of determining a detailed site-specific plan for minimizing the cost of achieving that objective. Level of protection planning models could again be used to help transform those objectives into a refined fire management plan.

Strategic models like LANIK, LEOPARDS, NFMAS, and KITRAL are aggregate models that will need to be enhanced or replaced as time passes. Fire managers can incorporate simple proposed strategies and tactics in such models in the form of rules that govern the use of fire suppression resources, but it will be difficult to translate those strategies and tactics into precise operational deployment, initial attack, and large fire suppression in each zone. They will have to develop simple operational guidelines for prevention, detection, initial attack, and large fire suppression that vary by fire danger as measured by, for example, fire danger rating indices. For example, they will have to specify daily detection size and initial attack response time objectives by zone given the fire danger.

VI. SOME CHALLENGES

The development and implementation of a fire exclusion policy is relatively simple and straightforward when compared with the challenges that fire managers will have to deal with in the 21st century. It is, of course, no mean feat to decide what fire-fighting resources are required and where they should be based, predict daily fire occurrence, deploy fire suppression resources close to areas where fires are expected to occur, find the fires that do occur while they are still small, dispatch initial attack resources to fight those fire and contain them quickly and at small sizes, and deal with any large escaped fires that result. These are enormous and potentially costly logistic problems, but they are manageable given the widespread acceptance that fire is destructive and that fire managers must minimize the cost of achieving specified fire control. But such challenges quickly fade when compared with the need to understand the social and ecological impacts of fire and how human intervention will shape forest landscapes and the animal populations that will struggle to survive there and to resolve the many conflicts that different constituencies will bring to the table as society attempts to develop site-specific fire management strategies for every significant parcel of forest land. We address some of those issues on the following pages.

A. RELATING OUTCOMES TO STRATEGIES

Fire management programs produce fire regimes that forest management clients and others affected can use to assess the impact of fire management on what they do and the things they value. It is, however, difficult to predict the fire regime consequences of specified fire management strategies, and even more difficult to determine how best to achieve a specified fire management regime. Consider, for example, the simplest case of a small forest management region that functions independently. It would indeed be difficult to predict what regime would result by using a specified set of fire management resources according to some specified sets of rules. Researchers can and have, of course, developed planning models that are designed to help resolve such problems. Examples include NFMAS, CFEES, and LEOPARDS, all of which have already been discussed. But these systems are primarily initial attack planning tools that focus on initial attack operations, and their ability to model large fires that escape initial attack is either nonexistent or, at best, very simplistic.

We can expect progress in this area, but the real challenge lies in the fact that fire management units are not independent units. Consider the case of two interdependent fire management units that lie adjacent to each other and suppose airtankers are used for initial attack purposes in both units. One form of fire management might call for specified initial attack time targets that vary by compartment and fire weather within each unit. If there is a shortage of airtankers in one unit, then fires will have to wait longer in the initial attack queue during which they will grow and demand even longer service times. That will reduce the likelihood that the busy unit can share its resources with the adjacent unit which will increase the likelihood that fires escape initial attack in the adjacent unit. In turn, that will lead to an increase in the need for suppression resources in the adjacent unit and produce shortages in both units that may last for days or even weeks. Thus, a particularly troublesome fire or a shortage of fire suppression resources in one unit can have significant impacts that percolate across time and space. Furthermore, fire management agencies can borrow resources from other provinces, states, and countries. The availability of such resources will vary depending upon the weather in those jurisdictions. The first challenge is that the set of empirical data required to document what strategies were used in the past on some designated area is enormous. For example, if the fire regime is thought to be a function of the initial attack response time strategy, one needs to know how easy it was to borrow resources from other agencies to meet initial attack response time targets. Equally challenging is the fact that when a manager sets initial attack response time targets for one management unit, he or she needs to model the ability of that unit to borrow resources from other widely scattered agencies.

B. DEALING WITH UNCERTAINTY

Fire is a stochastic disturbance agent that introduces enormous uncertainty into forest land management planning. Forest managers recognized that fire threatened their ability to exploit forest resources and took measures to limit its extent and impact. They developed contingency plans that were designed to enhance their ability to cope with fire losses, but their understanding of fire and the lack of mathematical modeling techniques and computing resources made it impossible for them to incorporate fire loss explicitly in their planning systems in anything other than a rudimentary fashion. Reed and Errico (1986) developed a forest-level timber harvest scheduling model that makes it possible to account for fire loss in large forest management units if fire losses are not severe and some deterministic average fire loss is assumed to occur each year. Boychuk and Martell (1996) showed how stochastic programming methods could be used to deal with variability in fire processes, but stochastic modeling methodologies and computer technology limit the ability to do so. There is, of course, a need to enhance stochastic programming planning methods, but even more important is the need for forest-based societies to learn to live in harmony with the variability that is characteristic of natural forests.

Gunn (1996) has suggested that replanning after significant uncertain events have taken place is a reasonable strategy for coping with uncertain fire losses and, assuming forests are not "pushed to the limit" with respect to timber production, they may be robust enough to "respond." One cannot maintain both natural ecosystems and stable timber harvest flows simultaneously. Societies must develop site-specific compromises for such problems, and they have to learn to cope with the variability that will result. That might, for example, call for explicit recognition that harvest levels will rise and fall and the aggregation of relatively small independently owned forest management units into larger woodsheds in which forest landowners agree to share their resources with those that suffer loss so that each of them enjoys a measure of insurance concerning resource flows. We will also have to learn how to deal with the problems that will arise when we embed islands of flammable "natural" forest landscapes that are designed to provide habitats for rare and endangered species in larger landscapes that are not so well endowed with suitable habitat. To date the forest fire management research community has focused on timber production under uncertainty. It is essential that ecologists and land management planners work together to deal with the much more complex issues of wildlife conservation in managed forests that are expected to provide both resources and suitable wildlife habitats.

C. MOVING BEYOND FIRE EXCLUSION

Most modern wildland fire management agencies were established and developed in response to threats to public safety, property, and timber production. Despite the growing recognition that fire is a natural component of many forest ecosystems and that it is neither economically nor ecologically sound to attempt to exclude it completely from forested landscapes, there has not been a significant shift away from traditional fire exclusion practices in most jurisdictions. Fire has, of course, been "reintroduced" into some areas, primarily parks and wilderness areas where the desire to maintain or restore natural ecosystem processes is paramount and the land managers responsible believe that they can increase the amount of fire on the landscape without undermining public safety and property significantly.

Choosing not to fight all fires aggressively does, however, pose significant challenges to the managers who must develop and implement such policies, the fire fighters who must deliver them, and the public who must live with the consequences. The primary motivation for fire exclusion is the desire to reduce uncertainty and minimize the risk that significant losses will be incurred in the near future. While fire exclusion may lead to hazardous fuel buildups and potentially even more destructive fires in the distant future in some biomes, every wildfire that is not extinguished represents a potential short-term threat to public safety, property, and timber. Historical accounts of wildland fire disasters in North America during the early decades of the 1900s describe how hundreds of lives were lost as fires swept across the landscape and engulfed entire communities and the rural homesteads between them, almost without warning (see, for example, Holbrook, 1943; Lambert and Pross, 1967; Pyne, 1982). There is no simple common explanation for all such tragedies, but, in many cases, small land clearing and lightning-caused fires had been left to burn largely unattended across the landscape. That posed no significant threat as long as benign fire weather conditions prevailed, but from time to time nature conspired to produce several good drying days and one or more days with high temperatures, low relative humidities, and high winds. The heretofore small benign fires then began to spread, joined up, and raced across the landscape on wide fronts pushed by strong winds. The surveillance, telecommunications, and transportation systems were not up to the task, and the resulting damage precipitated calls for fire exclusion that largely persist to the present.

Modern forest fire management agencies have the capability to monitor ongoing fires, and they can rely on meteorologists to forecast weather so that they can predict potential fire behavior some days in advance. They can, in principle, predict when fires are about to make a run, intervene, and extinguish them before that happens. If and when they are unable to do so, they can use modern transportation technology to evacuate threatened communities. The irony is that, since such policies would not produce the large burned areas characteristic of natural fire regimes, it is questionable what environmental benefits would be gained by such practices.

There are no simple answers. Martell (1984) addressed this issue as it pertains to the boreal forest region of Canada and advocated the adoption of what he described as "fire impact management policies" whereby decisions concerning wildfire suppression and the use of prescribed fire are based on sound social, economic, and ecological principles. He also discussed some of the practical problems that would complicate the development and implementation of such policies.

The simple truth is that fire cannot be reintroduced into fire-prone landscapes without significant threat to public safety, property, and other values. A fire manager that cannot control what proves to be a destructive fire will not be criticized if he or she is judged to have done his or her best to deal with the situation in a professional manner. However, if he or she decides to let what ultimately proves to be a destructive fire burn with little or no significant suppression, he or she and the agency that employs him or her will no doubt be liable for civil litigation. The issue is not unlike the concern Howard *et al.* (1972) identified when they realized that decisions to seed hurricanes to diminish their destructive potential should be tempered by the possibility that a change in the storm's characteristics might, rightly or wrongly, be attributed to the seeding operation and could open the government to civil litigation.

Mitchell (1995, p. 4) explains that the precautionary principle that emerged from the Earth Summit in Rio de Janeiro in 1992 recognizes that even though "uncertainty creates a serious dilemma for resource and environmental managers. . . . incomplete understanding should not be used as an excuse for delaying action when environmental degradation appears imminent." Some might well be tempted to use the precautionary principle to support calls for nonexclusion policies, but that must be tempered by concern for public safety and property damage and the possibility that, although fire is natural, some forest ecosystems may have been so perturbed by humans that "natural" fire regimes would simply hasten their demise. Fire management is ecosystem management over massive spatial and temporal landscapes, and it is fraught with considerable uncertainty concerning what will happen in the next few days, weeks, decades, or centuries. Given the importance of fire in natural ecosystem processes, we cannot afford to simply continue on with the way we in North America have behaved for the last century. It should all make for an interesting future.

ACKNOWLEDGMENTS

J. Beverly and A. Tithecott provided helpful comments on earlier versions of this chapter. The Aviation, Flood and Fire Management Branch of the Ontario Ministry of Natural Resources provided a digital file of their 1976-1999 fire report data that was used to compute the Ontario fire statistics reported in this chapter.

FURTHER READING

Winston, W. L. (1994). "Operations Research: Applications and Algorithms." Duxbury Press, Belmont.

REFERENCES

Baumgartner, D. C., and Simard, A. J. (1982). "Wildland Fire Economics: A State of the Art Review and Bibliography." Gen. Tech. Rep. NC-72. USDA Forest Service, North Central Forest Experiment Station, St. Paul.

Blake, I. F. (1979). "An Introduction to Applied Probability." John Wiley and Sons, New York.

Boxall, P. C., Watson, D. O., and Englin, J. (1996). Backcountry recreationists' valuation of forest and park management features in wilderness parks of the western Canadian Shield. Can. J. For. Res. 26, 982–990 .

Boychuk, D. B., and Martell, D. L. (1996). A multistage stochastic programming model for sustainable forest-level timber supply under risk of fire. For. Sci. 42, 10–26.

Clough, D. J. (1963). "Concepts in Management Science." Prentice-Hall, Inc., Englewood Cliffs.

Cunningham, A. A., and Martell, D. L. (1973). A stochastic model for the occurrence of man-caused forest fires. Can. J. For. Res. 3, 282–287.

Fried, J. S., and Gilless, J. K. (1988). "The California Fire Economics Simulator Initial Attack Model (CFES-IAM): MS-DOS Version 1.11 User's Guide." Bulletin 1925. Division of Agriculture and Natural Resources, University of California, Oakland.

Fried, J. S., and Fried, B. D. (1996). Simulating wildfire containment with realistic tactics. For. Sci. 42, 267–281.

Gass, S. I. (1994). Public sector analysis and operations research/management science. In "Operations Research and the Public Sector" (S. M. Pollock, M. H. Rothkopf, and A. Barnett, Eds.), Handbooks in Operations Research and Management Science Vol. 6, pp. 23–46. North-Holland, New York.

Gorte, J. K., and Gorte, R. W. (1979). "Application of Economic Techniques to Fire Management—A Status Review and Evaluation." Gen. Tech. Rep. INT-53. USDA Forest Service, Intermountain Forest and Range Experiment Station, Ogden.

Greulich, F. E. (1976). "A Model for the Seasonal Assignment of Airtankers to Home Bases Under Optimal Daily Transfer Rules." PhD dissertation. University of California, Berkeley.

Gunn, E. A. (1996). Hierarchical planning processes in forestry: A stochastic programming-decision analytic perspective. In "Workshop on Hierarchical Approaches to Forest Management in Public and Private Organizations" (D. L. Martell, L. S. Davis, and A. Weintraub, Eds.), pp. 85–95. Inf. Rep. PI-X-124. Canadian Forest Service, Petawawa National Forestry Institute, Chalk River.

Holbrook, S. H. (1943). "Burning an Empire." The Macmillan Company, New York.

Howard, R. A., Matheson, J. E., and North, D. W. (1972). The decision to seed hurricanes. Science 176, 1191–1202.

Islam, K. M. S. (1998). "Spatial Dynamic Queueing Models for the Daily Deployment of Airtankers for Forest Fire Control." PhD dissertation. University of Toronto, Toronto.

Islam, K. M. S., and Martell, D. L. (1998). Performance of initial attack airtanker systems with interacting bases and variable initial attack ranges. Can. J. For. Res. 28, 1448–1455.

Johnson, K. N., and Scheurman, H. L. (1977). Techniques for prescribing optimal timber harvest and investment under different objectives—Discussion and synthesis. For. Sci. Monogr. 18, 1–31.

Kourtz, P. H., and Mroske, B. (1991). "Routing Forest Fire Detection Aircraft: A Multiple-Salesman, Travelling Salesman Problem." Canadian Forest Service, Petawawa National Forestry Institute, Chalk River.

Kourtz, P. H., and O'Regan, W. G. (1968). A cost-effectiveness analysis of simulated forest fire detection systems. *Hilgardia* 39, 341–366.

Kourtz, P. H., and Todd, B. (1992). "Predicting the Daily Occurrence of Lightning-Caused Forest Fires." Inf. Rep. PI-X-112. Canadian Forest Service, Petawawa National Forestry Institute, Chalk River.

Lambert, R. S., and Pross, P. (1967). "Renewing Nature's Wealth." Ontario Department of Lands and Forests, Toronto.

MacLellan, J. I., and Martell, D. L. (1996). Basing airtankers for forest fire control in Ontario. *Oper. Res.* 44, 677–686.

Martell, D. L. (1982). A review of operational research studies in forest fire management. *Can. J. For. Res.* 12, 119–140.

Martell, D. L. (1984). Fire impact management in the boreal forest region of Canada. *In* "Resources and Dynamics of the Boreal Zone, Proceedings of a Conference held at Thunder Bay, Ontario, August 1982" (R. W. Wein, R. R. Riewe, and I. R. Methven, Eds.), pp. 526–533. Association of Canadian Universities for Northern Studies, Ottawa.

Martell, D. L. (1994). The impact of fire on timber supply in Ontario. *For. Chron.* 70, 164–173.

Martell, D. L., and Boychuk, D. (1994). "Levels of Fire Protection for Sustainable Forestry in Ontario: Final Report." Report prepared for the Canada-Ontario Northern Ontario Development Agreement, Sustainable Forestry Development/Decision Support Project: Developing Analytical Procedures for Establishing the Level of Protection for Forest Fire Management to Support Sustainable Forestry in Ontario.

Martell, D. L., and Tithecott, A. (1991). Development of daily airtanker deployment models. *In* "Proceedings of the 1991 Symposium on Systems Analysis in Forest Resources" (M. A. Buford, Ed.), pp. 366–368. Gen. Tech. Rep. SE-74. USDA Forest Service, Southeastern Forest Experiment Station, Asheville.

Martell, D. L., Bevilacqua, E., and Stocks, B. J. (1989). Modeling seasonal variation in daily people-caused forest fire occurrence. *Can. J. For. Res.* 19, 1555–1563.

Martell, D. L., Boychuk, D., MacLellan, J. I., Sakowicz, B. M., and Saporta, R. (1995). Decision analysis of the level of forest fire protection in Ontario. *In* "Symposium on Systems Analysis in Forest Resources: Management Systems for a Global Economy with Global Resource Concerns" (J. Sessions and J. D. Brodie, Eds.), pp. 138–149. Pacific Grove, California.

Martell, D. L., Drysdale, R. J., Doan, G. E., and Boychuk, D. (1984). An evaluation of forest fire initial attack resources. *Interfaces* 14(5), 20–32.

Martell, D. L., Gunn, E. A., and Weintraub, A. (1998). Forest management challenges for operational researchers. *Euro. J. Oper. Res.* 104, 1–17.

Martell, D. L., Kourtz, P. H., Tithecott, A., and Ward, P. C. (1999). The development and implementation of forest fire management decision support systems in Ontario, Canada. *In* "Proceedings of the symposium on fire economics, planning, and policy: Bottom Lines" (P. N. Omi and A. Gonzalez-Caban, technical coordinators), pp. 131–142. Gen. Tech. Rep. PSW-GTR-173. USDA Forest Service, Pacific Southwest Research Station, Albany, California.

Martell, D. L., Otukol, S., and Stocks, B. J. (1987). A logistic model for predicting daily people-caused forest fire occurrence in Ontario. *Can. J. For. Res.* 17, 394–401.

McAlpine, R. S., and Hirsch, K. G. (1999). An overview of LEOPARDS: The level of protection analysis system. *For. Chron.* 75, 615–621.

Mees, R. M. (1976). "Computer Evaluation of Existing and Proposed Fire Lookouts." Gen. Tech. Rep. PSW-19. USDA Forest Service, Pacific Southwest Forest Experiment Station, Berkeley.

Merrill, D. F., and Alexander, M. E., Eds. (1987). "Glossary of Forest Fire Management Terms."

Fourth Edition. Canadian Committee on Forest Fire Management, National Research Council of Canada. Ottawa.

Mitchell, B. (1995). Addressing conflict and uncertainty. In "Resource and Environmental Management in Canada: Addressing Conflict and Uncertainty" (B. Mitchell, Ed.), pp. 1–8. Oxford University Press, Toronto.

Parks, G. M. (1964). Development and application of a model for suppression of forest fires. Manage. Sci. 10, 760–766.

Pedernera. P., and Julio, G. (1999). Improving the economic efficiency of combatting forest fires in Chile: The KITRAL system. In "Proceedings of the symposium on fire economics, planning, and policy: Bottom lines" (P. N. Omi and A. Gonzalez-Caban, technical coordinators), pp. 149–155. Gen. Tech. Rep. PSW-GTR-173. USDA Forest Service, Pacific Southwest Research Station, Albany, California.

Plonski, W. L. (1974). Normal yield tables (metric) for major forest species of Ontario. Division of Forests, Ontario Ministry of Natural Resources, Toronto.

Pollock, S. M., and Maltz, M. D. (1994). Operations research in the public sector: An introduction and a brief history. In "Operations Research and the Public Sector" (S. M. Pollock, M. H. Rothkopf, and A. Barnett, Eds.), Handbooks in Operations Research and Management Science, Volume 6, pp. 1–22. North-Holland, New York.

Poulin-Costello, M. (1993). "People-Caused Forest Fire Prediction Using Poisson and Logistic Regression." MSc thesis. University of Victoria, Victoria.

Pyne, S. J. (1982). "Fire in America: A Cultural History of Wildland and Rural Fire." Princeton University Press, Princeton.

Pyne, S. J. (1997). "World Fire: The Culture of Fire on Earth." University of Washington Press, Seattle.

Reed, W. J., and Errico, D. (1986). Optimal harvest scheduling at the forest level in the presence of the risk of fire. Can. J. For. Res. 16, 266–278.

Ross, S. M. (1985). "Introduction to Probability Models," 3rd ed. Academic Press, Orlando.

Sparhawk, W. N. (1925). The use of liability ratings in planning forest fire protection. J. Agric. Res. 30, 693–762.

Van Wagner, C. E. (1979). The economic impact of individual fires on the whole forest. For. Chron. 55, 47–50.

Van Wagner, C. E. (1987). "Development and Structure of the Canadian Forest Fire Weather Index System." Forestry Technical Report 35. Canadian Forestry Service, Ottawa.

Whalen, R. J. (1995). "The Ecology of Fire." Cambridge University Press, Cambridge.

SUBJECT INDEX

Printed in the United States
122079LV00003B/76-84/A